T0291630

LONDON MATHEMATICAL SOCIETY STUDENT TEXTS

59 Frobenius algebras and 2D topological quantum field theories, JOACHIM KOCK
60 Linear operators and linear systems, JONATHAN R. PARTINGTON
61 An introduction to noncommutative Noetherian rings (2nd Edition), K. R. GOODEARL & R. B. WARFIELD, JR
62 Topics from one-dimensional dynamics, KAREN M. BRUCKS & HENK BRUIN
63 Singular points of plane curves, C. T. C. WALL
64 A short course on Banach space theory, N. L. CAROTHERS
65 Elements of the representation theory of associative algebras I, IBRAHIM ASSEM, DANIEL SIMSON & ANDRZEJ SKOWROŃSKI
66 An introduction to sieve methods and their applications, ALINA CARMEN COJOCARU & M. RAM MURTY
67 Elliptic functions, J. V. ARMITAGE & W. F. EBERLEIN
68 Hyperbolic geometry from a local viewpoint, LINDA KEEN & NIKOLA LAKIC
69 Lectures on Kähler geometry, ANDREI MOROIANU
70 Dependence logic, JOUKU VÄÄNÄNEN
71 Elements of the representation theory of associative algebras II, DANIEL SIMSON & ANDRZEJ SKOWROŃSKI
72 Elements of the representation theory of associative algebras III, DANIEL SIMSON & ANDRZEJ SKOWROŃSKI
73 Groups, graphs and trees, JOHN MEIER
74 Representation theorems in Hardy spaces, JAVAD MASHREGHI
75 An introduction to the theory of graph spectra, DRAGOŠ CVETKOVIĆ, PETER ROWLINSON & SLOBODAN SIMIĆ
76 Number theory in the spirit of Liouville, KENNETH S. WILLIAMS
77 Lectures on profinite topics in group theory, BENJAMIN KLOPSCH, NIKOLAY NIKOLOV & CHRISTOPHER VOLL
78 Clifford algebras: An introduction, D. J. H. GARLING
79 Introduction to compact Riemann surfaces and dessins d'enfants, ERNESTO GIRONDO & GABINO GONZÁLEZ-DIEZ
80 The Riemann hypothesis for function fields, MACHIEL VAN FRANKENHUIJSEN
81 Number theory, Fourier analysis and geometric discrepancy, GIANCARLO TRAVAGLINI
82 Finite geometry and combinatorial applications, SIMEON BALL
83 The geometry of celestial mechanics, HANSJÖRG GEIGES
84 Random graphs, geometry and asymptotic structure, MICHAEL KRIVELEVICH *et al*
85 Fourier analysis: Part I - Theory, ADRIAN CONSTANTIN
86 Dispersive partial differential equations, M. BURAK ERDOĞAN & NIKOLAOS TZIRAKIS
87 Riemann surfaces and algebraic curves, R. CAVALIERI & E. MILES
88 Groups, languages and automata, DEREK F. HOLT, SARAH REES & CLAAS E. RÖVER
89 Analysis on Polish spaces and an introduction to optimal transportation, D. J. H. GARLING
90 The homotopy theory of $(\infty,1)$-categories, JULIA E. BERGNER
91 The block theory of finite group algebras I, M. LINCKELMANN
92 The block theory of finite group algebras II, M. LINCKELMANN
93 Semigroups of linear operators, D. APPLEBAUM
94 Introduction to approximate groups, M. C. H. TOINTON
95 Representations of finite groups of Lie type (2nd Edition), F. DIGNE & J. MICHEL
96 Tensor products of C^*-algebras and operator spaces, G. PISIER
97 Topics in cyclic theory, D. G. QUILLEN & G. BLOWER
98 Fast track to forcing, M. DŽAMONJA

London Mathematical Society Student Texts 99

A Gentle Introduction to Homological Mirror Symmetry

RAF BOCKLANDT
University of Amsterdam

CAMBRIDGE
UNIVERSITY PRESS

University Printing House, Cambridge CB2 8BS, United Kingdom

One Liberty Plaza, 20th Floor, New York, NY 10006, USA

477 Williamstown Road, Port Melbourne, VIC 3207, Australia

314–321, 3rd Floor, Plot 3, Splendor Forum, Jasola District Centre,
New Delhi – 110025, India

103 Penang Road, #05–06/07, Visioncrest Commercial, Singapore 238467

Cambridge University Press is part of the University of Cambridge.

It furthers the University's mission by disseminating knowledge in the pursuit of education, learning, and research at the highest international levels of excellence.

www.cambridge.org
Information on this title: www.cambridge.org/9781108483506
DOI: 10.1017/9781108692458

First published 2021

A catalogue record for this publication is available from the British Library.

ISBN 978-1-108-48350-6 Hardback
ISBN 978-1-108-72875-1 Paperback

Contents

Preface

This book grew out of an advanced masters course that I teach biannually at the University of Amsterdam. The course is aimed at students who are doing a masters in algebra and geometry or mathematical physics. In this course I try to give them a feeling of what homological mirror symmetry is and how it ties together many different areas of mathematics. The focus of the course is to explain the main concepts and results and to illustrate them with examples, without getting too technical. In this way the students will be better prepared to delve into the primary literature if they want to understand the theory at a deeper and more detailed level.

As there are many different topics to cover, it is not an easy task to decide what to include and what to omit. Both symplectic geometry and algebraic geometry come with a powerful toolbox that enables researchers to deal with many different situations and levels of generality, but make the fields quite hard to penetrate for outsiders and newcomers. My own background is in representations of quivers and I am an expert in neither symplectic geometry nor algebraic geometry. Therefore, I decided to approach the book more from a representation-theoretic perspective instead of geometrical, and to use mainly examples in complex or symplectic dimension 1, in other words surfaces. This is not perfect because the one-dimensional perspective misses some important features that are crucial in understanding the higher-dimensional cases, but the main advantage is that these examples can easily be visualized. Moreover, they retain some of the key aspects of homological mirror symmetry while from a complex and symplectic view avoid many extra difficulties. Finally, these examples are often of tame type, which means that their representation theory is very well understood. Some of these examples, such as the affine and projective line or the linear quiver, are familiar to many mathematicians and therefore they offer a good hook to enter the field of homological mirror symmetry.

The book is split into three parts offering different looks at the subject.

- The first part sets up the A_∞-formalism. We first have a brief look at the representation theory of categories, then we study complexes and homology and merge the two subjects into the theory of A_∞-algebras and categories. In the final chapter of this part we apply these ideas to the representation theory of some quivers and show how we can interpret the results in two different geometrical ways: intersection theory of curves on certain surfaces, and sheaves on certain one-dimensional algebraic geometrical objects. These two interpretations offer a first glimpse at homological mirror symmetry.
- The second part takes a more classical approach. We start with a chapter of motivation from theoretical physics that explains the origins of homological mirror symmetry by seeing quantum physics as representation theory. After that we have a closer look at the A-model and the B-model and introduce various A_∞-categories that describe these models. The last chapter of this part looks at mirror symmetry for the complex torus and how to extend it using techniques from toric and tropical geometry.
- The third part of the book focuses on surfaces. First we give an explicit construction of the Fukaya category of a surface using gentle algebras and look at how these categories can be constructed by gluing simpler categories together. This perspective also allows us to construct many different mirrors in the $\mathbb{Z}_2$-graded case. Next we move to the $\mathbb{Z}$-graded setting using line fields. The final two chapters are devoted to two concepts that are important in the study of mirror symmetry but also have many applications in other fields: stability and deformation theory.

While the first part introduces some machinery concerning A_∞-categories that is needed in the other two parts, the latter two stand on their own and can be read independently.

Writing a book about a fast evolving and rapidly expanding field of research is not an easy task and keeping up with all the exciting developments is nigh impossible, let alone giving a concise overview of them. Therefore, these notes form only an incomplete and personal view on the subject and I hope that my limited understanding has not resulted in grave omissions and mistakes or that I have cut too many corners in presenting the material. Nevertheless, I hope that these notes may serve as an entrance guide for mathematicians interested in the subject and prepare them for their own forays into homological mirror symmetry.

Naturally I want to thank some people for their support while writing this book. First of all my thanks go to the students at the University of Amsterdam who took my mirror symmetry class and whose questions and comments were very helpful for organizing the material. In particular I want to thank Edward Berengoltz, Sam van den Brink, Niek Lamoree, Peter Spacek, Okke van Garderen, Ronen Brilleslijper and my PhD student Jasper van de Kreeke for the illuminating and puzzling moments in front of the blackboard when we worked out examples, stumbled on difficulties and tried to clear them up.

I am also grateful to Pieter Belmans, Simon Cramer, Casper Loman, Philip Schlösser, Jan Marten Sevenster, Lucas Smits, Leonard Tokic and Mick van Vliet for the helpful suggestions and comments on various drafts of the manuscript.

Many thanks go to Hessel Posthuma, Sergey Shadrin, Han Peters, Lenny Taelman and all my other colleagues at the University of Amsterdam for the little conversations at the coffee machine and entertaining discussions during lunch, which make the Korteweg-de Vries Institute such a pleasant working environment.

I thank Steffen Opperman for inviting me to give a mini-course on A_∞-structures in Njordfordeid, and the other lecturers, Alice Rizzardo and Julian Külshammer, and the students at this summer school for the nice interactions and stimulating atmosphere. I also fondly remember my visits to London with Yankı Lekili, Travis Schedler, Ed Segal and Michael Wong where we covered many napkins in coffee bars with drawings of arc systems, line fields, dimer models, twisted complexes and much more.

Finally, these acknowledgements would be incomplete without thanking Ella, Idun and Jorun for the fine family moments that offered welcome diversions during the writing process.

PART ONE

TO A_∞ AND BEYOND

1
Categories

In this chapter we review some basics of category theory and representation theory. We will extend the representation theory of algebras to categories and discuss Morita equivalence in this context. Most of the material covered can be found in every standard book about category theory or representation theory [176, 148, 258]. We include it to fix notation and as a reference because in later chapters we will expand this formalism to the world of A_∞-categories.

1.1 Categories

Definition 1.1 A *category* C consists of a collection of objects $\mathrm{Ob}(\mathsf{C})$ and for each pair of objects $A, B \in \mathrm{Ob}(\mathsf{C})$ there is a set $\mathsf{C}(A, B)$. This set is the *hom-space* from A to B and its elements are called *morphisms*. They satisfy the following properties:

- morphisms can be composed: if $\phi \in \mathsf{C}(A, B)$ and $\psi \in \mathsf{C}(B, C)$ then $\psi\phi \in \mathsf{C}(A, C)$,
- the composition is associative: $(\phi\psi)\chi = \phi(\psi\chi)$ whenever it is defined,
- for each object A there is an identity morphism $\mathbb{1}_A \in \mathsf{C}(A, A)$ such that for $\phi \in \mathsf{C}(A, B)$ we have $\phi = \phi\mathbb{1}_A = \mathbb{1}_B\phi$.

A morphism $\phi \in \mathsf{C}(A, B)$ is called an *isomorphism* if there is a $\psi \in \mathsf{C}(B, A)$ such that $\phi\psi = \mathbb{1}_B$ and $\psi\phi = \mathbb{1}_A$. In that case, A and B are called *isomorphic*.

Example 1.2 The standard example of a category is `Sets`: its objects are sets and morphisms are maps between sets. Two sets are isomorphic if they have the same cardinality.

Example 1.3 (Examples from algebra) For every algebraic structure we can construct a corresponding category; its objects are sets equipped with this algebraic structure and its morphisms are maps preserving this structure. Examples of this construction are $\mathtt{Groups}$, $\mathtt{Fields}$ and $\mathtt{Rings}$.

Two important categories we will often use are $\mathtt{VECT}(\mathbb{k})$ and $\mathtt{vect}(\mathbb{k})$, which are the categories whose objects are all $\mathbb{k}$-vector spaces and all finite-dimensional $\mathbb{k}$-vector spaces and whose morphisms are linear maps.

Example 1.4 (Examples from geometry) In topology the category $\mathtt{Top}$ consists of all topological spaces with continuous maps as morphisms, while $\mathtt{Top}_*$ consists of pairs of a topological space and a point in this space, together with morphisms that map the selected points to each other.

Similarly, in differential geometry we can construct $\mathtt{Man}$ (and $\mathtt{Man}_*$) whose objects are (pointed) manifolds and whose morphisms are smooth maps (that identify the selected points).

Every topological space $\mathbb{X}$ can also be considered as a category $\mathtt{Open}(\mathbb{X})$, whose objects are the open subsets and whose morphisms are the inclusions.

All these examples are categories for which the objects are sets with a special structure and the morphisms are maps between these sets preserving the structure. There are also other interesting categories that do not fall into this class.

Example 1.5 The *fundamental groupoid* of a topological space $\Pi_1(\mathbb{X})$ is the category for which the objects are the points in $\mathbb{X}$ and the morphisms are homotopy classes of paths between points. Note that if $\mathbb{X}$ is connected then $\Pi_1(\mathbb{X})(p, p)$ is by definition equal to the fundamental group $\pi_1(\mathbb{X}, p)$.

1.2 Functors

Definition 1.6 A *covariant functor* $\mathcal{F}: \mathtt{C} \to \mathtt{D}$ consists of maps $\mathcal{F}: \mathrm{Ob}(\mathtt{C}) \to \mathrm{Ob}(\mathtt{D})$ and $\mathcal{F}: \mathtt{C}(A, B) \to \mathtt{D}(\mathcal{F}(A), \mathcal{F}(B))$ such that $\mathcal{F}(\psi\phi) = \mathcal{F}(\psi)\mathcal{F}(\phi)$ and $\mathcal{F}(\mathbb{1}_A) = \mathbb{1}_{\mathcal{F}(A)}$.

Example 1.7 Examples of covariant functors are *forgetful functors*, which consider the same objects and maps but forget some of the structure: $\mathcal{F}: \mathtt{Groups} \to \mathtt{Sets}: (G, *) \mapsto G$ or $\mathcal{F}: \mathtt{Rings} \to \mathtt{Groups}: (R, +, \times) \mapsto (R, +)$.

Example 1.8 Other functors create more structure, such as $\mathcal{U}: \mathtt{Sets} \to \mathtt{VECT}(\mathbb{k}): S \mapsto \langle S \rangle$ which associates to each set a vector space with as basis that set. Because a linear map is determined by the images of the basis, the functor turns each map between generators into a linear map between vector spaces.

The group ring construction is also an example of this: it associates to each group its group ring, consisting of all linear combinations of group elements. This functor turns each group morphism into a ring morphism by linearizing it.

Example 1.9 A more fancy example of a covariant functor is the fundamental group, which goes from pointed topological spaces to groups,

$$\pi_1 \colon \mathtt{Top}_* \to \mathtt{Groups} \colon (X, p) \mapsto \pi_1(X, p).$$

If $f \colon (X, p) \to (Y, q)$ is a continuous map, we set $\pi_1(f) \colon \pi_1(X, p) \to \pi_1(Y, q) \colon [\ell] \mapsto [f\ell]$ where $\ell \colon [0, 1] \to X$ is a loop starting at $p \in X$ and $[\ell]$ is its corresponding element in the fundamental group.

Definition 1.10 A *contravariant functor* $\mathcal{F} \colon \mathtt{C} \to \mathtt{D}$ consists of maps $\mathcal{F} \colon \mathrm{Ob}(\mathtt{C}) \to \mathrm{Ob}(\mathtt{D})$ and $\mathcal{F} \colon \mathtt{C}(A, B) \to \mathtt{D}(\mathcal{F}(B), \mathcal{F}(A))$ such that

$$\mathcal{F}(\psi\phi) = \mathcal{F}(\phi)\mathcal{F}(\psi) \quad \text{and} \quad \mathcal{F}(\mathbb{1}_A) = \mathbb{1}_{\mathcal{F}(A)}.$$

If we define the *opposite category* $\mathtt{C}^{\mathrm{op}}$ to be the category with the same objects but

$$\mathtt{C}^{\mathrm{op}}(V, W) := \mathtt{C}(W, V)$$

and the multiplication reversed, then a contravariant functor from $\mathtt{C} \to \mathtt{D}$ is the same as a covariant functor from $\mathtt{C}^{\mathrm{op}} \to \mathtt{D}$.

Example 1.11 The basic example of a contravariant functor is the dual of a vector space

$$-^* \colon \mathtt{vect}(\Bbbk) \to \mathtt{vect}(\Bbbk) \colon V \mapsto V^* := \mathrm{Hom}_{\Bbbk}(V, \Bbbk).$$

This functor is contravariant because if $f \colon V \to W$ is a linear map, the corresponding dual map $f^* \colon W^* \to V^* \colon \phi \mapsto \phi \circ f$ goes in the opposite direction.

Example 1.12 Another example is the construction that associates to every manifold its ring of smooth functions

$$C_\infty \colon \mathtt{Man} \to \mathtt{Rings} \colon \mathbb{M} \mapsto C_\infty(\mathbb{M}) := \{f \colon \mathbb{M} \to \mathbb{R} \mid f \text{ is smooth}\}.$$

Definition 1.13 A covariant or contravariant functor $\mathcal{F} \colon \mathtt{C} \to \mathtt{D}$ is called *full*, *faithful* or *fully faithful* if all the maps $\mathcal{F} \colon \mathtt{C}(A, B) \to \mathtt{D}(\mathcal{F}(A), \mathcal{F}(B))$ are surjective, injective or bijective. A fully faithful covariant or contravariant functor is called an *equivalence* or *antiequivalence* if for each $B \in \mathrm{Ob}(\mathtt{D})$ there is an $A \in \mathrm{Ob}(\mathsf{C})$ with $\mathcal{F}(A)$ isomorphic to B. This last property is also called *essentially surjective*.

Example 1.14 Define the category $\mathtt{mat}(\Bbbk)$ as follows:

$$\mathrm{Ob}(\mathtt{mat}(\Bbbk)) := \{0, 1, 2, \dots\} \quad \text{and} \quad \mathtt{mat}(n, m) := \mathrm{Mat}_{m\times n}(\Bbbk).$$

This category is equivalent to the category of vector spaces through the functor

$$\mathcal{F}\colon \mathtt{mat}(\Bbbk) \to \mathtt{vect}(\Bbbk)\colon n \mapsto \Bbbk^n,$$
$$\mathcal{F}\colon \mathtt{mat}(n, m) \to \mathtt{vect}(\Bbbk^n, \Bbbk^m)\colon A \mapsto (\phi\colon x \mapsto Ax),$$

which is fully faithful because every linear map between $\Bbbk^n$ and $\Bbbk^m$ is represented by a unique matrix. It is also essentially surjective because every finite-dimensional vector space is isomorphic to $\Bbbk^n$ for some n.

The main idea is that equivalent categories describe two types of mathematical objects which behave the same. All constructions between one of these types of objects and their morphisms can be translated to the other setting and vice versa. In our particular situation it means that working with vector spaces and linear maps is the same as working with matrices.

Example 1.15 The functor $-^*$ is an antiequivalence for the category of all finite-dimensional vector spaces $\mathtt{vect}(\Bbbk)$, but not for the category $\mathtt{VECT}(\Bbbk)$ of all vector spaces over $\Bbbk$. This is because the dual of an infinite-dimensional vector space never has a countable basis, so $-^*$ is not essentially surjective.

1.3 Natural Transformations

The final main ingredient of category theory is natural transformations. They can be seen as morphisms between functors.

Definition 1.16 A *natural transformation* $\eta\colon \mathcal{F} \to \mathcal{G}$ between two covariant functors $\mathcal{F}, \mathcal{G}\colon \mathsf{C} \to \mathsf{D}$ is a collection of D-morphisms $(\eta_X\colon \mathcal{F}(X) \to \mathcal{G}(X))_{X\in \mathrm{Ob}\,\mathsf{C}}$ such that for every morphism $f\colon X \to Y$ in C we have $\eta_Y \circ \mathcal{F}(f) = \mathcal{G}(f) \circ \eta_X$.

Example 1.17 The standard example of a natural transformation is the one between the identity functor $\mathbb{1}\colon \mathtt{vect}(\Bbbk) \to \mathtt{vect}(\Bbbk)$ and the double dual of a vector space $-^{**}\colon \mathtt{vect}(\Bbbk) \to \mathtt{vect}(\Bbbk)$:

$$\eta_X\colon X \to X^{**}\colon v \mapsto (\mathrm{ev}_v\colon X^* \to \Bbbk\colon \phi \mapsto \phi(v)).$$

This is the mathematical way of saying that a finite-dimensional vector space and its double dual are canonically isomorphic: there is a standard isomorphism between a vector space and its double dual. This is not the case for the single dual: although a finite-dimensional vector space is isomorphic to its dual, to construct an actual isomorphism we need some extra data (like a basis).

Example 1.18 It is also easy to check that every natural transformation of the identity functor $\mathbb{1}: \mathtt{vect}(\Bbbk) \to \mathtt{vect}(\Bbbk)$ to itself is a global rescaling by a common factor $\lambda \in \Bbbk$:

$$(\nu_\lambda)_X : X \to X : v \mapsto \lambda v.$$

Indeed, if $\nu: \mathbb{1} \to \mathbb{1}$ is a natural transformation then it acts by rescaling on the one-dimensional vector space $\Bbbk$: $\nu_\Bbbk : \Bbbk \to \Bbbk : x \to \lambda x$. If v is an element of a vector space X then there is a linear map $f: \Bbbk \to X: 1 \mapsto v$, so $\nu_X(v) = \nu_X(f(1)) = f(\nu_\Bbbk(1)) = f(\lambda) = \lambda v$.

1.4 Linear Categories

If we fix a field $\Bbbk$ we can consider $\Bbbk$-linear categories.

Definition 1.19 A category C is called a $\Bbbk$-*linear category* if all hom-spaces are $\Bbbk$-vector spaces and the multiplication is bilinear. A functor between two $\Bbbk$-linear categories is called $\Bbbk$-*linear* if the maps between the hom-spaces are $\Bbbk$-linear. An object X in a $\Bbbk$-linear category C is called a *zero object* if for all $Y \in \mathrm{Ob}(\mathsf{C})$ we have $\mathsf{C}(X, Y) = \mathsf{C}(Y, X) = 0$.

Unless it is specifically stated otherwise, a functor between two $\Bbbk$-linear categories will always be assumed to be $\Bbbk$-linear. This is not a big restriction because almost all natural functors are $\Bbbk$-linear.

Example 1.20 There are many examples of $\Bbbk$-linear categories: $\mathtt{VECT}(\Bbbk)$, $\mathtt{vect}(\Bbbk)$, $\mathtt{mat}(\Bbbk)$. For any $\Bbbk$-algebra A, we have $\Bbbk$-linear categories

- $\mathtt{MOD}\text{-}A$: the category of all *right A-modules*,
- $\mathtt{Mod}\text{-}A$: the category of all *finitely generated right A-modules*, and
- $\mathtt{mod}\text{-}A$: the category of all *finite-dimensional right A-modules*.

The zero object in these categories is the zero-vector space with trivial A-action. The hom-spaces in these categories are often denoted by $\mathrm{Hom}_A(M, N)$ instead of $\mathtt{MOD}\text{-}A(M, N)$. The same triplet exists for left modules as well: $A\text{-}\mathtt{MOD}$, $A\text{-}\mathtt{Mod}$ and $A\text{-}\mathtt{mod}$.

Example 1.21 Any $\mathbb{k}$-algebra A can also be considered as a $\mathbb{k}$-linear category $\mathtt{A}$ with one object O and $\mathtt{A}(\mathsf{O},\mathsf{O}) = A$. On the other hand, it is also possible to consider a category C with more than one object as an algebra: we can take the direct sum of all hom-spaces

$$C := \bigoplus_{X,Y\in \mathrm{Ob}(\mathsf{C})} \mathsf{C}(X,Y).$$

The product of two morphisms is composition if they are composable and zero otherwise. Note that for each object X, the unit $\mathbb{1}_X$ gives an idempotent in the algebra. If there is a finite number of objects the sum of all these idempotents is the unit in the algebra, but if there are infinitely many objects this will be an algebra without a unit.

Note that these operations are not inverses. If we start with a category with a finite number of objects, turn it into an algebra and then back into a category, this new category is not equivalent to the original. There is however a weaker sense in which they are equivalent. This will be explored in the next sections.

1.5 Modules

Given a $\mathbb{k}$-algebra A, a left module consists of a vector space V and an algebra morphism $\rho\colon A \to \mathrm{Hom}_{\mathbb{k}}(V,V)$. If we consider A as a category $\mathtt{A}$ with one object, this is precisely a covariant functor from $\mathtt{A}$ to $\mathtt{Vect}(\mathbb{k})$. A morphism between two modules is a linear map $f\colon V \to W$ such that for all $a \in A$ we have $f \circ \rho_V(a) = \rho_W(a) \circ f$, in which we recognize a natural transformation between the functors ρ_V and ρ_W.

Definition 1.22 If C is a ($\mathbb{k}$-linear) category we define its *left module category* $\mathsf{C}\text{-}\mathtt{MOD}$ as the category with as objects the ($\mathbb{k}$-linear) functors $\rho\colon \mathsf{C} \to \mathtt{VECT}(\mathbb{k})$ and as morphisms the natural transformations between them. An object of $\mathsf{C}\text{-}\mathtt{MOD}$ is called a *left* C*-module*. The *right module category* $\mathtt{MOD}\text{-}\mathsf{C}$ is the category of contravariant functors from C to $\mathtt{VECT}(\mathbb{k})$. The objects are *right* C*-modules*.

Example 1.23 If C is a $\mathbb{k}$-linear category with a finite number of objects and C is the corresponding algebra then we can turn every classical left C-module M into a functor $\mathcal{F}_M\colon \mathsf{C} \to \mathtt{VECT}(\mathbb{k})$ that maps an object X to the vector space $\mathcal{F}_M := \mathbb{1}_X M$ and every morphism $\phi\colon X \to Y$ to $\mathcal{F}_M(\phi)\colon \mathbb{1}_X M \to \mathbb{1}_Y M\colon m \mapsto \phi m$.

Vice versa, every C-module $\mathcal{F}\colon \mathsf{C} \to \mathtt{VECT}(\mathbb{k})$ can be seen as a classical module of the algebra $C = \bigoplus_{X,Y\in \mathrm{Ob}\,\mathsf{C}} \mathsf{C}(X,Y)$ by taking the direct sum $M_{\mathcal{F}} := \bigoplus_{X\in \mathrm{Ob}\,\mathsf{C}} \mathcal{F}(X)$ and letting $\mathcal{F}(\phi)$ act on the appropriate components.

Example 1.24 Many classical constructions in geometry are actually modules of categories.

- Let $C_\infty(\mathbb{M})$ denote the space of smooth functions on a manifold $\mathbb{M}$. If we forget the ring structure on $C_\infty(\mathbb{M})$, the functor $C_\infty\colon \mathtt{Man} \to \mathtt{VECT}(\mathbb{R})$ can be seen as a right module of $\mathtt{Man}$. Similarly, the differential n-forms Ω^n and the de Rham cohomology H^n_{dR} can also be seen as right modules of $\mathtt{Man}$.
- If $\mathscr{V}$ is a vector bundle over a topological space $\mathbb{X}$ then the section functor $\Gamma(\mathscr{V}, -)$, which maps every open $\mathbb{U}$ to its space of sections $\Gamma(\mathscr{V}, \mathbb{U})$, is a contravariant functor from $\mathtt{Open}(\mathbb{X})$ to $\mathtt{VECT}(\mathbb{R})$ and hence a right module of $\mathtt{Open}(\mathbb{X})$. In general, a right module of $\mathtt{Open}(\mathbb{X})$ is also called a *presheaf* of $\mathbb{X}$.
- A *local system* on a topological space $\mathbb{X}$ can be seen as a module of the fundamental groupoid $\Pi_1(\mathbb{X})$: it assigns to each point in $\mathbb{X}$ a vector space and to each path a linear map that only depends on the homotopy class. In other words, it is a vector bundle with a flat connection.

Modules of categories behave in all respects like modules of rings and all familiar concepts hold. We say that a morphism between C-modules is *injective*, *surjective* or *bijective* if its natural transformation is respectively injective, surjective or bijective in every object of C.

For every morphism we can also define its *kernel* and *cokernel* by looking at the kernel and cokernel of the natural transformation in every object. Other notions such as short exact sequences and direct sums and summands can also be defined objectwise.

Example 1.25 The exterior derivative can be viewed as a $\mathtt{Man}$-module morphism $d_n\colon \Omega^n \to \Omega^{n+1}$ and $H^n_{\mathrm{dR}} = \operatorname{Ker} d_n / \operatorname{Im} d_{n-1}$ as right $\mathtt{Man}$-modules. For more details see Section 2.2.3.

Just as we can consider an algebra as a module over itself, we can see every object in a $\mathbb{k}$-linear category as a module of this category: to $X \in \mathrm{Ob}(\mathsf{C})$ we associate the contravariant functor $\mathsf{C}(-, X)$. A morphism between two objects $\phi\colon X \to Y$ will give a natural transformation $\phi \circ -\colon \mathsf{C}(-, X) \to \mathsf{C}(-, Y)\colon f \mapsto \phi \circ f$. Therefore we have a covariant functor $\mathcal{Y}\colon \mathsf{C} \to \mathtt{MOD}\text{-}\mathsf{C}$. This functor is called the *Yoneda embedding*.

Lemma 1.26 (Yoneda lemma) *The functor* $\mathcal{Y}\colon \mathsf{C} \to \mathtt{MOD}\text{-}\mathsf{C}$ *is fully faithful.*

Yoneda's lemma implies that C can be seen as a full subcategory of $\mathtt{MOD}\text{-}\mathsf{C}$. In a similar way C^{op} embeds in the left module category $\mathsf{C}\text{-}\mathtt{MOD}$.

1.6 Morita Equivalence

Sometimes nonisomorphic algebras can have the equivalent module categories. This phenomenon is called *Morita equivalence* and has been well studied [9, 151].

Example 1.27 If A is an algebra we say that $e \in A$ is an *idempotent* if $e^2 = e$ and it is a *full idempotent* if additionally $AeA = A$. For a full idempotent e we can construct an algebra $B = eAe = \mathrm{Hom}_A(eA, eA)$ and a functor

$$\mathcal{E}\colon \mathtt{MOD}\text{-}A \to \mathtt{MOD}\text{-}B\colon M \mapsto Me \quad \text{with } \mathcal{E}(f\colon M \to N) = f|_{Me}.$$

This functor is fully faithful because we can reconstruct f from $f|_{Me}$ as $M = MAeA = (Me)A$. It is also essentially surjective because

$$\mathcal{E}(N \otimes_B eA) = (N \otimes_B eA)e = N \otimes_B eAe = N \otimes_B B \cong N.$$

Therefore, $\mathcal{E}$ is an equivalence between $\mathtt{MOD}\text{-}A$ and $\mathtt{MOD}\text{-}B$.

Example 1.28 The elementary matrix E_{11} with a 1 in the upper-left corner is a full idempotent for $\mathrm{Mat}_n(A) = \mathrm{Hom}_A(A^{\oplus n}, A^{\oplus n})$, so the latter is Morita equivalent with A.

The previous examples indicate that there are two processes to make Morita equivalences: taking endomorphism rings of certain direct summands (eA) and direct sums ($A^{\oplus n}$). These processes are sufficient:

Theorem 1.29 (Morita [177]) *Two algebras A and B are Morita equivalent if and only if $B \cong e\,\mathrm{Mat}_n(A)e$ for some full idempotent matrix $e \in \mathrm{Mat}_n(A)$.*

These ideas generalize to Morita equivalences for $\mathbb{k}$-linear categories as well.

Definition 1.30 Two categories C and D are called *Morita equivalent* if $\mathtt{MOD}\text{-}\mathsf{C}$ and $\mathtt{MOD}\text{-}\mathsf{D}$ are equivalent categories.

Definition 1.31 The *additive completion* $\mathtt{Add}\,\mathsf{C}$ of a $\mathbb{k}$-linear category C is the category with objects that are finite formal direct sums of objects in C, including the zero sum. We write such objects as $(A_1 \oplus \cdots \oplus A_n)$ with $A_i \in \mathrm{Ob}(\mathsf{C})$ and

$$\mathtt{Add}\,\mathsf{C}(A_1 \oplus \cdots \oplus A_n, B_1 \oplus \cdots \oplus B_m) := \bigoplus_{i,j} \mathsf{C}(A_i, B_j),$$

where the right-hand side is a direct sum of $\mathbb{k}$-vector spaces. The composition of morphisms is the bilinear extension of the composition in C. Note that C is a full subcategory of $\mathtt{Add}\,\mathsf{C}$.

Example 1.32 The *additive completion* of an algebra A (viewed as a category with one object) is equivalent to the category $\mathtt{mat}(A)$. In the case that $A = \mathbb{k}$ this category is equivalent to $\mathtt{vect}(\mathbb{k})$.

Example 1.33 A module category $\mathtt{MOD}\text{-}\mathsf{C}$ is equivalent to its additive completion because the direct sum of modules is again a module, so the embedding $\mathtt{MOD}\text{-}\mathsf{C} \subset \mathtt{Add}\,\mathtt{MOD}\text{-}\mathsf{C}$ is essentially surjective.

Definition 1.34 The *split closure* (or *Karoubi completion*) $\mathtt{Split}\,\mathsf{C}$ is the category for which the objects are pairs (X, e), where $X \in \mathrm{Ob}\,\mathsf{C}$ and $e \in \mathsf{C}(X, X)$ is an idempotent. The morphism space between (X, e) and (X', e') is $e'\mathsf{C}(X, X')e$. Again note that C is a full subcategory of $\mathtt{Split}\,\mathsf{C}$ if we identify X with $(X, \mathbb{1}_X)$.

Example 1.35 The category $\mathtt{Split}\,\mathtt{vect}(\mathbb{k})$ is equivalent to $\mathtt{vect}(\mathbb{k})$ because if (V, e) is a vector space with an idempotent endomorphism then (V, e) is equivalent to $(\mathrm{Im}(e), \mathbb{1}_{\mathrm{Im}(e)})$.

This also holds for module categories: if M is an A-module and $e\colon M \to M$ is an idempotent then $M = eM \oplus (1 - e)M$ and we can identify (M, e) with $(eM, \mathbb{1}_{eM})$. So all objects in $\mathtt{Split}\,(\mathtt{MOD}\text{-}A)$ can be seen as direct summands of modules and therefore the embedding $\mathtt{MOD}\text{-}A \subset \mathtt{Split}\,(\mathtt{MOD}\text{-}A)$ is an equivalence. Furthermore, if C is a subcategory of a module category then $\mathtt{Split}\,\mathsf{C}$ will be equivalent to the subcategory that contains all direct summands of objects in C.

Definition 1.36 The combination of both operations is denoted by $\mathtt{Split}\,\mathtt{Add}$ and it is called the *split-add completion*.

Example 1.37 If A is an algebra, viewed as a category with one object, then the objects in $\mathtt{Split}\,\mathtt{Add}\,A$ can be seen as pairs $(A^{\oplus n}, U)$, where U is an idempotent matrix with coefficients in A. We can see U as an idempotent endomorphism of the right module $A^{\oplus n}$ and identify $(A^{\oplus n}, U)$ with the right A-module $\mathrm{Im}(U)$. This gives a fully faithful functor $\mathtt{Split}\,\mathtt{Add}\,A \to \mathtt{MOD}\text{-}A$. The modules in the image of this functor are called the *finitely generated (f.g.) projective right A-modules* and denoted by $\mathtt{Proj}\,A$.

Example 1.38 If $\mathbb{M}$ is a manifold and we view $C_\infty(\mathbb{M})$ as a category with one object, then the objects in $\mathtt{Split}\,\mathtt{Add}\,C_\infty(\mathbb{M})$ can be identified with finite-dimensional vector bundles over $\mathbb{M}$ in the following way:

$$(C_\infty(\mathbb{M})^{\oplus n}, U) \mapsto \mathscr{V} = \{(p, U(p)x) \mid p \in \mathbb{M},\ x \in \mathbb{R}^n\},$$

so at each point $p \in \mathbb{M}$ we put the vector space $\mathrm{Im}\,U(p)$. The *Serre–Swan theorem* [222, 235] states that this gives an equivalence between the category of finitely generated projective $C_\infty(\mathbb{M})$-modules and the category of finite-dimensional vector bundles over $\mathbb{M}$:

$$\mathtt{Split\,Add}\,C_\infty(\mathbb{M}) \cong \mathtt{vect}\,\mathbb{M}.$$

Theorem 1.39 *For a* $\Bbbk$*-linear category* C *the categories* $\mathtt{Split}\,\mathsf{C}$, $\mathtt{Add}\,\mathsf{C}$ *and* $\mathtt{Split\,Add}\,\mathsf{C}$ *are Morita equivalent to* C.

Sketch of the proof Given a linear functor $\mathcal{F}\colon \mathsf{C} \to \mathtt{VECT}(\Bbbk)$ we can construct functors

$$\mathcal{F}_{\mathtt{Add}}\colon \mathtt{Add}\,\mathsf{C} \to \mathtt{VECT}(\Bbbk)\colon A_1 \oplus \cdots \oplus A_n \mapsto \mathcal{F}(A_1) \oplus \cdots \oplus \mathcal{F}(A_n),$$
$$\mathcal{F}_{\mathtt{Split}}\colon \mathtt{Split}\,\mathsf{C} \to \mathtt{VECT}(\Bbbk)\colon (A, e) \mapsto \operatorname{Im}\mathcal{F}(e).$$

These operations induce functors between $\mathtt{MOD}\text{-}\mathsf{C}$ and $\mathtt{MOD}\text{-}\mathtt{Add}\,\mathsf{C}$, respectively $\mathtt{MOD}\text{-}\mathsf{C}$ and $\mathtt{MOD}\text{-}\mathtt{Split}\,\mathsf{C}$, which turn out to be fully faithful and essentially surjective. □

So if we want to do representation theory of a category C, it is better to look at $\mathtt{Split\,Add}\,\mathsf{C}$. Usually this category is much larger than the original category, but in some sense it is easier to work with because it contains all the objects that *should* be there. Just like for ordinary algebras $\mathtt{Split\,Add}\,\mathsf{C}$ can be seen as a subcategory of $\mathtt{MOD}\text{-}\mathsf{C}$. It is the smallest subcategory of $\mathtt{MOD}\text{-}\mathsf{C}$ that contains C and is closed under direct sums and summands. This gives a tower of categories

$$\mathsf{C} \subset \mathtt{Add}\,\mathsf{C} \subset \mathtt{Split\,Add}\,\mathsf{C} \subset \mathtt{MOD}\text{-}\mathsf{C},$$

which generalizes the following inclusions of modules for an algebra A:

$$\{A\} \subset \{\text{f.g. free } A\text{-modules}\} \subset \{\text{f.g. projective } A\text{-modules}\} \subset \{\text{all } A\text{-modules}\}.$$

1.7 Exercises

Exercise 1.1 Show that the abelianization $\mathcal{Ab}\colon \mathtt{Groups} \to \mathtt{Groups}\colon G \mapsto \frac{G}{[G,G]}$ can be made into a covariant functor, but the center $\mathcal{Z}\colon \mathtt{Groups} \to \mathtt{Groups}\colon G \mapsto Z(G)$ cannot. (Hint: factorize $\mathbb{1}_{\mathbb{Z}/2\mathbb{Z}}$ over S_3 and look at the images of these morphisms under $\mathcal{Z}$.)

Exercise 1.2 Let G be a group considered as a category with one object. Show that the natural transformations of the identity functor $1\colon \mathsf{G} \to \mathsf{G}$ to itself form a group that is isomorphic to $Z(\mathsf{G})$.

Exercise 1.3 Show that $\mathcal{T}_p^*\colon \mathtt{Man}_* \to \mathtt{VECT}(\mathbb{R})\colon (\mathbb{M}, p) \mapsto \mathrm{T}_p^*\mathbb{M}$ and $\Omega^1\colon \mathtt{Man}_* \to \mathtt{VECT}(\mathbb{R})\colon (\mathbb{M}, p) \mapsto \Gamma(\mathrm{T}^*\mathbb{M})$ are both contravariant functors and

$$\eta_{(\mathbb{M},p)}\colon \Gamma(\mathrm{T}^*\mathbb{M}) \to \mathrm{T}_p^*\mathbb{M}\colon \phi \mapsto \phi_p$$

is a natural transformation between them. Does there exist a nontrivial natural transformation in the opposite direction?

Exercise 1.4 Let $\mathtt{C}$ be any $\Bbbk$-linear category. Show that $\mathtt{Add}\,(\mathtt{Split}\,\mathtt{Add}\,\mathtt{C})$ is equivalent to $\mathtt{Split}\,\mathtt{Add}\,\mathtt{C}$.

Exercise 1.5 Let $\mathtt{C}$ be any $\Bbbk$-linear category and $\mathcal{F}\colon \mathtt{Add}\,\mathtt{C} \to \mathtt{VECT}(\Bbbk)$ a $\Bbbk$-linear functor.

(i) Show that

$$\mathcal{F}(X \oplus Y) \cong \mathcal{F}(X) \oplus \mathcal{F}(Y),$$

where the left-hand $\oplus$ is the formal direct sum in $\mathtt{Add}\,\mathtt{C}$ and the right-hand $\oplus$ is the standard direct sum of vector spaces.

(ii) Use this to complete the first part of Theorem 1.39.

(iii) Show that $\mathtt{MOD}\,\mathtt{vect}(\Bbbk)$ is equivalent to $\mathtt{VECT}(\Bbbk)$.

Exercise 1.6 Show that the category of presheaves over a discrete topological space with k points is equivalent to the category of $\mathbb{C}^k$-modules.

Exercise 1.7 Let G be a finite group with k irreducible representations. Show that $\mathtt{Split}\,\mathtt{Add}\,\mathbb{C}G$ is equivalent to $\mathtt{Split}\,\mathtt{Add}\,\mathbb{C}^{\oplus k}$. (Hint: use the idempotents $e_\rho = \frac{1}{|G|}\sum \operatorname{Tr}\rho(g)g$ where $\rho\colon G \to \mathrm{Mat}_n(\mathbb{C})$ is an irreducible representation.)

Exercise 1.8 Let $A = \mathbb{C}[X]$ and $B = \mathbb{C}[X, X^{-1}]$.

(i) Show that every finite-dimensional A-module (B-module) can be written as a direct sum of modules of the form $A/(X-\lambda)^k$ and that the hom-spaces between modules with different λ are zero.

(ii) Use this to show that A-$\mathtt{mod}$ and B-$\mathtt{mod}$ are equivalent categories. (Hint: use a bijection between $\mathbb{C}$ and $\mathbb{C} \setminus \{0\}$.)

(iii) Show that A and B are not Morita equivalent.

Exercise 1.9 Let A be the ring $\mathbb{C}[X]$ considered as a category with one object. Show that $\mathtt{Add}\,A$ and $\mathtt{Split}\,\mathtt{Add}\,A$ are equivalent. (Hint: use the classification theorem for finitely generated modules of a principal ideal domain.)

Exercise 1.10 Let $A = C_\infty(\mathbb{S}^1)$ and write elements in A as smooth 2π-periodic functions in θ. Find an idempotent matrix $U = \mathrm{Mat}_{2\times 2}(A)$ such that $(A^{\oplus 2}, U)$ corresponds to a Moebius strip viewed as a vector bundle over $\mathbb{S}^1$. (Hint: construct a projection matrix that projects $\mathbb{R}^2$ onto a line that makes an angle $\theta/2$ with the X-axis.) Use $(A^{\oplus 2}, U)$ to show that $\mathtt{Add}\,A$ and $\mathtt{Split}\,\mathtt{Add}\,A$ are not equivalent.

2
Cohomology

In this chapter we have a look at (co)homology theories. These were first developed in the context of geometry and topology, but have subsequently spread to many other areas in mathematics. We will first set out the general framework and then look at some examples in topology and algebra. For a more in-depth treatment of these subjects we refer to [111, 172] and [253].

2.1 Complexes

2.1.1 Definition

Definition 2.1 A *cochain complex of* $\mathbb{k}$*-vector spaces* is a sequence of objects in $\mathsf{VECT}(\mathbb{k})$, $(C^i)_{i\in\mathbb{Z}}$, together with linear maps $d_i\colon C^i \to C^{i+1}$ such that $d_{i+1}d_i$ is zero. Equivalently, we can see $C^\bullet = \oplus_i C^i$ as a $\mathbb{Z}$-graded vector space with a degree 1 linear map d such that $d^2 = 0$. The map d is called the *differential*.

The *cohomology* of a chain complex $C^\bullet$ is the graded vector space

$$\mathsf{H}C^\bullet := \frac{\operatorname{Ker} d}{\operatorname{Im} d}.$$

The elements in the image d are called *exact* and the elements in the kernel are called *closed*; the cohomology measures the difference between closedness and exactness. A complex is called *bounded* if $C^i = 0$ for $|i| \gg 0$.

The *nth shift* of a complex $C^\bullet$ is the complex $C[n]^\bullet$ with $C[n]^i := C^{i+n}$ and $(d_i)_{C[n]} := (-1)^n (d_{i+n})_C$. In other words, a shift by 1 decreases the degrees of the elements by 1 and adds an extra minus sign in front of the differential.

Remark 2.2 Similarly, we can look at *chain complexes*. These are graded vector spaces $C_\bullet$ with a differential of degree -1. Exact elements in chain complexes are called *boundaries* and closed elements are called *cycles*.

The degrees in cochain complexes are usually written as superscripts, while chain complexes are indexed by subscripts. One can easily switch from chain complexes to cochain complexes by putting a minus sign in front of the index:

$$C^\bullet := C_{-\bullet}.$$

In what follows we will always work with cochain complexes.

Remark 2.3 Sometimes a complex does not have a full $\mathbb{Z}$-grading but only a $\mathbb{Z}_2$-grading, for which d has degree $1 \in \mathbb{Z}_2$. In that case, we speak of $\mathbb{Z}_2$-complexes and $\mathbb{Z}_2$-homology and we will write $C^\pm$ instead of $C^\bullet$. This notation uses the multiplicative notation for $\mathbb{Z}_2$, so the degree 0 component is written as C^+ and the degree 1 component as C^-. Every $\mathbb{Z}$-complex is also a $\mathbb{Z}_2$-complex with $C^+ = \oplus_i C^{2i}$ and $C^- = \oplus_i C^{2i+1}$.

Remark 2.4 Cochain complexes can also be defined in any category of modules MOD-A or even more generally an abelian category (this is a linear category in which it makes sense to talk about kernels, images and cokernels).

2.1.2 Morphisms

Definition 2.5 A *morphism between cochain complexes* $C^\bullet, D^\bullet$ is a degree 0 linear map $f\colon C^\bullet \to D^\bullet$ such that $d_D f - f d_C = 0$.

This condition ensures that f will induce a map between the cohomologies of the complexes. Indeed, if $d_C x = 0$ then $d_D f(x) = 0$ and $f(d_C y) = d_D(f(y))$, so the map

$$\mathsf{H}f\colon \mathsf{H}C^\bullet \to \mathsf{H}D^\bullet\colon x + \operatorname{Im} d_C \mapsto f(x) + \operatorname{Im} d_D$$

is well defined and it is also compatible with composition: $\mathsf{H}(f \circ g) = \mathsf{H}f \circ \mathsf{H}g$.

Definition 2.6 A morphism $f\colon C^\bullet \to D^\bullet$ is called

(i) a *quasi-isomorphism* if $\mathsf{H}f$ is an isomorphism,
(ii) *null-homotopic* if it is equal to $d_D g + g d_C$, where $g\colon C^\bullet \to D^\bullet$ is any degree -1 linear map. In that case, $\mathsf{H}f = 0$.

Two morphisms are called *homotopic* if their difference is null-homotopic.

Given two complexes $C^\bullet$ and $D^\bullet$, one can construct the complex of graded module morphisms of arbitrary degree $\operatorname{Hom}^\bullet(C^\bullet, D^\bullet)$:

$$\operatorname{Hom}^j(C^\bullet, D^\bullet) := \bigoplus_i \operatorname{Hom}_{\mathbb{k}}(C^i, D^{i+j}) \quad \text{with } df = d_D f - (-1)^{\deg f} f d_C.$$

This is again cochain complex because the differential has degree +1. The cohomology of this complex in degree 0 will give us the morphisms of complexes up to homotopy.

The cohomology in other degrees can also be considered as morphisms of complexes but between shifted versions of the complex because

$$\mathrm{Hom}^\bullet(C[i]^\bullet, D[j]^\bullet) \cong \mathrm{Hom}^\bullet(C^\bullet, D^\bullet)[j-i],$$

where the isomorphism adds in an appropriate sign.

2.1.3 A Structure Theorem for Complexes

Definition 2.7 Let $C^\bullet$ be a cochain complex.

(i) $C^\bullet$ is *minimal* if its differential is zero.
(ii) $C^\bullet$ is *contractible* if its homology is zero.

The homology of a complex can be seen as a minimal complex if we equip it with the zero differential. This idea leads to the following structure theorem.

Theorem 2.8 *Every complex of* $\mathbb{k}$*-vector spaces* $C^\bullet$ *is a direct sum of a contractible and a minimal complex.*

Sketch of the proof Because the C^i are vector spaces we can choose splits

$$C^i = H^i(C) \oplus I^i(C) \oplus R^i(C)$$

such that $\mathrm{Im}\, d = I^\bullet(C)$, $\mathrm{Ker}\, d = H^\bullet(C) \oplus I^\bullet(C)$ and $d \colon R^i(C) \to I^{i+1}(C)$ is an isomorphism. Note that we can see $C^\bullet_m := H^\bullet(C)$ and $C^\bullet_c := I^\bullet(C) \oplus R^\bullet(C)$ as subcomplexes of $C^\bullet$. The former is minimal, while the latter is contractible. □

This split allows us to decompose every linear map $\phi \colon C^\bullet \to D^\bullet$ into nine components: $\phi^{IH} \colon H^\bullet(C) \to I^\bullet(D)$, etc.

Lemma 2.9 *Let* $f \colon C^\bullet \to D^\bullet$ *be a morphism of chain complexes of* $\mathbb{k}$*-vector spaces.*

(i) f *is a quasi-isomorphism if and only if* f^{HH} *is an isomorphism.*
(ii) f *is null-homotopic if and only if* f^{HH} *is zero.*
(iii) *The map*

$$\pi \colon \mathsf{H}\,\mathrm{Hom}^\bullet(C^\bullet, D^\bullet) \to \mathrm{Hom}^\bullet(\mathsf{H}C^\bullet, \mathsf{H}D^\bullet) \colon f \mapsto \bar{f} = f^{HH}$$

is an isomorphism.
(iv) *A morphism* $f \colon C^\bullet \to D^\bullet$ *is a quasi-isomorphism if and only if there is a morphism* $f^\dagger \colon D^\bullet \to C^\bullet$ *such that* $f^\dagger f$ *and* $f f^\dagger$ *are homotopic to* $\mathbb{1}_C$ *and* $\mathbb{1}_D$ *(*$f^\dagger$ *is called a quasi-inverse of* f*).*

Sketch of the proof The first statement is obvious because by construction $H^\bullet(-)$ is naturally identified with the cohomology of the complex. To prove the second statement, note that according to the split the differential has the form

$$d = \begin{pmatrix} 0 & 0 & \delta \\ 0 & 0 & 0 \\ 0 & 0 & 0 \end{pmatrix},$$

where δ is invertible.

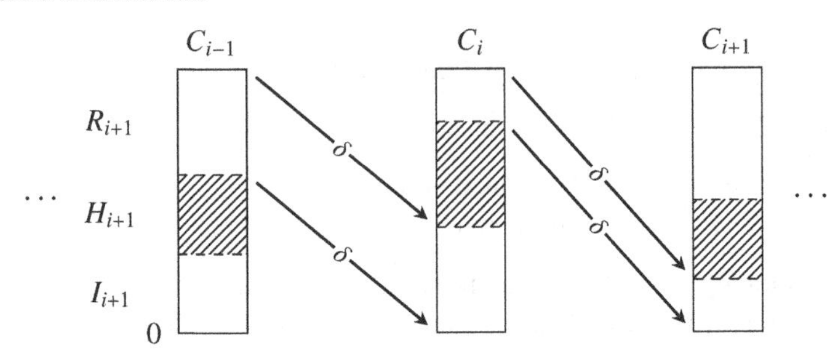

If we set

$$h = \begin{pmatrix} 0 & 0 & 0 \\ 0 & 0 & 0 \\ \delta^{-1} & 0 & 0 \end{pmatrix}$$

It is easy to see that dh, dh and $\mathbb{1} - dh - hd$ are the idempotents e_I, e_R and e_H that project onto the three components. Moreover, we have the following identities: $h^2 = 0$, $dhd = d$, $hdh = h$.

If f is a morphism of complexes and $f^{HH} = 0$ then

$$\begin{aligned}
0 = f^{HH} &= e_H f e_H \\
&= f - dhf - hdf - fdh + dhfdh + hdfdh - fhd + dhfhd + hdfhd \\
&= f - dhf - hdf - dfh + dfh + 0 - fhd + dhfhd + hfd \\
&= f - dhf - fhd + dhfhd \\
&= f - \left(d \underbrace{(hf - dhfh)}_{g} + \underbrace{(hf - dhfh)}_{g} d \right).
\end{aligned}$$

So $f = dg + gd$ is null-homotopic.

For the third part the injectivity of π follows directly from (b) and π is surjective because any map f for which only the component f^{HH} is nonzero is a morphism of complexes. To prove (d) we take for $f^\dagger$ any function in $\pi^{-1}((f^{HH})^{-1})$. □

2.1.4 A Categorical Point of View

From a more abstract point of view we can introduce the category $\mathtt{K}^\bullet(\Bbbk)$. The objects of $\mathtt{K}^\bullet(\Bbbk)$ are the cochain complexes of $\Bbbk$-vector spaces and the morphisms are all $\Bbbk$-linear maps: $\mathtt{K}^\bullet(C^\bullet, D^\bullet) = \mathrm{Hom}^\bullet(C^\bullet, D^\bullet)$. This category is in fact a *differentially graded or dg-category*. This is a special type of $\Bbbk$-linear category for which the hom-spaces are cochain complexes and the differentials satisfy the *graded Leibniz rule*: $d(f \circ g) = df \circ g + (-1)^{\deg f} f \circ dg$.

If we want to study complexes up to quasi-isomorphism, we have to work in a category where the quasi-isomorphisms are invertible. This can be achieved by looking at

$$\mathtt{HK}^0(\Bbbk)$$

whose objects are complexes of $\Bbbk$-vector spaces and whose hom-spaces are $\mathsf{H}\,\mathrm{Hom}^0(C^\bullet, D^\bullet)$. Lemma 2.9 implies that this category is equivalent to the category of graded $\Bbbk$-vector spaces with degree 0 homogeneous linear maps because the functor

$$\pi\colon \mathtt{HK}^0(\Bbbk) \to \mathtt{grVECT}(\Bbbk)\colon \begin{cases} C^\bullet & \mapsto \mathsf{H}C^\bullet, \\ f & \mapsto \mathsf{H}f \ \ (\cong f^{HH}) \end{cases}$$

is an equivalence of categories.

Remark 2.10 The discussion in the last two sections holds for complexes of vector spaces. As soon as the vector spaces have extra structure, like a ring structure or a module structure over some $\Bbbk$-algebra A, these structures usually do not survive the split and the quasi-inverse we constructed will not be compatible with the extra structure. Therefore we need to adapt the construction of $\mathtt{HK}^0(A)$ in case A is an algebra. This will be studied in detail in Chapter 3.

2.2 Cohomology in Topology

2.2.1 Gluing Simplices

Many topological spaces, like polyhedra, can be made by gluing smaller pieces together. If we want to describe a polyhedron made up of triangles, we need to give a set of triangles, a set of edges and a set of points. Moreover, for each of the triangles we need to assign three edges, and for each edge two points. There is a neat way to do this by means of $\boldsymbol{\Delta}$-sets [209].[1]

[1] In the literature these are also known as Δ-complexes [111]. We prefer to call them $\boldsymbol{\Delta}$-sets because the term complexes is used for cochain complexes. The $\boldsymbol{\Delta}$-sets are closely related to the more common notion of simplicial sets [87].

Definition 2.11 The *strictly increasing simplex category* Δ has as objects all sets $[n] := \{0, \dots, n\}$ with n any nonnegative integer and as morphisms all strictly increasing maps:

$$\Delta([m], [n]) = \{\phi\colon [m] \to [n] \mid x > y \implies \phi(x) > \phi(y)\}.$$

All strictly increasing maps are injections and they can be constructed by composing certain special maps: the *face maps*. These are injections that map $[n-1]$ into $[n]$ by skipping one element:

$$\delta_i\colon [n-1] \to [n]\colon x \mapsto \begin{cases} x, & x < i, \\ x+1, & x \geq i. \end{cases}$$

We can also realize this category in a geometrical way. The standard n-simplex is the topological space

$$\mathsf{Simp}_n = \{(x_0, \dots, x_n) \in [0,1]^n \mid \textstyle\sum x_i = 1\}.$$

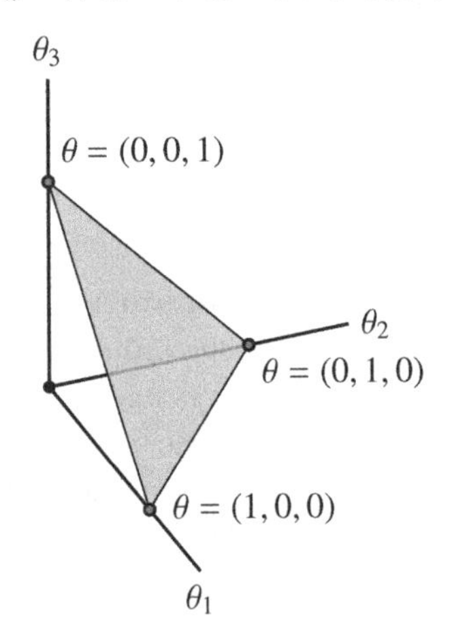

So Simp_0 is a point, Simp_1 a line segment, Simp_2 a triangle, etc. The face maps can be realized as the embeddings of the faces:

$$\delta_i\colon \mathsf{Simp}_{n-1} \to \mathsf{Simp}_n\colon (x_1, \dots, x_{n-1}) \mapsto (x_1, \dots, x_{i-1}, 0, x_i, \dots, x_{n-1}).$$

The geometrical strict simplex category has as objects the standard simplices, and its morphisms are generated by the face maps. By construction it is equivalent to Δ.

Suppose that we have a topological space built up of points, line segments and triangles, etc., and we have identified each of these objects with a standard n-simplex. We can describe these data as follows: for each n-simplex we have a set of objects that are identified with it. If we have a map $\phi\colon \mathsf{Simp}_n \to \mathsf{Simp}_m$

and an object that is identified with the standard m-simplex, then composing ϕ with this identification will give a map that identifies a standard n-simplex with another object. So ϕ gives us a map from objects that are identified with m-simplices to objects that are n-simplices. This warrants the following definition.

Definition 2.12 A Δ-*set* is a contravariant functor $\mathcal{S}\colon \Delta \to \texttt{Sets}$. To a Δ-set we associate its *geometrical realization*

$$|\mathcal{S}| = \left(\bigcup_i \mathcal{S}[i] \times \mathsf{Simp}_i\right) / \sim$$

where $(a, \phi(x)) \sim (\mathcal{S}(\phi)(a), x)$ for every morphism $\phi \in \Delta$.

Note that if ϕ is a face map δ_i, this equivalence relation implies that the simplex associated to the ith face of a (i.e. points of the form $(\mathcal{S}(\delta_i)(a), x)$) is identified with its embedding into the simplex of a (i.e. points of the form $(a, \delta_i(x))$).

Example 2.13

(i) The torus $\mathbb{S}^1 \times \mathbb{S}^1$ can be realized by the functor $\mathcal{S}$ with $\mathcal{S}[0] = \{P\}$, $\mathcal{S}[1] = \{\ell_0, \ell_1, \ell_2\}$ and $\mathcal{S}[2] = \{a, b\}$. The action on the hom-spaces is $\mathcal{S}\delta_i(\ell_j) = p$, $\mathcal{S}(\delta_i)(a) = \ell_i$ and $\mathcal{S}(\delta_i)(b) = \ell_{2-i}$.

(ii) The sphere $\mathbb{S}^2$ can be realized by the functor $\mathcal{S}$ with $\mathcal{S}[0] = \{p, q, r\}$, $\mathcal{S}[1] = \{\ell_0, \ell_1, \ell_2\}$ and $\mathcal{S}[2] = \{a, b\}$. The action on the hom-spaces is $\mathcal{S}(\delta_i)(a) = \ell_i$ and $\mathcal{S}(\delta_i)(b) = \ell_i$, while $\mathcal{S}\delta_i(\ell_j)$ is the head of ℓ_j if $i = 1$ and its tail if $i = 0$ (see the picture below).

Torus: Sphere:

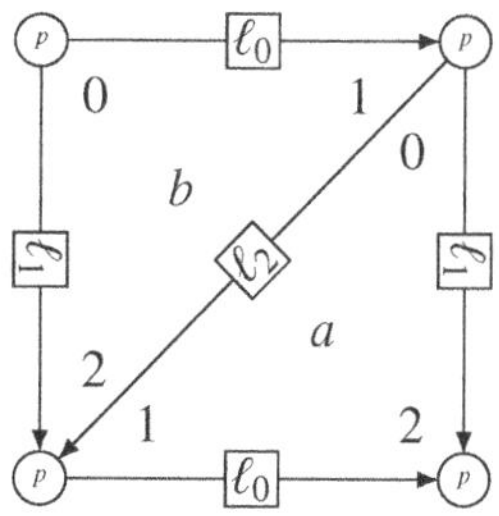

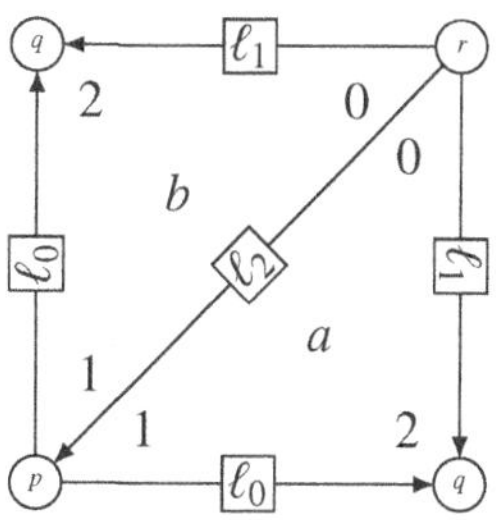

Different Δ-sets can give rise to homeomorphic topological spaces. In order to control this we want to define invariants that only depend on the topological realization.

2.2.2 Simplicial Cohomology

Definition 2.14 If $\mathcal{S}$ is a Δ-set and $\Bbbk$ is a field, we define a cochain complex of vector spaces,

$$\mathsf{C}\Delta^i(\mathcal{S}, \Bbbk) = \mathtt{Sets}(\mathcal{S}(i), \Bbbk),$$

with differential

$$df = \sum_j (-1)^j f \circ \delta_j.$$

One can easily check that d^2 is zero. The cohomology of $\mathsf{C}\Delta^\bullet(\mathcal{S}, \Bbbk)$ is called the *simplicial cohomology* of $\mathcal{S}$ and denoted by $\mathsf{H}^\bullet(\mathcal{S}, \Bbbk)$.

Remark 2.15 The construction can be split into two steps:

(i) We transform $\mathcal{S}$ into a left Δ-module $\mathcal{M}\colon \Delta \to \mathtt{VECT}(\Bbbk)$ by composing it with the functor $\mathtt{Sets}(-, \Bbbk)$.
(ii) We turn the left Δ-module $\mathcal{M}$ into a cochain complex $\mathcal{M}^\bullet$ by $\mathcal{M}^i = \mathcal{M}[i]$ and $d = \sum_j (-1)^j \mathcal{M}(\delta_j)$.

In some cases the Δ-set already has the structure of a Δ-module and then the first step can be omitted.

How can we interpret the simplicial cochain complex? Note that $f \in \mathsf{C}\Delta^i(\mathcal{S}, \Bbbk)$ associates to each k-simplex a number, which can be thought of as an amount that flows through that simplex. From this point of view df measures for each $(k+1)$-simplex how much flows in/out through its boundary. If $df = 0$ then the flow does not accumulate in some $(k+1)$-simplex so it represents a closed flow.

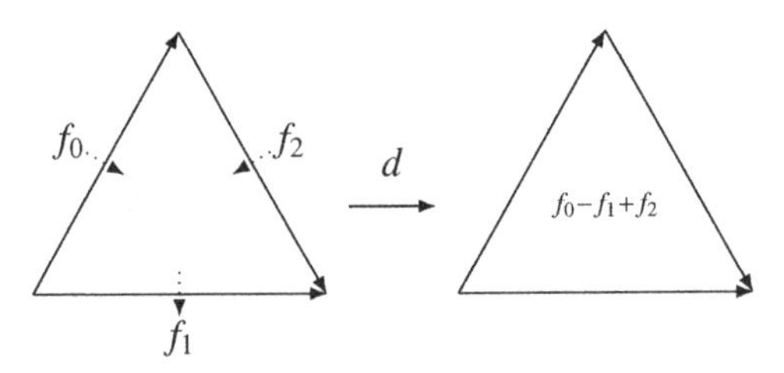

But suppose that the Δ-set has a hole (e.g. a missing $(k+1)$-simplex); then closed flows are allowed to accumulate in that missing simplex. These flows will however not be exact, so they end up in the cohomology. Therefore we can say that the cohomology of $\mathcal{S}$ measures holes in the topological space $|\mathcal{S}|$.

Definition 2.16 The cochain complex $\mathsf{C\Delta}^\bullet(\mathcal{S}, \Bbbk)$ comes with a natural product. If $f \in \mathsf{C\Delta}^p(\mathcal{S}, \Bbbk)$ and $g \in \mathsf{C\Delta}^q(\mathcal{S}, \Bbbk)$ we define the *cup product* $f \smile g \in \mathsf{C\Delta}^{p+q}(\mathcal{S}, \Bbbk)$ as

$$f \smile g(a) := f(\mathcal{S}(\iota_p)(a)) \cdot g(\mathcal{S}(\upsilon_q)(a)),$$

where ι_p is the embedding $[p] = \{0, \dots, p\} \subset \{0, \dots, p+q\} = [p+q]$ and $\upsilon_q \colon [q] \to [p+q] \colon x \mapsto x + p$.

One can check that this product is associative and satisfies the graded Leibniz rule:

$$d(f \smile g) = df \smile g + (-1)^p f \smile dg.$$

This means that we can also define a product on the cohomology because if $df = dg = 0$ then $d(f \smile g) = 0$ so products of closed elements are closed. Also

$$(f + dh) \smile g = f \smile g - d(h \smile g),$$

so adding something exact does not change the cohomology class of the product.

Definition 2.17 The space $\mathsf{HC\Delta}^\bullet(\mathcal{S}, \Bbbk)$ equipped with the cup product is called the *cohomology ring* of $\mathcal{S}$ (with coefficients in $\Bbbk$) and denoted by $\mathsf{H}^\bullet(\mathcal{S}, \Bbbk)$.

Example 2.18 We calculate the cohomology ring of the torus with the Δ-set of 2.13. Let P, L_0, L_1, L_2, A, B be the functions that map $p, \dots, b$ to 1 and the rest to 0. The differential satisfies $dP = dA = dB = 0$ and $dL_i = (-1)^{i+1}(A+B)$, so we can use $[P], [L_0+L_1], [L_2+L_1], [A] = -[B]$ as a basis for the cohomology. We have that $P \smile X = X$ because the corners of all simplices are p. Furthermore $\mathcal{S}\iota_1(a) = \ell_2$ and $\mathcal{S}\upsilon_1(a) = \ell_0$, while for b we have the opposite $\mathcal{S}\iota_1(b) = \ell_0$ and $\mathcal{S}\upsilon_1(b) = \ell_2$. Therefore

$$L_0 \smile L_2 = B \text{ and } L_2 \smile L_0 = A,$$

while all other products between L_i are zero. The cohomology ring is isomorphic to

$$\mathsf{HC\Delta}^\bullet(\mathcal{S}, \Bbbk) \cong \frac{\Bbbk\langle \xi, \eta \rangle}{\langle \xi\eta + \eta\xi, \xi^2, \eta^2 \rangle}$$

where we identified

$$\begin{aligned} 1 &= [P], \\ \xi &= [L_0 + L_1], \\ \eta &= [L_2 + L_1], \text{ and} \\ \xi\eta &= [B] = -[A] = -\eta\xi. \end{aligned}$$

Different Δ-sets with homeomorphic topological realizations will result in different cochain complexes with the same cohomology ring. One way to establish this result is to relate them to a cohomology ring that only depends on the topology of $\mathbb{X} = |\mathcal{S}_1| = |\mathcal{S}_2|$.

Definition 2.19 For a topological space $\mathbb{X}$ define the Δ-set $\mathcal{X}$ whose k-simplices are all continuous maps from Simp_k to $\mathbb{X}$:

$$\mathcal{X}[i] = \mathsf{Top}(\mathsf{Simp}_k, \mathbb{X}) \quad \text{and} \quad \mathcal{X}[\delta_i](f) = f \circ \delta_i.$$

The cohomology ring $\mathsf{HC\Delta}^\bullet(\mathcal{X}, \Bbbk)$ is called the *singular cohomology ring* of $\mathbb{X}$ and denoted by

$$\mathsf{H}^\bullet(\mathbb{X}, \Bbbk) := \mathsf{HC\Delta}^\bullet(\mathcal{X}, \Bbbk).$$

Theorem 2.20 (Topological invariance) *If $\mathcal{S}$ is a Δ-set with realization $\mathbb{X} = |\mathcal{S}|$ then $\mathsf{HC\Delta}^\bullet(\mathcal{S}, \Bbbk) \cong \mathsf{H}^\bullet(\mathbb{X}, \Bbbk)$ as graded rings.*

Sketch of the proof The simplicial complex $\mathcal{S}$ naturally embeds into $\mathcal{X}$ and this gives a restriction map

$$\mathsf{C\Delta}^\bullet(\mathcal{X}, \Bbbk) \to \mathsf{C\Delta}^\bullet(\mathcal{S}, \Bbbk),$$

which is a morphism of dg-algebras. One can show that it is a quasi-isomorphism, so on the level of cohomology it induces an isomorphism of graded rings. More information can be found in [111]. □

Remark 2.21 If we have two Δ-sets $\mathcal{S}_1$ and $\mathcal{S}_2$ with $|\mathcal{S}_1| \cong |\mathcal{S}_2| = \mathbb{X}$ then we have the diagram

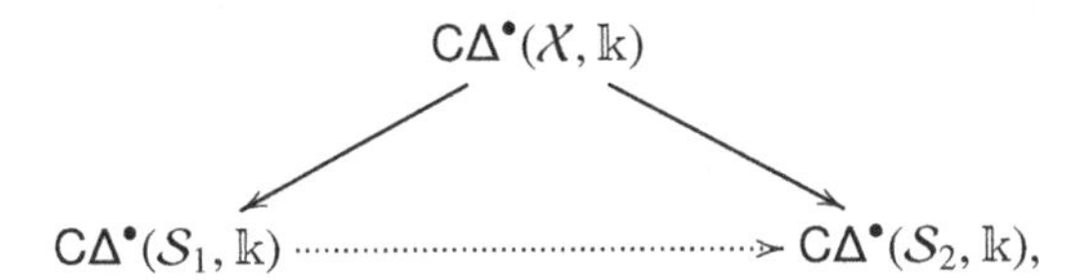

where the two solid arrows are quasi-isomorphisms of rings. By Lemma 2.9 there exists a quasi-isomorphism of vector spaces that makes the diagram commute: compose the second projection with a quasi-inverse of the first. Unfortunately, this quasi-isomorphism is in general not a quasi-isomorphism of rings.

2.2.3 De Rham Cohomology

Let $\mathbb{M}$ be a smooth compact orientable manifold. Recall that a k-form is a section of the antisymmetric product $\Lambda^k T^*\mathbb{M}$. In local coordinates we can write such a form as

$$\omega = \sum_{i_1<\cdots<i_k} \omega_{i_1\ldots i_k} dx_{i_1} \wedge \cdots \wedge dx_{i_k},$$

and the differential is given by

$$d\omega = \sum_{i_1<\cdots<i_k} \sum_j \frac{\partial \omega_{i_1\ldots i_k}}{\partial x_j} dx_j \wedge dx_{i_1} \wedge \cdots \wedge dx_{i_k},$$

where we used the convention that $dx_i \wedge dx_j = -dx_j \wedge dx_i$. The space of k-forms is denoted by $\Omega^k(\mathbb{M})$.

Definition 2.22 The triple $(\Omega^\bullet(\mathbb{M}), d, \wedge)$ forms a dg-algebra which is called the *de Rham complex*. The cohomology ring of this complex is called the *de Rham cohomology*:

$$H^\bullet_{\mathrm{dR}}(\mathbb{M}) := \mathsf{H}\Omega^\bullet(\mathbb{M}).$$

If $f\colon \mathsf{Simp}_k \to \mathbb{M}$ is a smooth map, we can pull back any k-form $\omega \in \Omega^k(\mathbb{M})$ and integrate it. If $\mathcal{S}$ is a Δ-set such that $|\mathcal{S}| \cong \mathbb{M}$ this procedure gives a map

$$\phi\colon \Omega^\bullet(\mathbb{M}) \to \mathsf{C}\Delta^\bullet(\mathcal{S}, \mathbb{R})\colon \omega \mapsto f \quad \text{with } f(a) = \int_a \omega,$$

where $\int_a$ stands for integration over the simplex $\{a\} \times \mathsf{Simp}_k$ embedded in $|\mathcal{S}| \cong \mathbb{M}$.

Stokes' theorem implies that this map is a morphism of complexes:

$$\phi(d\omega)(a) = \int_a d\omega = \int_{\partial a} \omega = \sum_i \pm \int_{\delta_i a} \omega = d\phi(\omega)(a).$$

This morphism is also compatible with the products and we have the following theorem. Its proof and more details can be found in [45].

Theorem 2.23 (De Rham) *The map $\phi\colon \Omega^\bullet(\mathbb{M}) \to \mathsf{C}\Delta^\bullet(\mathcal{S}, \mathbb{R})$ is a quasi-isomorphism.*

2.2.4 Massey Products

Going from the product on $\mathsf{C}\Delta^\bullet(\mathcal{S}, \mathbb{R})$ or $\Omega^\bullet(\mathbb{M})$ to the product on the cohomology tends to lose information. To remedy this one can introduce higher products, called *Massey products*. A Massey product is a way to associate new cohomology classes to a sequence of cohomology classes that satisfy certain additional constraints.

The ternary Massey product is constructed as follows. Consider three cohomology classes $[u]$, $[v]$, $[w]$ and suppose that $[u \smile v] = [v \smile w] = 0$.

This means that there are elements $s, t \in \mathsf{C}\Delta^i(\mathcal{S}, \Bbbk)$ such that $u \smile v = ds$ and $v \smile w = dt$. Using the Leibniz rule we see that

$$d(u \smile t) = du \smile t \pm u \smile dt = \pm u \smile v \smile w,$$
$$d(s \smile w) = ds \smile w \pm s \smile dw = u \smile v \smile w.$$

Therefore, if we choose the appropriate sign, we have that $d(u \smile t \pm s \smile w) = 0$ and it makes sense to look at the cohomology class $[u \smile t \pm s \smile w]$. Note that this definition depends on the choice of s, t and the Massey product $\langle [u], [v], [w] \rangle$ is defined as the set of all classes we can get in this way by varying s, t.

Example 2.24 Let $\mathcal{S}$ be a Δ-set that realizes the unit cube $[0, 1]^3$. Let $\gamma_1, \gamma_2, \gamma_3 \colon \mathbb{S}^1 \to (0, 1)^3$ be three disjoint circles embedded in the unit cube such that each simplex intersects at most one curve and only in one interval. Define $\mathcal{S}_\gamma$ as the Δ-set containing all simplices disjoint from the γ_i.

Fix a disk $\mathbb{D}_i \subset [0, 1]^3$ bounding γ_i and define $\lambda_i \in \mathsf{H}^1(\mathcal{S}_\gamma, \Bbbk)$ as

$$\forall\, a \in \mathcal{S}(1), \text{ we have } \lambda_i(a) = \begin{cases} 1 & \text{if } a \text{ crosses } \mathbb{D}_i \text{ in the positive direction,} \\ -1 & \text{if } a \text{ crosses } \mathbb{D}_i \text{ in the negative direction,} \\ 0 & \text{if } a \text{ does not cross } \mathbb{D}_i. \end{cases}$$

Although this definition depends on the choice of $\mathbb{D}_i$, its cohomology class only depends on γ_i.

If γ_i and γ_j are not linked then one can show that $\lambda_i \smile \lambda_j = 0$ in $\mathsf{H}^2(\mathcal{S}_\gamma, \Bbbk)$, while if they are linked like a Hopf link then $\lambda_i \smile \lambda_j \neq 0$.

Hopf link: 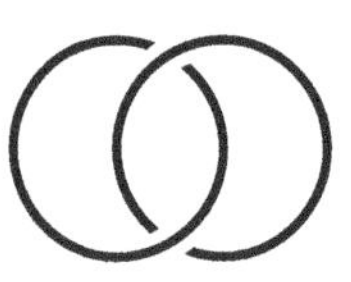 Borromean rings: 

The Borromean rings are three rings embedded in three-dimensional space such that if we ignore one ring the other two can be pulled apart without intersecting, but all three together cannot be pulled apart. Therefore we have that for all i, j, $\lambda_i \smile \lambda_j = 0$, so this is a situation where we can define the Massey product. In [171] Massey calculated that $\langle \lambda_1, \lambda_2, \lambda_3 \rangle$ contains a nonzero cohomology class, while if we do the same calculations for three circles that are completely unlinked then $\langle \lambda_1, \lambda_2, \lambda_3 \rangle$ contains only zero.

The ring structure of the cohomology ring does not see the difference between the Borromean rings and the loose rings, but the Massey products do. In the next chapter we will see a way to enhance the cohomology ring that incorporates these higher products.

2.3 Cohomology in Algebra

The idea of cutting a complicated space into simpler pieces and looking at how they are glued together can be imitated in many different areas of mathematics. Here we will investigate this idea in representation theory for algebras.

2.3.1 Resolutions

Let A be an algebra over $\Bbbk$ and let M be a right A-module. If M is generated by a set S_0 we can make a surjective morphism from a free module to A:

$$M \xleftarrow{m\cdot} \bigoplus_{m \in S_0} A =: P^0.$$

This map has a kernel with generating set S_1, so we can construct a map

$$P^0 \xleftarrow{m\cdot} \bigoplus_{m \in S_1} A =: P^{-1}.$$

If we continue like this (and set $P^i = 0$ for $i > 0$) we end up with a cochain complex of free modules $P^\bullet$. By construction, the homology of this complex is M in degree 0 and 0 in all other degrees.

Definition 2.25 A *free resolution* of a right A-module M is a complex of free A-modules $P^\bullet$ such that $\mathsf{H}P^\bullet = M^\bullet$, where $M^\bullet$ is the complex with M in degree 0 and all other components 0.

Note that this construction does not give a unique free resolution, because it depends on the choice of generators. But one can show that any two free resolutions of M are quasi-isomorphic. Even better, we have the following lemma.

Theorem 2.26 (Comparison theorem) *If $\varphi\colon M \to N$ is a morphism and $P^\bullet$, $Q^\bullet$ are free resolutions of M, N, then there exists a morphism of complexes $\bar{\varphi}$ such that the induced morphism on the homology is φ. This lift is unique up to homotopy.*

Sketch of the proof The idea is to construct $\bar{\varphi}$ stepwise: Because $Q^0 \to N$ is surjective we can find $q_i \in Q^0$ that map to the images of the generators of P^0 under the map $P^0 \to M \to N$. If we map the generators of P^0 to these q_i we have constructed a lift $\bar{\varphi}^0$ for P_0.

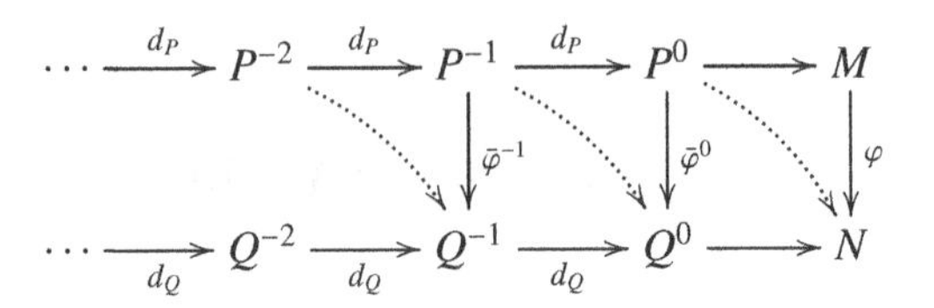

Now look at the images of the generators of P^{-1} under $\bar{\varphi}^0 d_P$. Because of the lifting property they sit in the kernel of $Q_0 \to N$, so we can find preimages for them under d_Q and we map the generators of P^{-1} to these preimages to obtain $\bar{\varphi}^{-1}$. Then we look at the generators of P^{-2} and keep on performing the same procedure to obtain the full $\bar{\varphi}$. The uniqueness can be shown with a similar argument; see [253, Theorem 2.2.6]. □

In particular if we take the identity morphism $\mathbb{1}_M$ for ϕ we get a quasi-isomorphism between each pair of free resolutions $P^\bullet$, $Q^\bullet$ of a module M.

Remark 2.27 Instead of working with free resolutions, it is also possible to work with the more general notion of a projective resolution, in which all components are projective A-modules (i.e. direct summands of free modules). Just like in the free case, all projective resolutions of a module are quasi-isomorphic.

2.3.2 Ext-Groups

Suppose that we have two A-modules M, N with resolutions $P^\bullet$, $Q^\bullet$. Between these two complexes we can look at the hom-complex

$$\mathrm{Hom}_A^j(P^\bullet, Q^\bullet) := \bigoplus_i \mathrm{Hom}_A(P^i, Q^{i+j}) \quad \text{with} \quad df = d_Q f - (-1)^{\deg f} f d_P.$$

Although this complex depends on the specific resolutions, the cohomology will only depend on M and N. In the case of the zeroth cohomology this follows straight from Theorem 2.26 as

$$\mathrm{H}\,\mathrm{Hom}_A^0(P^\bullet, Q^\bullet) \cong \mathrm{Hom}_A(M, N).$$

For the other cohomology groups this is a consequence of the fact that all free resolutions of the same module are quasi-isomorphic.

Definition 2.28 If M is an A-module with free resolution $P^\bullet$, and N an A-module with free resolution $Q^\bullet$, we define the ith *Ext-group* as

$$\mathrm{Ext}^i_A(M,N) := \mathsf{H}\,\mathrm{Hom}^i_A(P^\bullet, Q^\bullet).$$

Remark 2.29 By the comparison theorem this definition does not depend on the choices of free resolutions for M and N. Moreover, using a similar argument as in Theorem 2.26 one can also show that $\mathsf{H}\,\mathrm{Hom}^i_A(P^\bullet, Q^\bullet) = \mathsf{H}\,\mathrm{Hom}^i_A(P^\bullet, N)$; this often makes the calculation of the $\mathrm{Ext}^i_A(M,N)$ much easier.

Example 2.30 Consider the algebra $\Bbbk[X]$ and let S_λ be the simple one-dimensional module on which X acts as λ. This module has a resolution

$$P^\bullet_\lambda := \Bbbk[X] \xrightarrow[d_{-1}]{(X-\lambda)} \Bbbk[X]$$

and therefore the complex $\mathrm{Hom}^\bullet_{\Bbbk[X]}(P^\bullet_\lambda, P^\bullet_\mu)$ takes the form

$$\Bbbk[X] \xrightarrow[d_{-1}]{\begin{pmatrix} X-\mu \\ -X+\lambda \end{pmatrix}} \Bbbk[X]^{\oplus 2} \xrightarrow[d_0]{(X-\lambda\ \ X-\mu)} \Bbbk[X]\ ,$$

where the middle term has degree 0. Because $\Bbbk[X]$ is a domain we have $\mathrm{Ext}^{-1}(S_\lambda, S_\mu) = 0$. This is a general feature: for any algebra and modules $\mathrm{Ext}^{<0}_A(M,N) = 0$.

For Ext^1 the answer depends on the values of λ, μ. The image of d_0 is the ideal $(X-\lambda, X-\mu) \lhd \Bbbk[X]$. If $\lambda \neq \mu$ this ideal is the whole ring and $\mathrm{Ext}^1(S_\lambda, S_\mu) = 0$, while if $\lambda = \mu$ we get

$$\mathrm{Ext}^1(S_\lambda, S_\lambda) = \frac{\Bbbk[X]}{(X-\lambda)} = \Bbbk.$$

There is also a nice representation-theoretic interpretation of $\mathrm{Ext}^1_A(M,N)$.

Definition 2.31 An *extension* is a short exact sequence $N \hookrightarrow E \twoheadrightarrow M$. We say that two extensions are *equivalent* if there is a commutative diagram

$$\begin{array}{ccccc} N & \hookrightarrow & E_1 & \twoheadrightarrow & M \\ \downarrow{\scriptstyle \mathbb{1}_N} & & \downarrow{\scriptstyle \phi} & & \downarrow{\scriptstyle \mathbb{1}_M} \\ N & \hookrightarrow & E_2 & \twoheadrightarrow & M. \end{array}$$

Theorem 2.32 *The space* $\mathrm{Ext}^1_A(M,N)$ *measures all extensions from* M *to* N *up to equivalence. In particular, if* $\mathrm{Ext}^1_A(M,N) = 0$ *then every short exact sequence* $N \hookrightarrow E \twoheadrightarrow M$ *splits:* $E \cong M \oplus N$.

Sketch of the proof Given an extension we can construct a morphism by combing back the identity on M:

$$\begin{array}{ccccccc} \cdots & \longrightarrow & P^{-1} & \longrightarrow & P^0 & \longrightarrow & M \\ & & \downarrow{\scriptstyle \phi_1} & & \downarrow{\scriptstyle \phi_0} & & \downarrow{\scriptstyle \mathbb{1}_M} \\ & & N & \hookrightarrow & E_2 & \twoheadrightarrow & M. \end{array}$$

This induces a map $[\phi_1] \in \mathrm{Ext}^1_A(M, N)$. One can show that this identification is an isomorphism; see [253, Theorem 3.4.3]. If $\mathrm{Ext}^1_A(M, N) = 0$ then every extension is isomorphic to the trivial extension and hence splits. □

Example 2.33 In the previous example one can easily check that the sequence

$$\frac{\Bbbk[X]}{(X-\lambda)} \xhookrightarrow{X-\lambda} \frac{\Bbbk[X]}{(X-\lambda)^2} \twoheadrightarrow \frac{\Bbbk[X]}{(X-\lambda)}$$

does not split and this gives rise to the nonzero Ext-group.

2.3.3 The Ext-Ring

Given an A-module M and a free resolution $P^\bullet$, the hom-space $\mathrm{Hom}^\bullet_A(P^\bullet, P^\bullet)$ has a ring structure coming from the composition of morphisms. Because this ring structure is compatible with the differential it also induces a ring structure on the cohomology. This gives us a graded ring $\mathrm{Ext}^\bullet_A(M, M)$, which is independent of the choice of resolutions. In this section we will look at the structure of the Ext-ring in some special cases.

Example 2.34 The simplest case happens when $A = \Bbbk[X]$ and $M = \Bbbk[X]/(X)$. In this case, M has a resolution $F^\bullet = A \xrightarrow{X} A$ and

$$\mathrm{Hom}^\bullet_A(F^\bullet, F^\bullet) = \left(\mathrm{Mat}_2(A), df = f\left(\begin{smallmatrix} 0 & X \\ 0 & 0 \end{smallmatrix}\right) \pm \left(\begin{smallmatrix} 0 & X \\ 0 & 0 \end{smallmatrix}\right) f\right).$$

The homology is a two-dimensional graded vector space

$$\Bbbk\left(\begin{smallmatrix} 1 & 0 \\ 0 & 1 \end{smallmatrix}\right) \oplus \Bbbk\left(\begin{smallmatrix} 0 & 1 \\ 0 & 0 \end{smallmatrix}\right) =: \Bbbk\iota \oplus \Bbbk\xi,$$

the generator ξ has degree 1 and $\xi^2 = 0$, so

$$\mathrm{Ext}^\bullet_A(M, M) \cong \frac{\Bbbk[\xi]}{\langle \xi^2 \rangle}.$$

This ring is called the *ring of dual numbers*.

Example 2.35 Let us look at the case where $A = \Bbbk[X, Y]$ and the module is the simple module on which X, Y act trivially: $S = \Bbbk[X, Y]/(X, Y)$. In this case, we have a very nice resolution, called the Koszul resolution,

$$K^\bullet := A \xrightarrow[d_{-2}]{\binom{Y}{-X}} A^{\oplus 2} \xrightarrow[d_{-1}]{(X\ Y)} A.$$

Keeping Remark 2.29 in mind, we see that the Ext-groups are given by

$$\begin{aligned}\operatorname{Ext}^\bullet_A(S,S) &= \mathsf{H}\operatorname{Hom}^\bullet_A(K^\bullet, S)\\ &= S \xrightarrow{\binom{0}{0}} S^{\oplus 2} \xrightarrow{(0\ 0)} S\\ &= \Bbbk\iota \oplus (\Bbbk\xi \oplus \Bbbk\eta) \oplus \Bbbk\rho,\end{aligned}$$

where ι, ξ, η and ρ represent the map $A \to S : a \to a \mod (X, Y)$ applied to each of the four copies of A in the resolution. We can lift these four maps to complexes of morphisms $K^\bullet \to K^\bullet$. Clearly, ι lifts to the identity map and

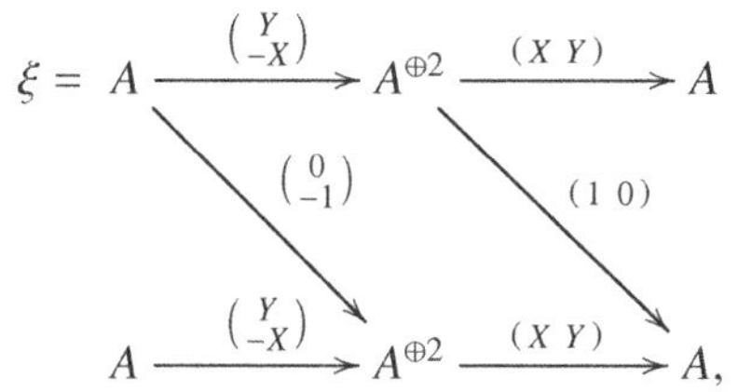

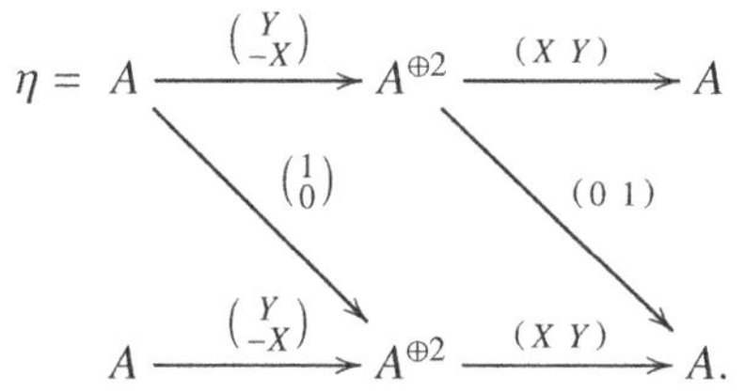

$$\begin{array}{cccccc} \xi = & A & \xrightarrow{\binom{Y}{-X}} & A^{\oplus 2} & \xrightarrow{(X\ Y)} & A \\ & & \searrow^{\binom{0}{-1}} & & \searrow^{(1\ 0)} & \\ & A & \xrightarrow{\binom{Y}{-X}} & A^{\oplus 2} & \xrightarrow{(X\ Y)} & A, \end{array}$$

$$\begin{array}{cccccc} \eta = & A & \xrightarrow{\binom{Y}{-X}} & A^{\oplus 2} & \xrightarrow{(X\ Y)} & A \\ & & \searrow^{\binom{1}{0}} & & \searrow^{(0\ 1)} & \\ & A & \xrightarrow{\binom{Y}{-X}} & A^{\oplus 2} & \xrightarrow{(X\ Y)} & A. \end{array}$$

These diagrams anticommute instead of commute because $\deg \xi = \deg \eta = 1$. Their composition

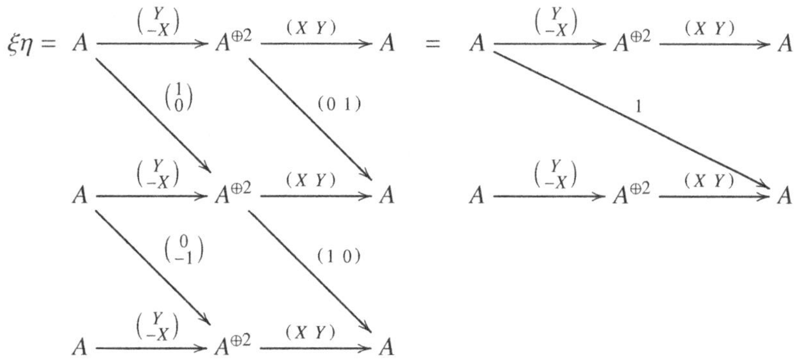

$$\begin{array}{cccccccccccc} \xi\eta = & A & \xrightarrow{\binom{Y}{-X}} & A^{\oplus 2} & \xrightarrow{(X\ Y)} & A & = & A & \xrightarrow{\binom{Y}{-X}} & A^{\oplus 2} & \xrightarrow{(X\ Y)} & A \\ & & \searrow^{\binom{1}{0}} & & \searrow^{(0\ 1)} & & & & & \searrow^{1} & & \\ & A & \xrightarrow{\binom{Y}{-X}} & A^{\oplus 2} & \xrightarrow{(X\ Y)} & A & & A & \xrightarrow{\binom{Y}{-X}} & A^{\oplus 2} & \xrightarrow{(X\ Y)} & A \\ & & \searrow^{\binom{0}{-1}} & & \searrow^{(1\ 0)} & & & & & & & \\ & A & \xrightarrow{\binom{Y}{-X}} & A^{\oplus 2} & \xrightarrow{(X\ Y)} & A & & & & & & \end{array}$$

gives a lift of ρ, while the composition in the other direction gives a lift of $-\rho$. The composition of ξ with itself is zero and so is the composition of η with itself, so we can conclude that

$$\mathrm{Ext}^\bullet_A(S,S) \cong \frac{\Bbbk\langle \xi, \eta\rangle}{\langle \xi^2, \eta^2, \xi\eta + \eta\xi\rangle}.$$

In general we obtain the following result; see [253, Section 4.5].

Theorem 2.36 *If $A = \Bbbk[X_1, \dots, X_n]$ and $S = \Bbbk[X_1, \dots, X_n]/(X_1, \dots, X_n)$ then*

$$\mathrm{Ext}^\bullet_A(S,S) \cong \frac{\Bbbk\langle \xi_1, \dots, \xi_n\rangle}{\langle \xi_i\xi_j + \xi_j\xi_i \mid 1 \le i, j \le n\rangle} =: \Lambda_n.$$

This algebra is also known as the exterior algebra with n variables.

We can generalize this construction to more complicated algebras. Suppose that we have an algebra of the form

$$A = \frac{\Bbbk\langle X_1, \dots, X_k\rangle}{\langle r_1, \dots, r_l\rangle},$$

where the relations are all linear combinations of monomials of degree at least 2. If we denote $\langle X_1, \dots, X_k\rangle$ by $\mathcal{X}$ and $\langle r_1, \dots, r_l\rangle$ by $\mathcal{R}$, we can write this condition as

$$\mathcal{R} \subset \mathcal{X}^2.$$

Furthermore, we will assume that the r_i give a basis for the vector space $\mathcal{R}/\mathcal{X}\mathcal{R} + \mathcal{R}\mathcal{X}$. If this is the case we call $\Bbbk\langle X_1, \dots, X_k\rangle/\langle r_1, \dots, r_l\rangle$ a *minimal presentation* of A. Just like for the polynomial ring, we can consider the simple module $S = A/\mathcal{X}$ and look at its Ext-ring.

Lemma 2.37 *The module S has a free resolution that starts like*

$$K^\bullet := \cdots \longrightarrow A^{\oplus l} \xrightarrow{(\partial_i r_j)_{ij}} A^{\oplus k} \xrightarrow{(X_1 \dots X_k)} A\,,$$

where ∂_i is a $\Bbbk$-linear operator such that $\partial_i(X_{s_1} \dots X_{s_u}) = X_{s_2} \dots X_{s_u}$ if $X_i = X_{s_1}$ and zero otherwise.

Sketch of the proof First remark that the ∂_i-operator removes an X_i at the front of each monomial of r_j and deletes all monomials for which this is impossible. This means that for every $f \in \Bbbk\langle X_1, \dots, X_k\rangle$ we have the identity $f = \sum_i X_i(\partial_i f)$.

Because the X_i are generators of the ring A, the map $A^{\oplus k} \xrightarrow{(X_1 \dots X_k)} A$ is surjective. Now suppose that $f_i \in \Bbbk\langle X_1, \dots, X_k\rangle$ are such that $\sum_i X_i f_i = r \in R$. We can write each r as $\sum_j r_j g_j + u$, where u is a linear combination of expressions of the form $v r_j w$ with v at least degree 1. If we apply ∂_i to r we get

$$f_i = \partial_i r = \sum (\partial_i r_j) g_j + \partial_i u.$$

The last term is zero in A because ∂_i does not touch the relations inside. In A we have $f_i = \sum(\partial_i r_j) g_j$ and the vector $(f_i)_i$ is a linear combination of the $(\partial_i r_j)_i$. Therefore, the second map in the resolution maps onto the kernel of the first. For the rest of the maps we proceed as usual: find generators of the kernel of d_i and map a free module onto them. □

This resolution allows us to describe a significant part of the Ext-algebra and its product structure.

Theorem 2.38 *If A and S are as above then*

- $\mathrm{Ext}^0_A(S,S) = \Bbbk$,
- $\mathrm{Ext}^1_A(S,S) = \Bbbk^{\oplus k} := \Bbbk\xi_1 \oplus \cdots \oplus \Bbbk\xi_k$,
- $\mathrm{Ext}^2_A(S,S) = \Bbbk^{\oplus l} := \Bbbk\rho_1 \oplus \cdots \oplus \Bbbk\rho_l$,

and the product in $\mathrm{Ext}^\bullet_A(S,S)$ *satisfies the following identities:*

$$\xi_i\xi_j = \sum_u \lambda^u_{ij}\rho_u \iff r_u = \sum_{ij} \lambda^u_{ij} X_j X_i + \textit{higher-degree terms.}$$

Sketch of the proof If we map the resolution $K^\bullet$ to S we get

$$S \xrightarrow{0} S^{\oplus k} \xrightarrow{0} S^{\oplus l} \xrightarrow{0} \cdots .$$

The first map is zero because all generators act as zero on S, and the second is zero because no $(\partial_i r_j)$ contains a constant term as $R \subset \langle X_1, \dots, X_k\rangle^2$. The third map is zero because the r_i form a basis for $\mathcal{R}/\mathcal{X}\mathcal{R} + \mathcal{R}\mathcal{X}$.

Because these maps are zero we have $\mathrm{Ext}^1_A(S,S) = \mathrm{Hom}(A^{\oplus k}, S) \cong \Bbbk^{\oplus k}$ and we can choose basis elements ξ_i that map the 1 in the ith summand of $A^{\oplus k}$ to $1 \in S$. This basis can be considered as dual to the generators $X_1, \dots, X_k$. Similarly, we have a basis $\rho_1, \dots, \rho_l$ for $\mathrm{Ext}^2_A(S,S) = \mathrm{Hom}(A^{\oplus l}, S) \cong \Bbbk^{\oplus l}$ that is dual to the relations $r_1, \dots, r_l$. The structure constants are determined in a completely analogous way to the calculation for the polynomial ring. For more details we refer to [168]. □

Remark 2.39 This theorem tells us that we can recover the quadratic part of the relations from the multiplication in the Ext-ring. A natural question that pops up is whether we can also recover the cubic or higher-degree parts. This is not possible using the product alone, but this result seems to suggest that we may need to define a triple product to recover the cubic part, a quadruple product for the quartic part and so forth. In the next chapter we will see that this is indeed the way to go.

Remark 2.40 In the section on cohomology in geometry we found out that the cohomology ring of a torus is the exterior algebra $\Lambda_2 = \mathbb{C}\langle\xi,\eta\rangle/\langle\xi^2,\eta^2,\xi\eta+\eta\xi\rangle$, while in the section on cohomology in algebra we found out that the same ring can be interpreted as the Ext-ring of a simple $\mathbb{C}[X,Y]$-module. So two seemingly unrelated objects give rise to the same cohomology ring. This coincidence is one of the basic facts that form the foundation of mirror symmetry.

2.4 Exercises

Exercise 2.1 Lemma 2.9 works for vector spaces but for more general modules the situation is more complicated.

(i) Show that Lemma 2.9 fails for complexes of $\mathbb{Z}$-modules. (Hint: look at $\mathbb{Z} \xrightarrow{2\cdot} \mathbb{Z}$.)
(ii) Show that Lemma 2.9 holds for complexes of $\mathbb{C}G$-modules if G is a finite group.

Exercise 2.2 The *simplex category* $\underline{\Delta}$ has as objects $[n] = \{0,\dots,n\}$ and as morphisms all nondecreasing maps $\underline{\Delta}([m],[n]) = \{\phi\colon [m] \to [n] \mid x \geq y \implies \phi(x) \geq \phi(y)\}$. A simplicial set is a contravariant functor from $\underline{\Delta}$ to sets. Note that the strictly increasing simplex category Δ is a subcategory of $\underline{\Delta}$, so every simplicial set can also be seen as a Δ-set. To go in the opposite direction we need to do some work.

(i) Show that if $\mathcal{S}$ is a nontrivial simplicial set then $\mathcal{S}[i]$ is nonempty for all $i \geq 0$.
(ii) If $\mathcal{S}$ is a simplicial set and $a \in \mathcal{S}[i]$ then we call a degenerate if it is in the image of $\mathcal{S}(f)$ for f some noninjective map. Define $\mathcal{S}^{\mathrm{nd}}[i]$ to be the set of nondegenerate i-simplices. Show that $\mathcal{S}^{\mathrm{nd}}$ is a Δ-set.
(iii) Show that $\mathsf{C}\Delta^\bullet(\mathcal{S},\Bbbk)$ and $\mathsf{C}\Delta^\bullet(\mathcal{S}^{\mathrm{nd}},\Bbbk)$ are quasi-isomorphic complexes.

Exercise 2.3 Take a two-dimensional Δ-set $\mathcal{S}$ and subdivide the triangles by introducing one extra point at their centers.

Show that the new simplicial cochain complex is quasi-isomorphic to the old cochain complex.

Exercise 2.4 Let $\mathbb{X} = |\mathcal{S}|$ be a topological space coming from a Δ-set $\mathcal{S}$ and $\mathbb{S}^1$ be the circle. Construct a Δ-set for $\mathbb{X} \times \mathbb{S}^1$ and use it to show that

$$\mathsf{H}^k(\mathbb{X} \times \mathbb{S}^1, \mathbb{k}) \cong \mathsf{H}^{k-1}(\mathbb{X}, \mathbb{k}) \oplus \mathsf{H}^k(\mathbb{X}, \mathbb{k}).$$

Exercise 2.5 Calculate the cohomology ring of a compact orientable surface with genus g using an appropriate Δ-set.

Exercise 2.6 Let $\mathbb{T} = \mathbb{R}^2/\mathbb{Z}^2$ be the torus and consider $\Omega^\bullet(\mathbb{T})$. As a $C_\infty(\mathbb{T})$-module this is generated by $1, dX, dY, dX \wedge dY$, where X, Y are the standard coordinates on $\mathbb{R}^2$. Show that these elements also form a basis for the de Rham cohomology ring. Divide the torus into two triangles and use this division to describe the isomorphism between simplicial cohomology and de Rham cohomology.

Exercise 2.7 Consider the ring $R = \mathbb{C}\langle X_1, Y_1, \ldots, X_g, Y_g\rangle/\langle \sum_i X_iY_i - Y_iX_i\rangle$ and let $S = R/\langle X_1, Y_1, \ldots, X_g, Y_g\rangle$. Calculate the Ext-ring of S.

Exercise 2.8 Let $R = \mathbb{C}[\epsilon]/(\epsilon^3)$ and $S = R/\langle\epsilon\rangle$. Calculate the Ext-ring of S and show that it is possible to define a nonzero triple Massey product in this ring.

Exercise 2.9 Let $R = \mathbb{C}\langle e, u\rangle/\langle e^2 - e, eu - u, ue\rangle$ and $S = R/\langle u\rangle$. Show that the Ext-ring of S is isomorphic to R itself.

Exercise 2.10 Let G be a finite group and let V, W be two $\mathbb{C}G$-modules. Show that $\mathrm{Ext}^i_{\mathbb{C}G}(V, W) = 0$ if $i > 0$.

3

Higher Products

In the previous chapter we saw that it is interesting to define higher products on the cohomology ring to capture information that got lost in the process of taking the cohomology. In this chapter we will introduce the A_∞-formalism, originally developed by Stasheff [229], which deals with such higher products in a coherent way. After that we will merge this formalism with the category theory and representation theory from the first chapter. Finally, we will look at some useful extensions of these ideas.

For more details and background material we refer to the overview articles by Keller [133, 135], the first part of Seidel's book [218] and the book by Kontsevich and Soibelman [150].

3.1 Motivation and Definition

3.1.1 Generalizing Associativity

Let us have a look at three algebraic objects we have encountered before. A cochain complex is a graded vector space V with a degree 1 unary operation δ that satisfies the differential law $\delta^2 = 0$. A graded algebra R is a graded vector space with a degree 0 binary operation "$\cdot$" that satisfies the associativity law. The third object, a dg-algebra, has both operations. Each operation satisfies its own identity and there is an additional law: the graded Leibniz rule. We will now give a uniform interpretation of these three seemingly different laws, but to avoid being sidetracked by signs, we will temporarily assume that *our working field* $\Bbbk$ *has characteristic* 2.

If we write $\mathbb{1}$ for the identity map on V, set $\mu_1 \colon V \to V \colon v \mapsto \delta v$ and define $\mu_2 \colon V^{\otimes 2} \to V \colon v \otimes w \mapsto v \cdot w$, the three laws become

[DIF] $\mu_1 \circ \mu_1 = 0,$
[LEI] $\mu_1 \circ \mu_2 + \mu_2 \circ (\mu_1 \otimes \mathbb{1}) + \mu_2 \circ (\mathbb{1} \otimes \mu_1) = 0,$
[ASS] $\mu_2 \circ (\mu_2 \otimes \mathbb{1}) + \mu_2 \circ (\mathbb{1} \otimes \mu_2) = 0.$

The three laws each say that a certain operator (unary, binary or ternary) is zero. This operator is the sum of all possible (unary, binary or ternary) operators that can be made by composing two operators, tensored with additional identity maps to make them composable.

If we represent each product by a node in a rooted tree, we can graphically represent each identity by a sum of all trees with two nodes and $k = 1, 2, 3$ leaves (we do not consider the output as a leaf):

[DIF]  $= 0,$

[LEI] $+ \quad + \quad = 0,$

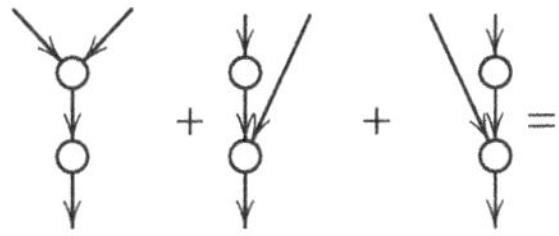

[ASS] $+ \quad = 0.$

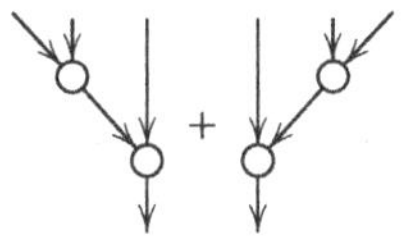

This pictorial interpretation can be generalized to all instances of k.

3.1.2 A_∞-Algebras

Definition 3.1 An A_∞-*algebra* is a graded vector space $A^\bullet$ with a collection of maps $b_i \colon A[1]^{\otimes i} \to A[1]$ of degree 1 with $i = 1, 2, \ldots$, such that the identities

$$[\mathtt{M_n}] \qquad \sum_{\substack{r+s+t=n \\ r,t \geq 0, s>0}} b_{r+t+1} \circ (\mathbb{1}^{\otimes r} \otimes b_s \otimes \mathbb{1}^{\otimes t}) = 0$$

hold for all $n \geq 1$. The formula can be graphically represented as

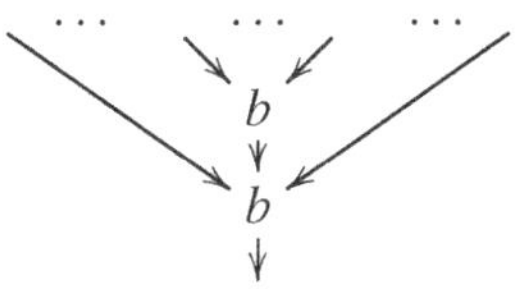

$$\sum \qquad = 0,$$

where the sum is over all rooted trees with n leaves and two nodes.

Remark 3.2 The degree 1 map $b_i\colon A[1]^{\otimes i} \to A[1]$ can also be seen as a map $\mu_i\colon A^{\otimes i} \to A$ of degree $2-i$. From this point of view the unary product has degree 1 and the binary has degree 0, just like a differential and an ordinary product. This degree convention is also in line with the definition of the triple Massey product $\langle u, v, w\rangle = [ud^{-1}(vw) \pm d^{-1}(uv)w]$, which has degree $-1 = 2-3$. Furthermore, the discussion after Theorem 2.38 suggested that the higher products should remember the coefficients of the relations $\mu_k(\xi_1 \otimes \cdots \otimes \xi_k) = \sum \lambda_i \rho_i$ with $\xi_i \in \mathrm{Ext}^1_A(S,S)$ and $\rho_i \in \mathrm{Ext}^2_A(S,S)$, so here the degree is again $\deg \rho_i - \deg \xi_1 - \cdots - \deg \xi_k = 2-k$.

Remark 3.3 Condition [$\mathtt{M}_1$] is the same as [DIF] so $(A[1], b_1)$ is always a cochain complex. Condition [$\mathtt{M}_2$] is the same as [LEI], while [$\mathtt{M}_3$] reduces to [ASS] if $b_1 = 0$ or $b_3 = 0$, as each of the extra terms contains both b_1 and b_3.

3.1.3 A_∞-Morphisms

Naturally, we also want to define morphisms between A_∞-algebras. Naively one would think that these are just maps $f\colon A \to B$ such that $b_k^B \circ f^{\otimes k} = f \circ b_k^A$. Unfortunately, this notion is far too strict because it does not allow us to mix products. We can only relate the k-ary product of A with the k-ary product of B, but the reason we want to introduce higher products is that they remember features of the binary product if we go to homology. Therefore we need a notion of morphisms that can mix products.

Definition 3.4 An A_∞*-morphism* $f_\bullet\colon A \to B$ consists of a sequence of maps $f_i\colon A[1]^{\otimes i} \to B[1]$ of degree 0 that satisfy axioms

$$[\mathrm{F}_n] \qquad \sum_{\substack{r+s+t=n\\ r,t\geq 0, s>0}} f_{r+t+1} \circ (\mathbb{1}^{\otimes r} \otimes b_s^A \otimes \mathbb{1}^{\otimes t}) = \sum_{\substack{i_1+\cdots+i_k=n\\ k>0, i_j>0}} b_k^B \circ (f_{i_1} \otimes \cdots \otimes f_{i_k}).$$

The *composition of two* A_∞*-morphisms* $f_\bullet\colon A \to B$ and $g_\bullet\colon B \to C$ is defined as

$$(g \circ f)_n := \sum_{\substack{i_1+\cdots+i_k=n\\ k>0, i_j>0}} g_k \circ (f_{i_1} \otimes \cdots \otimes f_{i_k}).$$

Note that [F_1] implies that f_1 is a morphism of cochain complexes between $(A[1], b_1^A)$ and $(B[1], b_1^B)$. We say that $f_\bullet$ is an A_∞*-isomorphism* if f_1 is an isomorphism of cochain complexes and an A_∞*-quasi-isomorphism* if f_1 is a quasi-isomorphism.

Remark 3.5 The axioms for the morphisms can be graphically represented as

$$\sum f(\ldots, b^A(\ldots), \ldots) = \sum b^B(f(\ldots), \ldots, f(\ldots)).$$

If the higher-order maps are all zero ($f_{\geq 2} = 0$) we recover the naive definition of a morphism. Such morphisms are called *strict A_∞-morphisms*. We say that $g_\bullet \colon B_\bullet \to A_\bullet$ is a (quasi-)inverse of $f_\bullet \colon A_\bullet \to B_\bullet$ if their compositions are strict and g_1 is a (quasi-)inverse of f_1. In this way $(f \circ g)_1$ and $(g \circ f)_1$ are (homotopic to) the identity map.

Lemma 3.6 *Let $f_\bullet \colon A^\bullet \to B^\bullet$ be an A_∞-morphism; then*

(i) *$f_\bullet$ has an inverse if and only if $f_\bullet$ is an A_∞-isomorphism,*
(ii) *$f_\bullet$ has a quasi-inverse if and only if $f_\bullet$ is an A_∞-quasi-isomorphism.*

Sketch of the proof The idea is to build up the (quasi-)inverse $g_\bullet$ step by step. We start with g_1. Because f_1 is a (quasi-)isomorphism we can find a (quasi-)inverse g_1 using Lemma 2.9. The higher g_i can be deduced inductively from the condition $(g \circ f)_{>0} = 0$. For example, if g_1 is an isomorphism and $(g \circ f)_2 = 0$ then $g_2(f_1 \otimes f_1) + g_1 f_2 = 0$ so $g_2 = g_1 f_2(f_1^{-1} \otimes f_1^{-1}) = g_1 f_2(g_1 \otimes g_1)$. See also [218, I 1.14]. □

The power of A_∞-morphisms is that just like morphisms of complexes, quasi-isomorphisms can be quasi-inverted. This is not true if we only allow strict morphisms. A quasi-inverse of a strict morphism is in general not strict anymore because higher components are needed to make it satisfy the $[\mathrm{F_n}]$-axioms.

3.1.4 Getting the Signs Right

If we work over a field with characteristic 0, the product b_2 does not give us an ordinary product because the associativity rule only holds up to sign. Therefore, extra signs are needed to make things work. There are many approaches to this problem; for a concise overview we refer to [202].

The starting point is the Koszul sign convention, which tells us how to apply operators to operands.

Definition 3.7 (Koszul sign convention) If $\phi \colon V^{\otimes k} \to V$ and $\psi \colon V^{\otimes l} \to V$ are two graded linear operators, we define the left action of their tensor product on $V^{\otimes k+l}$ by

$$\phi\otimes\psi(a_1\otimes\cdots\otimes a_k\otimes b_1\otimes\cdots\otimes b_l) = (-1)^{|\psi|(|a_1|+\cdots+|a_k|)}\phi(a_1\otimes\cdots\otimes a_k)\otimes\psi(b_1\otimes\cdots\otimes b_l),$$

where $|a|$ is shorthand for the degree of a.

In words, the Koszul rule states that we must add a minus sign every time we move an odd-degree operator past an odd-degree element. This rule of thumb also lets us work out the formula when the operators act from the right:

$$(a_1\otimes\cdots\otimes a_k\otimes b_1\otimes\cdots\otimes b_l)\phi\otimes\psi = (-1)^{|\phi|(|b_1|+\cdots+|b_l|)}\phi(a_1\otimes\cdots\otimes a_k)\otimes\psi(b_1\otimes\cdots\otimes b_l).$$

From here onwards there are two main conventions, one mainly used in symplectic geometry and one mainly used in representation theory.

Symplectic Geometry

This convention is used in Seidel's book [218] and in most papers on Fukaya categories. Let the shifted products $b_\bullet$ and A_∞-morphism $f_\bullet$ act from the right and define

$$\mu_k(a_1,\ldots,a_k) := ((a_1[1]\otimes\cdots\otimes a_k[1])b_k)[-1],$$
$$\phi_k(a_1,\ldots,a_k) := ((a_1[1]\otimes\cdots\otimes a_k[1])f_k)[-1],$$

where $a[1]$ stands for the element $a \in A$ considered in $A[1]$ where it has one degree less. This gives the following version of the A_∞-axioms:

$$[\mathtt{SM_n}] \qquad \sum_{r+s+t=n} (-1)^{\dagger}\mu(a_1,\ldots,a_r,\mu(a_{r+1},\ldots,a_{r+s}),a_{r+s+1},\ldots,a_{r+s+t}) = 0,$$

with $\dagger = (|a_{r+s+1}|-1)+\cdots+(|a_{r+s+t}|-1)$. Note that there is a $(|a_{r+s+1}|-1)$ in the exponent because we have shifted the degree of the elements by 1.

For an A_∞-algebra with $\mu_{\geq 3}=0$ the rules $[\mathtt{SM_2}]$ and $[\mathtt{SM_3}]$ become

$$\mu_1(\mu_2(a,b)) + (-1)^{|b|-1}\mu_2(\mu_1(a),b) + \mu_2(a,\mu_1(b)) = 0,$$
$$(-1)^{|c|-1}\mu_2(\mu_2(a,b),c) + \mu_2(a,\mu_2(b,c)) = 0,$$

which are unfortunately neither the Leibniz nor the associativity rule. This can be solved by additionally defining

$$da := (-1)^{|a|}\mu_1(a) \quad \text{and} \quad a\cdot b := (-1)^{|b|}\mu_2(a,b).$$

In this way we get

$$(-1)^{|a|+|b|+|b|}d(a\cdot b) + (-1)^{|b|-1+|a|+|b|}da\cdot b + (-1)^{|b|+|b|+1}a\cdot db = 0,$$
$$(-1)^{|a|}(d(a\cdot b) - da\cdot b - (-1)^{|a|}a\cdot db) = 0,$$

$$(-1)^{|c|-1+|b|+|c|}(a\cdot b)\cdot c + (-1)^{|b|+|c|+|c|}a\cdot(b\cdot c) = 0,$$
$$(-1)^{|b|-1}((a\cdot b)\cdot c - a\cdot(b\cdot c)) = 0.$$

For A_∞-morphisms the sign rule becomes

$$[\mathtt{SF_n}] \quad \sum_{\substack{r+s+t=n\\ r,t\geq 0, s>0}} (-1)^{\dagger}\phi_{r+t+1} \circ (\mathbb{1}^{\otimes r} \otimes \mu_s^A \otimes \mathbb{1}^{\otimes t}) = \sum_{\substack{i_1+\cdots+i_k=n\\ k>0, i_j>0}} \mu_k^B \circ (\phi_{i_1} \otimes \cdots \otimes \phi_{i_k}).$$

Note that there are no extra signs for the ϕ's because their shifted versions have degree 0.

Representation Theory

This convention is used in Keller's overview articles [133, 135] and many higher algebra papers. First define the degree -1 map

$$\sigma\colon A \to A[1]\colon x \mapsto (-1)^{|x|}x[1].$$

Let the shifted products $b_\bullet$ and A_∞-morphism $f_\bullet$ act from the left and define

$$m_k(a_1, \ldots, a_k) := \sigma^{-1} \circ b_k \circ \sigma^{\otimes k}(a_1 \otimes \cdots \otimes a_k),$$
$$f_k(a_1, \ldots, a_k) := \sigma^{-1} \circ f_k \circ \sigma^{\otimes k}(a_1 \otimes \cdots \otimes a_k).$$

This gives the following version of the A_∞-axioms:

$$[\mathtt{RM_n}] \qquad \sum_{r+s+t=n} (-1)^{\epsilon} m(a_1, \ldots, a_r, m(a_{r+1}, \ldots, a_{r+s}), a_{r+s+1}, \ldots, a_{r+s+t}) = 0,$$

with $\epsilon = r + st + s(|a_1| + \cdots + |a_r|)$. Note that this time the Leibniz rule and associativity rule automatically drop out:

$$(-1)^{0+2\cdot 0} m_1(m_2(a, b)) + (-1)^{0+1\cdot 1} m_2(m_1(a), b) + (-1)^{1+1\cdot 0+|a|} m_2(a, m_1(b)) = 0,$$
$$(-1)^{0+2\cdot 1} m_2(m_2(a, b), c) + (-1)^{1+2\cdot 0+2|a|} m_2(a, m_2(b, c)) = 0.$$

Remark 3.8 We will always use the letter b for the shifted maps, a Latin m for the representation theory convention and a Greek μ for the symplectic convention. Because the latter will be important when we study Fukaya categories, we will mostly use the μ-products in what follows. We will write definitions in terms of the μ-products with appropriate signs. They come from definitions of b-products that only contain $+$-signs.

3.1.5 Examples

Example 3.9 Any ordinary algebra $R, \cdot$ can be seen as an A_∞-algebra $R^\bullet$ concentrated in degree 0 with $\mu_2(x, y) = x \cdot y$ and $\mu_{\neq 2} = 0$.

Example 3.10 If $(A, d, \cdot)$ is a dg-algebra, we will see it as an A_∞-algebra with μ-products such that

$$da := (-1)^{|a|}\mu_1(a) \quad \text{and} \quad a \cdot b := (-1)^{|b|}\mu_2(a, b).$$

Example 3.11 Consider the ordinary algebra $R = \Bbbk[X]/(X^2)$ and the dg-algebra $A = \Bbbk[X,Y]/(Y^2)$ with $|X| = 0$, $|Y| = -1$ and $dY = X^2$. Because $R^\bullet$ is the homology of $A^\bullet$ there is a quasi-isomorphism $\pi \colon A^\bullet \to R^\bullet$ of dg-algebras, which corresponds to a strict A_∞-isomorphism.

There is no quasi-isomorphism of dg-algebras in the other direction because X^2 is zero in R but there are no zero divisors in degree 0 in A. We can however construct an A_∞-quasi-isomorphism $\phi_\bullet \colon R^\bullet \to A^\bullet$ with $\phi_1(a + bX) = a + bX$, $\phi_2(1, X) = \phi_2(X, 1) = 0$ and $\phi_2(X, X) = Y$. All higher ϕ_i must be zero for reasons of degree. The A_∞-rule $[\mathrm{SF}_2]$ for (X, X) becomes

$$\begin{aligned}\phi_1\mu_2^R(X,X) &= \mu_2^A(\phi_1(X),\phi_1(X)) + \mu_1^A(\phi_2(X,X)),\\ 0 &= X^2 + (-1)^{|Y|}dY,\\ 0 &= X^2 - X^2.\end{aligned}$$

Example 3.12 Let $r \in \Bbbk[\![X]\!]$ be a Laurent series of degree at least 1:

$$r = r_1X + r_2X^2 + r_3X^3 + \cdots .$$

Set $A = \Bbbk\xi \oplus \Bbbk\rho$ with $|\xi| = 1$, $|\rho| = 2$. Define the following A_∞-structure:

$$\mu_k^r(\xi,\ldots,\xi) = r_k\rho$$

and all products involving a ρ are equal to zero. It is easy to check that the A_∞-axioms hold. Now take another $f = f_1X + f_2X^2 + \cdots \in \Bbbk[\![X]\!]$. This defines a ring morphism $\iota_f \colon \Bbbk[\![X]\!] \to \Bbbk[\![X]\!] \colon u(X) \mapsto u(f(X))$ and set $s = \iota_f(r)$. Define the A_∞-maps $\phi_k^f \colon A^{\otimes k} \to A$ by

$$\phi_k^f(\xi,\ldots,\xi) = f_k\xi$$

and let all expressions involving a ρ be equal to zero except for $\phi_1^f(\rho) = \rho$. This gives an A_∞-morphism from (A,μ^s) to (A,μ^r):

$$\begin{aligned}\sum_{i_1+\cdots+i_k=n} \mu^r(\phi_{i_1}^f(\xi,\ldots,\xi),\ldots,\phi_{i_k}^f(\xi,\ldots,\xi)) &= \sum_{i_1+\cdots+i_k=n} \mu^r(f_{i_1}\xi,\ldots,f_{i_k}\xi)\\ &= \sum_{i_1+\cdots+i_k=n} f_{i_1}\cdots f_{i_k}r_k\rho\\ &= \phi_1^f\mu_n^s(\xi,\ldots,\xi)\\ &= \sum_{i+j+k=n} \phi^f(\ldots,\mu_j^s(\xi,\ldots,\xi),\ldots).\end{aligned}$$

The composition of two such A_∞-morphisms corresponds to the composition of the two ring morphisms of $\Bbbk[\![X]\!]$:

$$\phi_\bullet^f \circ \phi_\bullet^g = \phi_\bullet^h \quad \text{with } \iota_f \circ \iota_g = \iota_h.$$

Finally, $\phi^f_\bullet$ is an A_∞-isomorphism if $f_1 \neq 0$ because then ι_f is invertible. So in a sense A_∞-structures on A are dual to elements in $\mathbb{k}[[X]]$.

Remark 3.13 We can enlarge the previous example by adding a unit in degree 0: $E^\bullet = \mathbb{k}\mathbb{1} \oplus \mathbb{k}\xi \oplus \mathbb{k}\rho$. For the products we add the rule that $\mu_2^r(u, \mathbb{1}) = (-1)^{|u|}\mu_2(\mathbb{1}, u) = u$ for any homogeneous $u \in E^\bullet$ and set all other products involving $\mathbb{1}$ to zero.

3.2 Minimal Models

3.2.1 A Structure Theorem

Just like for complexes of vector spaces we have a structure theorem for A_∞-algebras that allows us to split every A_∞-algebra into a direct sum of two parts.

Definition 3.14 An A_∞-structure (A, b) is

- *minimal* if $b_1 = 0$,
- *contractible* if $b_{>1} = 0$ and $H(A) = 0$.

Theorem 3.15 (Kadeishvili's minimal model theorem [131]) *Every A_∞-algebra is A_∞-isomorphic to the direct sum of a contractible and a minimal A_∞-algebra.*

Sketch of the proof We give the outline of the proof by Kontsevich and Soibelman [147] because it has a nice graphical interpretation. Other more detailed accounts by Merkulov and Markl can be found in [174, 170].

Suppose A is an A_∞-algebra. Choose a split $A^\bullet = I^\bullet \oplus H^\bullet \oplus R^\bullet$ as in Lemma 2.9 for the differential $-b_1$ on A. This gives rise to a map h and a projection $\pi = \mathbb{1} + b_1 h + h b_1$ onto $H^\bullet$.

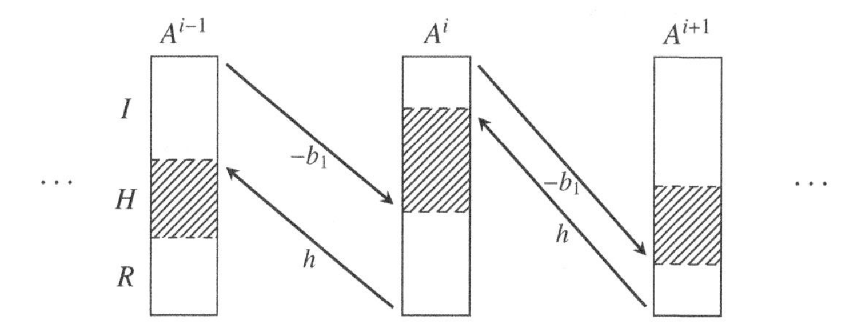

The cochain complex (A, b_1) decomposes as $A_{\text{contr}} \oplus A_{\text{min}}$ with $A_{\text{contr}} = I^\bullet \oplus R^\bullet = \operatorname{Ker} \pi$ and $A_{\text{min}} = H^\bullet = \operatorname{Im} \pi$.

The idea is to construct a new A_∞-structure $\bar{b}$ that respects the split and an A_∞-isomorphism $f_\bullet$ between (A,b) and $(A,\bar{b})$. We keep the unary multiplication the same: $\bar{b}_1 = b_1$. To calculate the higher $\bar{b}_k$ we will look at trees with k leaves. For each such tree $\mathcal{T}$ we can construct a product $b_{\mathcal{T}}$. Each leaf represents the map π, each node a map b_i and each internal branch a map h. The stem of the tree represents π.

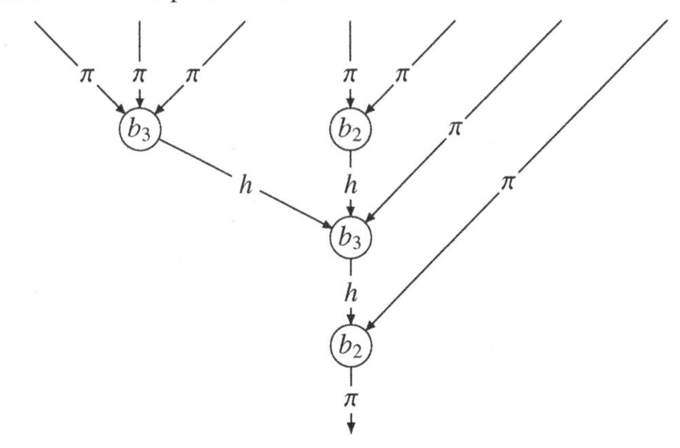

In the example above we get $b_{\mathcal{T}} = \pi b_2(hb_3(hb_3\pi^{\otimes 3} \otimes hb_2\pi^{\otimes 2} \otimes \pi) \otimes \pi)$. The multiplication is defined as the sum of all $b_{\mathcal{T}}$:

$$\bar{b}_k := \sum b_{\mathcal{T}}.$$

Because of all the π's in $b_{\mathcal{T}}$, the image of the new higher product $\bar{b}_\bullet$ is in A_{min} and it is zero as soon as one of the entries is in A_{contr}. On the other hand, the image of $\bar{b}_1$ is in A_{contr} and it is zero on A_{min}. This means that

$$(A,\bar{b}) = (A_{\mathrm{contr}}, \bar{b}_1) \oplus (A_{\mathrm{min}}, \bar{b}_{>1}).$$

To show that $\bar{b}$ satisfies the A_∞-relations, note that the operator $\bar{b}_k$ depends on h and we can take the partial derivative:

$$\partial_h \bar{b}_k[u] := \lim_{\epsilon \to 0} \frac{\bar{b}_k^{h+\epsilon u} - \bar{b}_k^h}{\epsilon}.$$

This derivative is the sum over all trees with one h substituted by a u. If we substitute u by π^2 and cut the tree between these two π's we get two smaller trees. Therefore we can rewrite $\partial_h \bar{b}_k[\pi^2]$ as a sum over all pairs of trees:

$$\partial_h \bar{b}_k[\pi^2] = \sum_{\mathcal{T}} \sum_h \underbrace{b_{\mathcal{T}}}_{h \leftarrow \pi^2} = \sum_{\mathcal{T}_1, \mathcal{T}_2} b_{\mathcal{T}_1}(\cdots \otimes b_{\mathcal{T}_2} \otimes \cdots).$$

The expression $\partial_h \bar{b}_k[\pi^2]$ is in fact the combination of all pairs of higher products $\bar{b}_{\geq 2}$. This is almost the A_∞-relation for $\bar{b}$ except for the terms containing a $\bar{b}_1$. Fortunately, all pairs containing a $\bar{b}_1$ and one $\bar{b}_{\geq 2}$ are zero because $b_1\pi = \pi b_1 = 0$, so the A_∞-relation $[\mathtt{M_k}]$ for $\bar{b}$ can be rewritten as $\partial_h \bar{b}_k[\pi^2] = 0$. If we fill in the formula $\pi^2 = \pi = \mathbb{1} + b_1 h + hb_1$ we get

$$\partial_h \bar{b}_k[\pi^2] = \partial_h \bar{b}_k[\mathbb{1}] + (\partial_h \bar{b}_k[b_1 h] + \partial_h \bar{b}_k[hb_1]).$$

The terms in brackets consist of a sum over all trees and edges where a b_1 has been inserted before or after the edge. Each term in the sum contains a $b_l(\cdots b_1 \cdots)$ or $b_1 b_l$ for some $l \geq 2$. Using the A_∞-relations for b we can rewrite these terms as minus the sum of all products $b_r(\cdots b_s \cdots)$ with $r, s \geq 2$. This product directly connects b_r with b_s without an h in between, so all terms are trees with one h substituted by $\mathbb{1}$ and

$$(\partial_h \bar{b}_k[b_1 h] + \partial_h \bar{b}_k[hb_1]) = -\partial_h \bar{b}_k[\mathbb{1}],$$

which means that $\partial_h \bar{b}_k[\pi^2] = 0$ and the A_∞-axioms hold for $\bar{b}$.

The proof that there is an A_∞-isomorphism $f_\bullet$ with $f_1 = \mathbb{1}$ between (A, b) and $(A, \bar{b})$ consists of similar calculations. The formula for f_i is the same as for $\bar{\mu}_i$ but with a map h at the root instead of a π. □

Remark 3.16 Note that if we apply the formula for $\bar{b}$ to a sequence $a_1 \otimes \cdots \otimes a_n$, no extra signs from the Koszul rule appear because all maps involved have even degree (the π's and the $h \circ b$'s). Therefore, the formula also holds for the μ-products without additional signs.

Because $(A, \bar{b})$ splits as a direct sum, the embedding $A_{\min} \hookrightarrow A$ and projection $A \to A_{\min}$ are strict morphisms. If we compose the embedding with the A_∞-isomorphism $f_\bullet \colon (A, b) \to (A, \bar{b})$ and the projection with its inverse then we get two maps $i_\bullet \colon A_{\min} \to (A, b)$ and $p_\bullet \colon (A, b) \to A_{\min}$.

If $g_\bullet \colon A^\bullet \to B^\bullet$ is an A_∞-quasi-isomorphism then $p^B_\bullet \circ \psi_\bullet \circ i^A_\bullet$ is an A_∞-isomorphism between the two minimal models. On the other hand, if $g_\bullet \colon A^\bullet_{\min} \to B^\bullet_{\min}$ is an A_∞-isomorphism between the minimal models of $A^\bullet$ and $B^\bullet$ then $i^B_\bullet \circ \psi_\bullet \circ p^A_\bullet$ is an A_∞-quasi-isomorphism between $A^\bullet$ and $B^\bullet$.

Corollary 3.17 *Two A_∞-algebras are A_∞-quasi-isomorphic if and only if their minimal models are A_∞-isomorphic.*

It seems natural to construct minimal models of A_∞-algebras and work only with those as they are uniquely defined up to A_∞-isomorphism. However, it is notoriously hard to work out all the products in the minimal model and given two minimal models it is also hard to check whether they are A_∞-isomorphic or not.

On the other side of the spectrum it is also possible to construct dg-models. These are A_∞-algebras without higher products: $b_{\geq 3} = 0$. Dg-models have the advantage that the algebra structure is rather easy, but the disadvantage that they contain many extra elements that are cut away when going to the homology.

Depending on the application, a dg-model might be more suitable than a minimal model or vice versa. The A_∞-formalism offers a framework to move between these different models with relative ease. This can be done using the *homological perturbation lemma*, which is a generalization of the minimal model theorem.

Lemma 3.18 (Homological perturbation lemma) *Let $\phi\colon A \to B$ and $\phi^\dagger\colon B \to A$ be quasi-inverses between two complexes (A, μ_1^A), (B, μ_1^B) and take a map $h\colon B \to B$ such that $\phi\phi^\dagger = \mathbb{1}_B + h\mu_1^B + \mu_1^B h$.*

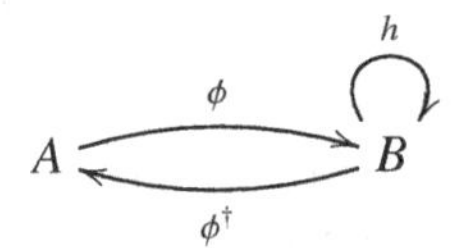

If $\mu_\bullet^B$ is an A_∞-structure on B extending μ_1^B then we can extend μ_1^A to an A_∞-structure $\mu_\bullet^A$ and construct an A_∞-quasi-isomorphism $\phi_\bullet\colon A \to B$ such that $\phi_1 = \phi$.

Sketch of the proof The idea is the same: we construct μ_n^A as a sum over all trees with μ^B at the nodes, ϕ for the leaves, $\phi^\dagger$ for the root and h for the internal edges, while $\phi_\bullet$ is constructed similarly but with an h at the root. This can also be expressed recursively as

$$\mu^A(a_1,\dots,a_n) = \sum_r \sum_{i_1+\cdots+i_r=n} \phi^\dagger\mu^B(\phi_{i_1}(a_1,\dots,a_{i_1}),\dots,\phi_{i_r}(a_{n-i_r+1},\dots,a_n)),$$

$$\phi(a_1,\dots,a_n) = \sum_r \sum_{i_1+\cdots+i_r=n} h\mu^B(\phi_{i_1}(a_1,\dots,a_{i_1}),\dots,\phi_{i_r}(a_{n-i_r+1},\dots,a_n)).$$

For more details see [170]. □

3.2.2 Examples

Example 3.19 If we return to Example 2.35 of the polynomial ring, $R = \mathbb{k}[X, Y]$ and $S = R/(X, Y)$, we know that

$$\operatorname{Ext}_R^\bullet(S, S) = \mathbb{k}\langle \xi, \eta\rangle / \langle \xi^2, \eta^2, \xi\eta + \eta\xi\rangle.$$

This calculation was done by looking at the cohomology of $\operatorname{Hom}_R^\bullet(K^\bullet, K^\bullet)$ with

$$K^\bullet := R \xrightarrow{\begin{pmatrix} X \\ -Y \end{pmatrix}} R^2 \xrightarrow{(X\ Y)} R\,.$$

Apart from the cohomological grading, we can give $K^\bullet$ a second grading by considering R as a graded ring where X, Y have degree 1. For this second grading the elements ι, ξ, η, ρ have degree 0 because the calculations in Example 2.35 show that as maps they only contain constants. The second degree of the differential b_1 is 1 because it is linear in X, Y. This implies that h will have degree -1 if we choose the split $H \oplus I \oplus R$ compatible with the new grading.

To construct the higher products we have to sum over all trees. Each node of a tree is an ordinary product, so it has second degree 0, while each internal edge has degree -1 because it represents an h. Therefore, the higher products all have negative degree. Because $\mathrm{Ext}^\bullet_R(S,S)$ only has elements with second degree 0, we can conclude that all higher products are 0.

Example 3.20 Let us look at the case where $A = \Bbbk\langle X, Y\rangle/(X^2Y)$ and the module is the simple right module on which X, Y act trivially: $S = A/(X,Y)$. The Koszul resolution for this module is

$$K^\bullet := A \xrightarrow{\binom{XY}{0}} A^{\oplus 2} \xrightarrow{(X\ Y)} A\ .$$

The Ext-algebra is given by

$$\mathrm{Ext}^\bullet_A(S,S) = \Bbbk\iota \oplus (\Bbbk\xi \oplus \Bbbk\eta) \oplus \Bbbk\rho,$$

with

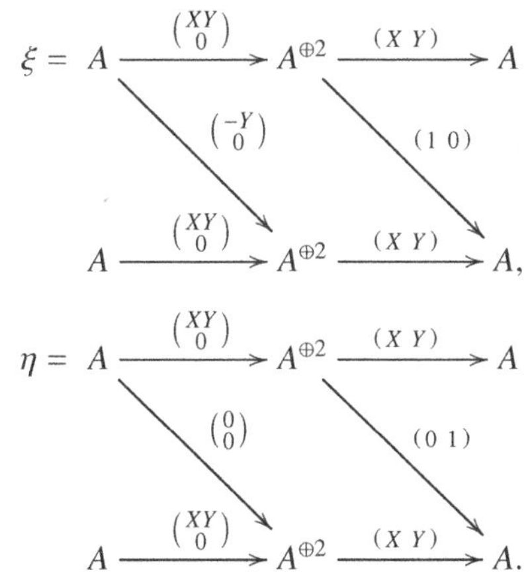

$$\begin{array}{ccccc} \xi = A & \xrightarrow{\binom{XY}{0}} & A^{\oplus 2} & \xrightarrow{(X\ Y)} & A \\ & \searrow^{\binom{-Y}{0}} & & \searrow^{(1\ 0)} & \\ A & \xrightarrow{\binom{XY}{0}} & A^{\oplus 2} & \xrightarrow{(X\ Y)} & A, \end{array}$$

$$\begin{array}{ccccc} \eta = A & \xrightarrow{\binom{XY}{0}} & A^{\oplus 2} & \xrightarrow{(X\ Y)} & A \\ & \searrow^{\binom{0}{0}} & & \searrow^{(0\ 1)} & \\ A & \xrightarrow{\binom{XY}{0}} & A^{\oplus 2} & \xrightarrow{(X\ Y)} & A. \end{array}$$

From these expressions we see that $\eta \cdot \xi = \eta \cdot \eta = \xi \cdot \eta = 0$ and

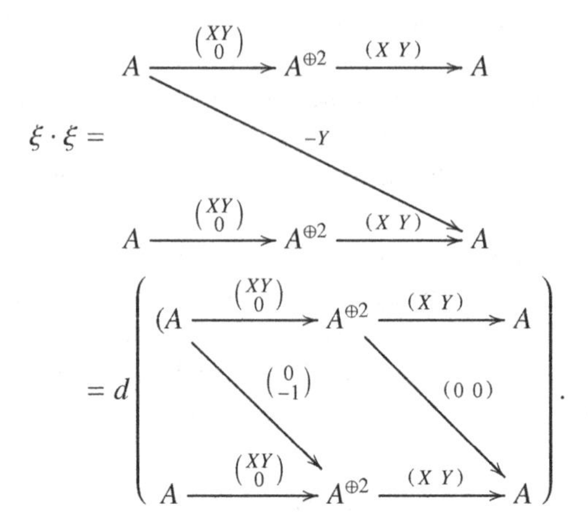

So while the product $\xi \cdot \xi$ is nonzero in $\mathrm{Hom}^\bullet_A(S, S)$, it is zero in homology: $\xi{\cdot}\xi = d\kappa$. Because ξ and κ both have degree 1 this identity becomes $-\mu_2(\xi, \xi) = -\mu_1(\kappa)$.

Construct a split of $\mathrm{Hom}^\bullet_A(K^\bullet, K^\bullet)$ into $R \oplus H \oplus I$ with a basis that contains $\iota, \xi, \eta, \rho \in H$, $\xi \cdot \xi \in I$ and $\kappa \in R$. In this way we have $h\mu_2(\xi, \xi) = -\kappa$. To calculate $\bar{\mu}_3$ we need to take the sum of two trees:

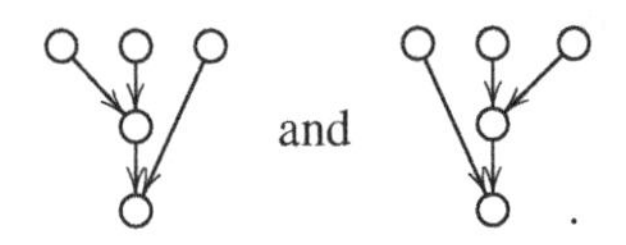

This gives a triple product

$$\begin{aligned}\bar{\mu}_3(\eta, \xi, \xi) &= \pi(\mu_2(h\mu_2(\eta, \xi), \xi) + \mu_2(\eta, h\mu_2(\xi, \xi))) \\ &= \pi(0 - \mu_2(\eta, \kappa)) \\ &= \pi(\eta \cdot \kappa) = -\rho,\end{aligned}$$

which is nonzero. On the other hand, we get

$$\begin{aligned}\bar{\mu}_3(\xi, \xi, \eta) &= \pi(\mu_2(h\mu_2(\xi, \xi) \otimes \eta) + \mu_2(\xi, h\mu_2(\xi, \eta))) \\ &= \pi(-\mu_2(\kappa, \xi) - 0) \\ &= \pi(\kappa \cdot \xi) = 0.\end{aligned}$$

This example generalizes to the following theorem.

Theorem 3.21 (Lu, Palmieri, Wu, Zhang [168]) *Let $A = \Bbbk\langle X_1, \ldots, X_n\rangle / \langle r_1, \ldots, r_m\rangle$ be a minimal presentation and let S be the simple module $A/\langle X_1, \ldots, X_n\rangle$. The minimal model $\mathrm{Ext}^\bullet_A(S, S)$ satisfies the following properties:*

(i) $\mathrm{Ext}^1(S,S) = \Bbbk\xi_1 \oplus \cdots \Bbbk\xi_n$.
(ii) $\mathrm{Ext}^2(S,S) = \Bbbk\rho_1 \oplus \cdots \Bbbk\rho_m$.
(iii) $\mu(\xi_{i_1},\ldots,\xi_{i_k}) = \sum \pm\lambda_j\rho_j$ *if the coefficient of* $X_{i_k}\cdots X_{i_1}$ *in* r_j *is* λ_j.

Remark 3.22 The signs and the order of the monomials depend on the particular sign convention and whether we work with right modules (reverse order) or left modules (original order).

3.3 A_∞-Categories

3.3.1 From Algebras to Categories

Just as we can see $\Bbbk$-linear categories as generalizations of algebras, we can also define the notion of an A_∞-category.

Definition 3.23 An *A_∞-category* $\mathtt{C}^\bullet$ consists of a collection of objects $\mathrm{Ob}\,\mathtt{C}^\bullet$, graded $\Bbbk$-vector spaces of morphism $\mathtt{C}^\bullet(A,B)$, and products

$$\mu_n\colon\ \mathtt{C}^\bullet(A_1,A_0)\otimes\cdots\otimes\mathtt{C}^\bullet(A_n,A_{n-1})\to\mathtt{C}^\bullet(A_n,A_0)$$

of degree $2-n$ satisfying the signed generalized associativity rules $[\mathtt{SM_n}]$ whenever they are defined.

An *A_∞-functor* $\mathcal{F}\colon\ \mathtt{A}^\bullet\to\mathtt{B}^\bullet$ consists of a map $\mathcal{F}\colon\ \mathrm{Ob}\,\mathtt{A}^\bullet\to\mathrm{Ob}\,\mathtt{B}^\bullet$ and maps

$$\mathcal{F}_n\colon\ \mathtt{A}^\bullet(A_1,A_0)\otimes\cdots\otimes\mathtt{A}^\bullet(A_n,A_{n-1})\to\mathtt{B}^\bullet(\mathcal{F}A_n,\mathcal{F}A_0)$$

of degree $1-n$ satisfying the signed generalized associativity rules $[\mathtt{SF_n}]$ whenever they are defined. The composition of A_∞-functors is the same as the composition of A_∞-morphisms.

Given an A_∞-category $\mathtt{C}^\bullet$ we again have the notion of a minimal model. We will denote this model by $\mathtt{HC}^\bullet$ and the injection and projection A_∞-functors by $\mathcal{I}\colon\ \mathtt{HC}^\bullet\to\mathtt{C}^\bullet$ and $\mathcal{P}\colon\ \mathtt{C}^\bullet\to\mathtt{HC}^\bullet$.

Remark 3.24 To construct the minimal model of an A_∞-category $\mathtt{C}^\bullet$ we have to make many choices: for each pair of objects we need a degree -1 map $h_{AB}\colon\ \mathtt{C}^\bullet(A,B)\to\mathtt{C}^\bullet(A,B)$. Different choices result in different but A_∞-isomorphic categories.

Remark 3.25 In the definition of A_∞-algebras, we have not talked about the notion of a unit. A unit can be defined in several ways. The easiest one is to say that $1\in A$ is a *strong unit* if $\mu_1(1)=0$, $\mu_2(a,1)=(-1)^{|a|}\mu_2(1,a)=a$ and $\mu_k(\ldots,1,\ldots)=0$ if $k>2$. A second possibility is to say that 1 is a *weak unit* if it becomes a unit in the cohomology ring $H(A)$. One can show that if A has a weak unit then there is an A_∞-isomorphic A_∞-algebra that has a strong unit. So without loss of generality we can assume that A has a strong unit. Similarly, for categories we can speak of A_∞-categories with strong units for each object or with weak units for each object. From now on we will implicitly assume that the A_∞-categories we discuss have strong units. For more details see [218, I 2.1].

The degree 0 part of minimal models of an A_∞-category $\mathtt{C}^\bullet$ with weak or strong units can be seen as an ordinary category if we use μ_2 for the ordinary product $x \cdot y = (-1)^{|y|}\mu_2(x \otimes y)$. We will denote this category by $\mathtt{HC}^0$. While the precise higher A_∞-structure on $\mathtt{HC}^\bullet$ depends on choices, the category $\mathtt{HC}^0$ is independent of these choices.

Definition 3.26 We say that a degree 0 morphism f in $\mathtt{C}^\bullet$ is a *quasi-isomorphism* if $\mu_1(f) = 0$ and $\mathcal{P}(f)$ is an isomorphism in $\mathtt{HC}^0$. Two objects are *quasi-isomorphic* if they are connected by a quasi-isomorphism.

We will say that an A_∞-functor $\mathcal{F}\colon \mathtt{A}^\bullet \to \mathtt{B}^\bullet$ is a *quasi-equivalence* if it gives rise to a functor between the minimal models $\bar{\mathcal{F}} = \mathcal{P}_B\mathcal{F}\mathcal{I}_A : \mathtt{HA}^0 \to \mathtt{HB}^0$, which is an ordinary equivalence. In that case, $\mathtt{A}^\bullet$ and $\mathtt{B}^\bullet$ are called *quasi-equivalent* A_∞-categories and this is denoted as $\mathtt{A}^\bullet \overset{\infty}{=} \mathtt{B}^\bullet$.

Example 3.27 Given an ordinary algebra A, we can consider its module category as an ordinary category, but we would also like to consider $\mathtt{MOD}\text{-}A$ as an A_∞-category. From the previous chapter we know there is a natural notion of higher-degree morphisms in this case: the Ext-groups. How can we construct an A_∞-category for which the objects are modules and the hom-spaces the Ext-groups?

The idea is to use resolutions. Choose for each module M a free resolution $F_M^\bullet$ and construct $\mathtt{RES}^\bullet A$ as the category of all A-modules with morphism spaces $\mathtt{RES}^\bullet A(M,N) := \mathrm{Hom}_A^\bullet(F_M^\bullet, F_N^\bullet)$. This is a dg-category and define $\mathtt{MOD}^\bullet A := \mathtt{HRES}^\bullet A$ as the minimal model of this dg-category. Because $\mathrm{Ext}_A^0(M,N) = \mathrm{Hom}_A(M,N)$ the degree 0 component of this category is equivalent to the original module category $\mathtt{MOD}\text{-}A$.

Note that this construction depends on many choices: a resolution for each module and splits of the hom-spaces for the minimal models. Different choices will result in A_∞-equivalent categories. Alternatively, one can add all resolutions of modules as separate but isomorphic objects.

3.3.2 Twisted Complexes

In the first chapter we saw that when doing representation theory, it is often better to make our category bigger. Going to the additive completion adds the free modules of an algebra, while going to the split-add completion also adds projective modules. In Chapter 2 we showed that other modules could be approximated by complexes of free (or projective) modules, so one could ask whether there is a way to include these complexes as well. Complexes of free modules can be seen as graded direct sums of copies of the algebra, together with a degree 1 element δ that satisfies $\delta^2 = 0$. In this section we will introduce an appropriate generalization of a complex for A_∞-algebras and categories.

Let $\mathsf{C}^\bullet$ be an A_∞-category. A direct sum of shifted objects is a finite formal expression of the form $\oplus_i A_i[k_i]$ where the A_i are objects in $\mathsf{C}^\bullet$ and the k_i are integers. We can group these objects in an A_∞-category $\mathtt{Add}\ \mathsf{C}^\bullet$ if we define the hom-space between two direct sums as

$$\mathtt{Add}\ \mathsf{C}^\bullet(\oplus_i A_i[k_i], \oplus_i B_i[l_i]) = \bigoplus_{i,j} \mathsf{C}^\bullet(A_i, B_j)[l_i - k_j],$$

and extend the A_∞-products to these hom-spaces using the matrix multiplication convention: e.g.

$$\mu_k(\alpha^1, \dots, \alpha^k)_{ij} := \sum_{i_1,\dots,i_{n-1}} \mu_k(\alpha^1_{ii_1}, \alpha^2_{i_1 i_2}, \dots, \alpha^k_{i_k j})$$

and extra signs for the shifts

$$\mu_k(a_1[i_1], \dots, a_k[i_k]) = (-1)^{\sum_{1\le u<v\le k} i_u(|a_v|-1)} \mu_n(a_1, \dots, a_k)[i_1 + \dots + i_k],$$

because the shift can be seen as a degree 1 operator that has to be pulled through from the right using the Koszul sign rule.

Definition 3.28 A *twisted complex* A over $\mathsf{C}^\bullet$ consists of a pair (M, δ) where $M = \oplus_i A_i[k_i] \in \mathtt{Add}\ \mathsf{C}^\bullet$ is a direct sum of shifted objects and δ is a degree 1 element in $\mathtt{Add}\ \mathsf{C}^\bullet(M, M)$. Additionally, we assume that δ is *strictly lower triangular*[1] and satisfies the *Maurer–Cartan equation*:

$$\mathsf{MC}(\delta) := \mu_1(\delta) + \mu_2(\delta, \delta) + \mu_3(\delta, \delta, \delta) + \dots = 0.$$

Because δ is strictly lower triangular, this infinite sum has only a finite number of nonzero terms.

Given a sequence of twisted complexes $(M_0, \delta_0), \dots, (M_k, \delta_k)$ we can introduce a twisted k-ary product

$$\tilde{\mu}_k\colon \operatorname{Hom}(M_1, M_0) \otimes \dots \otimes \operatorname{Hom}(M_k, M_{k-1}) \to \operatorname{Hom}(M_k, M_0)$$

[1] By this we mean that $\delta_{ij} = 0$ if $i < j$, but if we do not want to impose a particular order on the summands we can also demand that $\delta_{i_1 i_2} \otimes \dots \otimes \delta_{i_{k-1} i_k} = 0$ for $k \gg 0$.

by taking a sum over all possible ways to insert δ's between the entries:

$$\tilde{\mu}_k(a_1,\ldots,a_k) = \sum_{m_0,\ldots,m_k\geq 0} \mu_{k+m_0+\cdots+m_k}(\underbrace{\delta_0,\ldots,\delta_0}_{m_0},a_1,\ldots,a_k,\underbrace{\delta_k,\ldots,\delta_k}_{m_k}).$$

Again, although this sum is infinite, only a finite number of terms are zero because the δ are lower triangular.

Lemma 3.29 *The twisted products satisfy the A_∞-axioms.*

Sketch of the proof First of all, the fact that δ_i has degree 1 and $\mu_{k+m_0+\cdots+m_k}$ degree $2-(k+m_0+\cdots+m_k)$ implies that $\tilde{\mu}_k$ still has degree $2-k$. If we expand the expression $\sum \pm\tilde{\mu}(a_1,\ldots,\tilde{\mu}(\cdots),\ldots,a_n)$ from $[\mathtt{SM_n}]$ using the definition of $\tilde{\mu}$, we can group terms together according to the number of δ_i that are inserted for each i.

If we look at one group separately we see that we get all terms of an A_∞-axiom except those terms for which the inner μ only acts on δ's and not on a_i. If we add these terms for each group, every group will become zero. Regroup all the added terms according to the entries of the outer product. For each outer product, the terms that were added all had the same sign and their sum is zero:

$$\mu(\ldots,\mu_1(\delta_i),\ldots) + \mu(\ldots,\mu_2(\delta_i,\delta_i),\ldots) + \cdots = \mu(\ldots,\mathsf{MC}(\delta_i),\ldots) = 0.$$

Therefore, the original expression must also be zero and $\tilde{\mu}$ satisfies the A_∞-axioms. □

Definition 3.30 If $\mathsf{C}^\bullet$ is an A_∞-category, we define $\mathtt{Tw}\,\mathsf{C}^\bullet$ as the A_∞-category for which the objects are twisted pairs, the hom-spaces are the standard hom-spaces between direct sums and the multiplications are the $\tilde{\mu}$. This category comes with two natural functors:

- The *standard embedding* $\mathcal{I}\colon \mathsf{C}^\bullet \to \mathtt{Tw}\,\mathsf{C}^\bullet$ is the strict A_∞-functor for which $\mathcal{I}(A) = (A[0], 0)$ and $\mathcal{I}_1(\alpha) = \alpha$.
- The *shift functor* $\Sigma\colon \mathtt{Tw}\,\mathsf{C}^\bullet \to \mathtt{Tw}\,\mathsf{C}^\bullet$ is the strict A_∞-functor that shifts the degree in the direct summands by 1: $\Sigma(M,\delta) = (M[1], -\delta)$ and $\Sigma_1(\alpha) = \alpha$.

Remark 3.31 If $\mathsf{C}^\bullet$ has strong units then $\mathtt{Tw}\,\mathsf{C}^\bullet$ also has strong units because the identity map $\mathbb{1}_A \in \mathtt{Tw}\,\mathsf{C}^\bullet(A,A)$ acts as a strong unit for (A,δ_A). Indeed,

$$\begin{aligned}
\tilde{\mu}_1(\mathbb{1}_A) &= \mu(\mathbb{1}_A) + \mu(\delta_A,\mathbb{1}_A) + \mu(\mathbb{1}_A,\delta_A) + \mu_{\geq 3}(\ldots,\mathbb{1}_A,\ldots)\\
&= 0 + \delta_A - \delta_A + 0 = 0,\\
\tilde{\mu}_2(\mathbb{1}_A,u) &= \mu(\mathbb{1}_A,u) + m_{\geq 3}(\ldots,\mathbb{1}_A,\ldots)\\
&= (-1)^{|u|}u + 0,
\end{aligned}$$

and higher-order products with $\mathbb{1}_A$ only contain higher-order terms so they are also zero.

Remark 3.32 If $(M,\delta) = (A_1[i_1]^{\oplus e_1} \oplus \cdots \oplus A_k[i_k]^{\oplus e_k}, \delta)$ and $(g_1,\ldots,g_k) \in \mathsf{GL}_{e_1}(\mathbb{k}) \times \cdots \times \mathsf{GL}_{e_k}(\mathbb{k})$ is a k-tuple of matrices then we can define matrices

$$\gamma := \begin{pmatrix} g_1\otimes\mathbb{1}_{A_1} & & \\ & \ddots & \\ & & g_k\otimes\mathbb{1}_{A_k} \end{pmatrix} \quad \text{and} \quad \gamma^{-1} := \begin{pmatrix} g_1^{-1}\otimes\mathbb{1}_{A_1} & & \\ & \ddots & \\ & & g_k^{-1}\otimes\mathbb{1}_{A_k} \end{pmatrix}$$

in $\mathtt{Add}\ \mathsf{C}^\bullet(M,M)$. We can use these to define the conjugate twisted complex

$$(M,\delta^\gamma) := (M, \gamma\cdot\delta\cdot\gamma^{-1}) = \big(M, -\mu_2(\mu_2(\gamma,\delta),\gamma^{-1})\big),$$

which is isomorphic to the original because γ can be seen as an invertible element in $\mathtt{Tw}\,\mathsf{C}^\bullet((M,\delta),(M,\delta^g))$ with

$$\tilde{\mu}_1(\gamma) = \mu_1(\gamma) + \mu_2(\gamma,\delta) + \mu_2(\delta^\gamma,\gamma) = 0.$$

Example 3.33 Let R be an algebra and consider it as an A_∞-algebra concentrated in degree 0 with trivial higher products. This A_∞-algebra is also an A_∞-category $\mathsf{R}^\bullet$ with one object X. Because $\mathsf{R}^\bullet(X,X) = R$ we can consider a shifted direct sum as a free $\mathbb{Z}$-graded right R-module and the hom-space between two such $\mathbb{Z}$-graded sums is equal to the space of all right R-module morphisms. A degree 1 map δ satisfies the Maurer–Cartan equation if $\mu_2(\delta,\delta) = -\delta^2 = 0$, so a twisted complex (P,δ_P) is just a *bounded complex* $P^\bullet$ of free R-modules of finite rank.

If (P,δ_P), (Q,δ_Q) are twisted complexes then

$$\mathtt{Tw}\,\mathsf{R}^\bullet((P,\delta_P),(Q,\delta_Q)) = \mathrm{Hom}_R^\bullet(P^\bullet,Q^\bullet)$$

and if $f\colon P^\bullet \to Q^\bullet$ then

$$\begin{aligned}\tilde{\mu}_1(f) &= \mu_1(f) + \mu_2(\delta_Q,f) + \mu_2(f,\delta_P)\\ &= (-1)^{|f|}\delta_Q\cdot f - f\cdot\delta_P.\end{aligned}$$

As there are no higher products in $\mathsf{R}^\bullet$, we have $\tilde{\mu}_2 = \mu_2$ and $\tilde{\mu}_{>2} = 0$. If we put $df = (-1)^{|f|}\tilde{\mu}_1(f)$ and $f\cdot g = (-1)^{|g|}\mu_2(f,g)$, we see that the category of twisted complexes of $\mathsf{R}^\bullet$ is the ordinary dg-category of finite-rank complexes of free R-modules.

3.3.3 Derived A_∞-Categories

Definition 3.34 Let $\mathsf{C}^\bullet$ be an A_∞-category.

- The *derived A_∞-category* of $\mathtt{C}^\bullet$ is the minimal model of the twisted completion:

$$\mathtt{D}\,\mathtt{C}^\bullet := \mathtt{HTw}\,\mathtt{C}^\bullet.$$

- The *derived category* of $\mathtt{C}^\bullet$ is the degree 0 component of the derived A_∞-category, viewed as an ordinary $\mathbb{k}$-linear category: $\mathtt{D}^b\mathtt{C} := \mathtt{HTw}\,\mathtt{C}^0$.

Example 3.35 Take the ring $R = \mathbb{k}$ and construct $\mathtt{Tw}\mathbb{k}^\bullet$. As explained in Example 3.33, this category is equivalent to the dg-category of finitely generated (so in this case finite-dimensional) cochain complexes of $\mathbb{k}$-vector spaces:

$$\mathtt{Tw}\mathbb{k}^\bullet = \mathtt{fdK}^\bullet\,\mathbb{k}.$$

The minimal model $\mathtt{D}\mathbb{k}^\bullet$ can be identified with the category of finite-dimensional graded vector spaces: $\mathtt{grvect}^\bullet\,\mathbb{k}$ viewed as an A_∞-category with only $\mu_2 \neq 0$. The functor $\mathcal{F}\colon \mathtt{Tw}\mathbb{k}^\bullet \to \mathtt{grvect}^\bullet\,\mathbb{k}$ with

$$\mathcal{F}(C^\bullet) = \mathsf{H}C^\bullet \quad \text{and} \quad \mathcal{F}(f) = \mathsf{H}f$$

can be seen as a strict A_∞-functor and because of Lemma 2.9(c) it is a quasi-equivalence:

$$\mathtt{D}\mathbb{k}^\bullet \overset{\infty}{=} \mathtt{grvect}^\bullet\,\mathbb{k}.$$

Example 3.36 In Example 3.27 we upgraded the module category of an algebra A to an A_∞-category $\mathtt{MOD}^\bullet A$, so it makes sense to construct the twisted completion $\mathtt{Tw}\,\mathtt{Mod}^\bullet A$ and the derived category $\mathtt{D}\,\mathtt{MOD}^\bullet A$. The degree 0 component of this last category is better known in the literature [253] as the *bounded derived category* $\mathtt{D}^b\mathtt{MOD}\,A$.

Definition 3.37 Two A_∞-algebras or categories $\mathtt{A}^\bullet, \mathtt{B}^\bullet$ are *derived equivalent* if and only if $\mathtt{D}\,\mathtt{A}^\bullet$ and $\mathtt{D}\,\mathtt{B}^\bullet$ are A_∞-equivalent categories.

Lemma 3.38 *If $\mathtt{A}^\bullet$ and $\mathtt{B}^\bullet$ are quasi-equivalent then $\mathtt{D}\,\mathtt{A}^\bullet$ and $\mathtt{D}\,\mathtt{B}^\bullet$ are equivalent.*

Sketch of the proof If $\mathcal{F}_\bullet\colon \mathtt{A}^\bullet \to \mathtt{B}^\bullet$ is an A_∞-functor then we can construct a new A_∞-functor by $\mathtt{Tw}\mathcal{F}_\bullet\colon \mathtt{Tw}\,\mathtt{A}^\bullet \to \mathtt{Tw}\,\mathtt{B}^\bullet$,

$$\mathtt{Tw}\mathcal{F}(\oplus_i A_i[k_i], \delta) = (\oplus \mathcal{F}(A_i)[k_i], \mathcal{F}(\delta) + \mathcal{F}(\delta, \delta) + \cdots),$$

$$\mathtt{Tw}\mathcal{F}_k(a_1, \ldots, a_k) = \sum_{m_0,\ldots,m_k \geq 0} \mathcal{F}_{k+m_0+\cdots+m_k}(\underbrace{\delta_0, \ldots, \delta_0}_{m_0}, a_1, \ldots, a_k, \underbrace{\delta_k, \ldots, \delta_k}_{m_k})$$

with appropriate matrix and shift extensions like for the product. One can check that this definition satisfies the $[\mathtt{SF_n}]$-axioms and that $\mathtt{Tw}\mathcal{F}$ is a quasi-equivalence if $\mathcal{F}$ is (see [218, I 3m]). This quasi-equivalence induces an equivalence between the minimal models. □

Definition 3.39 If $\mathcal{F}_\bullet\colon \mathtt{A}^\bullet \to \mathtt{B}^\bullet$ is an A_∞-functor we define the derived functor

$$\mathtt{D}\mathcal{F}\colon \mathtt{D}\,\mathtt{A}^\bullet \to \mathtt{D}\,\mathtt{B}^\bullet := \mathcal{P}_{\mathtt{Tw}\,\mathtt{B}^\bullet} \circ \mathtt{Tw}\mathcal{F} \circ \mathcal{I}_{\mathtt{Tw}\,\mathtt{A}^\bullet},$$

where $\mathcal{I}_{\mathtt{Tw}\,\mathtt{A}^\bullet}\colon \mathtt{D}\,\mathtt{A}^\bullet \to \mathtt{Tw}\,\mathtt{A}^\bullet$ and $\mathcal{P}_{\mathtt{Tw}\,\mathtt{B}^\bullet}\colon \mathtt{Tw}\,\mathtt{B}^\bullet \to \mathtt{D}\,\mathtt{B}^\bullet$ are the standard injections and projections of the minimal models.

Example 3.40 If M is a right A-module then $\mathrm{Hom}_A(-, M)$ is a dg-functor from $\mathtt{RES}^\bullet A$ to $\mathtt{K}^\bullet\,\Bbbk$ that maps a module N with resolution $P^\bullet$ to the complex of vector spaces $\mathrm{Hom}(P^\bullet, M)$. The corresponding derived functor $\mathtt{D}\,\mathrm{Hom}(-, M)$ goes from $\mathtt{D}\,\mathtt{RES}^\bullet A \cong \mathtt{D}\,\mathtt{MOD}^\bullet A$ to $\mathtt{D}\,\mathtt{K}^\bullet\,\Bbbk \cong \mathtt{grVECT}^\bullet\,\Bbbk$ and it maps N to the graded vector space $\mathtt{Ext}^\bullet(N, M)$. This functor is the A_∞-version of the derived functor $\mathbf{R}\,\mathrm{Hom}(-, M)$.

3.3.4 Cones and Generators

Let $\mathtt{C}^\bullet$ be an A_∞-category and let $\mathtt{Tw}\,\mathtt{C}^\bullet$ be its twisted completion. In this section we will investigate how to build twisted complexes out of others.

Definition 3.41 Let (A, δ_A) and (B, δ_B) be two twisted complexes over $\mathtt{C}^\bullet$. To any homogeneous morphism $f \in \mathtt{Tw}\,\mathtt{C}^\bullet(A, B)$ of degree 1 we assign its *cone*:

$$\mathrm{Cone}(f) := \left(A \oplus B, \begin{pmatrix} \delta_A & 0 \\ f & \delta_B \end{pmatrix}\right).$$

Lemma 3.42

(i) *If* $\tilde{\mu}_1(f) = 0$ *then* $\mathrm{Cone}(f)$ *is a twisted complex.*

(ii) *If* $f, g \in \mathtt{Tw}\,\mathtt{C}^\bullet(A, B)$ *are in the same homotopy class then* $\mathrm{Cone}(f)$ *and* $\mathrm{Cone}(g)$ *are isomorphic objects in the derived category* $\mathtt{D^b C}$.

(iii) *If* f *has a quasi-inverse in* $\mathtt{Tw}\,\mathtt{C}^\bullet$ *then* $\mathrm{Cone}(f)$ *is a zero object in* $\mathtt{D^b C}$.

(iv) *If* $u\colon A \to A'$ *and* $v\colon B' \to B$ *are quasi-isomorphisms in* $\mathtt{Tw}\,\mathtt{C}^\bullet$ *then* $\mathrm{Cone}(f)$ *and* $\mathrm{Cone}(ufv)$ *are quasi-isomorphic.*

Sketch of the proof We only work out the first two and leave the rest as an exercise. To prove (a) we just write out the Maurer–Cartan equation

$$\mathsf{MC}\begin{pmatrix} \delta_A & 0 \\ f & \delta_B \end{pmatrix} = \sum_i \mu\left(\begin{pmatrix} \delta_A & 0 \\ f & \delta_B \end{pmatrix}^{\otimes i}\right) = \begin{pmatrix} \mathsf{MC}(\delta_A) & 0 \\ \tilde{\mu}_1(f) & \mathsf{MC}(\delta_B) \end{pmatrix} = 0.$$

For (b) let $f - g = \tilde{\mu}(h)$ and construct the map

$$\phi = \begin{pmatrix} \mathbb{1}_A & 0 \\ h & \mathbb{1}_B \end{pmatrix} \in \mathrm{Hom}(\mathrm{Cone}(f), \mathrm{Cone}(g)).$$

This map satisfies

$$\tilde{\mu}(\phi) = \sum_{k,l} \mu\left(\begin{pmatrix} \delta_A & 0 \\ g & \delta_B \end{pmatrix}^{\otimes k} \otimes \begin{pmatrix} \mathbb{1}_A & 0 \\ h & \mathbb{1}_B \end{pmatrix} \otimes \begin{pmatrix} \delta_A & 0 \\ f & \delta_B \end{pmatrix}^{\otimes l} \right)$$
$$= \begin{pmatrix} \tilde{\mu}(\mathbb{1}_A) & 0 \\ \tilde{\mu}(g, \mathbb{1}_A) + \tilde{\mu}(\mathbb{1}_B, f) + \tilde{\mu}(h) & \tilde{\mu}(\mathbb{1}_B) \end{pmatrix} = \begin{pmatrix} 0 & 0 \\ g - f + \tilde{\mu}(h) & 0 \end{pmatrix} = 0$$

and it has an inverse $\phi^{-1} = \begin{pmatrix} \mathbb{1}_A & 0 \\ 0 & \mathbb{1}_B \end{pmatrix}$ because

$$\tilde{\mu}(\phi, \phi^{-1}) = \sum_{k,l,m} \mu\left(\begin{pmatrix} \delta_A & 0 \\ g & \delta_B \end{pmatrix}^{\otimes k}, \begin{pmatrix} \mathbb{1}_A & 0 \\ h & \mathbb{1}_B \end{pmatrix}, \begin{pmatrix} \delta_A & 0 \\ f & \delta_B \end{pmatrix}^{\otimes l}, \begin{pmatrix} \mathbb{1}_A & 0 \\ -h & \mathbb{1}_B \end{pmatrix}, \begin{pmatrix} \delta_A & 0 \\ g & \delta_B \end{pmatrix}^{\otimes m} \right)$$
$$= \begin{pmatrix} \tilde{\mu}(\mathbb{1}_A, \mathbb{1}_A) & 0 \\ u & \tilde{\mu}(\mathbb{1}_B, \mathbb{1}_B) \end{pmatrix} = \begin{pmatrix} \mathbb{1}_A & 0 \\ 0 & \mathbb{1}_B \end{pmatrix},$$

with $u = \tilde{\mu}(h, \mathbb{1}_A) + \tilde{\mu}(\mathbb{1}_B, h) + \tilde{\mu}(g, \mathbb{1}_A, \mathbb{1}_A) - \tilde{\mu}(\mathbb{1}_B, f, \mathbb{1}_A) + \tilde{\mu}(\mathbb{1}_B, \mathbb{1}_B, g) = 0$. The last three terms are zero because $\mathbb{1}_A$ and $\mathbb{1}_B$ are strong units. □

Example 3.43 If M is an A-module and N a submodule of M then we have a short exact sequence $N \to M \to M/N$. This short exact sequence can be seen as an element in $f \in \mathrm{Ext}^1(N, M/N) = \mathtt{DMOD}^1 A(N, M/N)$. One can easily check that $\mathrm{Cone}(f)$ is isomorphic to M in $\mathtt{D^bMOD}\text{-}A$.

Example 3.44 If $f\colon M \to N$ is a morphism between A-modules then we can consider f as a degree 1 morphism between $M[1]$ and N in $\mathtt{Tw}\,\mathtt{MOD}^\bullet A$. If f is surjective the $\mathrm{Cone}(f)$ is quasi-isomorphic with the shift of the kernel of f. If f is injective then $\mathrm{Cone}(f)$ is quasi-isomorphic with the cokernel of f.

Lemma 3.45 (Higher products and cones) *If $A_1 \xleftarrow{f_1} \cdots \xleftarrow{f_{k-1}} A_k$ is a sequence of morphisms in $\mathtt{C}^\bullet$ and $1 < i < k-1$ then*

$$\mu(f_1, \ldots, f_{k-1}) = \tilde{\mu}(f_1, \ldots, f_{i-2}, \tilde{f}_{i-1}, \tilde{f}_{i+1}, f_{i+2}, \ldots, f_k),$$

where $\tilde{f}_{i+1}\colon A_{i+2} \to \mathrm{Cone}(f_i)$ and $\tilde{f}_{i-1}\colon \mathrm{Cone}(f_i) \to A_{i-1}$ are the induced cone morphisms of f_{i-1} and f_{i+1} in $\mathtt{Tw}\,\mathtt{C}^\bullet$.

Sketch of the proof This follows from a straightforward calculation using the facts that $\mathrm{Cone}(f_i) = \left(A_{i+1} \oplus A_i, \begin{pmatrix} 0 & 0 \\ f_i & 0 \end{pmatrix}\right)$, $\tilde{f}_{i+1} = \begin{pmatrix} f_{i+1} \\ 0 \end{pmatrix}$ and $\tilde{f}_{i-1} = \begin{pmatrix} 0 & f_{i-1} \end{pmatrix}$. □

Lemma 3.46 *If $\mathtt{C}^\bullet$ is closed under taking cones and shifts then $\mathtt{Tw}\,\mathtt{C}^\bullet$ and $\mathtt{D}\,\mathtt{C}^\bullet$ are quasi-equivalent to $\mathtt{C}^\bullet$.*

Sketch of the proof The condition of being closed under shifts and cones means that for any object $A \in \mathtt{C}^\bullet \subset \mathtt{Tw}\,\mathtt{C}^\bullet$ and for any morphism f, the objects

ΣA and Cone(f) are quasi-isomorphic to objects in $\mathsf{C}^\bullet \subset \mathtt{Tw}\,\mathsf{C}^\bullet$. The lemma follows from the fact that every twisted object is a cone of smaller twisted objects:

$$(A_1[i_1] \oplus \cdots \oplus A_k[i_k], \delta)$$
$$= \mathrm{Cone}\left((A_1[i_1] \oplus \cdots \oplus A_k[i_{k-1}], \delta|_{1\cdots k-1 \times 1 \cdots k-1}) \xrightarrow{\delta_{k \times 1 \cdots k-1}} (A_k, 0) \right).$$

□

Definition 3.47 If S is a set of objects in $\mathtt{D}\,\mathsf{C}^\bullet$ we denote by $\langle S \rangle$ the smallest subcategory containing S that is closed under taking cones and shifts.[2] If the embedding $\langle S \rangle \subset \mathtt{D}\,\mathsf{C}^\bullet$ is an equivalence then we call S a *set of generators*. Note that in that case $\mathtt{D}\,\mathsf{S}^\bullet \overset{\infty}{=} \mathtt{D}\,\mathsf{C}^\bullet$, where $\mathsf{S}^\bullet$ is the full subcategory of $\mathtt{D}\,\mathsf{C}^\bullet$ containing the objects of S.

Example 3.48 Take the completed polynomial ring $R = \Bbbk[[X]]$. Because R is a principal ideal domain, every finitely generated R-module is a direct sum of copies of R and modules of the $R/(X^i)$. The latter have resolutions like

$$R \xrightarrow{X^i} R \;= \mathrm{Cone}(X^i).$$

So all *finitely generated modules* sit in $\langle R \rangle$ and R is a generator of $\mathtt{D}\,\mathtt{Mod}^\bullet R$.

The derived category of finite-dimensional modules $\mathtt{D}\,\mathtt{mod}^\bullet R$ is contained in $\langle R \rangle$ but R cannot be seen as a generator because it is not an object of $\mathtt{D}\,\mathtt{mod}^\bullet R$. Instead, the set of all simple R-modules will form a set of generators for $\mathtt{D}\,\mathtt{mod}^\bullet R$. Indeed, if M is a module and N is a submodule then by Example 3.43, M is in $\langle N, M/N \rangle$. Continuing decomposing N and M/N we see that every finite-dimensional module is generated by the simple modules. Because R is commutative and completed there is only one simple module $S = R/(X)$. Therefore,

$$\mathtt{D}\,\mathtt{mod}^\bullet R = \langle S \rangle \overset{\infty}{=} \mathtt{D}\,\mathsf{S}^\bullet,$$

where $\mathsf{S}^\bullet$ is an A_∞-category with one object with endomorphism ring $\mathtt{Ext}^\bullet{}_R(S, S)$. This gives the following tower of categories:

$$\langle S \rangle = \mathtt{D}\,\mathtt{mod}^\bullet R \subset \mathtt{D}\,\mathtt{Mod}^\bullet R = \langle R \rangle \subset \mathtt{D}\,\mathtt{MOD}^\bullet R.$$

Example 3.49 Now take the ring $R = \Bbbk[X]/(X^n)$. Again, every finitely generated module is a direct sum of copies of R and modules of the form $R/(X^i)$. But now the resolutions of the latter modules are infinite:

$$\cdots \xrightarrow{X^{n-i}} R \xrightarrow{X^i} R \xrightarrow{X^{n-i}} R \xrightarrow{X^i} R\,.$$

[2] Although the cones and shifts are defined as objects in the twisted completion, we can also consider them as objects in the minimal model or the derived category.

Therefore, they do not sit inside $\langle R\rangle$. On the other hand, because R itself is finite-dimensional, all finitely generated modules are generated by the unique simple module $S = R/(X)$. This gives the following tower of categories:

$$\langle R\rangle \subset \mathtt{D}\,\mathtt{mod}^\bullet R = \mathtt{D}\,\mathtt{Mod}^\bullet R = \langle S\rangle \subset \mathtt{D}\,\mathtt{MOD}^\bullet R.$$

3.3.5 Split-Derived Categories

In some cases $\mathtt{D}\,\mathtt{C}^\bullet$ is not closed under direct summands and then it makes sense to look at the *split-derived category* $\mathtt{D}^\pi\,\mathtt{C}^\bullet$. This is an enlargement of $\mathtt{D}\,\mathtt{C}^\bullet$ that is closed under shifts, cones and direct summands. To define it formally as an A_∞-category we would have to talk about idempotents in A_∞-categories and just like units there are strict and weak versions of this concept. For further details we refer to [218].

Example 3.50 For a Noetherian algebra A the category $\mathtt{D}^\pi A$ is the category that consists of all finitely generated complexes of finitely generated projective modules (remember that these are direct summands of finitely generated free modules). These complexes are also known as perfect complexes and therefore $\mathtt{D}^\pi A$ is sometimes denoted by $\mathtt{Perf}^\bullet A$.

Definition 3.51 A set of objects S is called a *split-generator* if the smallest subcategory of $\mathtt{D}^\pi\mathtt{C}$ closed under shifts, cones and direct summands containing S is equivalent to $\mathtt{D}^\pi\mathtt{C}$.

Example 3.52 Take $R = \mathbb{C}[X]/(X^2 - 1)$; then $P_1 = R/(X - 1)$ and $P_2 = R/(X + 1)$ are two projective modules that do not have a finite free resolution because

$$\cdots \longrightarrow R \xrightarrow{X-1} R \xrightarrow{X+1} R \xrightarrow{X-1} R \longrightarrow P_1\ .$$

Therefore, $\langle R\rangle \subset \mathtt{D}\,\mathtt{Mod}^\bullet R$ does not contain P_1, but it is isomorphic to $(R, \frac{1+X}{2}) \in \mathtt{D}^\pi R$. In fact, the object R is a split-generator of $\mathtt{D}\,\mathtt{Mod}^\bullet R$ but not a generator.

To check whether a certain object X is split-generated by S we can use the following criterion (see also [54, Chapter 6] and [6]).

Lemma 3.53 *An object $X \in \mathtt{D}^\pi\,\mathtt{C}^\bullet$ is split-generated by $S \subset \mathtt{C}$ if we can write $\mathbb{1}_X$ as a sum of morphisms that factor through objects in S. This means that*

$$\mathbb{1}_X = \sum_r \mu_k(f_{r,k_r-1}, \ldots, f_{r,0}),$$

where $f_{r,i} \in \mathtt{C}^\bullet(M_{r,i}, M_{r,i+1})$ with $M_{r,0} = M_{r,k_r} = X$ and $M_i \in S$ if $i \neq 0, k_r$.

Sketch of the proof If all $k_r = 2$ then $\mathbb{1}_X = \sum_r f_{r,1} f_{r,0}$, so the matrix $e = (f_{r,0} f_{s,1})_{rs}$ is an idempotent and $X \cong (\oplus_r M_{r,1}, e)$. For $k_r > 2$ we can use Lemma 3.42 to reduce the order of the product by introducing a cone. □

3.3.6 The Relation with Triangulated Categories

Classically, the derived category is not constructed as an A_∞-category but as a triangulated category. We will briefly outline the connection between this concept and the A_∞-approach. For more details about the classical approach we refer to [253, Chapter 10].

The classical starting point is an abelian category $\mathtt{C}$. We can think of this as a $\Bbbk$-linear category that has finite direct sums, kernels and cokernels that behave as expected, and therefore it also has the notion of a short exact sequence. The prime example of such a category is the category of right modules of an algebra $\mathtt{MOD}A$, or the category of finitely generated modules of a Noetherian algebra $\mathtt{Mod}\,A$.

From an abelian category one can construct the category of bounded complexes $\mathtt{K}^\mathrm{b}\mathtt{C}$ with as morphisms the degree 0 morphisms of complexes up to homotopy, just like we did in the case of vector spaces in Chapter 2. This category has a special set of morphisms $\mathtt{Q}$ called the quasi-isomorphisms, but unfortunately not all quasi-isomorphisms are invertible in $\mathtt{K}^\mathrm{b}\mathtt{C}$. To solve this problem we add formal inverses to all morphisms in $\mathtt{Q}$. The result is the bounded derived category

$$\mathtt{D}^\mathrm{b}\mathtt{C} := \mathtt{K}^\mathrm{b}\mathtt{C} \cdot \mathtt{Q}^{-1}.$$

Unfortunately, working with the derived category $\mathtt{D}^\mathrm{b}\mathtt{C}$ is not easy because of these extra formal inverses, but in some cases we can get around this. In the case of the derived category of $\mathtt{MOD}A$, every bounded complex of A-modules is quasi-isomorphic to a (not necessarily bounded) complex of projective A-modules. Moreover, quasi-isomorphisms between complexes of projective modules have quasi-inverses. Therefore, $\mathtt{D}^\mathrm{b}\mathtt{MOD}\,A$ is equivalent to a subcategory of $\mathtt{KPROJ}\,A$ because we do not have to invert quasi-isomorphisms anymore.

This matches what we have done in the A_∞-case. First we substituted each module for a projective (or free) resolution, then we took the twisted completion to make for bounded complexes with these resolutions, after that we went to the minimal model by taking the cohomology and finally we restricted the zeroth degree to leave the A_∞-world again. This allows us to conclude that the classical bounded derived category is the degree 0 part of its A_∞-version:

$$\mathtt{D}^\mathrm{b}\mathtt{MOD}\,A \cong \mathtt{DMOD}^0 A.$$

Remark 3.54 This works fine for the category of A-modules but some categories do not have *enough projectives*. If that is the case one can either enlarge the category to include them (e.g. construct $\mathtt{D^b mod}\, A$ as a subcategory of $\mathtt{D^b MOD}\, A$) or work with resolutions that use different types of objects that behave similarly with respect to quasi-isomorphisms. One example of these are injective objects, which occur in the classical construction of $\mathtt{D^b QCoh}\, \mathbb{X}$; see [124].

The classical derived category is not an abelian category itself anymore, but it still has some special structure.

Definition 3.55 A $\Bbbk$-linear category $\mathtt{T}$ is *triangulated* if it has an equivalence $\Sigma\colon \mathtt{T} \to \mathtt{T}$ and a collection of diagrams $A \to B \to C \to \Sigma A$, called *distinguished triangles*, such that every morphism $f\colon A \to B$ can be extended to distinguished triangle

$$A \xrightarrow{f} B \to C \to \Sigma A.$$

The object C is unique up to isomorphism and is called the *mapping cone* of f. The distinguished triangles satisfy some additional axioms that can be found in [253, 10.2.1].

Definition 3.56 A set S of objects in a triangulated category $\mathtt{T}$ is called

- a *generator* if the smallest subcategory of $\mathtt{T}$ containing S and closed under shifts and mapping cones is $\mathtt{T}$,
- a *classical generator* if the smallest subcategory of $\mathtt{T}$ containing S and closed under shifts, mapping cones and direct summands is $\mathtt{T}$.

If $\mathtt{A}^\bullet$ is any A_∞-category then its derived A_∞-category $\mathtt{D^b A} = \mathtt{DA}^0$ can be given a triangulated structure:

- The shift functor is the shift of complexes $\Sigma(M, \delta) := (M[1], -\delta)$.
- The distinguished triangles come from the cone construction:

$$(M, \delta_M) \xrightarrow{\quad f \quad} (N, \delta_N) \xrightarrow{\begin{pmatrix} 0 & 0 \\ 0 & \mathbb{1}_N \end{pmatrix}} \mathrm{Cone}(f)$$

$$\xrightarrow{\begin{pmatrix} \mathbb{1}_M & 0 \\ 0 & 0 \end{pmatrix}} (\Sigma M, \delta_M).$$

So from the A_∞-perspective, being triangulated merely means being closed under taking twisted complexes. If S is a generating set for the A_∞-category $\mathtt{D\,A}^\bullet$ then S will also be a generator for the triangulated category $\mathtt{D^b A}$, and if S split-generates $\mathtt{D}^\pi\, \mathtt{A}^\bullet$ then it is a classical generator for $\mathtt{D}^\pi \mathtt{A}^0$.

Definition 3.57 An A_∞-*enhancement* of a triangulated category $\mathtt{T}$ is a derived A_∞-category $\mathtt{T}^\bullet$ such that its degree 0 part is equivalent to $\mathtt{T}$ as a triangulated category.

Note that a triangulated category may not always have an enhancement or it can have more than one non-A_∞-equivalent enhancement (see [211, 210]). In general though, for most naturally occurring triangulated categories there is a unique enhancement [159]. In the case of the module category of an algebra A the unique enhancement of $\mathtt{D}^\mathrm{b}\mathtt{MOD}\,A$ is precisely the A_∞-category $\mathtt{D}\,\mathtt{MOD}^\bullet A$ that we defined before. Finally, for classical derived categories coming from an abelian category, we do not have to worry about the difference between the derived and the split-derived versions.

Theorem 3.58 (Balmer–Schlichting [43]) *If $\mathtt{C}$ is an abelian (or more generally, exact) category then the derived category is split-complete.*

3.4 Bells and Whistles

In the final section of this chapter we briefly discuss some extra features that can be added to the theory of A_∞-categories.

3.4.1 Modules

Given an A_∞-algebra $A^\bullet$ we can talk of right A_∞-modules. Just like in the ordinary case this can be done in two ways.

Definition 3.59 (Algebraic version) A right A_∞-module $M^\bullet$ is a graded vector space with degree $2-k$ linear maps

$$\mu_k^M\colon M \otimes A^{\otimes k-1} \to M$$

such that the A_∞-axioms $[\mathtt{SM_n}]$ hold, where the μ's are appropriately interpreted as μ^A or μ^M, depending on whether they have an input from $M^\bullet$ or not.

Definition 3.60 (Categorical version) A right A_∞-module is an A_∞-functor $\mathcal{F}\colon \mathtt{A}^{\mathrm{op}\bullet} \to \mathtt{VECT}^\bullet\,\Bbbk$, where $\mathtt{A}^{\mathrm{op}\bullet}$ has one object O and $\mathtt{A}^{\mathrm{op}\bullet}(\mathsf{O},\mathsf{O}) = A^\bullet$ but the multiplications are reversed:

$$\mu^{\mathrm{op}}(a_1,\ldots,a_k) := (-1)^\dagger \mu(a_k,\ldots,a_1).$$

We can move between the two formalisms by setting $\mathcal{F}(\mathsf{O}) = M^\bullet$ and demanding that

$$\mathcal{F}(a_1,\ldots,a_k)(m) = \mu^M(m,a_k,\ldots,a_1).$$

The right A_∞-modules form an A_∞-category $\mathtt{MOD}^\bullet_\infty\,\mathtt{A}$ and $\mathtt{A}^\bullet$ embeds into it via an A_∞-version of the Yoneda lemma. We can even embed the whole of $\mathtt{Tw}\,\mathtt{A}^\bullet$ into $\mathtt{MOD}^\bullet{}_\infty\,\mathtt{A}$. This gives a tower of A_∞-categories that is reminiscent of the last paragraph of Chapter 1:

$$\mathtt{A}^\bullet \subset \mathtt{Add}\ \mathtt{A}^\bullet \subset \mathtt{Tw}\,\mathtt{A}^\bullet \ \subset \mathtt{MOD}^\bullet_\infty\,\mathtt{A}.$$

We also saw that Morita-equivalent algebras and categories have equivalent module categories. In the A_∞-setting (split-)derived equivalent A_∞-categories will have quasi-equivalent module categories. More information on A_∞-modules can be found in [135].

Example 3.61 Consider the A_∞-algebra $A = \Bbbk\xi + \Bbbk\rho$ from Example 3.12 with $r(X) = X^3$, so the only nontrivial product is $\mu(\xi,\xi,\xi) = \rho$. This algebra has a twisted complex

$$X = \left(\mathsf{O}\oplus\mathsf{O}, \left(\begin{smallmatrix} 0 & 0 \\ \xi & 0 \end{smallmatrix}\right)\right),$$

where O represents the unique object in $\mathtt{A}^\bullet$. This twisted complex becomes the A_∞-module

$$M^\bullet = \mathtt{Tw}\,\mathtt{A}^\bullet(\mathsf{O}, X) = A^\bullet \oplus A^\bullet = \Bbbk\xi_1 \oplus \Bbbk\rho_1 \oplus \Bbbk\xi_2 \oplus \Bbbk\rho_2$$

with nontrivial products

$$\mu(\xi_1,\xi) = \rho_2, \quad \mu(\xi_1,\xi,\xi) = \rho_1, \quad \mu(\xi_2,\xi,\xi) = \rho_2.$$

3.4.2 Curvature

In some cases ordinary A_∞-algebras are not general enough and we need the notion of a curved A_∞-algebra [60]. This algebraic gadget includes one extra product: a nullary product $\mu_0\colon \Bbbk \to A$.

Definition 3.62 A *curved A_∞-algebra* is graded vector space A with a collection of maps $\mu_i\colon A^{\otimes i} \to A$ of degree $2 - i$ with $i = 0, 1, 2, \ldots$, such that the identities

$$[\mathtt{CM_n}] \qquad \sum_{\substack{r+s+t=n \\ r,s,t\geq 0}} (-1)^\dagger \mu(a_1, \ldots, a_r, \mu(a_{r+1}, \ldots, a_{r+s}), a_{r+s+1}, \ldots, a_{r+s+t}) = 0$$

hold for all $n \geq 0$. In the terms with $s = 0$ we interpret $\mu(\,)$ as $\mu_0(1)$. This is an element of degree 2 in A. It is called the *curvature of the A_∞-algebra* and denoted by W. For a *curved A_∞-category* $\mathtt{C}^\bullet$ there is a map $\mu_0\colon \Bbbk \to \mathtt{C}^\bullet(M,M)$ for each object in the category, so every object X can have its own curvature W_X.

Note that there is one extra rule [$\mathtt{CM}_0$], which states that $\mu_1(W) = 0$. All the other rules become more complicated because we can splice in extra W's. Take for instance [$\mathtt{CM}_1$]:

$$\mu_1^2(a) + \mu_2(W, a) - \mu_2(a, W) = 0.$$

It implies that μ_1 does not necessarily square to zero, so (A, μ_1) does not need to be a complex anymore. Hence there is no notion of quasi-isomorphism or minimal model. This makes working with curved A_∞-algebras or curved A_∞-categories much harder. We can get around this problem by focusing on the twisted complexes instead of the algebra or category itself.

Definition 3.63 Let $\mathsf{C}^\bullet$ be a curved A_∞-category and assume $\mu_k = 0$ for $k \gg 0$. A *curved twisted complex* over a curved A_∞-category $\mathsf{C}^\bullet$ consists of a pair

$$M := (\oplus_i A_i[k_i], \delta),$$

where $\oplus_i A_i[k_i]$ is a direct sum of shifted objects and δ is an element of degree 1 in $\mathtt{Add}\ \mathsf{C}^\bullet(\oplus_i A_i[k_i], \oplus_i A_i[k_i])$. We can group the curved twisted complexes into a curved A_∞-category ${}^{\mathtt{cu}}\mathtt{Tw}\,\mathsf{C}^\bullet$ with the following products:

$$\tilde{\mu}_k(a_1, \dots, a_k) = \sum_{m_0, \dots, m_k \geq 0} \mu_{k+m_0+\cdots+m_k}(\underbrace{\delta_0, \dots, \delta_0}_{m_0}, a_1, \dots, a_k, \underbrace{\delta_k, \dots, \delta_k}_{m_k}).$$

If $\mathsf{C}^\bullet$ has strict units we can fix a $\lambda \in \mathbb{k}$ and look at all the curved twisted complexes with curvature λ:

$$\tilde{\mu}_0(1) = \mu_0(1) + \mu(\delta) + \mu(\delta, \delta) + \cdots = \lambda \mathbb{1}_M.$$

Because $\lambda \mathbb{1}_M$ commutes with everything and turns higher products into zero, the subcategory ${}^{\lambda}\mathtt{Tw}\,\mathsf{C}^\bullet$ containing these complexes together with $\tilde{\mu}_{>0}$ will be an ordinary uncurved A_∞-category. We will denote its minimal model by ${}^{\lambda}\mathtt{D}\,\mathsf{C}^\bullet$. Note that if $\lambda \neq 0$ then we have to work in the $\mathbb{Z}_2$-graded context because the curvature has to have degree 0.

Remark 3.64 We dropped the condition of being lower triangular. The reason for this is that it is not possible to find a lower triangular δ once μ_0 is nontrivial. In order to make sense of the $\tilde{\mu}_k$ we therefore have to assume that all μ_k are zero if $k \gg 0$. Alternatively we can work over a completed ring that allows us to take infinite sums; this will be discussed in the next subsection.

Example 3.65 Suppose that $R = \mathbb{k}[X_1, \dots, X_n]$ and $W \in R$ is an element called the potential. Put R in degree 0, set $\mu_0(1) = W$, use the ordinary product to define μ_2 and set all the other $\mu_i = 0$. Then $R_W := (R, \mu_\bullet)$ is a $\mathbb{Z}_2$-graded curved A_∞-algebra if and only if W is a central element because of [$\mathtt{CM}_1$].

We have to work in the $\mathbb{Z}_2$-graded setting because the curvature W has to have degree 2.

The objects of ${}^{\lambda}\mathtt{Tw}R_W$ can be seen as pairs (P, d) consisting of a $\mathbb{Z}_2$-graded finitely generated free R-module $P = P^0 \oplus P^1$, together with a degree 1 map d such that $d^2 = (W - \lambda)\mathbb{1}_P$. If we choose basis for P^0 and P^1, we can write d as a block matrix

$$\begin{pmatrix} 0 & \Psi_1 \\ \Psi_0 & 0 \end{pmatrix}$$

for which $\Psi_0\Psi_1 = \Psi_1\Psi_0 = (W - \lambda)\mathbb{1}$, or in other words we factored $W - \lambda$ as a product of two matrices. Therefore, the minimal model of ${}^{\lambda}\mathtt{Tw}R_W$ is also called the *category of matrix factorizations* of $W - \lambda$, and it is denoted by

$$\mathtt{MF}^{\pm}(R, W - \lambda) := {}^{\lambda}\mathtt{D}\,R_W = {}^{0}\mathtt{D}\,R_{W-\lambda}.$$

3.4.3 Filtrations

As we have seen above, it is sometimes necessary to take infinite sums. In order to make these sums converge we have to work in the filtered setting. This leads to the notion of a filtered A_∞-algebra. We will describe a slightly altered version of the theory developed by Fukaya, Oh, Ohta and Ono in [83, 84].

Fix a monoid $(G, +, 0)$ and an additive morphism $E\colon G \to \mathbb{R}_{\geq 0}$ such that $E^{-1}(0) = \{0\}$ and $E^{-1}([0, c])$ is finite for every $c \in \mathbb{R}_{\geq 0}$. To G we can assign a filtered ring

$$\mathbb{k}[\![G]\!] := \left\{\textstyle\sum_{g\in G} \lambda_g \hbar^g \mid \lambda_g \in \mathbb{k}\right\} \quad \text{with } F_r\mathbb{k}[\![G]\!] := \left\{\textstyle\sum_{E(g)\geq r} \lambda_g \hbar^g \mid \lambda_g \in \mathbb{k}\right\},$$

where $\hbar$ is just a formal placeholder for the exponents in G. Note that the product in $\mathbb{k}[\![G]\!]$ is well defined because of the finiteness condition on E. Here $\mathbb{k}[\![G]\!]$ is a domain and we will denote its field of fractions by $\mathbb{k}(\!(G)\!)$.

This finiteness condition also ensures that every $g \in G$ can only be split into a finite number of sums $g_1 + g_2$ and therefore the following definitions make sense.

Definition 3.66 A *G-filtered A_∞-algebra* $\mathtt{A}^{\hbar}$ consists of a $\mathbb{Z}$- or $\mathbb{Z}_2$-graded vector space A and morphisms $\mu_{k,g}\colon A^{\otimes k} \to A$ of degree $2 - k$ that satisfy the A_∞-axioms:

$$[\mathtt{M}_{\mathtt{n},\mathtt{g}}] \qquad \sum_{\substack{r+s+t=n \\ g_1+g_2=g}} (-1)^{\dagger}\mu_{r+t+1,g_1}(a_1, \ldots, a_r, \mu_{s,g_2}(a_{r+1}, \ldots, a_{r+s}), a_{r+s+1}, \ldots, a_{r+s+t}) = 0.$$

For the *curved* version we allow $\mu_{0,g}$ to be nonzero but we demand that $\mu_{0,0} = 0$. An element $\mathbb{1} \in A$ is a *strict unit* if $\mu_{2,0}(a, \mathbb{1}) = (-1)^{|a|}\mu_{2,0}(\mathbb{1}, a) = a$ and all other products involving $\mathbb{1}$ are zero.

Similarly, we also define *filtered A_∞-morphisms, categories and functors.* We say that a filtered A_∞-category is *minimal* if $\mu_{1,0} = 0$ and an A_∞-morphism is a quasi-isomorphism if $\mathcal{F}_{1,0}$ is a quasi-isomorphism for the differential $\mu_{1,0}$. Just like for products we allow $\mathcal{F}_{0,g}$ to be nonzero, as long as $\mathcal{F}_{0,0} = 0$.

Remark 3.67 Instead of keeping all products separate we could also have defined a G-filtered A_∞-algebra as an A_∞-structure on $A \otimes \mathbb{k}[[G]]$. This can be done by demanding that

$$\hat{\mu}_n(a_1\hbar^{g_1}, \ldots, a_n\hbar^{g_n}) = \sum_{g \in G} \mu_{n,g}(a_1, \ldots, a_n)\hbar^{g+g_1+\cdots+g_n}$$

and extending this linearly to power series. These A_∞-structures are *deformations* of $\mathtt{A}^\bullet = (A, \mu_{\bullet,0})$. In principle these approaches are equivalent and we can switch between the two ways of representing the structures.

The main reason to define filtered A_∞-algebras using $\mathbb{k}$-linear products $\mu_{n,g}$ on A instead of $\mathbb{k}[[G]]$-linear products on $A \otimes \mathbb{k}[[G]]$ is that this makes going to the cohomology more controllable. For complexes over $\mathbb{k}[[G]]$ we do not have the decomposition theorem (Lemma 2.9) because $\mathbb{k}[[G]]$ is not a field. For the filtered version we do have a minimal model if we keep working over $\mathbb{k}$ and use $\mu_{1,0}$ for the differential. For more details see [82].

Another thing we can do is view this structure as an A_∞-structure on $A \otimes \mathbb{k}((G))$ over the field $\mathbb{k}((G))$. In this way we can also talk about minimal models, but take care: this notion is different from the one considered in [82].

Example 3.68 Take for instance $G = \mathbb{N}$ and $\mathtt{A}^\hbar = \mathbb{k}x \oplus \mathbb{k}y$ with $\mu_{1,1}(x) = y$ and all other products zero.

- As a G-filtered A_∞-algebra it is minimal because $\mu_{1,0} = 0$.
- As a $\mathbb{k}[[G]]$-linear A_∞-algebra its minimal model is $\mathbb{k}y = \mathbb{k}[[G]]y/\hbar\mathbb{k}[[G]]$.
- As a $\mathbb{k}((G))$-linear A_∞-algebra its minimal model is 0.

We will denote a G-filtered A_∞-algebra by $\mathtt{A}^\hbar$ and its $\mathbb{k}[[G]]$- and $\mathbb{k}((G))$-linear versions by $\mathtt{A}^\hbar \otimes \mathbb{k}[[G]]$ and $\mathtt{A}^\hbar \otimes \mathbb{k}((G))$ respectively.

Definition 3.69 Let $\mathtt{A}^\hbar$ be a G-filtered A_∞-category. A *curved twisted complex* is a pair

$$(\oplus_i A_i[k_i], \delta) \quad \text{with } \delta = \sum_g \delta_g \hbar^g \in \oplus_{i,j} \mathtt{A}^1(A_i, A_j)[k_j - k_i] \otimes \mathbb{k}[[G]]$$

such that $(\oplus_i A_i[k_i], \delta_0)$ is an ordinary uncurved twisted complex over $\mathtt{A}^\bullet = (A, \mu_{\bullet,0})$. This implies that although δ need not be lower triangular, δ_0 must be.

The curved twisted complexes form a curved G-filtered A_∞-category ${}^{\mathrm{cu}}\mathtt{TwA}^\hbar$ with products

$$\hat{\mu}_n(a_1,\ldots,a_n) = \sum \mu(\delta_{M_0}^{\otimes e_0}, x_1,\ldots,x_n,\delta_{M_n}^{\otimes e_0}).$$

For any $\lambda \in \Bbbk[\![G]\!]$ the twisted complexes with curvature $\lambda\mathbb{1}$ form an uncurved G-filtered A_∞-category ${}^{\lambda}\mathtt{TwA}^\hbar$. If (M,δ) has curvature $\lambda\mathbb{1}$ then δ is also sometimes called a *weak bounding cochain*.

We can also consider ${}^{\lambda}\mathtt{TwA}^\hbar \otimes \Bbbk(\!(G)\!)$ by tensoring all the hom-spaces with the field $\Bbbk(\!(G)\!)$. This operation is not innocent because some objects can become zero objects (e.g. if $\hbar\mathbb{1}$ sits in the image of $\hat{\mu}_1$). We will denote the minimal model of ${}^{\lambda}\mathtt{TwA}^\hbar \otimes \Bbbk(\!(G)\!)$ viewed as $\Bbbk(\!(G)\!)$-linear A_∞-category by ${}^{\lambda}\mathtt{DA}^\hbar \otimes \Bbbk(\!(G)\!)$.

Example 3.70 Take $G = \mathbb{N}$ and $\mathtt{E} = \Bbbk\mathbb{1} \oplus \Bbbk\xi$ with $|\xi| = 1 \in \mathbb{Z}_2$. For every function $r(X,\hbar) = \sum_{n,m} r_{nm}X^n\hbar^m \in \Bbbk[\![X,\hbar]\!]\hbar$ we can define a filtered $\mathbb{Z}_2$-graded A_∞-structure with strict unit $\mathbb{1}$ and $\mu_{n,m}(\xi,\ldots,\xi) = r_{nm}\mathbb{1}$. This is a deformation of the $\mathbb{Z}_2$-graded algebra $\mathtt{E}^{\pm} = \mathbb{C}[\xi]/(\xi^2)$. Let us now focus on the deformation $\mathtt{E}^\hbar$ given by $r(X,\hbar) = X^n\hbar$.

Every twisted complex of $\mathtt{E}^{\pm}$ is isomorphic to a direct sum of complexes of the form

$$S_k := (E^{\oplus k}, \xi J) \quad \text{with } J = \begin{pmatrix} 0 & 0 \\ \mathbb{1}_{k-1} & 0 \end{pmatrix}.$$

If $k \leq n$ this complex stays uncurved so S_k is an object in ${}^{0}\mathtt{TwE}^{\pm}$. For $k = n$ we have $\hat{\mu}_1(\xi J^{\top n-1}) = \hbar\mathbb{1}$, so this becomes a zero object in ${}^{0}\mathtt{DE}^\hbar \otimes \Bbbk(\!(G)\!)$. If $k > n$ the complex S_k becomes curved with curvature $\hat{\mu}_0 = \hbar J^k$. It is impossible to make this curvature go away by adding an extra term to the twist.

Example 3.71 Take $\mathtt{A}^\hbar = \mathbb{C}[\![X]\!]$ with $\hat{\mu}_2$ the ordinary product, $\hat{\mu}_0 = X^n\hbar$ and all other products equal to 0. This is a $\mathbb{Z}_2$-graded curved deformation of $\mathbb{C}[\![X]\!]$. Every twisted complex of $\mathbb{C}[\![X]\!]$ is a direct sum of complexes of the form

$$S_k := \left(A[1] \oplus A, \begin{pmatrix} 0 & 0 \\ X^k & 0 \end{pmatrix}\right).$$

The complex S_k becomes curved with curvature $\hat{\mu}_0 = \hbar X^n\mathbb{1}$. For $k \leq n$ it is possible to make this curvature go away by adding an extra term to the twist:

$$\tilde{S}_k := \left(A[1] \oplus A, \begin{pmatrix} 0 & \hbar X^{n-k} \\ X^k & 0 \end{pmatrix}\right).$$

These twisted complexes sit in ${}^{0}\mathtt{TwA}^\hbar$. However, if $k = n$ then this procedure again has the consequence that $\hat{\mu}(\begin{pmatrix} 0 & 0 \\ 1 & 0 \end{pmatrix}) = \hbar\mathbb{1}$, so this becomes a zero object in ${}^{0}\mathtt{DA}^\hbar \otimes \Bbbk(\!(G)\!)$.

3.4.4 Transversality

In some cases it is not possible to define all hom-spaces and products in a category, but only those between certain collections of objects. To take care of this, Kontsevich and Soibelman introduced the notion of a precategory [146].

Definition 3.72 An A_∞-*precategory* $\underline{C}^\bullet$ consists of a set of objects $\mathrm{Ob}\,\underline{C}$ and for each $n \geq 1$ a set of n-tuples of objects

$$\underline{C}^n_{\mathrm{tr}} \subset \mathrm{Ob}\,\underline{C}^n.$$

The elements of $\underline{C}^n_{\mathrm{tr}}$ are called *transversal sequences*. We require that $\underline{C}^1_{\mathrm{tr}} = \mathrm{Ob}\,\underline{C}$ and every subsequence of a transversal sequence is transversal:

$$(A_1, \ldots, A_n) \in \underline{C}^n_{\mathrm{tr}}: \quad \forall\, i_1 < \cdots < i_k \leq n \text{ we have } (A_{i_1}, \ldots, A_{i_k}) \in \underline{C}^k_{\mathrm{tr}}.$$

For each transversal pair $(A_0, A_1) \in \underline{C}^2_{\mathrm{tr}}$ there is a graded $\Bbbk$-vector space $\underline{C}^\bullet(A_1, A_0)$ and for each transversal sequence $(A_0, \ldots, A_n) \in \underline{C}^{n+1}_{\mathrm{tr}}$ there is a degree $2 - n$ product

$$\mu_n\colon\ \underline{C}^\bullet(A_1, A_0) \otimes \cdots \otimes \underline{C}^\bullet(A_n, A_{n-1}) \to \underline{C}^\bullet(A_n, A_0).$$

These products satisfy the $[\mathtt{SM}_\Bbbk]$-axioms whenever they make sense.

For precategories it is still possible to talk about minimal models. We just take the homology of the morphism spaces that are defined and we construct new products for all old products that were defined. In general, the minimal model is not a category because some hom-spaces are missing and there is no notion of an identity element. This can be solved as follows.

Definition 3.73 A morphism $f \in \underline{C}^\bullet(B, A)$ is called a *(quasi-)isomorphism* if for every pair of transversal triples $(A, B, X), (Y, A, B) \in \underline{C}^3_{\mathrm{tr}}$ the maps

$$f^*\colon\ \underline{C}^\bullet(X, B) \to \underline{C}^\bullet(X, A)\colon\ u \mapsto \mu_2(f, u)$$
$$\text{and}\quad f_*\colon\ \underline{C}^\bullet(A, Y) \to \underline{C}^\bullet(B, Y)\colon\ u \mapsto \mu_2(u, f)$$

are (quasi-)isomorphisms. The objects A, B are called *quasi-isomorphic*.

The idea is to take the minimal model and add formal inverses to all isomorphisms. In this way we can define $\underline{HC}^\bullet(X, Y) := f^{-1}\, \underline{HC}^\bullet(A, B) g^{-1}$ where $f \in \underline{HC}^\bullet(X, B)$ and $g \in \underline{HC}^\bullet(A, Y)$ are isomorphisms and (X, B, A) and (B, A, Y) are transversal. For this to work we need to make sure that there are enough isomorphisms.

Definition 3.74 Let $\underline{C}^\bullet$ be a precategory. We say that $\underline{C}^\bullet$ is *unital* or has *enough quasi-isomorphisms* if for any object X and any finite collection $\{S_i\}_{i \in I}$

of transversal sequences, there exist quasi-isomorphisms $f_+ : X_+ \to X$ and $f_- : X \to X_-$ such that the sequences (X_+, S_i, X_-) are transversal.

Definition 3.75 An A_∞-functor $\mathcal{F} : \underline{\mathsf{C}}^\bullet \to \underline{\mathsf{D}}^\bullet$ between A_∞-precategories can be defined in the standard way, by requiring that $\mathcal{F}_0$ maps transversal sequences to transversal sequences. If the precategories are unital we demand that $\mathcal{F}_1$ maps quasi-isomorphisms to quasi-isomorphisms.

An A_∞-functor is a *quasi-equivalence* if $\mathcal{F}_1$ is a quasi-isomorphism for each transversal pair in $\underline{\mathsf{C}}^\bullet$ (fully faithful) and every object in $\underline{\mathsf{D}}^\bullet$ is quasi-isomorphic to one in $\mathcal{F}_0(\underline{\mathsf{C}}^\bullet)$ (essentially surjective).

Theorem 3.76 (Efimov) *There is a one-to-one correspondence between quasi-equivalence classes of unital A_∞-precategories and quasi-equivalence classes of A_∞-categories with strict units.*

Sketch of the proof The idea is first to go to the minimal model and prove the statement if there are no higher products by formally inverting the isomorphisms. If there are higher products present one first forgets the higher products and uses the first part. Then one can show that there is a bijection between higher product structures on the precategory and on the category up to quasi-equivalence. The full proof can be found in [70]. □

Corollary 3.77 *If $\underline{\mathsf{C}}^\bullet$ is a precategory with enough quasi-isomorphisms then one can construct a minimal A_∞-category $\mathsf{C}^\bullet$ with strict units that is quasi-equivalent to it.*

Definition 3.78 We will call $\mathsf{C}^\bullet$ the *categorification* of $\underline{\mathsf{C}}^\bullet$ and $\mathsf{D}\,\mathsf{C}^\bullet$ the *derived A_∞-category* of $\underline{\mathsf{C}}^\bullet$.

Example 3.79 Let $\underline{\mathtt{Lines}}^\bullet$ be the precategory of lines in the Euclidean plane. We say that $(\ell_1, \dots, \ell_k)$ is transversal if all pairs (ℓ_i, ℓ_j) intersect at different points: $\#\ell_i \cap \ell_j = 1$ and $\ell_i \cap \ell_j \cap \ell_k = \emptyset$. We define $\mathtt{Lines}(\ell_i, \ell_j) = \Bbbk p$ where p is the intersection point. If ℓ_1, ℓ_2, ℓ_3 form a triangle with corners p_1, p_2, p_3 we set $\mu_2(p_1, p_2) = p_3$. It is easy to check that this gives a minimal precategory concentrated in degree 0 without higher products.

Every nonzero morphism in this precategory is an isomorphism. The precategory has enough isomorphisms because given a transversal set $(\ell_1, \dots, \ell_k)$ and a line ℓ, we can easily find two more lines ℓ_+ and ℓ_- that intersect each other and all other lines. If we formally invert all isomorphisms, all objects become isomorphic and $\mathtt{Lines}^\bullet \cong \Bbbk$.

3.5 Exercises

Exercise 3.1 Count how many terms appear in the rules $[\mathtt{M_n}]$ and $[\mathtt{F_n}]$.

Exercise 3.2 Work out the inductive formulas for the homological perturbation lemma (Lemma 3.18) using trees like in the minimal model Theorem 3.15 and adapt the proof for the latter to the former.

Exercise 3.3 Assume $\Bbbk$ has characteristic 2 and let $A = \Bbbk\langle\xi_1, \dots, \xi_n\rangle$ where the ξ_i have degree $|\xi_i|$. Let $d\colon A \to A$ be a degree 1 map that satisfies the graded Leibniz rule. Now define multiplications μ_k on $B = \Bbbk X_1 \oplus \cdots \oplus \Bbbk X_n$ with $|X_i| = 1 - |\xi_i|$:

$$\mu_k(X_{i_1}, \dots, X_{i_k}) = \sum_j a^j_{i_1,\dots,i_k} X_j.$$

If $d\xi_j = \sum_k \sum_{i_1,\dots,i_k} a^j_{i_1,\dots,i_k} \xi_{i_k} \dots \xi_{i_1}$. Show that $d^2 = 0$ implies that $\mu_\bullet$ satisfies the A_∞-axioms.

Exercise 3.4 Let A be an A_∞-category. Define its A_∞-category of right A_∞-modules $\mathtt{MOD}^\bullet{}_\infty A$ and generalize Yoneda's lemma in this context. Show also that $\mathtt{Tw}A$ embeds into $\mathtt{MOD}^\bullet{}_\infty A$.

Exercise 3.5 Let $f\colon \mathtt{A}^\bullet \to \mathtt{B}^\bullet$ be an A_∞-morphism between two A_∞-algebras and let $\delta_A \in A^1 \otimes \mathbb{C}[[\epsilon]]$ be an element that satisfies the Maurer–Cartan equation (we tensored with $\mathbb{C}[[\epsilon]]$ to make the sum converge). Show that $f(\delta_A) + f(\delta_A, \delta_A) + \cdots$ satisfies the Maurer–Cartan equation for $B^\bullet \otimes \mathbb{C}[[\epsilon]]$.

Exercise 3.6 Assume that $\mathtt{A}^\bullet$ is a nonnegatively graded minimal A_∞-algebra and 1 is a weak unit. Pick a graded basis (b_i) for A with $b_0 = 1$ and define new products such that all higher products involving only b_i with $i > 0$ remain the same and all higher products with at least one b_0 are zero. Show that this is an A_∞-structure that is A_∞-isomorphic to $\mathtt{A}^\bullet$.

Exercise 3.7 Define the notion of a strict idempotent and construct the split-completion $\mathtt{Split}\ \mathtt{A}^\bullet$ as an A_∞-category for an A_∞-category $\mathtt{A}^\bullet$ where every idempotent is strict.

Exercise 3.8 Work out the proofs of Lemmas 3.42 and 3.45.

Exercise 3.9 Show that $\mathtt{MF}^\pm(\mathbb{C}[X, Y], XY)$ is quasi-equivalent with the category of finite-dimensional $\mathbb{Z}_2$-graded vector spaces $\mathtt{grvect}^\pm(\mathbb{C})$ (viewed as an A_∞-category).

Exercise 3.10 Let $f(X) \in \mathbb{C}[X]$. Show that $\mathtt{MF}^\pm(\mathbb{C}[X], f(X))$ is quasi-equivalent to the zero category unless there is a $\xi \in \mathbb{C}$ such that $\frac{df}{dX}(\xi) = 0$ and $f(\xi) = 0$.

4
Quivers

Representations of quivers have been studied intensively and there are many nice references such as [97, 213, 48, 47, 41]. The aim of this chapter is to view these classical results from the perspective of A_∞-categories. We will work out a couple of examples for which we can classify all representations. Each of these examples can be interpreted geometrically in two different ways. This will give us a first glimpse of homological mirror symmetry.

4.1 Representations of Quivers

4.1.1 The Path Algebra

Definition 4.1 A *quiver* Q is an oriented graph. We denote the set of vertices by Q_0, the set of arrows by Q_1 and the maps $h, t\colon Q_1 \to Q_0$ assign to each arrow its head and tail.

A *nontrivial path* p of length k is a sequence of arrows $a_1 \cdots a_k$ such that $t(a_i) = h(a_{i+1})$. A *trivial path* is just a vertex $v \in Q_0$. A path $a_1 \cdots a_k$ is called *cyclic* if $h(a_1) = t(a_k)$ and the equivalence class of a cyclic path under cyclic permutation is called an *oriented cycle*.

Definition 4.2 The *path algebra* $\mathbb{C}Q$ is a vector space spanned by all paths. The product of two paths is their concatenation if possible and 0 otherwise. Paths are concatenated from right to left: $pq = \circ \xleftarrow{p} \circ \xleftarrow{q} \circ$.

The trivial paths corresponding to the vertices v are a complete set of orthogonal idempotents:

$$\forall\, v, w \in Q_0 \text{ we have } vw = \begin{cases} v, & v = w, \\ 0, & v \neq w \end{cases} \quad \text{and} \quad \sum_{v \in Q_0} v = 1.$$

They span a semisimple subalgebra, which we will denote by $\mathbb{C}Q_0$. All nontrivial paths span a two-sided ideal $J = \langle a \mid a \in Q_1 \rangle \lhd \mathbb{C}Q$.

Example 4.3 If we take for example the quivers $\circ$, $\circlearrowleft$ and $\circ \leftarrow \circ$ then the corresponding path algebras are $\mathbb{C}$, $\mathbb{C}[X]$ and the nonstrictly upper-triangular (2×2)-matrices $\left(\begin{smallmatrix} \mathbb{C} & \mathbb{C} \\ 0 & \mathbb{C} \end{smallmatrix}\right)$. A quiver with one vertex and k loops gives rise to the free algebra $\mathbb{C}\langle X_1, \ldots, X_k \rangle$.

Remark 4.4 The path algebra can also be considered as a category $\mathbb{C}\mathtt{Q}$. The objects are the vertices, the hom-space between two vertices is $\mathbb{C}\mathtt{Q}(v, w) = w\mathbb{C}Qv$ and the vertex idempotents are the identities $\mathbb{1}_v = v \in v\mathbb{C}Qv$.

4.1.2 Simples and Projectives

We will denote the categories of finite-dimensional, finitely generated and all right $\mathbb{C}Q$-modules by $\mathtt{mod}\, Q$, $\mathtt{Mod}\, Q$ and $\mathtt{MOD}\, Q$ respectively and their A_∞-versions by $\mathtt{mod}^\bullet\, Q$, $\mathtt{Mod}^\bullet\, Q$ and $\mathtt{MOD}^\bullet\, Q$. There are two important sets of right modules parametrized by Q_0:

- the *basic projectives* $P_v := v\mathbb{C}Q$, which can also be viewed as contravariant functors $\mathcal{P}_v \colon \mathbb{C}\mathtt{Q} \to \mathtt{VECT}(\mathbb{C})$ with $\mathcal{P}_v(w) = v\mathbb{C}Qw$ and $\mathcal{P}_v(a)x = xa$;
- the *basic simples* $S_v := v\mathbb{C}Q/J$, which can also be viewed as contravariant functors $\mathcal{S}_v \colon \mathbb{C}\mathtt{Q} \to \mathtt{VECT}(\mathbb{C})$ with $\mathcal{S}_v(w) = \delta_{vw}\mathbb{C}$ and $\mathcal{S}_v(a) = 0$.

Because the former are projective there are no nontrivial extensions between them so

$$\mathrm{Ext}^i_{\mathbb{C}Q}(P_v, P_w) = \begin{cases} w\mathbb{C}Qv, & i = 0, \\ 0, & i > 0. \end{cases}$$

This implies that the full subcategory $\mathtt{P}^\bullet \subset \mathtt{D\,MOD}^\bullet\, Q$ containing the basic projectives is isomorphic to $\mathbb{C}\mathtt{Q}$ viewed as an A_∞-category concentrated in degree 0 with trivial higher products.

The simple module S_v has a projective resolution

$$\bigoplus_{h(a)=v} P_{t(a)} \xrightarrow{a\cdot} P_v.$$

From this we can easily calculate

$$\mathrm{Ext}^i_{\mathbb{C}Q}(S_v, S_w) = \begin{cases} \delta_{vw}\mathbb{C}, & i = 0, \\ \mathbb{C}^{\oplus \#\{a\colon v \leftarrow w\}}, & i = 1, \\ 0, & i > 1. \end{cases}$$

If we put $S = \bigoplus_{v\in Q_0} S_v$ then $\mathrm{Ext}^\bullet_{\mathbb{C}Q}(S,S)$ is a graded vector space spanned by a degree 0 element $v^* = \mathbb{1}_{S_v}$ for each vertex $v \in Q_0$ and a degree 1 element a^* for each arrow $a \in Q_1$. The product structure between the degree 0 and degree 1 elements is reversed, so we can consider v^* and a^* as vertices and arrows of the opposite quiver. The product of two degree 1 elements is zero because there are no degree 2 elements. This gives the following Ext-ring:

$$\mathrm{Ext}^\bullet_{\mathbb{C}Q}(S,S) = \frac{\mathbb{C}Q^{\mathrm{op}}}{\langle a^*b^* \mid a^*b^* \in Q_1^{\mathrm{op}}\rangle}.$$

In analogy to the dual numbers (Example 2.34), we will call this algebra the *dual quiver algebra* and denote it by ΛQ. Because there are no degree 2 elements the higher products on ΛQ are all zero. This implies that the full subcategory $\mathtt{S}^\bullet \subset \mathtt{D\,MOD}^\bullet\, Q$ of all basic simples is strictly isomorphic to ΛQ viewed as an A_∞-category with trivial higher products.

Theorem 4.5 *Let Q be any quiver; then we have the following inclusions of A_∞-categories:*

$$\mathtt{D\,S}^\bullet \subset \mathtt{D\,mod}^\bullet\, Q \subset \mathtt{D\,P}^\bullet.$$

When Q has no oriented cycles these inclusions are all equivalences.

Sketch of the proof The first inclusion is obvious because the simple modules are one-dimensional. The second inclusion follows from the fact that every module M has a standard projective resolution of the form

$$\bigoplus_{a\in Q_0} Mh(a) \otimes_{\mathbb{C}} P_{t(a)} \xrightarrow{\quad d \quad} \bigoplus_{v\in Q_0} Mv \otimes_{\mathbb{C}} P_v$$

with $d(\sum_a m_a \otimes x_a) = \sum_a m_a a \otimes x_a - m_a \otimes a x_a$. If M is finite-dimensional then both terms are finite direct sums of projectives, so the resolution is a twisted complex over $\mathtt{P}^\bullet$.

When Q has no oriented cycles then the basic projectives are finite-dimensional because there are only a finite number of paths. Furthermore, the only simple modules are the S_v, so $\mathbb{C}Q$ and hence all the basic projectives are generated by the basic simples. This implies that $\mathtt{D\,P}^\bullet \subset \mathtt{D\,S}^\bullet$. □

Remark 4.6 If Q has oriented cycles then all three categories are different: $\mathtt{D\,S}^\bullet$ will only contain the nilpotent modules (i.e. those that factor through $\mathbb{C}Q/J^k$ for some $k \geq 1$), while $\mathtt{D\,mod}^\bullet\, Q$ does not contain those P_v for which v sits in an oriented cycle.

4.1.3 Twisted Complexes

To move easily between the three categories $\mathtt{D\,mod}^\bullet\, Q$, $\mathtt{D\,P}^\bullet$, $\mathtt{D\,S}^\bullet$, we will use the notion of quiver representations.

Definition 4.7 A *dimension vector* of a quiver Q is a map $\alpha\colon Q_0 \to \mathbb{N}$ and we define the size of α as $|\alpha| = \sum_v \alpha_v$. An *$\alpha$-dimensional representation* of Q is an element

$$\rho \in \operatorname{Rep}(Q,\alpha) := \bigoplus_{a\in Q_1} \operatorname{Mat}_{\alpha_{h(a)}\times\alpha_{t(a)}}(\mathbb{C}).$$

A representation ρ can be interpreted in three different ways:

- as a right $\mathbb{C}Q$-module $\mathcal{M}_\rho\colon \mathbb{C}Q \to \mathtt{vect}(\mathbb{C})$ that maps the object v to the vector space $\mathcal{M}_\rho(v) := \mathbb{C}^{\alpha_v}$ and the morphism $a \in Q_1$ to the linear map $\mathcal{M}_\rho(a)\colon \mathbb{C}^{\alpha_{t(a)}} \to \mathbb{C}^{\alpha_{h(a)}}\colon x \mapsto x\rho(a)$, where x is considered as a row vector;
- as a twisted complex over $\mathtt{P}^\bullet$ which comes from the standard projective resolution of $\mathcal{M}_\rho$:

$$P_\rho = \bigoplus_{a\in Q_1} P_t(a)^{\oplus\alpha_{h(a)}} \xrightarrow{\rho(a)-a} \bigoplus_{v\in Q_0} P_v^{\oplus\alpha_v};$$

- as a twisted complex over $\mathtt{S}^\bullet$:

$$S_\rho := \left(\bigoplus S_v^{\oplus\alpha_v}, \delta = \sum_{a\in Q_0} \rho(a)^\top a^*\right),$$

  where $\rho(a)^\top a^*$ should be seen as a matrix block between $S_{h(a)}^{\oplus\alpha_{h(a)}}$ and $S_{t(a)}^{\oplus\alpha_{t(a)}}$. The Maurer–Cartan equation is trivially satisfied because all products between the a^*'s are zero, but S_ρ is not necessarily a twisted complex because δ might not be lower triangular. This is only the case when $\mathcal{M}_\rho$ is nilpotent.

Lemma 4.8 *If $\mathcal{M}_\rho$ is nilpotent then $S_\rho \cong \mathcal{M}_\rho \cong P_\rho \in \mathtt{D\,P}^\bullet$.*

Sketch of the proof We can transform S_ρ into P_ρ by using the standard resolution of the vertex simples; P_ρ can be transformed into $\mathcal{M}_\rho$ by taking the cokernel. □

Theorem 4.9 *Every object in $\mathtt{D\,S}^\bullet$ is isomorphic to a direct sum of shifts of S_ρ's.*

Sketch of the proof We first show that every (M,δ) is isomorphic to a complex without v^*'s in δ by induction on the number of summands in M. If there is only one summand then $\delta = 0$, so the statement is trivially true.

Take a twisted complex (M,δ) and look at the bottom row of δ. By the induction hypothesis we may assume that the other rows of δ only contain dual arrows. After a conjugation (see Remark 3.31) we can ensure that at most one entry of the bottom row is a vertex $v^* = \mathbb{1}_{S_v}$.

This vertex connects two summands $S_v[i+1] \xrightarrow{\mathbb{1}} S_v[i]$. If (M, δ) is not a direct sum of two smaller complexes, all other summands are connected to $S_v[i+1]$ or $S_v[i]$ via an entry containing dual arrows, but not to both for reasons of degree. The entries leaving $S_v[i]$ or arriving in $S_v[i+1]$ are zero because $\delta^2 = 0$. Hence, up to a shift the complex looks like

$$\delta_1 \circlearrowright N_1[1] \xleftarrow{r} S_v[1] \xrightarrow{\mathbb{1}} S_v \xleftarrow{s} N_2 \circlearrowleft \delta_2 .$$

This complex is quasi-isomorphic to $(M_{\mathrm{red}}, \delta_{\mathrm{red}}) = (N_1 \oplus N_2[1], \delta_1 + \delta_2)$ via the following pair of quasi-isomorphisms:

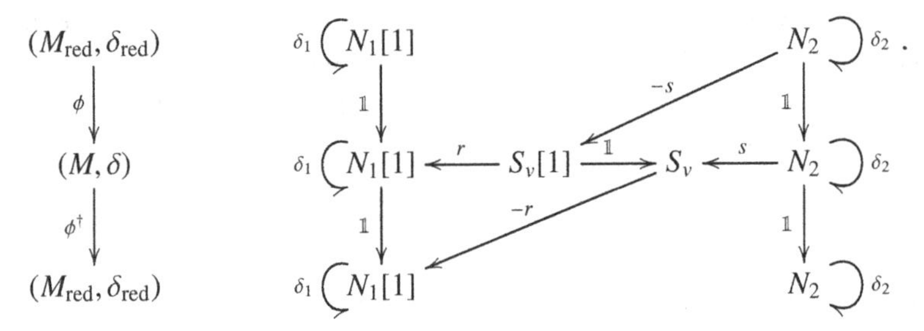

If a twisted complex (M, δ) has only dual arrows in δ then we can write it as a direct sum of complexes for which all summands are shifted by the same degree. Such a twisted complex is of the form $S_\rho[i]$ with $\rho(a)$ the matrix of coefficients of a^*. □

Remark 4.10 More generally one can prove that every object in $\mathtt{D\,MOD}^\bullet\, Q$ is isomorphic to a direct sum of shifts of modules (a complex with zero differential). This is a well-known theorem in representation theory that follows from the fact that $\mathbb{C}Q$ is a *hereditary algebra*: the submodules of projective $\mathbb{C}Q$-modules are also projective.

4.2 Strings and Bands

Using the formalism developed above we will now look at three special quivers and interpret their representation theory geometrically as the intersection theory of lines on a surface. For most quivers it is impossible to classify all indecomposable representations. Only if the underlying graph is a Dynkin diagram or an extended Dynkin diagram can one get a complete classification. The three quivers we will examine are of this type.

4.2.1 The Linear Quiver

Let $Q = L_n$ be the linear quiver with n vertices:

$$v_n \xleftarrow{a_{n-1}} \bigcirc \leftarrow \cdots \leftarrow v_2 \xleftarrow{a_1} v_1 .$$

We denote the vertices from left to right by $v_1, \ldots, v_n$ and the arrows by $a_1, \ldots, a_{n-1}$. The vertex simples will be denoted by S_i and the arrows in the dual quiver by $\alpha_i = a_i^*$:

$$S_n \xrightarrow{\alpha_{n-1}} S_{n-1} \cdots\cdots > S_2 \xrightarrow{\alpha_1} S_1.$$

For each pair $1 \leq i \leq j \leq n$, there is a twisted complex

$$S_{ij} = (S_i \oplus \cdots \oplus S_j, \delta = \alpha_i + \cdots + \alpha_{j-1}),$$

which corresponds to a module that maps the vertices $v_i, \ldots, v_j$ to a one-dimensional vector space and the arrows between them to a nonzero map, or a representation ρ_{ij} with dimension vector $\epsilon_{ij} = (0, \ldots, 0, 1, \ldots, 1, 0, \ldots, 0)$ and $\rho_{ij}(a_k) = 1$ if $i \leq k < j$.

Theorem 4.11 *All indecomposable objects in* $\mathrm{D\,S}^\bullet$ *are isomorphic to shifts of the* S_{ij}.

Sketch of the proof Theorem 4.9 tells us that every indecomposable object in $\mathrm{D\,S}^\bullet$ is a shift of an S_ρ. Because there is only one α_i leaving each dual vertex and one arriving, an indecomposable S_ρ must look like

$$\left(S_i^{\oplus m_i} \oplus S_{i+1}^{\oplus m_{i+1}} \oplus \cdots \oplus S_j^{\oplus m_j}, \delta = B_i\alpha_i + \cdots + B_{j-1}\alpha_{j-1}\right),$$

where B_j is an $(m_j \times m_{j+1})$-matrix with coefficients in $\mathbb{C}$.

Decompose $\mathbb{C}^{m_k}$ as $\mathrm{Im} B_k \oplus R$. If $R \neq 0$ then

$$\cdots \oplus B_{k-2}B_{k-1}R \otimes S_{k-2} \oplus B_{k-1}R \otimes S_{k-1} \oplus R \otimes S_k$$

is a direct summand, so B_k must be surjective. Similarly, B_k must be injective because otherwise

$$\mathrm{Ker}\, B_k \otimes S_{k-1} \oplus B_{k+1}^{-1}\, \mathrm{Ker}\, B_k \otimes S_k \oplus B_{k+2}^{-1}B_{k+1}^{-1}\, \mathrm{Ker}\, B_k \otimes S_{k+1} \oplus \cdots$$

is a summand. Therefore, all B_k are invertible and we can conjugate (see Remark 3.31) them into identity matrices. If each matrix is of size $r \times r$, the twisted complex decomposes into r subcomplexes of the form S_{ij}. □

Now that we have established the indecomposable objects, the next step is to study the morphisms between them. From the definition we can deduce that

$$\mathtt{Tw}\,\mathtt{S}^{\bullet}(S_{ij}, S_{kl}) = \bigoplus_{\substack{i \le u \le j \\ k \le u \le l}} \mathbb{C}\mathbb{1}_{S_u} \oplus \bigoplus_{\substack{i \le v+1 \le j \\ k \le v \le l}} \mathbb{C}\alpha_v$$

with $d\alpha_v = 0$ and $d\mathbb{1}_{S_u} = \alpha_{u-1} - \alpha_u$, unless α_{u-1} or α_u are not present in $\mathtt{Tw}\,\mathtt{S}^{\bullet}(S_{ij}, S_{kl})$. If that is the case, these terms are deleted from the sum. If we calculate the homology we get the following table.

Lemma 4.12 *Let S_{ij}, S_{kl} be twisted complexes as above.*

Case	$\mathtt{DS}^0(S_{ij}, S_{kl})$	$\mathtt{DS}^1(S_{ij}, S_{kl})$
$i-1 \le k-1 < j \le l$	$\mathbb{C}\iota$	0
$k-1 < i-1 \le l < j$	0	$\mathbb{C}\gamma$
Otherwise	0	0

The degree 0 *element is $\iota = \sum_u \mathbb{1}_{S_u}$ where the sum runs over all S_u in the overlap. The degree* 1 *element is $\gamma = \alpha_v$ (any of them works as they are all equal up to homology).*

This rule can be interpreted in a geometrical way by viewing the twisted complexes as lines in a polygon.

Definition 4.13 Consider the regular $(n+1)$-gon with corners $p_0, \ldots, p_n$ (we number the corners anticlockwise and assume the line $p_n p_0$ is the horizontal one at the bottom of the polygon).

A *graded line segment* is a pair $\mathbb{L} = (\overrightarrow{p_i p_j}, \theta)$ consisting of an oriented line segment that connects two corners of the polygon and a *phase* $\theta \in \mathbb{R}$. This is a number such that θ mod 2π is equal to the angle between the X-axis and $\overrightarrow{p_i p_j}$.

Every $\overrightarrow{p_i p_j}$ can be given different phases that all differ by a multiple of 2π. If we reverse the direction of a line segment we have to add π to the phase, so therefore we introduce the notation $\mathbb{L}[k] := (\overrightarrow{p_j p_i}, \theta + k\pi)$. Every graded line segment is a shift of one of the

$$\mathbb{L}_{ij} := (\overrightarrow{p_{i-1} p_j}, \theta) \quad \text{with } \theta \in (0, 2\pi].$$

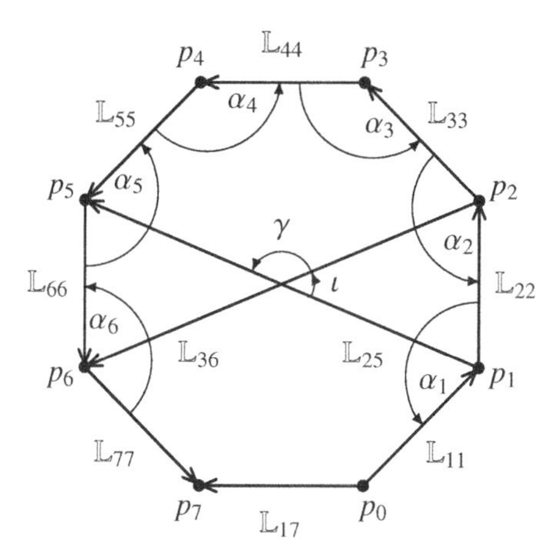

If we interpret the graded line segment $\mathbb{L}_{ij}[k]$ as the representation $S_{ij}[k]$, we see that the basic simples $S_i = S_{ii}$ form the edges of the polygon except for the horizontal edge, which corresponds to S_{17} (note that the latter is oriented differently). There is a degree 1 morphism (extension) between two simples if the line segments touch. From this point of view we can see the corresponding morphisms α_i as angles of the polygon.

This also holds for other representations. Between S_{25} and S_{36} there are two morphisms: a degree 0 morphism $\iota\colon S_{25} \to S_{36}$ and a degree 1 morphism $\gamma\colon S_{36} \to S_{25}$. If we look at the picture we see that the line segments $\mathbb{L}_{25}$ and $\mathbb{L}_{36}$ intersect in the interior of the polygon and there are two angles: one in every direction.

Observation 4.14 (Morphisms are angles) *Let L_1, L_2 be two twisted complexes corresponding to graded line segments $(\mathbb{L}_1, \theta_1)$ and $(\mathbb{L}_2, \theta_2)$. If the underlying line segments are different then we have $\mathrm{D\,S}^\bullet(L_1, L_2) \neq 0$ if and only if there is a positive angle $\beta \in [0, \pi)$ inside the polygon from $\mathbb{L}_1$ to $\mathbb{L}_2$. In that case, the morphism space is generated by one morphism of degree*

$$\frac{\beta - (\theta_2 - \theta_1)}{\pi}.$$

The degree measures the difference between the turning angle and the phase shift.

This interpretation is also useful for constructing cones. If we look again at the polygon above we notice the following:

- The cone of $\alpha_i\colon S_{i+1} \to S_i$ is S_{ii+1}, which can be interpreted as the line segment formed by stitching the line segments of S_i and S_{i+1} together using the angle α_i and straightening it.

- The cone of $\gamma\colon S_{36} \to S_{25}$ is

$$\left(\underbrace{S_2 \oplus S_3 \oplus S_4 \oplus S_5}_{S_{25}} \oplus \underbrace{S_3 \oplus S_4 \oplus S_5 \oplus S_6}_{S_{36}},\right.$$

$$\left.\underbrace{\alpha_3 + \alpha_4 + \alpha_5}_{S_{25}} + \underbrace{\alpha_4 + \alpha_5 + \alpha_6}_{S_{36}} + \underbrace{\alpha_3}_{\gamma}\right)$$

which we can reorder as

$$\left(\underbrace{S_2 \oplus \cdots \oplus S_6}_{S_{26}} \oplus \underbrace{S_3 \oplus S_4 \oplus S_5}_{S_{35}}, \underbrace{\alpha_3 + \cdots + \alpha_6}_{S_{26}} + \underbrace{\alpha_4 + \alpha_5}_{S_{35}} + \underbrace{\alpha_3}_{0 \in \mathrm{D\,S}^\bullet(S_{35}, S_{26})}\right)$$

$$\cong S_{26} \oplus S_{35}.$$

Geometrically we used the angle γ to stitch the pieces of the line segments together differently.

- If we take the cone over $\iota\colon S_{25} \to S_{36}$ viewed as a degree 1 morphism $\iota\colon S_{25}[1] \to S_{36}$ we get

$$\left(\underbrace{S_2[1] \oplus \cdots \oplus S_5[1]}_{S_{25}[1]} \oplus \underbrace{S_3 \oplus \cdots \oplus S_6}_{S_{36}},\right.$$

$$\left.\underbrace{-\alpha_3 - \alpha_4 - \alpha_5}_{S_{25}[1]} + \underbrace{\alpha_4 + \alpha_5 + \alpha_6}_{S_{36}} + \underbrace{\mathbb{1}_{S_3} + \mathbb{1}_{S_4} + \mathbb{1}_{S_5}}_{\iota}\right)$$

which, using the method in the proof of Theorem 4.9, can be seen as $S_2[1] \oplus S_6$. Again we stitched the segments together using the angle.

Observation 4.15 (Cones are stitches) *Assume that $\mathbb{L}_1$, $\mathbb{L}_2$ are two graded line segments with phases in $(0, 2\pi)$ and α is an angle that corresponds to a morphism of degree 1. Taking the cone of α can be seen as the line segment obtained by stitching the two line segments over α.*

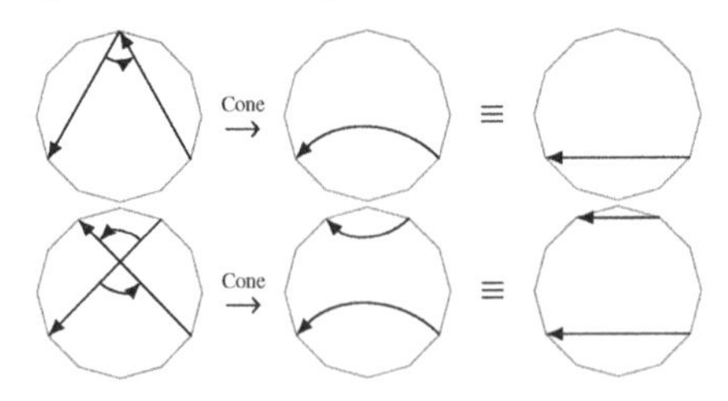

The final part of this geometrical interpretation is the products. It is easy to check that boundary angles that sit on the same corner of the polygon add up if we multiply them, but what about the internal angles? Let us look at two examples again.

- Take the twisted complexes $A = S_{36}$, $B = S_{25}$, $X = S_{47}$ and $Y = S_{35}$. There is a degree 0 morphism $\iota_{AX}\colon S_{36} \to S_{47}$ and a degree 1 morphism $\gamma_{XB}\colon S_{47} \to S_{25}$. Their composition is the degree 1 morphism $\gamma_{AB}\colon S_{36} \to S_{25}$. If we look in the picture we see that ι_{AX}, γ_{XB} are the internal angles of a triangle and γ_{AB} the external angle at the third corner.
- The product of $\iota_{BY}\colon S_{25} \to S_{35}$ and $\iota_{YA}\colon S_{35} \to S_{36}$ is $\iota_{BA}\colon S_{25} \to S_{36}$, which is the outer angle of the third corner of the triangle with inner angles ι_{BY} and ι_{YA}.

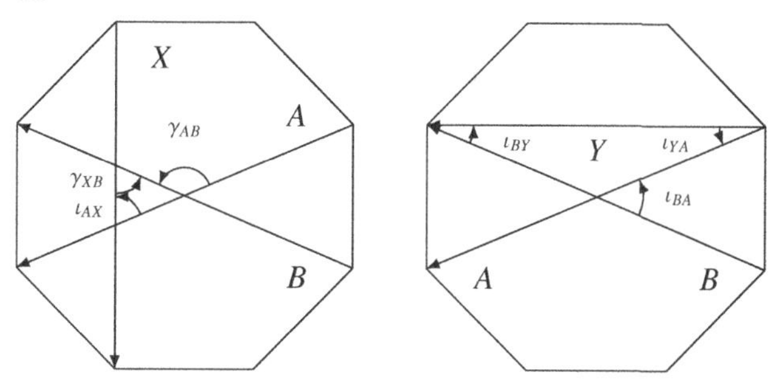

So the product sees triangles. This observation generalizes even to higher products. If we have an internal n-gon and we want to take a higher product of $n-1$ internal angles, then we can rewrite this as an $(n-2)$-ary product by taking a cone over one of the morphisms (see Lemma 3.42). The cone stitches two sides together to form an $(n-2)$-gon. Iterating this procedure we end up with a triangle that we can interpret using the binary product rule.

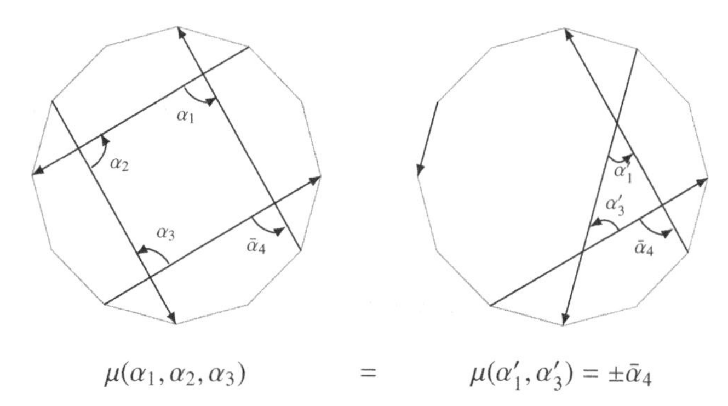

Observation 4.16 (Polygons induce products) *If $\mathbb{L}_1, \ldots, \mathbb{L}_k$ are line segments that bound a subpolygon then the higher product of all the connecting internal morphisms except one is the outer angle morphism at the remaining corner:*

$$\mu(\alpha_1, \ldots, \alpha_{k-1}) = \pm\bar{\alpha}_k.$$

The sign depends on the degrees of the morphisms.

Remark 4.17 Take care: not every nonzero product in $\mathtt{D\,S}^\bullet$ comes from a polygon, because sometimes these polygons can be degenerate (e.g. if three line segments run through the same point, or if some of the line segments are the same).

Observation 4.18 *The A_∞-category $\mathtt{D\,S}^\bullet$ describes the intersection theory of line segments in a polygon.*

Remark 4.19 Making this observation rigorous is not an easy task. One way to do this is to work with precategories. Define a precategory whose

- transversal sequences are graded line segments for which all pairwise intersection points are different,
- hom-spaces are graded $\mathbb{C}$-linear vector spaces spanned by the intersection angles,
- the only nonzero products are given by the rule in Observation 4.16.

Unfortunately, this precategory has no isomorphisms. To solve this problem we can add *graded curves*. These are pairs $\mathbb{L} = (\gamma\colon [0,1] \to \mathbb{R}^2, \theta\colon [0,1] \to \mathbb{R})$ where γ is a smooth embedding of $[0,1]$ into the polygon that connects two different corners and $\theta(x)$ is the angle between $\frac{d\gamma}{dt}(x)$ and the X-axis. In this precategory the angles

are quasi-isomorphisms and therefore the corresponding minimal model $\underline{\mathtt{Polygon}^\bullet_{n+1}}$ is a precategory with enough isomorphisms. If we categorify this precategory we get an equivalence

$$\mathtt{D\,S}^\bullet \overset{\infty}{\cong} \mathtt{Polygon}^\bullet_{n+1}.$$

4.2.2 The Kronecker Quiver

The second quiver we will have a look at is the Kronecker quiver $Q = K$. It consists of two vertices v_0, v_1 connected by two arrows a_0, a_1 in the same direction.

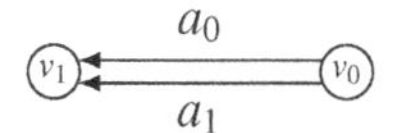

The category $\mathsf{P}^\bullet$ contains the two projective modules $P_i = v_i\mathbb{C}Q$ and the category $\mathsf{S}^\bullet$ contains two simple modules $S_i = v_i\mathbb{C}Q/\langle a_0, a_1\rangle$. The Ext-ring of those two simples is ΛQ. Because there are no paths of length 2, this algebra is isomorphic to $\mathbb{C}Q^{\mathrm{op}} \cong \mathbb{C}Q$. So the categories $\mathsf{S}^\bullet$ and $\mathsf{P}^\bullet$ look the same but they have different gradings. In $\mathsf{P}^\bullet$ all morphisms have degree 0 but in $\mathsf{S}^\bullet$ the dual arrows $\alpha_i = a_i^*$ have degree 1.

Theorem 4.20 *Every indecomposable object in* $\mathrm{D}\,\mathsf{S}^\bullet$ *is isomorphic to a shift of the following complexes:*

(i) $S(i, w)$ *is a linear subcomplex of the infinite string below, which starts with* S_i *on the right and has a total of* w *terms.*

$$\cdots \xleftarrow{\alpha_1} S_1 \xrightarrow{\alpha_0} S_0 \xleftarrow{\alpha_1} S_1 \xrightarrow{\alpha_0} S_0 \xleftarrow{\alpha_1} S_1 \xrightarrow{\alpha_0} \cdots$$

$S(0,4)$

(ii) $B(\lambda, n) = S_0^{\oplus n} \xleftarrow{\alpha_0 \mathbb{1}_n + \alpha_1 J(\lambda,n)} S_1^{\oplus n}$, *where* $J(\lambda, n)$ *is a Jordan block of size* n *with eigenvalue* $\lambda \in \mathbb{C}^*$.

Sketch of the proof Suppose (M, δ) is an indecomposable twisted complex. Using Theorem 4.9, we can assume that the twisted complex is of the form

$$\left(S_0[k]^{\oplus n} \oplus S_1[k]^{\oplus m}, B_0\alpha_0 + B_1\alpha_1\right),$$

where B_0, B_1 are two $(m \times n)$-matrices. Up to isomorphism we can multiply both matrices by the same invertible matrix on the left and on the right. If both B_0, B_1 are invertible we can bring B_0 to the identity and then conjugate B_1 to a Jordan block. This gives us a complex of the form $B(\lambda, n)$.

If either B_0 or B_1 is not invertible we will use induction on $n + m$ to show that (M, δ) is isomorphic to an $S(i, w)$. Clearly, if $n+m = 1$ then (M, δ) is either $S_0 = S(0, 1)$ or $S_1 = S(1, 1)$. If $n + m \geq 2$ we distinguish four cases depending on whether B_i is not injective or not surjective. We work out only one of these four cases.

Suppose B_0 has a kernel. Choose a basis vector v in that kernel and let $w = B_1 v$. The latter vector is nonzero because otherwise the twisted complex is decomposable. We can extend v and w to bases for $\mathbb{C}^n$ and $\mathbb{C}^m$ such that the matrices become

$$\begin{pmatrix} 0 & \vline & \\ \vdots & \vline & B'_0 \\ 0 & \vline & \end{pmatrix}, \quad \begin{pmatrix} 1 & \vline & \\ \vdots & \vline & B'_1 \\ 0 & \vline & \end{pmatrix}.$$

By multiplying both on the right by an invertible matrix we can assume that B'_1 starts with a zero row. The pair (B'_0, B'_1) satisfies the induction hypothesis because B'_1 is not invertible. Hence it describes an $S(i, w)$ and the total complex looks like

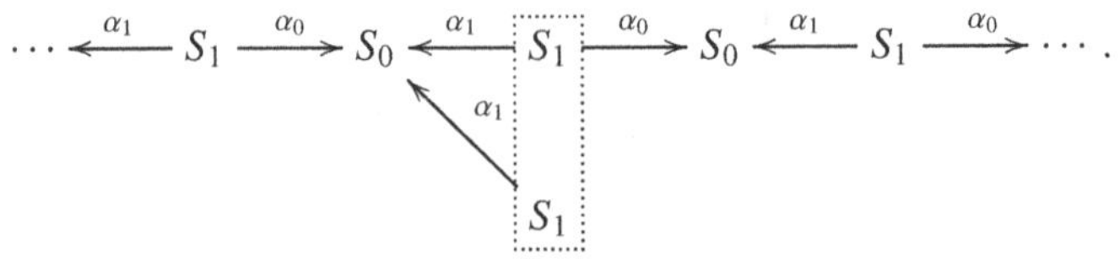

$$\cdots \xleftarrow{\alpha_1} S_1 \xrightarrow{\alpha_0} S_0 \xleftarrow{\alpha_1} \boxed{S_1} \xrightarrow{\alpha_0} S_0 \xleftarrow{\alpha_1} S_1 \xrightarrow{\alpha_0} \cdots .$$
$$\nwarrow^{\alpha_1} \; \boxed{S_1}$$

If the bottom S_1 connects somewhere in the middle of the chain then consider the kernels K_0, K_1 of B_0, B_1 restricted to the dotted $S_1 \oplus S_1$. If they coincide then there is a direct summand isomorphic to S_1. If they span $K_0 \oplus K_1 = S_1 \oplus S_1$ then we can split the total complex in two separate complexes:

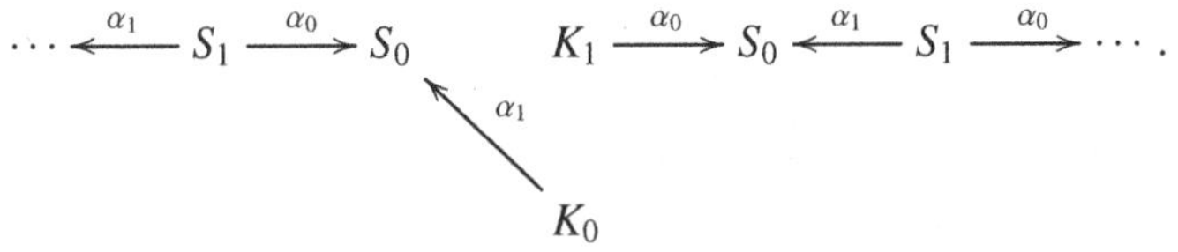

$$\cdots \xleftarrow{\alpha_1} S_1 \xrightarrow{\alpha_0} S_0 \qquad K_1 \xrightarrow{\alpha_0} S_0 \xleftarrow{\alpha_1} S_1 \xrightarrow{\alpha_0} \cdots .$$
$$\nwarrow^{\alpha_1} \; K_0$$

So if (M, δ) is indecomposable then that S_1 must connect to the end of the chain and hence it is a chain itself. □

Definition 4.21 We will call the $S(i, w)$ *string objects* and the $B(\lambda, n)$ *band objects*.

Remark 4.22 The string objects with even size can be seen as limits of band objects:

$$\lim_{\lambda \to 0} B(\lambda, n) \cong S(0, 2n) \quad \text{and} \quad \lim_{\lambda \to \infty} B(\lambda, n) \cong S(1, 2n).$$

The first limit can be seen directly from the diagrams of the complexes. For the second limit, observe that a base can turn the map $\alpha_0 \mathbb{1}_n + \alpha_1 J(\lambda, n)$ into $\alpha_0 J(\lambda^{-1}, n) + \alpha_1 \mathbb{1}_n$. Therefore we will sometimes indicate these string objects as $B(0, n)$ and $B(\infty, n)$.

Now we want to obtain a similar geometrical interpretation for the Kronecker quiver as we did for the linear quiver. Take the unit square and glue the left end to the right end. We obtain a cylinder with two marked points, one on each boundary circle, coming from the corners of the square. This cylinder is a flat surface and it is possible to measure angles in the standard way by pulling back to the square. As for the polygon we can define graded line segments between the marked points.

Let S_0 denote the graded line segment that corresponds to the vertical edge of the square with phase $\frac{\pi}{2}$ and let S_1 be the downward-pointing diagonal with phase $\frac{5\pi}{4}$. There are two angles going from S_1 to S_0, which we can identify with the two degree 1 morphisms between S_1 and S_0.

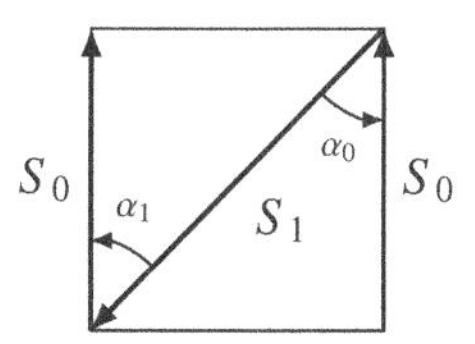

In the case of the polygon, we saw that we could interpret the cone construction as gluing line segments together along an angle. In the same way, every string object $S(i, w)$ can be seen as a line segment $S(i, w)$ in the universal cover of the cylinder.

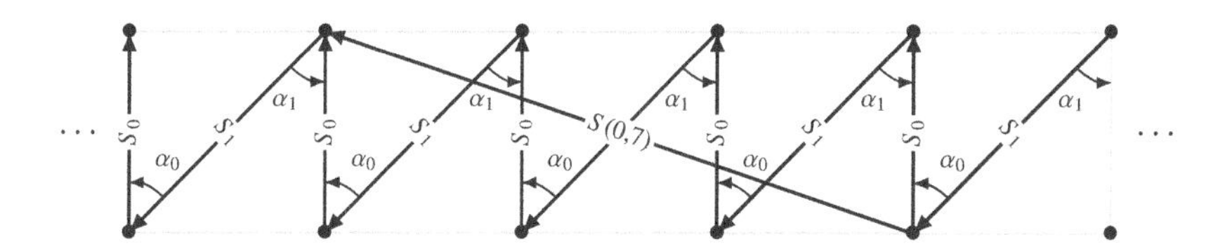

But what about the band objects? There is in fact a second type of curve with constant phase: a circle that goes around the cylinder. We would like to use these circles to interpret the band objects, but in order to do this we also need to attach some extra geometrical information: a local system.

Definition 4.23 A *local system* on a manifold $\mathbb{M}$ is a representation of its fundamental groupoid $\mathcal{L}\colon \Pi_1(\mathbb{M}) \to \mathtt{vect}(\mathbb{C})$.

A local system $\mathcal{L}$ assigns to each point of $\mathbb{L}$ a vector space $\mathcal{L}(p)$, called the fiber over p, and for each homotopy class of paths between two points there is an invertible map between the vector spaces. This map is sometimes called the transport along that path. If $\mathbb{M}$ is connected then all the vector spaces have the same dimension, which is called the *rank* of the local system.

Lemma 4.24 *The indecomposable local systems on a circle $\mathbb{S}^1$ are classified by invertible Jordan blocks.*

Sketch of the proof This is because the fundamental groupoid of the circle is equivalent to the group $\mathbb{Z}$. Up to isomorphism, a representation $\phi\colon \mathbb{Z} \to \mathsf{GL}_n$ is determined by the conjugacy class of the invertible matrix $\phi(1)$, which can be brought in Jordan normal form. □

If we fix one horizontal circle in the cylinder then each band object $B(n, \lambda)$ can be identified with a local system on that circle, for which the transport along the circle is a Jordan block of size n and eigenvalue λ. We denote the embedded circle together with the local system by $\mathbb{B}(n, \lambda)$.

The string objects can also be considered as line segments equipped with a rank 1 local system. Local systems on a contractible space are classified by their rank and a rank n local system is the direct sum of n copies of a rank 1 local system. Therefore we can interpret $S_0 \oplus S_0$ either as two line segments each with a rank 1 local system, or as one line segment with a rank 2 local system. So from now on we assume that a graded line segment $\mathbb{L}$ automatically comes with a local system and we let $\mathbb{L}(p)$ denote the fiber at p.

To work with these local systems, we also need to upgrade the morphism spaces. Every angle around an intersection point p should be seen as a morphism between the fibers of the two local systems at p. So if $\mathbb{L}_1$, $\mathbb{L}_2$ are local systems over two different line segments then we set

$$\operatorname{Hom}(\mathbb{L}_1, \mathbb{L}_2) = \bigoplus_{\alpha_p} \mathtt{vect}(\mathbb{L}_1(p), \mathbb{L}_2(p))\alpha_p,$$

where α_p represents an angle at the intersection point p. In other words, the angles come with coefficients that are linear maps.

This idea fits well with the cone construction. If we want to stitch two local systems together with a morphism $f\alpha_p$, where $f \in \mathtt{vect}(\mathbb{L}_1(p), \mathbb{L}_2(p))$ is invertible, we use f to identify the two fibers. If f is not invertible we break the local system in two (one where f is invertible and one where it is zero) and only stitch the bits where f becomes invertible.

The object $B(n, \lambda)$ can be seen as two local systems of rank n, one on S_0 and one on S_1, stitched together on one end by the identity and on the other end by the Jordan block. Together this gives a curve isotopic to the circle with a local system for which the transport is equal to the Jordan block: $\mathbb{B}(n, \lambda)$.

The final adaptation needed is the rule for the product. The coefficients for the angles must be included bilinearly in the products, but consecutive angles in a product do not connect fibers at the same intersection points, so the linear maps do not match up. Luckily this can easily be solved using the local system. If $e\colon p \to q$ is each edge of the polygon that lies on $\mathbb{L}_i$ then $\mathbb{L}_i(e)\colon \mathbb{L}(p) \to \mathbb{L}(q)$

is a homomorphism between the fibers on p and q. Therefore we can string all these maps together and define an extended product:

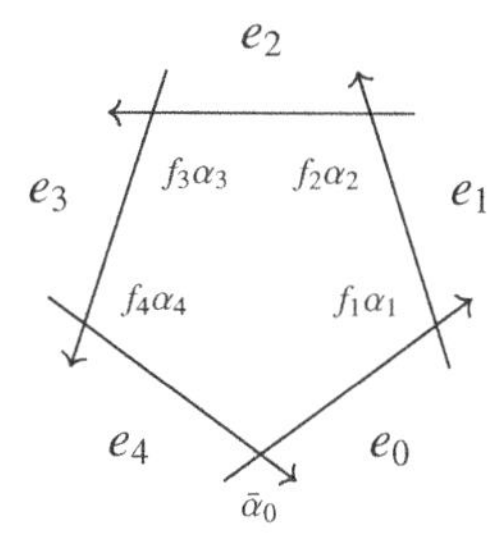

$$\mu(f_1\alpha_1, \ldots, f_4\alpha_4) = \mathbb{L}_1(e_0)f_1\mathbb{L}_1(e_1)\ldots\mathbb{L}_1(e_3)f_4\mathbb{L}_k(e_4)\cdot\bar{\alpha}_0 + \cdots.$$

If there is more than one polygon bounded by these angles, we take the sum of these expressions. With these extra features in operation one can check that the same observations we made for $\mathtt{D\,S}^\bullet$ in the case of the linear quiver also hold for the Kronecker quiver.

Observation 4.25 *The objects in $\mathtt{D\,S}^\bullet$ can be identified with direct sums of graded line segments with local systems. The morphisms are angles weighted by maps between the fibers of the local systems and polygons contribute to products weighted by the total transport around them.*

4.2.3 The Cyclic Quiver

The final quiver we will have a look at is the cyclic quiver $Q = C_n$. It has n vertices $v_1, \ldots, v_n$ and n arrows $a_i\colon v_i \to v_{i+1}$ where the index is considered modulo n.

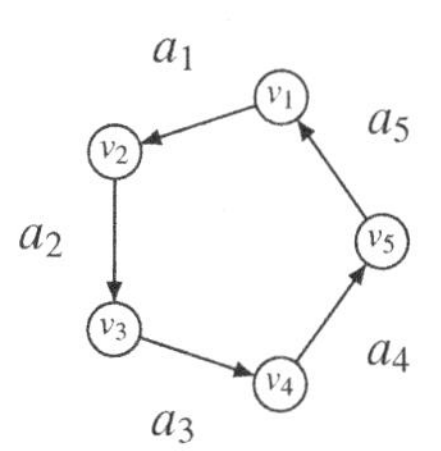

We allow the case $n = 1$ where the quiver has one vertex and one loop. There are n projective modules $P_i = v_i\mathbb{C}Q$ in $\mathtt{P}^\bullet$ and n simple modules $S_i = v_i\mathbb{C}Q/J$ in $\mathtt{S}^\bullet$. The dual arrows $\alpha_i = a_i^*$ are degree 1 morphisms from S_{i+1} to S_i.

The situation is slightly different from the two previous cases. The algebra $\mathbb{C}Q$ is infinite-dimensional because there are paths with arbitrary length. If we take the sum of all paths of length n we get a central element in $\mathbb{C}Q$ that we will denote by

$$\ell = a_n \dots a_1 + a_1 a_n \dots a_2 + \dots + a_{n-1} \dots a_1 a_n.$$

One can easily check that $\mathbb{C}[\ell]$ is a polynomial ring and $\mathbb{C}Q$ is a free $\mathbb{C}[\ell]$ module generated by all paths of length $< n$. Because $\mathbb{C}Q$ is infinite-dimensional, the three categories $\mathtt{D\,S}^\bullet \subset \mathtt{D\,mod}^\bullet\,Q \subset \mathtt{D\,P}^\bullet$ are all different and this is reflected in the classification theorem.

Theorem 4.26 *The indecomposable objects in* $\mathtt{D\,P}^\bullet$ *are shifts of object of the form*

- $S(i,w) = (S_i \oplus \dots \oplus S_{i+w-1}, \alpha_i + \dots + \alpha_{i+w-1})$,
- $B(k,\lambda) = \mathcal{M}_\rho$ *with* $\rho(a_n) = J(k,\lambda)$ *and* $\rho(a_i) = \mathbb{1}_k$ *for* $i \neq n$,
- P_i,

with $i \in \{1,\dots,n\}$, $\lambda \in \mathbb{C}$ *and* $w \in \mathbb{N}$. *We have that* $S(i,w) \in \mathtt{D\,S}^\bullet$, $B(n,\lambda) \in \mathtt{D\,mod}^\bullet\,Q$ *and* $P_i \in \mathtt{D\,P}^\bullet$.

Sketch of the proof Every indecomposable object in $\mathtt{D\,P}^\bullet$ is a shift of a finitely generated $\mathbb{C}Q$-module M. If M is infinite-dimensional then there is an injective morphism $P_v \to M$ which splits because P_v is projective. Therefore, $M \cong P_i$.

If M is finite-dimensional, it is of the form $\mathcal{M}_\rho$ for some representation ρ. Because ℓ is central we can split M as a direct sum according to the eigenvalues of ℓ. Therefore, ℓ can only have one eigenvalue λ. If λ is nonzero then all $\rho(a_i)$ are invertible and we can conjugate them to the identity matrix, except for $\rho(a_n)$ which can be conjugated to a Jordan block. If $\lambda = 0$ then the module is nilpotent and hence it sits in $\mathtt{D\,S}^\bullet$. We can use the same argument as in Theorem 4.20 to complete the classification. □

Again we can interpret this classification in terms of lines on a surface. Take a disk with one marked point at the center and n on the boundary. We identify the n simple objects with the n anticlockwise arcs on the boundary and the n projective objects with the spokes. The $B(n,\lambda)$ will correspond to curves that circle around the center equipped with a local system with transport equal to $J(n,\lambda)$.

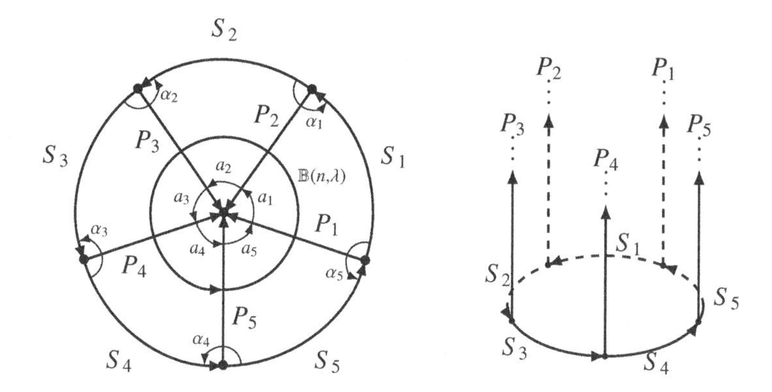

The arrows of the quiver Q correspond to morphisms between the projectives and can be seen as angles around the marked point at the center. The dual arrows run between the simples and they can be seen as angles around the marked points on the circle.

At first everything seems to work out except for the gradings; if we measure our angles in the Euclidean plane then the morphism $a_n \dots a_1$ from P_1 to itself must have degree 2 because it turns around 2π. The solution to this problem is to consider this surface as a cylinder with an infinite end instead of a disk. In that case, the line segments P_i all have phase $\frac{\pi}{2}$ and the S_i have phase 0. The triangle bounded by S_i, P_i and P_{i+1} has two angles of $\frac{\pi}{2}$ and one zero angle at infinity.

Observation 4.27 *We can interpret the representation theory of the cyclic quiver in terms of curves on a cylinder with one infinite end. The category* $\mathtt{D}\,\mathtt{S}^\bullet$ *contains the line segments that sit on the boundary circle and* $\mathtt{D}\,\mathtt{mod}^\bullet\, Q$ *the line segments that are compact (i.e. do not go to infinity), while* $\mathtt{D}\,\mathtt{P}^\bullet$ *also includes the noncompact line segments.*

4.3 Points and Sheaves

In this section we will review the three quivers from a different geometrical perspective: namely algebraic geometry. We will work our way through the list in reverse order.

4.3.1 The Cyclic Quiver

Let us start with the one-loop quiver. Its path algebra is the polynomial ring $R = \mathbb{C}[X]$. This ring can be viewed as the ring of polynomial functions on the affine line $\mathbb{A}^1 = \mathbb{C}$. For each point $p \in \mathbb{C}$ there is a morphism $\rho_p \colon R \to \mathbb{C} \colon f(X) \to f(p)$ that evaluates each function at the point p. This morphism has as kernel $\mathfrak{p} = \{f \in R \mid f(p) = 0\} = (X - p)$ and if we divide out this kernel we get the simple module $S_p = J(p, 1)$. All simple modules arise in this way so we can conclude that the simple R-modules correspond to the points on the affine line:

$$\{\text{points of } \mathbb{A}^1\} \leftrightarrow \{\text{maximal ideals in } R\} \leftrightarrow \{\text{simple } R\text{-modules}\}.$$

The higher Jordan blocks can be interpreted in a similar way: while ρ_p assigns to each function its value in p, the map $R \to J(p,n) = \mathbb{C}[X]/(X - p)^n$ assigns to each f its nth-order Taylor approximation in p, so the $J(p,n)$ explore infinitesimal surroundings of p.

Given a module M we say that it is supported at $p \in \mathbb{A}^1$ if $M/\mathfrak{p}M \neq 0$. Clearly, S_p is only supported at p and at no other $q \in \mathbb{A}^1$. The same holds for $J(p,n)$, but for R as a module over itself the situation is different. It is supported at all p. At every point $R/\mathfrak{p}$ is a one-dimensional vector space and we can group all these vector spaces together into a one-dimensional vector bundle (or line bundle) over $\mathbb{A}^1$. Each element $m \in R$ gives an element in $R/\mathfrak{p}$ for every $p \in \mathbb{A}^1$ so they can be seen as sections of the bundle. The line bundle is trivial because the element $1 \in R$ gives a section that forms a basis in each fiber. Because $\mathbb{A}^1$ is contractible every vector bundle over $\mathbb{A}^1$ is trivial and if it has rank n, it will correspond to the projective (even free) module $R^{\oplus n}$:

$$\{\text{vector bundles over } \mathbb{A}^1\} \leftrightarrow \{\text{projective } R\text{-modules}\}.$$

Modules that are neither projective nor simple have a more intricate geometrical interpretation that is called a coherent sheaf. We will discuss this concept in more detail in Chapter 7, but for now it is sufficient to know that a coherent sheaf on $\mathbb{A}^1$ is the same as a finitely generated $\mathbb{C}[X]$-module:

$$\mathtt{D}\,\mathtt{Coh}^\bullet\,\mathbb{A}^1 = \mathtt{D}\,\mathtt{Mod}^\bullet\,\mathbb{C}[X].$$

Observation 4.28 *The representations of the one-loop quiver can be interpreted as sheaves on the affine line $\mathbb{A}^1$. Under this correspondence the simple module S_p becomes the skyscraper sheaf $\mathscr{S}_p$, while the free module $\mathbb{C}[X]$ becomes the trivial line bundle $\mathscr{O}$, also known as the structure sheaf.*

We will now extend this geometrical interpretation to the higher cyclic quivers. Consider again the affine line $\mathbb{A}^1$. On this space we have a left action of the cyclic group $G = \langle g \mid g^n \rangle \cong \mathbb{Z}_n$:

$$g^k \cdot x = \zeta^k x \quad \text{where } \zeta = e^{\frac{2\pi i}{n}}.$$

The orbits for this action all have size n except for the zero orbit, which consists of one point. The action transfers to a right action on the coordinate ring $\mathbb{C}[X]$ such that $f\cdot g(x) = f(g\cdot x)$. If we want to quotient out the action, the coordinate ring of the quotient $\mathbb{A}^1/G$ should be seen as the ring of G-invariant functions $\mathbb{C}[X]^G$, because these are the functions that are constant on the orbits. This ring is $\mathbb{C}[X^n]$, which is isomorphic to $\mathbb{C}[X]$, so from this perspective $\mathbb{A}^1/G$ looks the same as $\mathbb{A}^1$.

However, this does not fully capture the geometry, because $\mathbb{A}^1/G$ has one special point (the zero), while all points of $\mathbb{A}^1$ are alike. Therefore, it is better to describe the quotient not as an affine variety but as something more delicate: the orbifold $[\mathbb{A}^1/G]$.

To do this we introduce the notion of a G-equivariant sheaf. This is a sheaf on $\mathbb{A}^1$ together with an action of G that is compatible with the action of G on $\mathbb{A}^1$. To be more precise, it is a $\mathbb{C}[X]$-module M with a G-action such that $(mf(x))\cdot g = (m\cdot g)(f(x)\cdot g)$. Both actions can be packaged together in one algebra

$$\mathbb{C}[X]\star G := \frac{\mathbb{C}\langle X, g\rangle}{\langle gXg^{-1} - \zeta^k X, g^n - 1\rangle},$$

and then M naturally gets the structure of a $\mathbb{C}[X]\star G$-module.

Therefore, it makes sense to define the derived category of G-equivariant coherent sheaves as

$$\mathtt{D}\,\mathtt{Coh}^\bullet[\mathbb{A}^1/\mathbb{Z}_n] := \mathtt{D}\,\mathtt{Mod}^\bullet\,\mathbb{C}[X]\star G.$$

This category is the right substitute for sheaves on $[\mathbb{A}^1/G]$, so $\mathtt{D}\,\mathtt{Coh}^\bullet[\mathbb{A}^1/G]$ should be considered as the derived category of coherent sheaves on the orbifold $[\mathbb{A}^1/G]$.

Theorem 4.29 *The algebra $\mathbb{C}[X]\star G$ is isomorphic to $\mathbb{C}Q$ with Q the cyclic quiver with n vertices.*

Sketch of the proof Put $\zeta = e^{\frac{2\pi i}{n}}$ and define

$$e_k = \frac{1}{n}\sum_{i=1}^{n} \zeta^{ki} g^i.$$

It is easy to check that $\sum_k e_k = 1$ and $e_i e_j = \delta_{ij} e_i$. In other words, they form a set of orthogonal idempotents. Moreover, we have $Xe_k = e_{k+1}X$ and

$$X = \sum_i e_{i+1} X e_i,$$

where the indices are taken modulo n. Now construct a morphism $\phi\colon \mathbb{C}Q \to \mathbb{C}[X] \star G$ that maps the vertices to the e_k and the arrows a_i to the $e_{i+1}Xe_i$. One can easily check that this is an isomorphism. □

Observation 4.30 *The representation theory of the cyclic quiver can be geometrically interpreted as coherent sheaves on an orbifold:*

$$\mathtt{D}\,\mathtt{Coh}^\bullet[\mathbb{A}^1/G] \overset{\infty}{=} \mathtt{D}\,\mathtt{Mod}^\bullet\,\mathbb{C}C_n.$$

4.3.2 The Kronecker Quiver

To get a similar interpretation for the Kronecker quiver, we have to look at the algebraic geometry of the projective line. This is the affine line with an extra point at infinity: $\mathbb{P}^1 = \mathbb{A}^1 \cup \{\infty\}$. Using homogeneous coordinates we can also see it as

$$\mathbb{P}^1 := \{(x : y) \mid (x, y) \in \mathbb{C}^2 \setminus \{(0, 0)\}\},$$

where $(x : y)$ is considered up to a multiple: $(x : y) = (\lambda x : \lambda y)$. We can see it as the union of two affine lines, $\{(x : 1) \mid x \in \mathbb{C}\}$ and $\{(1 : y) \mid y \in \mathbb{C}\}$, whose intersection is $\mathbb{C}^*$.

The coordinate rings of the two affine lines are $R_X = \mathbb{C}[X]$ and $R_Y = \mathbb{C}[Y]$. Both X and Y can be considered as functions on the overlap and there they satisfy the relation $Y = \frac{1}{X}$. The coordinate ring on the overlap can be seen as $R_{XY} = \frac{\mathbb{C}[X,Y]}{(XY-1)}$. On the other hand, the only functions in $\mathbb{C}[X]$ and $\mathbb{C}[Y]$ that extend to $\mathbb{P}^1$ are the constant ones. We can summarize this in two diagrams: one of spaces and one of algebras.

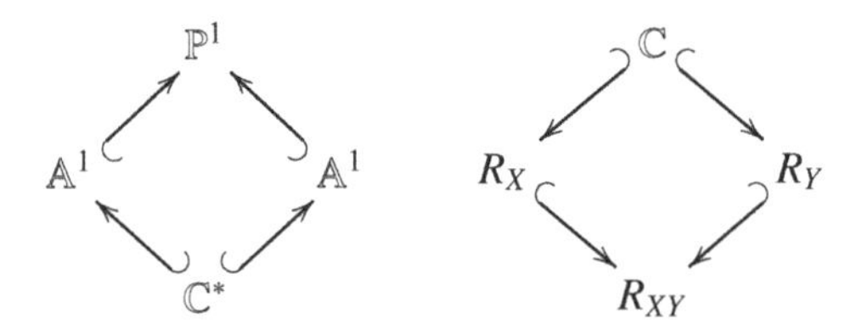

The ring of functions on $\mathbb{P}^1$ is just $\mathbb{C}$, which does not fully capture the geometry of the projective line. Therefore we cannot view sheaves on $\mathbb{P}^1$ as modules over its coordinate ring.

The solution to this problem is to view a coherent sheaf over $\mathbb{P}^1$ as two coherent sheaves over the two $\mathbb{A}^1$'s glued together. This means that we need three ingredients: a finitely generated R_X-module M_X, a finitely generated

R_Y-module M_Y and a gluing isomorphism between them on the overlap. The latter is an isomorphism of R_{XY}-modules:

$$\phi_M\colon \underbrace{M_X \otimes_{R_X} R_{XY}}_{M_{XY}} \to \underbrace{M_Y \otimes_{R_Y} R_{XY}}_{M_{YX}}.$$

We call such a triple $M = (M_X, M_Y, \phi_M)$ a coherent sheaf on $\mathbb{P}^1$.

The morphisms in $\mathsf{Coh}\,\mathbb{P}^1$ are the expected ones: pairs of module morphisms $(\psi_X\colon M_X \to N_X, \psi_Y\colon M_Y \to N_Y)$ that commute with the gluing maps. There is a neat way to see these morphisms as the zeroth homology of the complex

$$\begin{pmatrix} \mathrm{Hom}_{R_X}(M_X, N_X) \\ \oplus \\ \mathrm{Hom}_{R_Y}(M_Y, N_Y) \end{pmatrix} \xrightarrow{d_0} \mathrm{Hom}_{R_{XY}}(M_{XY}, N_{XY})$$

with $d_0(\psi_X \oplus \psi_Y) = \psi_X \otimes_{R_X} R_{XY} - \phi_N^{-1} \circ (\psi_Y \otimes_{R_Y} R_{XY}) \circ \phi_M$.

Theorem 4.31 *Every coherent sheaf is a direct sum of the following types:*

(i) $\mathscr{S}(\lambda, n) := (\frac{\mathbb{C}[X]}{(X-\lambda)^n}, \frac{\mathbb{C}[Y]}{(Y-\lambda^{-1})^n}, \mathbb{1})$,
(ii) $\mathscr{S}(0, n) := (\mathbb{C}[X]/(X^n), 0, 0)$,
(iii) $\mathscr{S}(\infty, n) := (0, \mathbb{C}[Y]/(Y^n), 0)$,
(iv) $\mathscr{O}(k) := (\mathbb{C}[X], \mathbb{C}[Y], Y^k)$,

where $\lambda \in \mathbb{C}^*$, $n \in \mathbb{N}$ and $k \in \mathbb{Z}$.

Sketch of the proof Decompose M_X as a direct sum of indecomposable modules. Its components are either of the form $R_X/(X - \lambda)^n$ with $\lambda \in \mathbb{C}$ or the free module R_X. If we tensor with R_{XY} the latter stays free, while $R_X/(X - \lambda)^n$ becomes $R_{XY}/(X - \lambda)^n$. In the case $\lambda \neq 0$ this module remains n-dimensional, but for $\lambda = 0$ it becomes zero. We can reason the same way for M_Y but note that $R_{XY}/(Y - \lambda)^n$ can be rewritten as $R_{XY}/(X - \lambda^{-1})^n$.

Pairing up isomorphic components we see that every $\mathbb{C}[X]/(X - \lambda)^n$ in M_X must be balanced by a $\mathbb{C}[Y]/(Y - \lambda^{-1})^n$ in M_Y and every $\mathbb{C}[X]$ by a $\mathbb{C}[Y]$. Only the $\mathbb{C}[X]/(X^n)$ in M_X and the $\mathbb{C}[Y]/(Y^n)$ in the M_Y do not need to be balanced and they can be split off as irreducible sheaves of type $\mathscr{S}(0, n)$ or $\mathscr{S}(\infty, n)$.

The rest needs to be glued together by an isomorphism. For example, if $M_{XY} = R_{XY}^{\oplus n}$ then this automorphism is an invertible $(n \times n)$-matrix G with coefficients in R_{XY}. By applying automorphisms of $M_X = R_X^{\oplus n}$ and $M_Y = R_Y^{\oplus n}$ this matrix is only defined up to multiplication on the right with invertible matrices over R_X and on the left with invertible matrices over R_Y. A classical theorem by Dedekind and Weber tells us that we can bring G into diagonal form with integer powers of Y on the diagonal, so in that case the sheaf is a direct sum of $\mathscr{O}(k)$'s.

On the other hand, morphisms between R_{XY}-modules of the form $R_{XY}/(X - \lambda)^n$ all come from R_X-morphisms. Therefore we can assume that the gluing between these components is just the identity, so we get direct sums of $\mathscr{S}(\lambda, n)$'s. For more details see [98, 228, 109, 118, 231]. □

Remark 4.32 The $\mathscr{S}_\lambda := \mathbb{S}(\lambda, 1)$ are the *skyscraper sheaves* and the $\mathscr{O}(k)$ are the *line bundles* over $\mathbb{P}^1$. The sheaf $\mathscr{O} := \mathscr{O}(0)$ is the trivial line bundle or structure sheaf.

In this way we have constructed an ordinary category of coherent sheaves, but we want to upgrade it to an A_∞-category. Just as we substituted every module with a free resolution, we can do the same here but now we need locally free resolutions. These are sheaves of the form (M_X, M_Y, ϕ_M) where both M_X, M_Y are free. It is easy to check that every coherent sheaf has a locally free resolution, which we can use to make dg-hom-spaces.

There is however one extra feature: as we saw earlier the sheaf homomorphisms themselves form the zeroth homology of a complex. Therefore we have two differentials and by taking their sum we get one big complex. The morphism spaces in $\mathsf{Coh}^\bullet\,\mathbb{P}^1$ are the cohomologies of these complexes. The degree 0 parts are the ordinary morphisms of sheaves, while the higher-degree part measures both the algebraic extensions between the modules as the geometrical obstructions in the gluing process.

Example 4.33 The dg-hom-space between $\mathscr{O}(r)$ and $\mathscr{O}(s)$ is

$$\underbrace{\mathrm{Hom}_{R_X}(R_X, R_X) \oplus \mathrm{Hom}_{R_Y}(R_Y, R_Y)}_{\text{degree } 0} \oplus \underbrace{\mathrm{Hom}_{R_{XY}}(R_{XY}, R_{XY})}_{\text{degree } 1}$$

with $d(f(X) \oplus g(Y)) = f(X) - X^{s-r} g(Y)$.

If $r > s$ then the second term can never cancel the first because it only consists of negative powers of X. Therefore, the zeroth homology is zero. If $r \le s$ then we have the following basis for the zeroth homology $(f, g) = (X^i, Y^{s-r-i})$ with $0 \le i \le s - r$:

$$\mathsf{Coh}^0(\mathscr{O}(r), \mathscr{O}(s)) = \begin{cases} \mathbb{C}(1, Y^{s-r}) + \cdots + \mathbb{C}(X^{s-r}, 1), & r \le s, \\ 0, & r > s. \end{cases}$$

The first homology on the other hand consists of $\mathbb{C}[X, X^{-1}]$ divided by the vector space spanned by X^i and X^{s-r-i} with $i \ge 0$. If $s < r - 1$ this space is nonzero and has dimension $r - s - 1$:

$$\mathsf{Coh}^1(\mathscr{O}(r), \mathscr{O}(s)) = \begin{cases} 0, & r \le s + 1, \\ \mathbb{C}Y + \cdots + \mathbb{C}Y^{r-s-1}, & r > s + 1. \end{cases}$$

Theorem 4.34 *The derived A_∞-category of coherent sheaves* $\mathtt{D\,Coh}^\bullet\,\mathbb{P}^1$ *is A_∞-equivalent to* $\mathtt{D\,mod}^\bullet\,\mathbb{C}Q$ *where $Q = K_2$ is the Kronecker quiver.*

Sketch of the proof We can construct all sheaves using $\mathcal{O}(0)$ and $\mathcal{O}(1)$:

- $\mathcal{O}(i-1)$ is the kernel of $\mathcal{O}(i)^{\oplus 2} \xrightarrow{(1\ Y)} \mathcal{O}(i+1)$,
- $\mathcal{O}(i+1)$ is the cokernel of $\mathcal{O}(i-1) \xrightarrow{\binom{1}{Y}} \mathcal{O}(i)^{\oplus 2}$,
- $\mathscr{S}(\frac{\lambda}{\mu}, n)$ is the cokernel of $\mathcal{O}(0) \xrightarrow{(\lambda+\mu Y)^i} \mathcal{O}(n)$.

Therefore, $\mathtt{D\,Coh}^\bullet\,\mathbb{P}^1 = \langle \mathcal{O}(0), \mathcal{O}(1)\rangle$. Take $\mathcal{O}(0) \oplus \mathcal{O}(1)$ and look at its A_∞-endomorphism algebra. Using the calculation in Example 4.33 we can conclude that

$$\mathtt{Coh}\,(\mathcal{O}(0) \oplus \mathcal{O}(1), \mathcal{O}(0) \oplus \mathcal{O}(1)) = \begin{pmatrix} \mathbb{C} & 0 \\ \mathbb{C} + \mathbb{C}Y & \mathbb{C} \end{pmatrix}$$

concentrated in degree 0. This is isomorphic to the path algebra of the Kronecker quiver, so $\langle \mathcal{O}(0), \mathcal{O}(1)\rangle = \mathtt{D}\mathbb{C}Q = \mathtt{D\,mod}^\bullet\,\mathbb{C}Q$. □

Remark 4.35 In this identification the string objects of odd length correspond to the line bundles $\mathcal{O}(k)$, the string objects of even length with the $\mathscr{S}(0, n)$ and $\mathscr{S}(\infty, n)$ and the band objects with the $\mathscr{S}(\lambda, n)$.

Observation 4.36 *The representation theory of the Kronecker quiver can be geometrically interpreted as coherent sheaves on the projective line:*

$$\mathtt{D\,Coh}^\bullet\,\mathbb{P}^1 \overset{\infty}{=} \mathtt{D\,Mod}^\bullet\,K_2.$$

4.3.3 The Linear Quiver

For the interpretation of the linear quiver we need the notion of a *Landau–Ginzburg model*. This consists of a pair $(\mathbb{X}, f)$ where $\mathbb{X}$ is a space (an affine variety or an orbifold) and $f \colon \mathbb{X} \to \mathbb{C}$ is a function. To this pair we can associate a curved algebra (R, f) where R is the coordinate ring of $\mathbb{X}$ and f is interpreted as an element in R.

The *category of singularities* of the Landau–Ginzburg model $(\mathbb{X}, f)$ is defined as the category of matrix factorizations. We will see in Chapter 7 that it gives a description of the geometry of the singular locus of $\mathbb{Y} = f^{-1}(0)$. Therefore, it is also denoted by

$$\mathtt{DSing}^{\pm}\mathbb{Y} = \mathtt{DMF}^{\pm}(R, f).$$

We will consider the case of the orbifold $\mathbb{X} = [\mathbb{A}^1/G]$. As we saw previously the correct notion for its coordinate ring is the algebra $\mathbb{C}[X] \star G$, which is isomorphic to the path algebra $\mathbb{C}Q$ of the cyclic quiver. This algebra has a central element $\ell = X^n$ which acts on each projective P_v as multiplication with the path of length n that starts and ends at v. This action gives an element $\mu_0(1) \in \mathrm{Hom}(P_v, P_v)$ which we can use to turn the $\mathbb{Z}_2$-graded version of $\mathtt{P}^\bullet$ into a curved category $\mathtt{P}^\pm_\ell$. The objects in $\mathtt{DP}^\pm_\ell$ consists of pairs (P, d) of a $\mathbb{Z}_2$-graded projective $\mathbb{C}Q$-module and a degree 1 map $d\colon P \to P$ such that d^2 acts as multiplication by ℓ.

Theorem 4.37 *Every indecomposable object in* $\mathtt{DP}^\pm_\ell$ *is a twisted complex of the form*

$$M_{ij} = \left(P_i \oplus P_j[1], \begin{pmatrix} 0 & p_{ij} \\ p_{ji} & 0 \end{pmatrix}\right),$$

where p_{uv} *is the shortest path from* v *to* u.

Sketch of the proof Let (P, d) be any matrix factorization. We split d into two parts $d_{10} + d_{01}$ that run between the degree 0 and degree 1 parts. Because $\mathrm{Hom}(P_v, P_w) = p_{wv}\mathbb{C}[\ell]$ and $\mathbb{C}[\ell]$ is a principal ideal domain, we can choose a basis in P^0 and P^1 such that d_{01} becomes a diagonal matrix. The fact that $d_{01}d_{10} = \ell$ implies that d_{10} is also diagonal. If (P, d) is indecomposable, d_{01} is a (1×1)-matrix and hence of the form M_{ij}. □

Now let us have a brief look at the hom-spaces:

$$\mathtt{TwP}^\pm_\ell(M_{ij}, M_{kl}) = \begin{pmatrix} p_{ki}\mathbb{C}[\ell] & p_{kj}\mathbb{C}[\ell] \\ p_{li}\mathbb{C}[\ell] & p_{lj}\mathbb{C}[\ell] \end{pmatrix}$$

with $d\begin{pmatrix} a & b \\ c & d \end{pmatrix} = \begin{pmatrix} 0 & p_{kl} \\ p_{lk} & 0 \end{pmatrix}\begin{pmatrix} a & b \\ c & d \end{pmatrix} - \begin{pmatrix} a & -b \\ -c & d \end{pmatrix}\begin{pmatrix} 0 & p_{ij} \\ p_{ji} & 0 \end{pmatrix}$. When $i = j$ or $k = l$ the homology is always zero, so M_{ii} is a zero object. Furthermore, we may assume that $i < j$ and $k < l$ because $M_{ij} = M_{ji}[1]$. Note that the indices are in fact cyclic but now we assume that they run from 0 to $n-1$. With these assumptions we get the following table, with $\iota = \begin{pmatrix} p_{ki} & 0 \\ 0 & p_{lj} \end{pmatrix}$ and $\gamma = \begin{pmatrix} 0 & p_{kj} \\ p_{li} & 0 \end{pmatrix}$.

Case	$\mathtt{DP}_\ell{}^0(M_{ij}, M_{kl})$	$\mathtt{DP}_\ell{}^1(M_{ij}, M_{kl})$
$i \le k < j \le l$	$\mathbb{C}\iota$	0
$k < i \le l < j$	0	$\mathbb{C}\gamma$
Otherwise	0	0

This is exactly the same table as for the linear quiver of length $n-1$ if we make the identifications $M_{ij} \leftrightarrow S_{i+1j}$.

Observation 4.38 *The representation theory of the linear quiver can be geometrically interpreted as the category of singularities of a Landau–Ginzburg model*

$$\mathtt{DMF}^{\pm}([\mathbb{A}^1/\mathbb{Z}_n], X^n) \overset{\infty}{\cong} \mathtt{DMod}^{\pm} L_{n-1}.$$

4.4 Picturing the Categories

In this section we will draw the categories $\mathtt{D\,Mod}^\bullet\, Q$ for the three quivers we have studied. In each case we draw all the indecomposable objects and we order them on the X-axis according to their phase. We also draw the indecomposable degree 0 angle morphisms. These are those angles that cannot be written as a product of smaller angle morphisms. Because angles are always positive they will point in the positive X-direction.

The resulting picture is also known as the Auslander–Reiten quiver of the category and has been studied in detail in the representation theory of finite-dimensional algebras [13]. The picture for $\mathtt{D\,Coh}^\bullet\,\mathbb{P}^1$ can be found in [136].

For each category we will also discuss its symmetries. These symmetries are a consequence of the geometric interpretations and can be seen as A_∞-functors of the categories. The different geometrical interpretations will often give rise to different symmetries.

4.4.1 The Linear Quiver

Using the interpretation as graded line segments, the diagram below shows the category $\mathtt{D\,S}^\bullet = \mathtt{D\,P}^\bullet$.

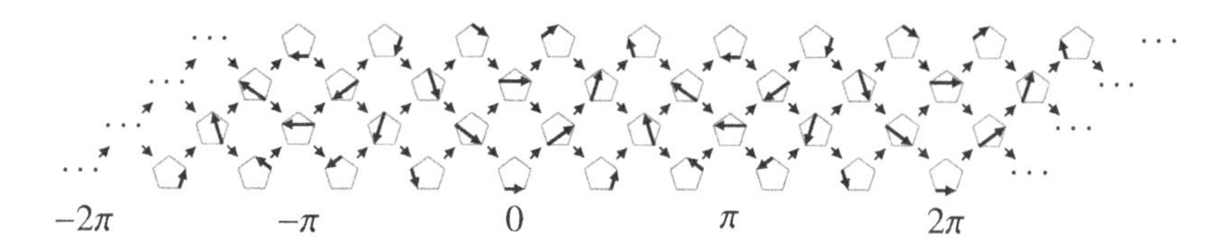

We ordered the objects in a grid using the phase for the horizontal axis and the anticlockwise difference between the source and target of the graded line segment for the vertical axis. Therefore, the shift operator is a glide reflection: it adds π to the phase and swaps source and target so it is a translation over π followed by a horizontal reflection.

We can draw the same picture for the derived category of representations of the linear quiver and in it we can easily identify the subcategories $\mathtt{S}^\bullet$, $\mathtt{mod}^\bullet\, Q$ and $\mathtt{P}^\bullet$.

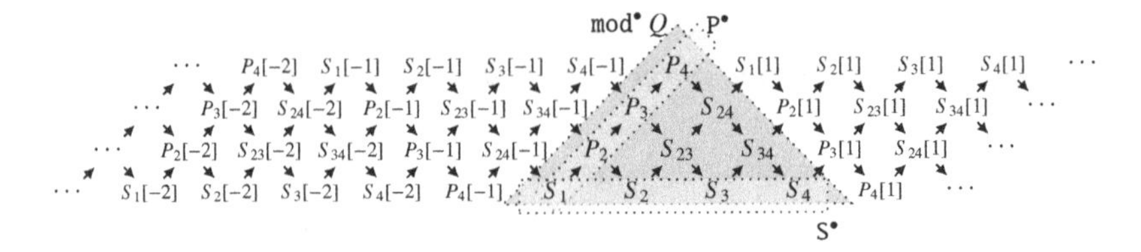

The geometrical interpretation shows that this category has a special autoequivalence $\mathcal{R}$. This autoequivalence rotates all the line segments over $\frac{2\pi}{n+1}$ and adds the same angle to all the phases. This autoequivalence is obvious from the geometric point of view but highly nontrivial if we interpret the category as $\mathrm{D}\,\mathtt{mod}^\bullet Q$. This functor maps some representations to others like S_i, which is mapped to S_{i-1} if $i > 2$. However, S_1 is mapped to the shift of a representation: $P_n[1]$. In representation theory this functor is also known as the *Auslander–Reiten translate* (or its inverse depending on the direction of the rotation) and it corresponds to a horizontal translation over $\frac{2\pi}{n+1}$ in the diagram.

Remark 4.39 For the second geometrical interpretation we only used the $\mathbb{Z}_2$-graded version. This means that we can forget about the phases of the line segments, and the category $\mathtt{DS}^{\pm}$ is a rolled up version of $\mathrm{D}\,\mathtt{S}^\bullet$ that only contains two copies of $\mathtt{mod}\,Q$, which are glued together to form a cylinder.

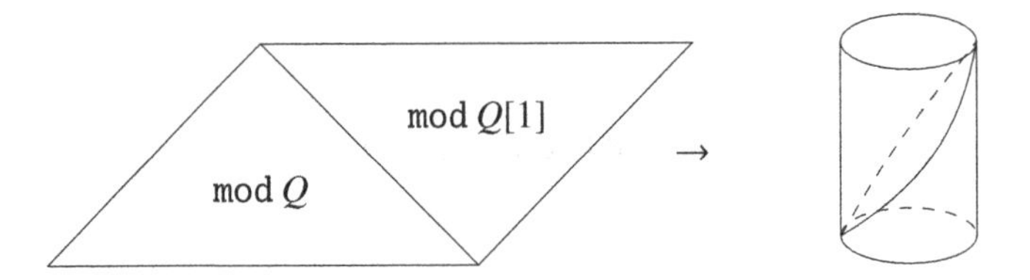

4.4.2 The Kronecker Quiver

For the Kronecker quiver all curves with a horizontal phase are of the form $B(-, n)$, so if we set them on a diagram with as horizonal axis the phase, they all sit on the same spot. To distinguish them we will put them in a box with width parametrized by λ and height by n. In this way we get a tower of objects on the locations where the phase is a multiple of π. Each object in the tower is connected to the one above it by the injection $B(\lambda, n) \subset B(\lambda, n+1)$ and to the one below it by the projection $B(\lambda, n) \to B(\lambda, n-1)$. There are no degree 0 morphisms between band objects with different λ.

The string objects with odd size all have different phases: for $S(0, 2k+1)$ the phase is $\frac{\pi}{2} - \tan^{-1}(k)$, while for $S(1, 2k+1)$ it is $-\frac{\pi}{2} + \tan^{-1}(k+1)$. Each of these objects is connected to the next by two degree 0 morphisms coming from the two embeddings (one on the left and one on the right). The subcategory $\mathtt{mod}^\bullet\, Q$ consists of everything with phases in $(-\frac{\pi}{2}, \frac{\pi}{2}]$.

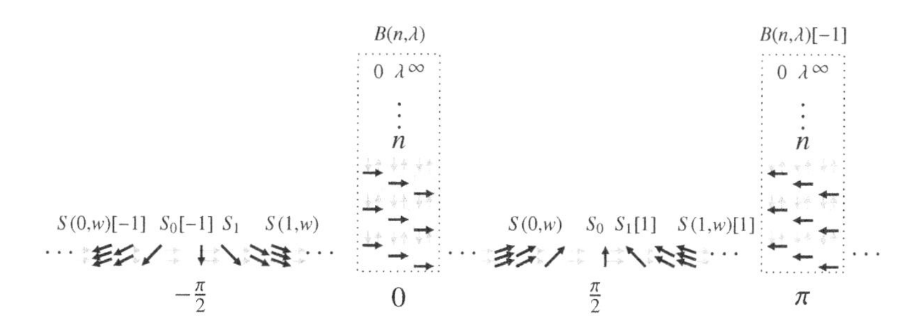

The picture indicates a symmetry of the category: we can shift all the odd string objects 1 to the right. On the cylinder this corresponds to a Dehn twist: we turn the upper boundary circle one full turn to the left, as if trying to open a jar. This gives every curve that crosses the cylinder an extra twist: $S(1, w)$ becomes $S(1, w + 2)$, while $S(0, w)$ becomes $S(0, w - 2)$ and S_0 turns into $S_1[1]$. Again we have an equivalence that is natural from the curve perspective but highly nontrivial from the quiver point of view.

A second symmetry comes from the geometrical interpretation as sheaves on $\mathbb{P}^1$. From this point of view every element $g \in \mathsf{PGL}_2(\mathbb{C})$ induces a transformation of the projective line that changes the sheaf $\mathscr{S}(p, n)$ to $\mathscr{S}(g \cdot p, n)$. The locally free sheaves are unaffected by these transformations. While obvious from the $\mathbb{P}^1$ point of view, this symmetry is very unnatural from the marked cylinder point of view because it mixes strings and band objects.

4.4.3 The Cyclic Quiver

In this picture there are only two phases up to shift. At $\frac{\pi}{2}$ we find all the projectives connected cyclically by the morphisms a_i. At 0 we find all the finite-dimensional modules, which we will order vertically according to their dimension. The string modules $S(i, w)$ can be ordered cyclically according to their starting point. In this way we get a tube with, at each height n, modules connected to the next and previous levels by a rhombic mesh. At heights that are multiples of n we find a floor of band objects.

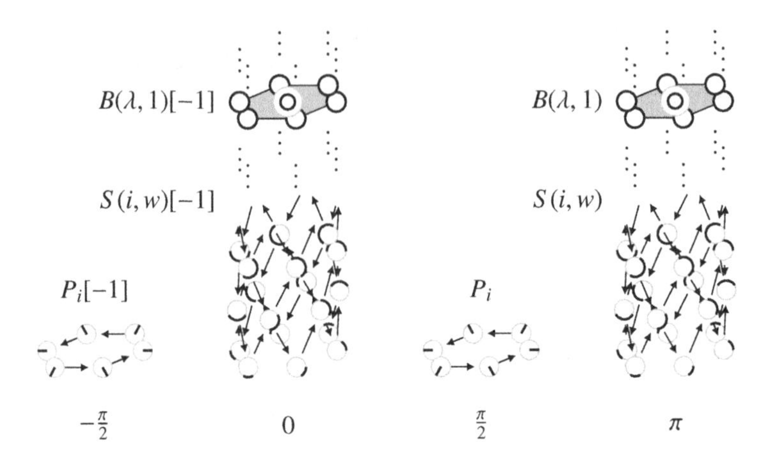

Again there are symmetries. The first one maps $S(i, w) \to S(i + 1, w)$ and $P_i \to P_{i+1}$ but fixes all the $B(\lambda, k)$. It corresponds to a rotation of the marked cylinder over $\frac{2\pi}{n}$. The second symmetry fixes the $S(i, w)$ and P_i but rescales the $B(\lambda, k)$ to $B(r\lambda, k)$. This corresponds to the rescaling of the orbifold $[\mathbb{A}^1/G]$ by a factor r.

4.5 A First Glimpse of Homological Mirror Symmetry

In this chapter we have seen that the representation theory of certain quivers can be interpreted geometrically in two different ways.

- The first interpretation has to do with the intersection theory of curves on a surface and gives rise to an A_∞-category where the objects are embedded curves and the morphisms are linear combinations of (angles at) intersection points. These categories are baby examples of a larger class of categories called Fukaya categories.

  In general, Fukaya categories describe the intersection theory of certain n-dimensional submanifolds, called Lagrangian submanifolds, of $2n$-dimensional symplectic manifolds. Just like for curves on a surface, two Lagrangian submanifolds will in general intersect at points and hence we can take the intersection points as a basis for the morphisms. There are many different versions of such categories: Fukaya categories, wrapped Fukaya categories, Fukaya–Seidel categories. Such constructions are often called *A-models* because they are related to type IIA superstring theory in physics.
- The second interpretation has to do with coherent sheaves on complex algebraic varieties. These are generalizations of vector bundles and the

morphisms measure maps between them (in degree 0), obstructions to the existence of maps and extensions between sheaves (in higher degrees). Again there are many different flavors of such categories: the derived category of coherent sheaves, categories of matrix factorizations, categories of singularities, etc. These constructions are called *B-models* because they appear in type IIB string theory.

The different models (quiver, A-model and B-model) give A_∞-categories that are not precisely equivalent but they become equivalent after we derive them. This procedure adds more objects to the categories and makes them nicer in a certain way without changing their representation theory.

Below is a list of models we have matched so far. This seems a small list but it is just the tip of the iceberg. It turns out that, with some creativity, it is possible to come up with a suitable B-model for almost any A-model and vice versa. This phenomenon was predicted by superstring theory and now goes under the name of *Homological Mirror Symmetry*. In the next part we are going to examine this idea in detail.

A-model	Quiver	B-model
Filled n-gon	Linear quiver	Curved orbifold
	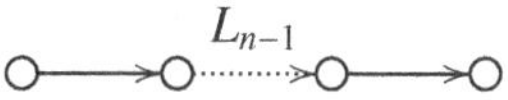	$([\mathbb{A}^1/\mathbb{Z}_n], X^n)$
Doubly marked cylinder	Kronecker quiver	Projective line
	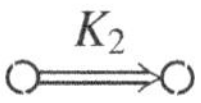	$\mathbb{P}^1$
Punctured 1-marked disk	One-loop quiver	Affine line
		$\mathbb{A}^1$
Punctured n-marked disk	Cyclic quiver	Orbifold
	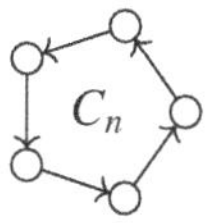	$[\mathbb{A}^1/\mathbb{Z}_n]$

4.6 Exercises

Exercise 4.1 Let Q be a quiver with n vertices and $n-1$ arrows connecting the vertices in a linear way, but not all oriented in the same direction. Show that up to equivalence $\mathtt{D\,mod}^\bullet\, \mathbb{C}Q$ does not depend on the orientation. (Hint: Choose lines in the $(n+1)$-gon such that their endomorphism ring is $\mathbb{C}Q$.)

Exercise 4.2 Let Q be the linear quiver with n vertices and let $X_1, \dots, X_k \in \mathtt{D\,mod}^\bullet\, Q$ be objects corresponding to graded line segments $\mathbb{L}_1, \dots, \mathbb{L}_k$ in the $(n+1)$-gon. Show that $X_1, \dots, X_k$ are split-generators for $\mathtt{D\,mod}^\bullet\, Q$ if and only if the endpoints cover all corners of the polygon.

Exercise 4.3 An full strong exceptional sequence for a derived category $\mathtt{D\,C}^\bullet$ is an ordered sequence of objects $X_1, \dots, X_k$ that split-generates the category and

$$\mathtt{DC}^d(X_i, X_j) = \begin{cases} 0, & d \neq 0 \text{ or } i > j, \\ \mathbb{C}, & i = j,\ d = 0. \end{cases}$$

In other words, all nontrivial morphisms are in degree 0 and between objects with increasing indices. Determine all full strong exceptional sequences in $\mathtt{D\,mod}^\bullet\, Q$, where Q is the linear quiver with three vertices.

Exercise 4.4 Determine the group of autoequivalences of $\mathtt{D\,mod}^\bullet\, Q$, where Q is the linear quiver with n vertices.

Exercise 4.5 Look at a cylinder with two marked points on one boundary circle and one on the other. Try to interpret the intersection theory of line segments on this surface as the derived category of the quiver $\circ \rightrightarrows \circ \rightarrow \circ$. Describe all indecomposable objects in this category.

Exercise 4.6 An object X in $\mathtt{D\,C}^\bullet$ is called n-spherical if its endomorphism ring is the cohomology ring of an n-sphere:

$$\mathtt{D\,C}^\bullet(X, X) \cong \mathbb{C} \oplus \mathbb{C}[n].$$

Determine all 1-spherical objects in $\mathtt{D\,Coh}^\bullet\, \mathbb{P}^1$.

Exercise 4.7 Let $G = \langle g \mid g^{kl} = 1 \rangle$ act on $\mathbb{C}[X]$ by $gX = e^{2\pi i/k}X$. Show that $\mathtt{D\,Coh}^\bullet[\mathbb{A}^1/G]$ can be described by the intersection theory of l disjoint one-punctured disks with each k marked points on the boundary.

Exercise 4.8 Show that the A-model of a sphere with two punctures can be described by $\mathtt{D}\mathbb{C}[X, X^1]$, while the B-model for this category is $\mathbb{A}^1 \setminus \{0\} = \mathbb{P}^1\{0, \infty\}$. Hence the A- and B-models have the same underlying topological space.

Exercise 4.9 Let $G = \langle g \mid g^n = 1 \rangle$ act on $\mathbb{C}[X, X^{-1}]$ by $gX = e^{2\pi i/n}X$. Show that the B-models for $[\mathbb{A}^1 \setminus \{0\}/G]$ and $\mathbb{A}^1 \setminus \{0\}$ are equivalent.

Exercise 4.10 Show that the category $\mathtt{MF}^{\pm}(\mathbb{C}[X], X^n)$ can be interpreted as the A-model of a disk with one point on the boundary and an order n orbifold point at the center. To make sense of this consider the n-fold cover $\pi\colon \mathbb{D} \to \mathbb{D}\colon z \mapsto z^n$. The objects in the A-model are curves downstairs but the products come from polygons upstairs.

PART TWO

A GLANCE THROUGH THE MIRROR

5
Motivation from Physics

The aim of this chapter is to give some idea of the physical motivation behind the theory of mirror symmetry. We will briefly sketch how one goes from classical physics to superstring theory and explain the origins of the A- and B-models in superstring theory. For more details about the underlying physics we refer to the lecture notes [62, 251] and the books [64, 114].

After that we will look at the physics from the perspective of category theory. The focus will be on the underlying algebraic structure and we assume tacitly that all technical issues, such as those arising in the path integral formalism, can be resolved.

5.1 The Path Integral Formalism

5.1.1 The Least Action Principle

Many aspects in classical physics can be described using a least action principle: the state of the system will change in such a way that it minimizes a certain quantity called the *action*.

Suppose for instance that we are looking at a particle moving on a manifold $\mathbb{M}$ and this can be described by a smooth curve $\gamma\colon [a,b] \to \mathbb{M}$. Denote the space of all such paths by $\mathcal{X}$. The action S can be seen as a function $S : \mathcal{X} \to \mathbb{R}$ and if we specify the start and end positions $x_a, x_b \in \mathbb{M}$ of the particle the path that the particle will take will be one that minimizes the action among all paths in $\mathcal{X}(x_a \to x_b) = \{\gamma \in \mathcal{X} \mid \gamma(a) = x_a,\ \gamma(b) = x_b\}$.

Every observable in the system can be thought of as a function $\phi\colon \mathcal{X} \to \mathbb{R}$. For example, the velocity at time $u \in [a,b]$ is the function $v_u\colon \mathcal{X} \to \mathbb{R}\colon \gamma \mapsto \frac{d\gamma}{dt}(u)$. To find the value of the observable we just plug in the solution we found using the least action principle. Together all the observables form an algebra O.

Observation 5.1 *Classical physics consists of finding critical points in a configuration space and evaluating functions (observables) at these critical points.*

5.1.2 Quantum Theories

The main idea in quantum mechanics is that particles do not follow a specified path, but all possible paths contribute to our observation with a certain amplitude. In the *path integral formalism* of quantum mechanics this is done by converting $\mathcal{X}$ into a measurable space $(\mathcal{X}, [d\gamma])$ over which we can integrate. In general, this is a highly nontrivial procedure but we will assume such a measure exists. The behavior of the quantum system is determined by the *partition function* of the theory and the expectation values of the observables, which are defined as the integrals

$$Z = \int_{\mathcal{X}} e^{iS(\gamma)/\hbar}[d\gamma] \quad \text{and} \quad \langle F \rangle = \int_{\mathcal{X}} F(\gamma) e^{iS(\gamma)/\hbar}[d\gamma],$$

where S is the classical action and $\hbar$ is the Planck constant.

The connection between quantum physics and classical physics is seen if we let the Planck constant $\hbar$ go to zero. If $S(\gamma)$ varies a lot, $e^{iS(\gamma)/\hbar}$ will oscillate very fast and the contributions of those γ will cancel out in the integral. In the limit, only those γ for which $\delta S = 0$ will contribute: the classical solutions.

This formalism is very general and can be used for many different situations. All we need is a space $\mathcal{X}$ of all possible configurations of the system, an algebra of observables O and a way to integrate expressions in these observables over $\mathcal{X}$. Once these are fixed we can choose an action $S \in O$ and study the dynamics of the system.

- We can start with a space-time manifold $\mathbb{M}$ and a vector bundle $\mathbb{E} \to \mathbb{M}$. The space $\mathcal{X}$ can be the space of sections $\Gamma(\mathbb{E})$. Such theories are basic examples of *quantum field theories*.
- We can start with a space-time manifold $\mathbb{M}$ and a principal bundle $\mathbb{G} \to \mathbb{M}$. The space $\mathcal{X}$ can be the space of all connections on $\mathbb{M}$ and the action can be the total curvature of the connection. This is known as *Yang–Mills theory*.
- We can start with a Riemann surface Σ and a Riemannian manifold $\mathbb{X}$. The space $\mathcal{X}$ can be the space of all smooth maps $\sigma\colon \Sigma \to \mathbb{X}$ and the action the area of the surface $\sigma(\Sigma)$. This can be seen as a rudimentary version of

string theory because if Σ is a cylinder, the map σ describes a closed string sweeping through the space $\mathbb{X}$. In this setup, Σ is called the worldsheet and $\mathbb{X}$ the target space.

Observation 5.2 *Quantum physics consists of integrating functions (observables) over a configuration space with a special measure.*

5.2 Symmetry

5.2.1 Symmetry and Cohomology

A physical theory has symmetries if there is a group G acting on the space $\mathcal{X}$ such that the action and the measure are invariant:

$$\forall\, g \in G \text{ we have } S(g\phi) = S(\phi) \quad \text{and} \quad [dg\phi] = [d\phi].$$

There are two possibilities for symmetries to emerge. First of all it can happen that $g\phi$ and ϕ really describe two different physical states of the system. The second case is that $g\phi$ and ϕ represent the same physical state and the difference between them came from choices we made in our description.

In the latter case the real space of states is not $\mathcal{X}$ but the orbit space $\mathcal{X}/G = \{G\phi \mid \phi \in \mathcal{X}\}$. The observables in this theory should not depend on the choices in our description so they should be invariant under the group action. Therefore, the observables over $\mathcal{X}/G$ are the G-invariant observables over $\mathcal{X}$.

There is also an infinitesimal version of symmetry. This means that each symmetry is a vector field $Q \in \Gamma(\mathsf{T}\mathcal{X})$. We say that $\phi \in \mathcal{X}$ is invariant if $Q\phi = 0$ or, equivalently, $\exp(tQ)\phi = \phi$. A vector field on $\mathcal{X}$ induces a derivation d_Q on the ring of functions on $\mathcal{X}$,

$$d_Q F = \lim_{\epsilon \to 0} \frac{F[\exp(\epsilon Q)\phi] - F[\phi]}{\epsilon}.$$

We say that a function is invariant under Q if $d_Q F = 0$. The observables of the theory are those that sit in the kernel of d_Q. An observable of the form $d_Q F$ will evaluate to zero for the path integral:

$$\begin{aligned}
\langle d_Q F \rangle &= \int (d_Q F)[\phi] e^{iS[\phi]/\hbar}\, d\phi = \int d_Q(F e^{iS/\hbar})[\phi]\, d\phi \\
&= \lim_{\epsilon \to 0} \frac{\int F e^{iS/\hbar}[\exp(\epsilon Q)\phi]\, d\phi - \int F e^{iS/\hbar}[\phi]\, d\phi}{\epsilon} = 0.
\end{aligned}$$

The two integrals in the last line are the same because the measure is invariant under $\exp(\epsilon Q)$ and therefore $[d\phi] = [d\exp(\epsilon Q)\phi]$. The physics cannot distinguish between observables that differ by something in the image of d_Q so, if we can ensure that $d_Q^2 = 0$, the relevant space of observables is in fact

$$\frac{\operatorname{Ker} d_Q}{\operatorname{Im} d_Q},$$

which is the cohomology of d_Q. A more in-depth approach to this idea leads to the notion of BRST-cohomology [122].

Observation 5.3 *Symmetries in quantum physics lead to cohomology theories on the algebra of observables.*

Finally, if we assume that the action is of the form $S = d_Q V$ for some observable V then the path integral for an observable $F \in \operatorname{Ker} d_Q$ does not depend on $\hbar$:

$$\begin{aligned}\frac{d}{d\hbar}\langle F\rangle &= \int F \frac{d}{d\hbar} e^{id_Q V/\hbar}\, d\phi = \frac{-i}{\hbar^2}\int F d_Q V e^{id_Q V/\hbar}\, d\phi \\ &= \frac{-i}{\hbar^2}\int d_Q(FV) e^{id_Q V/\hbar}\, d\phi = \frac{-i}{\hbar^2}\langle d_Q(FV)\rangle = 0.\end{aligned}$$

This means that we can take the limit of $\hbar \to 0$ without changing the path integral, so the path integral can be restricted to the space of classical solutions for which $\delta S = 0$. This makes the quantum mechanical system much easier and therefore physicists like to study theories where such a symmetry is present.

5.2.2 Supersymmetry

One way to create more symmetries is by making a theory *supersymmetric*. In a supersymmetric theory there are two types of fields: bosonic and fermionic. The former are described by ordinary variables that commute and the latter by Grassmann variables that anticommute.

Let us study a baby example: a zero-dimensional theory (see [62, Chapter 1]). The space $\mathcal{X}$ is the space of maps from a point to $\mathbb{R}$ so the space of field configurations is $\mathcal{X} = \mathbb{R}$ with the standard measure. We assume that the action is $S = \frac{1}{2}V(x)^2$ where V is a function. The action is minimal at the zeros of V, so these points are the classical solutions. To make things a bit easier, we will also work with the conventions of statistical mechanics. This means that we drop the factor $\frac{i}{\hbar}$ and add a minus sign to make the integrals converge. With this convention the partition function becomes

$$Z = \int_{\mathbb{R}} e^{-S}\, dx = \int_{\mathbb{R}} e^{-V(x)^2/2}\, dx,$$

which is usually not so easy to calculate.

We can alter this theory by introducing two extra fermions ψ_+, ψ_-. This means that the algebra of observables becomes

$$O = C_\infty(\mathbb{R}) \otimes \Lambda_2 = \frac{C_\infty(\mathbb{R})\langle\psi_+, \psi_-\rangle}{\langle\psi+^2, \psi_-^2, \psi_+\psi_- + \psi_-\psi_+\rangle}$$

and we add an extra term to the action

$$S = \frac{1}{2}V(x)^2 - \psi_+\psi_-\frac{dV(x)}{dx}.$$

If we work out the exponential of the action using a Taylor series, we get

$$\exp(-S) = \left(1 + \psi_+\psi_-\frac{dV(x)}{dx}\right)e^{-V(x)^2/2}$$

because the Grassmann variables ψ_+, ψ_- square to zero. To calculate the integral we also need to specify how to integrate over Grassmann variables. The correct rules are $\int d\psi = 0$ and $\int \psi\, d\psi = 1$. With this in mind we can integrate out ψ_+, ψ_- and with the extra factor $\frac{dV(x)}{dx}$ present, we can apply a change of variables from x to V. This results in the following expression for the supersymmetric (SUSY) partition function:

$$Z_{\text{SUSY}} = \int_{\mathbb{R}} \frac{dV(x)}{dx}e^{-V(x)^2/2}\,dx = \int_{V(-\infty)}^{V(+\infty)} e^{-V^2/2}\,dV.$$

If V is an increasing bijection then the integral is just a Gaussian integral and equal to $\sqrt{2\pi}$. If V is decreasing we get an extra minus sign. In general, as long as the limits of V are both infinite we can reinterpret the integral as follows: count the number of times V crosses the zero in the positive direction and subtract the number of times it crosses the zero in the negative direction. If the answer is 1 then V is overall increasing and we get $Z = \sqrt{2\pi}$. If the result is -1 then V is overall decreasing and $Z = -\sqrt{2\pi}$. We can also write this as

$$Z_{\text{SUSY}} = \sqrt{2\pi}\sum_{V(x)=0}\frac{V'(x)}{|V'(x)|}.$$

So while the original theory has a complicated partition function, the supersymmetric one is just a signed count of the classical solutions.

This theory also has two extra infinitesimal symmetries that swap bosons and fermions:

$$Q_+x = \epsilon\psi_+, \quad Q_+\psi_+ = 0, \quad Q_+\psi_- = \epsilon V(x),$$
$$Q_-x = \epsilon\psi_+, \quad Q_-\psi_- = 0, \quad Q_-\psi_+ = \epsilon V(x).$$

These symmetries are called *supersymmetries* and because there are two of them the theory is called $\mathcal{N} = 2$-supersymmetric.

Observation 5.4 *A supersymmetric theory is a $\mathbb{Z}_2$-graded theory with one or more degree 1 operators that square to zero. Adding supersymmetries to a theory makes calculations more manageable.*

5.3 Superstrings

5.3.1 From Fields to Strings

Let us now apply this philosophy to string theory. Fix a Riemann surface Σ and a target space $\mathbb{X}$. The space of maps $\{\sigma : \Sigma \to \mathbb{X}\}$ is equal to the space of sections of the trivial n-dimensional bundle, so this theory can be seen as a field theory of n scalar fields on a two-dimensional space-time Σ. From this point of view it makes sense to add extra fields to the theory to make it supersymmetrical. Let us start with the case where $\Sigma = \mathbb{C}$ and $\mathbb{X} = \mathbb{R}^n$.

Like in the zero-dimensional example we have to add two fermion fields for every scalar. Let us denote the scalar fields by ϕ^I and the fermion fields by $\psi^I_\pm$. At first glance each coordinate will have its own two-parameter family of supersymmetries but coordinate changes in $\mathbb{X}$ will mix up different coordinate fields, so all fields must share the same two-parameter family of supersymmetries. Such a theory is called a $d = 2$, $\mathcal{N} = 2$ theory, where d represents the dimension of the worldsheet Σ and $\mathcal{N}$ is the number of supersymmetries.

It is possible to get more supersymmetries if we additionally assume $\mathbb{X} = \mathbb{C}^n$. In that case, we get two sets of n scalar fields: the ϕ^i and their conjugates $\phi^{\bar{i}}$. There are four sets of fermions $\psi^i_\pm$, $\psi^{\bar{i}}_\pm$. Because real and imaginary parts move alongside in a coordinate transformation we have extra freedom for two more supersymmetries. Such a theory is called a $d = 2$, $\mathcal{N} = (2, 2)$ theory. In flat space the set of transformations looks like

$$\delta\phi^i = i\alpha_-\psi^i_+ + i\alpha_+\psi^i_-, \quad \delta\psi^i_\pm = -\tilde{\alpha}_\pm\partial_\pm\phi^i,$$
$$\delta\phi^{\bar{i}} = i\tilde{\alpha}_-\psi^{\bar{i}}_+ + i\tilde{\alpha}_+\psi^{\bar{i}}_-, \quad \delta\psi^{\bar{i}}_\pm = -\alpha_\pm\partial_\pm\phi^{\bar{i}},$$

where $\alpha_\pm, \tilde{\alpha}_\pm$ are four parameters and the ∂_+, ∂_- are the derivatives with respect to the complex coordinate z and its conjugate.

Observation 5.5 *Quantum field theories with complex scalars can be made supersymmetric with $\mathcal{N} = (2, 2)$ supersymmetries.*

Up to now we have treated $\mathbb{X}$ just as a complex vector space but we would like to have a theory where we can plug in any complex manifold with a compatible metric. Such a manifold is called a *Kähler manifold*. It is a real manifold $\mathbb{X}$ of dimension $2n$ and it has three structures on it:

- a Riemannian metric $g\colon \mathrm{T}\mathbb{X} \times \mathrm{T}\mathbb{X} \to \mathbb{R}$,
- a complex structure $J\colon \mathrm{T}\mathbb{X} \to \mathrm{T}\mathbb{X}$ with $J^2 = -1$,
- a symplectic form $\omega \in \Omega^2(\mathbb{X})$.

These three structures are tied together by the identity $g(X, JY) = \omega(X, Y)$, which can be used to deduce the third from the other two.

Using these structures it is possible to upgrade the supersymmetry from $\mathbb{C}^n$ to $\mathbb{X}$ by making the derivatives covariant using the Levi-Civita connection of the metric. For the sake of completeness we include the full supersymmetry transformation rules and the supersymmetric action:

$$\begin{aligned}
\delta\phi^i &= i\alpha_-\psi_+^i + i\alpha_+\psi_-^i, \quad \delta\psi_\pm^i = -\tilde{\alpha}_\pm\partial_\pm\phi^i - i\alpha_\pm\psi_\mp^k\Gamma^i_{km}\psi_\pm^m, \\
\delta\phi^{\bar{i}} &= i\tilde{\alpha}_-\psi_+^{\bar{i}} + i\tilde{\alpha}_+\psi_-^{\bar{i}}, \quad \delta\psi_\pm^{\bar{i}} = -\alpha_\pm\partial_\pm\phi^{\bar{i}} - i\tilde{\alpha}_\pm\psi_\mp^{\bar{k}}\Gamma^{\bar{i}}_{\bar{k}\bar{m}}\psi_\pm^{\bar{m}}, \\
S &= \int_\Sigma (g_{i\bar{j}}\partial_z\phi^i\partial_{\bar{z}}\phi^{\bar{j}} + ig_{i\bar{j}}\psi_+^i D_{\bar{z}}\psi_+^{\bar{j}} + ig_{i\bar{j}}\psi_-^i D_z\psi_-^{\bar{j}} \\
&\quad + g_{\bar{i}j}\partial_z\phi^{\bar{i}}\partial_{\bar{z}}\phi^j + ig_{\bar{i}j}\psi_+^{\bar{i}} D_{\bar{z}}\psi_+^j + ig_{\bar{i}j}\psi_-^{\bar{i}} D_z\psi_-^j \\
&\quad + \frac{1}{2}R_{i\bar{j}k\bar{l}}\psi_+^i\psi_+^{\bar{j}}\psi_-^k\psi_-^{\bar{l}})\, dz\, d\bar{z}.
\end{aligned}$$

Here the Γ^i_{km} are the Christoffel symbols of the metric, R_{ijkl} is the Riemann curvature tensor and D_z, $D_{\bar{z}}$ are the covariant derivatives on Σ coming from the pullback along the map $\Phi\colon \Sigma \to \mathbb{X}$ that is locally described by the scalar fields ϕ^i.

To go from $\mathbb{C}$ to an arbitrary Σ we have to consider the fermion fields as sections of a bundle over Σ. Because ϕ^i is a scalar, the equation $\delta\phi^i = \alpha_-\psi_+^i + i\alpha_+\psi_-^i$ tells us that the ψ_+^i are sections of the trivial bundle. On the other hand, $\delta\psi_+^i = -\tilde{\alpha}_+\partial_+\phi^i + \cdots$ tells us that the ψ_+^i are sections of the canonical bundle because they transform like in a 1-form: $d\phi^i = \partial_z\phi^i\, dz$. These cannot both be true unless Σ is very special and hence $\tilde{\alpha}_+$ and α_- cannot be nonzero at the same time. The same is true for α_+ and $\tilde{\alpha}_-$. To make a sensible theory we have to put some parameters to zero, and choosing which ones leads to different models of superstring theory.

Observation 5.6 *For string theories with arbitrary Riemann surfaces we have to choose which of the $\mathcal{N} = (2, 2)$ supersymmetries we want to keep.*

5.3.2 The A-Model

In the A-model we set $\alpha_+ = \tilde{\alpha}_- = -i$ and $\alpha_- = \tilde{\alpha}_+ = 0$ and we call the supersymmetry Q_A. In this case, ψ_-^i and $\psi_+^{\bar{i}}$ are sections of a trivial vector bundle on Σ and we have $d_{Q_A}\phi^i = \psi^i_-$ and $d_{Q_A}\phi^{\bar{i}} = \psi^{\bar{i}}_+$. Note that ψ^i_- and $\psi^{\bar{i}}_+$ still anticommute so if we rewrite these fields for instance as $d\phi^i$ and $d\phi^{\bar{i}}$ then at each point $x \in \Sigma$ expressions in the ϕ^i, $\phi^{\bar{i}}$, $d\phi^i$ and $d\phi^{\bar{i}}$ can be interpreted as differential forms on $\mathbb{X}$ and d_{Q_A} is just the exterior derivative. Therefore, the local algebra of observables can be seen as the de Rham cohomology of $\mathbb{X}$.

To calculate path integrals, we first need to have a closer look at the action. One can rewrite the action as

$$S = S_0 + S_1 = \int_\Sigma d_{Q_A} V + \int_\Sigma \Phi^* \omega,$$

where is $V = g_{i\bar{j}}(\psi^i_+ \partial_{\bar{z}}\phi^{\bar{j}} + \psi^{\bar{j}}_- \partial_z \phi^i)$ and ω is the symplectic form $g_{i\bar{j}}\, dx^i \wedge dx^{\bar{j}}$. The term S_1 is the integral of a closed 2-form over a surface, so it only depends on the homology class of the surface $\beta = [\Phi] \in H(\mathbb{X})$.

The configuration space X can be seen as the disjoint union of components, one for each homology class of the image of the map $\Phi\colon \Sigma \to \mathbb{X}$. The path integral is a sum over all homology classes of path integrals over each component separately. In each term of the sum we can bring the factor $\exp(\frac{i}{\hbar}S_1(\beta))$ to the front:

$$\begin{aligned}\langle F\rangle &= \int_X F e^{\frac{i}{\hbar}\int d_{Q_A}V + \frac{i}{\hbar}S_1(\beta)} \\ &= \sum_{\beta \in \mathrm{H}_2(\mathbb{X})} e^{\frac{i}{\hbar}S_1(\beta)} \int_{\{\Phi \in X | [\Phi] = \beta\}} F e^{\frac{i}{\hbar}\int_\Sigma d_{Q_A}V}\, d\phi.\end{aligned}$$

The path integrals in this sum do not depend on $\hbar$ because the action is d_{Q_A}-exact. This means that we can restrict the path integrals to the solutions of the classical equations for the action $S_0 = \int_\Sigma d_{Q_A} V$.

The classical equations δS_0 imply that $\partial_{\bar{z}}\phi^i = \partial_z \phi^{\bar{i}} = 0$, which means that $\Phi\colon \Sigma \to \mathbb{X}$ is in fact a holomorphic map. In the simplest situation there are only finitely many holomorphic maps and in that case the path integral is a weighted count of these holomorphic embeddings. Although the specific maps $\Sigma \to \mathbb{X}$ depend on the complex structure J on $\mathbb{X}$, their number and their area measured by ω are independent of J.

Observation 5.7 *The partition function of the A-model only depends on $(\mathbb{X}, \omega)$ as a symplectic manifold.*

5.3.3 The *B*-Model

In the *B*-model we define Q_B to be the supersymmetry operator with $\tilde{\alpha}_+ = \tilde{\alpha}_- = 1$. Now the ψ with the barred indices are sections of a trivial Σ-bundle and we perform a change of variables $d\phi^i := g_{i\bar{j}}(\psi^{\bar{j}}_+ - \psi^{\bar{j}}_-)$ and $d\phi^{\bar{i}} = \psi^{\bar{i}}_+ + \psi^{\bar{i}}_-$.

The complex algebra of local observables is the complexified algebra of differential forms on X: $\Omega_{\mathbb{C}}(\mathbb{X})$. We can check that $d_{Q_B}\phi^i = 0$ and $d_{Q_B}\phi^{\bar{i}} = d\phi^{\bar{i}}$ while $d_{Q_B}d\phi^i = d_{Q_B}d\phi^{\bar{i}} = 0$.

This means that in degree 0 the observables are expressions in the ϕ^i only, and not in the conjugate variables $\phi^{\bar{i}}$. In other words, these observables are precisely the holomorphic functions on $\mathbb{X}$. Further investigation shows that the cohomology of $(\Omega_{\mathbb{C}}(\mathbb{X}), d_{Q_B})$ can be identified with the Dolbeault cohomology [66, 91], which is an invariant of the complex structure on $\mathbb{X}$.

To calculate path integrals, one can rewrite the action as

$$S = S_0 + S_1 = \int_{\Sigma} d_{Q_B} V + \int_{\Sigma} \mathcal{Z},$$

where $V = -g_{i\bar{j}}(\psi^i_+ \partial_{\bar{z}}\phi^{\bar{j}} + \partial_z \phi^{\bar{j}} \psi^i_-)$ and $\mathcal{Z}$ is a complicated formula that does not depend on the metric $g_{i\bar{j}}$ or on the symplectic form ω, but only on the complex structure J.

Observation 5.8 *The partition function of the B-model only depends on $(\mathbb{X}, J)$ as a complex manifold.*

5.4 Categorical Interpretations

The path integral formalism is just one way to look at quantum mechanics. In this section we will approach quantum mechanics from a different angle. We will investigate how we can interpret it from the perspective of category theory.

5.4.1 Physics as a Functor

In classical physics the state of a system is described as a point in phase space, the space of all possible configurations of the system. Phase space has the structure of a symplectic manifold $(\mathbb{X}, \omega)$. To introduce dynamics in this system one needs to specify a Hamiltonian $H: \mathbb{X} \to \mathbb{R}$ and this will define a vector field X_H on $\mathbb{X}$ such that $\omega(X_H, -) = dH$. The system will follow the flow of this vector field: there is a one-parameter family of automorphisms $\phi_t: \mathbb{X} \to \mathbb{X}$ such that $\frac{d\phi_t}{dt} = X_H$ and if $v \in \mathbb{X}$ is the position of the system at time a then $\phi_t(v)$ is the state of the system at time $a + t$.

We mold this idea in a categorical framework. Define the category `Part`, which has just one object $\bullet$ and the morphism space $\texttt{Part}(\bullet, \bullet) = \mathbb{R}$.

Composition is given by the sum $t \circ u = t + u$ and the identity is zero. This easy category is a toy model for the time evolution of a particle:

$$\bullet \xrightarrow{t} \bullet \in \texttt{Part}(\bullet, \bullet).$$

A classical dynamical system can be seen as a functor from the particle category to the category of symplectic manifolds $\mathcal{E}\colon \texttt{Part} \to \texttt{Symp}$. It assigns to the unique object $\bullet$ the phase space of the system and $\mathcal{E}(t)$ becomes the time-t flow of the evolution of the system. The fact that this is a functor follows from the observation that $\phi_{t+u} = \phi_t \phi_u$.

In quantum mechanics the state of a particle at a given time is described as a vector in a Hilbert space, which is a complex vector space with a hermitian inner product, and the time evolution of the system is described by a unitary operator. Again this can be seen as a functor $Q\colon \texttt{Part} \to \texttt{Hilb}$, or in other words it is a special representation of the category $\texttt{Part}$.

Example 5.9 To construct that representation for a particle moving on a manifold $\mathbb{X}$, we take for $Q(\bullet) = L^2(\mathbb{X}, \mathbb{C})$ the space of complex-valued square-integrable functions on $\mathbb{X}$ with inner product $\langle f, g \rangle = \int_{\mathbb{X}} f\bar{g}\, dx$. We can describe the linear map $Q(t)$ by convolution:

$$Q(t)f(y) = \int_{\mathbb{X}} G_t(x, y) f(x)\, dx.$$

The convolution factor $G_t(x, y)$ is the weight of all paths starting with $\gamma(0) = x$ and ending with $\gamma(t) = y$. In the path integral formalism this factor can be thought of as

$$G_t(x, y) = \frac{\int_{\mathcal{X}(x,y)} e^{iS(\gamma)/\hbar}[d\gamma]}{\int_{\mathcal{X}} e^{iS(\gamma)/\hbar}[d\gamma]},$$

where $\mathcal{X}(x, y) = \{\gamma \in \mathcal{X} \mid \gamma(0) = x,\ \gamma(t) = y\}$.

Observation 5.10 *Ordinary quantum mechanics can be seen as a representation of the particle category* $\texttt{Part}$.

5.4.2 Topological Quantum Field Theories

If we want to introduce fields into this picture we have to do some more work. From a mathematical point of view a field is a section of some vector bundle $\mathbb{E}$ over a manifold $\mathbb{M}$. Instead of keeping time separately as we did for particles we will now include it in the base manifold $\mathbb{M}$, which will function as our space-time.

Again the dynamics is governed by some least action principle. The space of all possible field configurations is the space of sections $\mathcal{F} = \Gamma(\mathbb{E})$ and the action is a function $S : \mathcal{F} \to \mathbb{R}$. The classical solutions are those that minimize the action. To pick out one solution we can specify the field near a codimension-1 subspace of $\mathbb{M}$ and try to propagate the field to the whole of $\mathbb{M}$.

Just like with particles we can try to quantize this theory using a path integral formalism. Imagine a piece $\mathbb{U} \subset \mathbb{M}$ with an incoming boundary $\mathbb{B}_i$ and an outgoing boundary $\mathbb{B}_o$. To each of the boundaries we can associate a Hilbert space $Q(\mathbb{B}_{i/o})$ spanned by all possible states of the field restricted to that boundary and we can construct a linear map $Q(\mathbb{U}) \colon Q(\mathbb{B}_i) \to Q(\mathbb{B}_o)$ for which the coefficient between two states is the sum over all fields with those states as boundary conditions of a factor depending on the action:

$$\langle \phi_i, Q(t)\phi_o \rangle = \frac{\int_{\mathcal{F}(\phi_i,\phi_o)} e^{\frac{i}{\hbar}S(\phi)}[d\phi]}{\int_{\mathcal{F}} e^{\frac{i}{\hbar}S(\phi)}[d\phi]},$$

where $\mathcal{F}(\phi_i, \phi_o) = \{\phi \in \mathcal{F} \mid \phi|_{\mathbb{B}_{i/o}} = \phi_{i/o}\}$.

Now suppose that we have two pieces $\mathbb{U}, \mathbb{V} \subset \mathbb{M}$ such that $\mathbb{B}_{o,\mathbb{U}} = \mathbb{B}_{i,\mathbb{V}}$; then we can glue these pieces together as one piece $\mathbb{U} \cup \mathbb{V}$ with boundaries $\mathbb{B}_{i,\mathbb{U}\cup\mathbb{V}} = \mathbb{B}_{i,\mathbb{U}}$ and $\mathbb{B}_{o,\mathbb{U}\cup\mathbb{V}} = \mathbb{B}_{o,\mathbb{V}}$. If we further suppose that the action is additive, $S_{\mathbb{U}\cup\mathbb{V}}(\phi) = S_{\mathbb{U}}(\phi|_{\mathbb{U}}) + S_{\mathbb{V}}(\phi|_{\mathbb{V}})$, then we can show that $Q(\mathbb{U} \cup \mathbb{V}) = Q(\mathbb{U}) \circ Q(\mathbb{V})$. This suggests that Q can again be seen as a representation of an appropriate category.

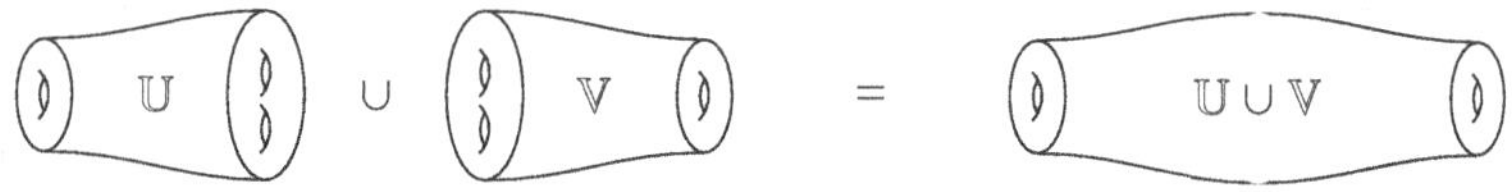

$\mathbb{U}$ $\cup$ $\mathbb{V}$ $=$ $\mathbb{U} \cup \mathbb{V}$

The *cobordism category* $\mathsf{Cob_n}$ has as objects oriented $(n-1)$-manifolds and as morphisms oriented n-manifolds with boundaries. The boundary of a morphism $\mathbb{U}$ is split into two disjoint parts $\mathbb{B}_i$ and $\mathbb{B}_o$, where the latter inherits the orientation of $\mathbb{U}$ and the former the opposite orientation. In this way $\mathbb{U}$ is seen as a morphism from $\mathbb{B}_i$ to $\mathbb{B}_o$. The composition of two morphisms is done by gluing them together over the common boundary.

Depending on what structure the manifolds have (topological, smooth, Riemannian, etc.) there are different versions of the cobordism category. A quantum field theory can now be seen as a functor from an appropriate cobordism category to the category of Hilbert spaces satisfying some additional properties. The main additional condition is that the functor has to be monoidal: it has to turn the disjoint union of boundary components into the tensor product of vector spaces. If the cobordism category is the topological one, we speak of a topological quantum field theory (TQFT). For more information we refer to [246, 18].

Observation 5.11 *Quantum field theory studies monoidal representations of cobordism categories.*

5.4.3 Closed Strings

This point of view is also helpful in string theory. A closed string is a circle embedded into a target manifold $\sigma\colon \mathbb{S}_1 \to \mathbb{X}$. If the string moves over time it will sweep out a cylinder in $\mathbb{X}$. Strings can also show more complicated behavior: they can appear or disappear, or two strings can fuse together. These dynamics can be identified with the following cobordisms:

$$\in \mathsf{Cob}_2(\emptyset, \mathbb{S}^1), \qquad \in \mathsf{Cob}_2(\mathbb{S}^1, \emptyset), \qquad \in \mathsf{Cob}_2(\mathbb{S}^1 \sqcup \mathbb{S}^1, \mathbb{S}^1).$$

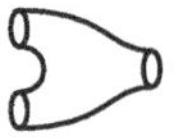

To construct a quantum theory for strings we associate to each n-tuple of strings a Hilbert space and the worldsheets will act as linear maps between those spaces with coefficients coming from an appropriate action principle. This can naturally be seen as a monoidal representation of a two-dimensional cobordism category. A topological string theory will give a functor from Cob_2 to $\mathtt{Hilb}$ (or $\mathtt{Vect}(\Bbbk)$). Because every topological surface can be made by pairs of pants and caps, the functor will be determined by $V = \mathcal{F}(\mathbb{S}^1)$ and the images of the three operations above: a unit $\Bbbk \to V$, a trace $V \to \Bbbk$ and a multiplication $V \otimes V \to V$. Together these data turn V into a *Frobenius algebra* [8].

Observation 5.12 *There is a one-to-one correspondence between 2d-TQFTs and Frobenius algebras.*

Most string theories are not topological but conformal, which means that we have to take the complex structure on the worldsheet into account [215]. The corresponding conformal cobordism category CFT will have an object n for every n-tuple of circles and $\mathsf{CFT}(n, m)$ will be the space of all Riemann surfaces with $n + m$ boundary components, together with identifications of $\mathbb{S}_1$ with the boundaries (framings). This set splits as a disjoint union of subspaces according to the genus of the surfaces: $\mathsf{CFT}(n, m) := \sqcup_g \mathcal{N}_{g,n+m}$.

Once a string theory is specified, we can try to calculate the corresponding functor $\mathtt{CFT} \to \mathtt{Hilb}$. This functor will map the object n to a Hilbert space $\mathcal{H}^{\otimes n}$, where $\mathcal{H}$ describes all states of a single string. If we fix one point in $\Sigma \in \mathsf{CFT}(n, m)$ the path integral formalism will give rise to a map $\mathcal{H}^{\otimes n} \to \mathcal{H}^{\otimes m}$ by integrating over all possible field configurations over the Riemann surface Σ.

This is not all however. We could also imagine integrating over all configurations where Σ is restricted to a subspace of $\mathtt{CFT}(n, m)$ instead of just one point. This gives a map $\mathcal{H}^{\otimes n} \to \mathcal{H}^{\otimes m}$ for each chain in the singular chain complex of $\mathtt{CFT}(n, m)$. We can make sense of this by defining a category $\mathtt{HCFT}^\bullet$ whose objects are $n \in \mathbb{N}$ and $\mathtt{HCFT}^\bullet(n, m)$ is a suitable complex that calculates the homology of $\mathtt{CFT}(n, m)$. From this point of view we can interpret the path integral formalism as a monoidal dg-functor $\mathcal{F}\colon \mathtt{HCFT}^\bullet \to \mathtt{dgVect}^\bullet$. This is known as a *topological conformal field theory* [90, 58, 214].

Example 5.13 In the A-model this idea is closely related to the notion of *Gromov–Witten invariants* [85]. Let $\mathbb{X}$ be a smooth projective variety, fix two numbers $n, g \in \mathbb{N}$ and a 2-cycle $\beta \in \mathsf{H}_2(\mathbb{X}, \mathbb{Z})$. Given n cohomology classes $\psi_1, \ldots, \psi_n \in \mathsf{H}^\bullet(\mathbb{X}, \mathbb{Q})$ one can define the Gromov–Witten invariant

$$\mathrm{GW}^{\mathbb{X}}_{\beta,g,n}(\psi_1, \ldots, \psi_n) \in \mathsf{H}^\bullet(\overline{\mathcal{M}}_{g,n}, \mathbb{Q}).$$

This is an element in the cohomology ring of $\overline{\mathcal{M}}_{g,n}$, which is a compactification of the moduli space of Riemann surfaces of genus g with n marked points (instead of boundary circles, so without framings). Intuitively these invariants should count the number of holomorphic maps $\sigma\colon \Sigma \to \mathbb{X}$ that map a Riemann surface with genus g and n marked points such that the homology class of $\sigma(\Sigma)$ is β and the marked points lie on homology classes dual to the ψ_i.

We briefly sketch how these data might be turned into a representation of $\mathtt{HCFT}^\bullet$. Fix a basis for $\beta_1, \ldots, \beta_k$ for $\mathsf{H}_2(\mathbb{X}, \mathbb{Z})$ and let $\mathbb{k} = \mathbb{C}(\!(t_1, \ldots, t_k)\!)$. If $\beta = b_1\beta_1 + \cdots + b_k\beta_k$ then we set $t^\beta = t_1^{b_1} \cdots t_k^{b_k}$. Now look at the vector space $V = \mathsf{H}^\bullet(\mathbb{X}, \mathbb{k})$. Because $\mathbb{X}$ is smooth and compact this vector space comes with a bilinear pairing $\langle\, ,\, \rangle\colon V \times V \to \mathbb{k}$, which allows us to identify V with its $\mathbb{k}$-dual and any linear map $f\colon V^{\otimes_{\mathbb{k}} n} \to V^{\otimes_{\mathbb{k}} m}$ with a map $f\colon V^{\otimes_{\mathbb{k}} n+m} \to \mathbb{k}$.

With these conventions fixed we can turn the Gromov–Witten invariants into a representation of $\mathtt{HCFT}^\bullet$. Set $\mathcal{GW}^{\mathbb{X}}(n) = V^{\otimes_{\mathbb{k}} n}$ and for any $\eta \in \mathtt{HCFT}^\bullet(n, m)$ let $\mathcal{GW}^{\mathbb{X}}(\eta)$ correspond to the map

$$V^{\otimes_{\mathbb{k}} n+m} \to \mathbb{k}\colon \psi_1 \otimes \cdots \otimes \psi_{n+m} \mapsto \sum_\beta t^\beta \int_\eta \mathrm{GW}^{\mathbb{X}}_{\beta,g,n+m}(\psi_1, \ldots, \psi_{n+m}).$$

In other words, the coefficients of the map $\mathcal{GW}^{\mathbb{X}}(\eta)$ are weighted sums of the Gromov–Witten invariants integrated over the cycle η, reinterpreted as a cycle in $\overline{\mathcal{M}}_{g,n+m}$. For further details we refer to [58]. It is also possible to go in the opposite direction: starting from a topological conformal field theory one can compute certain numerical invariants that behave like Gromov–Witten invariants [61, 50].

Observation 5.14 *Closed string theories give rise to representations of* CFT *or* HCFT$^\bullet$.

5.4.4 Open Strings

Closed strings are embedded circles moving through space-time, but one can also imagine open strings. These are embedded line segments and to describe their dynamics we also have to study what happens at their endpoints. In general, these endpoints are confined to submanifolds of the target space $\mathbb{X}$ and the fields at the endpoints sit on certain bundles on these submanifolds.

These submanifolds equipped with bundles should be considered as dynamical objects in their own right. In physics they are known as *branes* and depending on the model of string theory they come in different flavors. Just like for closed string theory we can try to wrap this into a categorical formalism [16]:

- the branes are the objects of a category,
- intersections between branes give rise to morphisms between them,
- the multiplication is dictated by the way the strings can move between these intersections.

For a fixed target space $(\mathbb{X}, J, \omega)$ we get two A_∞-categories of branes, one for the A-model and one for the B-model.

- In the *A-model* one works with symplectic geometry and the branes will be Lagrangian submanifolds. These are maximal subspaces $\mathbb{L} \subset \mathbb{X}$ such that $\omega_{\upharpoonright\mathbb{L}} = 0$. The corresponding category is called the *Fukaya category* and its mathematical definition will be the subject of Chapter 6.
- In the *B-model* the submanifolds will be complex submanifolds. Further considerations show that one also must allow for certain bundles on these submanifolds. This leads to the notion of the *derived category of coherent sheaves* and its mathematical definition will be the subject of Chapter 7.

How do these brane categories relate to the conformal field theories we discussed before? From a physical point of view it is natural to stick open strings together to form closed strings, so in principle it should be possible to construct the closed string theory from the open string version. In [58], Costello shows that this is indeed the case. For each set Λ he constructs three dg-categories: a category $\mathtt{O}^\bullet_\Lambda$ that models open strings with endpoints on objects in Λ, a mixed version $\mathtt{OC}^\bullet_\Lambda$ that includes both open and closed strings, and a subcategory $\mathtt{C}^\bullet$ that only models the closed strings. The last one is quasi-equivalent to $\mathtt{HCFT}^\bullet$ and does not depend on Λ:

$$\mathtt{O}^\bullet_\Lambda \subset \mathtt{OC}^\bullet_\Lambda \supset \mathtt{C}^\bullet .$$

If we have a representation of $\mathtt{O}^\bullet_\Lambda$, it can be extended to a representation of $\mathtt{OC}^\bullet_\Lambda$ and then restricted to a representation of $\mathtt{C}^\bullet$. The latter can then be interpreted as a cohomological conformal field theory. Costello also shows that specifying a representation of $\mathtt{O}^\bullet_\Lambda$ is the same as constructing a (cyclic) A_∞-category whose underlying set of objects is Λ. Furthermore, one can show that derived equivalent A_∞-categories will give equivalent representations of $\mathtt{C}^\bullet$.

Observation 5.15 *Open string theories are determined by (cyclic) A_∞-categories and derived equivalent A_∞-categories give rise to equivalent closed string theories.*

5.5 What Is Mirror Symmetry?

In the 1980s and 1990s, physicists noticed that certain calculations in the A-model on a Kähler manifold $\mathbb{X}$ matched calculations for the B-model on a different manifold $\mathbb{X}^\vee$. Therefore, they conjectured that for every A-model there should be a B-model on a different manifold that gives a physically equivalent theory. This new phenomenon became known as mirror symmetry and the pair $(\mathbb{X}, \mathbb{X}^\vee)$ is called a mirror pair. The origin of this name lies in the fact that a certain set of numerical invariants, the *Hodge diamond*, of a mirror pair are each others mirror image. This can be seen in the example below, which shows the Hodge diamonds of the quintic threefold and its mirror.

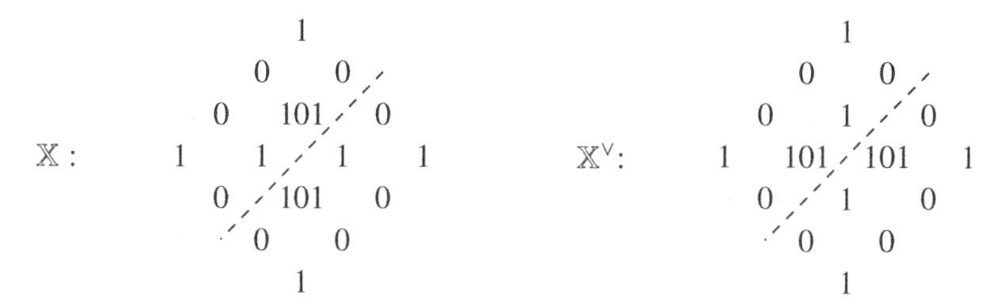

From a mathematical point of view this implied that certain questions in symplectic geometry could be answered by doing complex geometry on a different manifold and vice versa. At first this idea seemed crazy to most mathematicians, but when the physicists Candelas, de la Ossa, Green and

Parker [51] used it to count curves on the quintic threefold, they started to pay attention. These numbers, which can be defined rigorously using the theory of Gromov–Witten invariants, were notoriously hard to determine [75], but using mirror symmetry Candelas et al. could relate them to certain integrals – called periods – on the mirror manifold and find a generating series for them.

Many different ways were proposed to formalize mirror symmetry. Initially these formulations were framed in a classical geometrical framework using the standard tools from differential, symplectic and complex and enumerative geometry [92, 158]. In 1994, Kontsevich [141] proposed to look at it from a categorical point of view and this gave rise to the idea of *homological mirror symmetry*.

The idea is that mirror symmetry should be a derived equivalence between two A_∞-categories: one for the A-model, which goes by the name of the Fukaya category, and one for the B-model, which is the derived category of coherent sheaves. The rationale behind this idea is that because a superstring theory can be seen as representation theory of a certain A_∞-category, derived equivalent A_∞-categories will lead to the same physics. Mathematically this equivalence of categories then would imply that certain numerical invariants of these categories are the same and these invariants could then be interpreted in the context of symplectic or algebraic geometry, thus recovering the more classical statements of mirror symmetry:

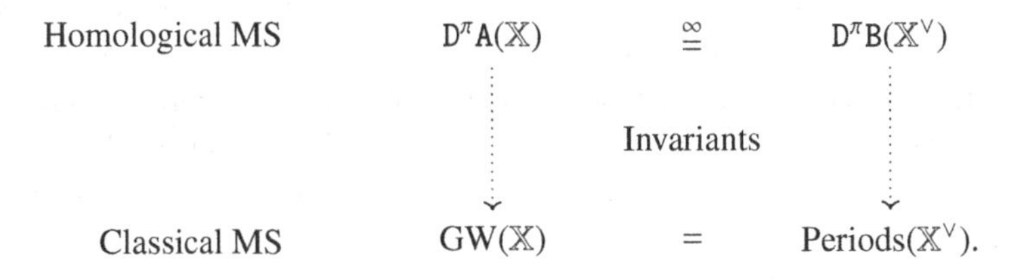

Originally mirror symmetry was only formulated for a special class of Kähler manifolds: compact Calabi–Yau manifolds. It gradually became clear that the principle could also be extended to other geometrical settings where the spaces involved could be non-Calabi–Yau [5, 139, 49], noncompact [1], nonsmooth [199] or even noncommutative [11]. To cope with these new spaces, several generalizations of the A-model and the B-model have been constructed.

In the next two chapters we will look at the constructions of these A_∞-categories. First we will discuss the A-model and look at various versions of Fukaya categories. For the B-model we will define the derived category of coherent sheaves as an A_∞-category and look at categories of matrix factorizations. In the final chapter of this part we will give some examples of equivalences between A-models and B-models that illustrate the phenomenon of homological mirror symmetry. We will discuss how to find mirror pairs $(\mathbb{X}, \mathbb{X}^\vee)$ using ideas from toric and tropical geometry and we will briefly sketch how this ties into two major research directions in this area: the SYZ-conjecture [237] and the Gross–Siebert program [104, 105].

5.6 Exercises

Exercise 5.1 Let $\mathbb{M} = \mathbb{R}^2$ and $\gamma\colon [0,1] \to \mathbb{M}$ be a path. The classical action for a free moving particle is $S(\gamma) = \int_0^1 \dot{\gamma}^2(t)\, dt$. Show that the curve $\gamma \in \mathcal{X}(x_0 \to x_1)$ that minimizes the action moves at a constant speed in a straight line from x_0 to x_1.

Exercise 5.2 Let $\mathbb{M} = \mathbb{R}^2$ and let $I_3 = \{0, \frac{1}{2}, 1\}$. The space of all discrete paths is

$$\mathcal{X}_N(x_0 \to x_1) := \{\gamma\colon I_3 \to \mathbb{M} \mid \gamma(0) = x_0,\ \gamma(1) = x_1\} = \mathbb{M},$$

and the action on this space is

$$S(\gamma) = r\Big(\big(x_1 - \gamma(\tfrac{1}{2})\big)^2 + \big(\gamma(\tfrac{1}{2}) - x_0\big)^2\Big),$$

where r is a constant. Show that the expectation value of $\gamma(\frac{1}{2})$ in the path integral formalism is $(x_0 - x_1)/2$. You may assume that $r \in \mathbb{C}$ is chosen in such a way that the relevant integrals converge.

Exercise 5.3 Let $\mathbb{M} = \mathbb{C}$ and look at the space of closed curves $\mathcal{X} = \{\gamma\colon [0,1] \to \mathbb{M} \mid \gamma(0) = \gamma(1)\}$. For every vector field On $\mathcal{X}$ we have a one-parameter family of symmetries $\exp(\epsilon Q)(\gamma)(t) = \gamma(t{+}\epsilon \bmod 1)$ that correspond to changing the starting point of the curve. Let $X_u := \gamma(u)$ be the observable that maps the curve γ to its location at time u. What is $d_Q X_u$?

Exercise 5.4 Let $\mathfrak{g} = \mathbb{R}g_1 + \cdots + \mathbb{R}g_k$ be a Lie algebra with an action on a space $\mathcal{X}$. Consider the tensor product

$$K^\bullet := C_\infty(\mathcal{X}) \otimes \Lambda^\bullet(\mathfrak{g}) = C_\infty(\mathcal{X}) \otimes \frac{\mathbb{C}\langle g_1, \ldots, g_k\rangle}{\langle g_i g_j + g_j g_i \mid 1 \le i, j \le k\rangle}$$

with the differential $d\phi\, g_{i_1} \ldots g_{i_r} = \sum_j (g_i \cdot \phi) g_i \wedge g_{i_1} \wedge \cdots \wedge g_{i_r}$. Show that this is a complex and

$$H^0(K) = \{f \in C_\infty(\mathcal{X}) \mid \forall\, g \in \mathfrak{g},\ g \cdot f = 0\}.$$

Show that if the action on $\mathcal{X}$ is free (i.e. at every point $x \in \mathcal{X}$ the vector fields corresponding to the g_i are linearly independent) then $H^1(K) = 0$ (in fact all higher cohomology is zero).

Exercise 5.5 Show that the action $S = \frac{1}{2}V(x)^2 + \psi_+\psi_-\frac{d}{dx}V(x)$ is indeed invariant under the supersymmetries $Q_\pm$.

Exercise 5.6 Consider a system with two bosonic fields x, H and two fermionic fields ψ_+, ψ_-. Show that the action

$$S = \frac{1}{2}H^2 + iHV(x) - \psi_+\psi_-\frac{d}{dx}V(x)$$

has a supersymmetry of the form

$$\delta x = \psi_+ + \psi_-, \quad \delta\psi_\pm = \pm iH, \quad \delta H = 0,$$

and show that $\delta^2 = 0$ and that $S = \delta F$ for some expression F.

Exercise 5.7 Show that every two-dimensional orientable Riemannian manifold can be seen as a Kähler manifold, where J corresponds to anticlockwise rotation of 90°.

Exercise 5.8 The topological cobordism category Cob_1 has as objects signed sets of points. These are finite sets of points each with a + sign or a − sign attached to it. A morphism consists of a set of oriented open curves that start at a positive point of S_1 or a negative point of S_2 and end at a positive point of S_2 or a negative point of S_1, together with a number of closed curves. Two morphisms are the same if there is a homeomorphism between them that preserves the endpoints.

(i) Let $S = \{\,\}$. Show that $\mathsf{Cob}_1(S, S), \circ \cong \mathbb{N}, +$.
(ii) Let $S = \{p_{1+}, \ldots, p_{k+}\}$. Show that $\mathsf{Cob}_1(S, S), \circ$ is isomorphic to the product of $\mathbb{N}$ with the symmetric group S_k.
(iii) Let $S = \{p_+, q_-\}$. Describe the semigroup $\mathsf{Cob}_1(S, S), \circ$.

Exercise 5.9 A functor $\mathcal{F}\colon \mathsf{Cob}_1 \to \mathtt{vect}(\Bbbk)$ is monoidal if it transforms the disjoint union of pointed sets naturally into the tensor product:

$$\mathcal{F}(S_1 \cup S_2) \cong \mathcal{F}(S_1) \otimes \mathcal{F}(S_2).$$

Similarly, it transforms disjoint unions of curves into tensor products of maps. For a precise definition see [246].

(i) Show that $\mathcal{F}$ is determined by the images of $^{+}_{-}\!\big)$ and $\big(\,^{+}_{-}$.

(ii) Show that the image of $\overset{+}{\underset{-}{)}}$ is a nondegenerate pairing between $V_+ :=$ $\mathcal{F}(+)$ and $V_- := \mathcal{F}(-)$. Conclude that there is precisely one 1d-TQFT for every dimension $n \in \mathbb{N}$.

(iii) In the unoriented cobordism category the objects are just sets of points and the morphisms unoriented curves between these points. Describe this category and its monoidal representations.

Exercise 5.10 Let $\mathcal{F}$ be a 2d-TQFT. Show that the multiplication in the corresponding Frobenius algebra is associative and commutative. Furthermore, show that the pairing $(a, b) \mapsto \mathrm{Tr}(a \cdot b)$ is nondegenerate.

6
The A-Side

The A-model in homological mirror symmetry deals with symplectic geometry, and its main features are captured by an A_∞-category that describes the intersection theory of Lagrangian submanifolds in a symplectic manifold.

This category is called the *Fukaya category* and its construction is quite technical. Therefore we will start with a warm-up on Morse theory. Then we will outline a rough road map to the construction of a basic version of the Fukaya category and describe what kind of problems may be encountered along the way. We will not describe in detail how all these problems can be resolved, but only indicate how they can be approached.

After that we will show how one can extend this basic version to include extra objects and how to adapt it to more complicated symplectic manifolds. We will give some examples for each of the different flavors of Fukaya categories that can be constructed and illustrate their differences in the most simple case: surfaces. We end the chapter with a discussion on how to find generators for these categories.

For a nice introduction to Morse homology we refer to the book by Banyaga and Hurtubise [28] and the lecture notes by Hutchings [123]. The lecture notes by Salamon [212] offer a good introduction to Floer homology. For an overview of Fukaya categories we refer to the beginner's guide by Auroux [20], the survey article by Smith [227], lecture notes by Chantraine [54] and the book by Seidel [218]. More technical details concerning the definitions of Fukaya categories are covered in the books by Fukaya, Oh, Ohta and Ono [83, 84].

6.1 Morse Theory

6.1.1 The Morse Complex

Our starting point is a compact oriented n-dimensional Riemannian manifold $(\mathbb{M}, g)$ and a smooth function $f\colon \mathbb{M} \to \mathbb{R}$. We can use the metric to define the *gradient* of f. This is the vector field ∇f with the property that $g(\nabla f, -) = df(-)$. Geometrically this vector field points in the direction where f increases the most. The points of $\mathbb{M}$ where $\nabla f = 0$ are called the *critical points* of f. To each critical point p we can associate its Hessian, which in local coordinates is the matrix of second derivatives

$$H(f)_p = \begin{pmatrix} \frac{\partial^2 f(p)}{\partial x_1 \partial x_1} & \cdots & \frac{\partial^2 f(p)}{\partial x_1 \partial x_n} \\ \vdots & & \vdots \\ \frac{\partial^2 f(p)}{\partial x_n \partial x_1} & \cdots & \frac{\partial^2 f(p)}{\partial x_n \partial x_n} \end{pmatrix}.$$

Under a coordinate change this matrix changes as JHJ^T, so we can diagonalize it to a matrix with $0, \pm 1$ on the diagonal.

Definition 6.1 A critical point is called *simple* if there are no zeros on the diagonal and the *Morse index* of a simple critical point $\operatorname{ind} p$ is the number of -1's on the diagonal. Note that the index of a local maximum is n, while that of a local minimum is 0.

The vector field ∇f induces a flow $\phi_t\colon \mathbb{M} \to \mathbb{M}$ such that $\frac{d\phi_t}{dt} = \nabla f$. To each singular point p we can assign its *ascending* and *descending* submanifolds

$$\begin{aligned} \mathsf{Asc}(p) &= \{x \in \mathbb{M} \mid \lim_{t \to +\infty} \phi_t(x) = p\} \\ \text{and} \quad \mathsf{Desc}(p) &= \{x \in \mathbb{M} \mid \lim_{t \to -\infty} \phi_t(x) = p\}. \end{aligned}$$

With this notation, $\mathsf{Desc}(p) \cap \mathsf{Asc}(q)$ is the union of all flow paths from p to q.

Lemma 6.2 *The Morse index of a simple critical point p is equal to the dimension of the ascending manifold or the codimension of the descending manifold.*

Sketch of the proof Let p be a simple critical point of f with index i. We can find a coordinate transformation such that locally around p the function looks like $f = -x_1^2 + \cdots - x_i^2 + x_{i+1}^2 + \cdots + x_n^2$. For this function the ascending manifold is an i-dimensional disk containing all points for which the coordinates $x_{i+1}, \ldots, x_n$ are zero. □

Definition 6.3 A function is called a *Morse function* if all its critical points are simple. It is called a *Morse–Smale function* if additionally every ascending manifold intersects every descending manifold transversally:

$$\dim \mathsf{Desc}(p) \cap \mathsf{Asc}(q) = \dim \mathsf{Desc}(p) + \dim \mathsf{Asc}(q) - n = \operatorname{ind} q - \operatorname{ind} p.$$

We denote the set of critical points with index i by $\mathsf{Crit}_i(f)$.

Remark 6.4 The Morse–Smale functions form a dense subset of all smooth functions, so every function can be tweaked such that it becomes Morse–Smale.

Example 6.5 Take a torus lying flat on the table and use the height function f. The critical points are the top and bottom circles, so f is not a Morse function. If we put it upright on the table then the height function is Morse and it has four critical points: the bottom with index 0, the top with index 2, and the top and the bottom of the inner circle with index 1.

This function is however not Morse–Smale because although the last two points both have index 1, there are two trajectories between them – the two halves of the inner circle – so $\dim \mathsf{Desc}(p) \cap \mathsf{Asc}(q) = 1 \neq 1 - 1$. If we tilt the torus a little, these two trajectories will disappear and the height function will become Morse–Smale.

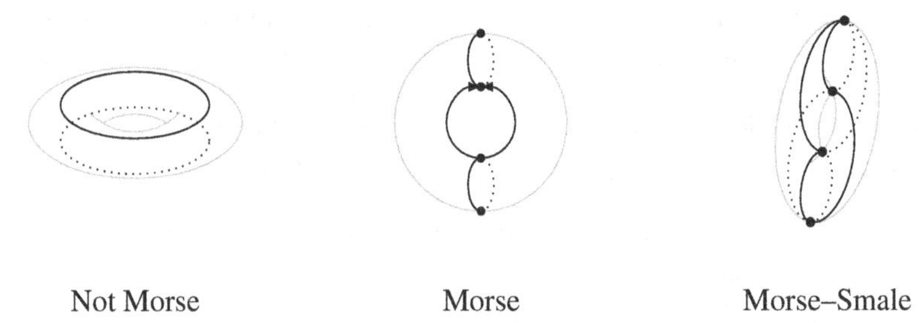

Not Morse Morse Morse–Smale

Now suppose $f: \mathbb{M} \to \mathbb{R}$ is Morse–Smale. If the indices of two critical points p, q differ by 1 then there are only finitely many flow trajectories between them because $\dim \mathsf{Desc}(p) \cap \mathsf{Asc}(q) = 1$. Each can be given a sign in the following way. Choose orientations for all the $\mathsf{Desc}(p)$. This automatically gives orientations to the $\mathsf{Asc}(p)$ if we demand that the split $\mathsf{T}_p\mathbb{M} = \mathsf{T}_p\mathsf{Desc}(p) \oplus \mathsf{T}_p\mathsf{Asc}(p)$ preserves the orientation. In a similar way we can also orient $\mathsf{Desc}(p) \cap \mathsf{Asc}(q)$ by demanding that there is a positive basis $b_1, \ldots, b_n$ in $\mathsf{T}_x\mathbb{M}$ for which the first vectors give a positive basis on $\mathsf{T}_x\mathsf{Desc}(p)$, the last on $\mathsf{T}_x\mathsf{Asc}(q)$ and the overlap on $\mathsf{T}_x\mathsf{Desc}(p) \cap \mathsf{Asc}(q)$.

Definition 6.6 For a Morse–Smale function f on a compact orientable Riemannian manifold $(\mathbb{M}, g)$, the *Morse complex* $\mathsf{CM}^\bullet(\mathbb{M}, g, f)$ is the vector space spanned by the critical points and graded by the index.

The differential of p is the signed sum of all endpoints of trajectories from p to a critical point with index one higher:

$$\mathsf{CM}^i(M, g, f) = \Bbbk\mathsf{Crit}_i(f) \quad \text{and} \quad dp = \sum_{\substack{\gamma:\, p \to q \\ |q|=|p|+1}} (-1)^\gamma q.$$

The sign $(-1)^\gamma$ is $+1$ if γ runs in the positive direction of $\mathsf{Desc}(p) \cap \mathsf{Asc}(q)$.

Theorem 6.7 *The Morse complex is a cochain complex and its cohomology is the cohomology of* $\mathbb{M}$.

Sketch of the proof The differential clearly has degree 1 and it also squares to zero. If we work out the formula for d^2 we get

$$d^2(p) = \sum_{\substack{\gamma_1:\, p \to q \\ |q|=|p|+1}} \sum_{\substack{\gamma_2:\, q \to r \\ |r|=|q|+1}} (-1)^{\gamma_1}(-1)^{\gamma_2} r,$$

so the square of the differential counts the ways to go to r via an intermediate.

Because the difference between the indices of p and r is 2, the intersection between $\mathsf{Asc}(p)$ and $\mathsf{Desc}(q)$ will be two-dimensional. This means that the trajectories from r to p form one-parameter families. At the boundaries of these one-parameter families, the trajectories break into two.

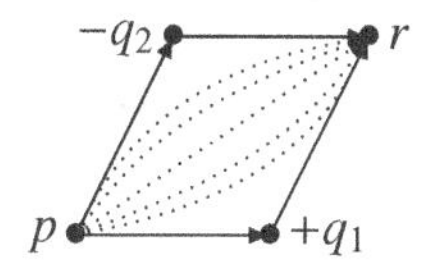

Vice versa, if we have two trajectories of degree 1 it is locally possible to deform them to a trajectory of degree 2. Each one-parameter family that breaks into two has two ends, so there are always two contributions with opposite signs that cancel each other.

To show that the cohomology of the Morse complex equals the cohomology of $\mathbb{M}$, associate to each critical point with index i its ascending manifold. Because all the ascending manifolds are contractible and cover $\mathbb{M}$, this gives a cellular decomposition, which we can use to define cellular cohomology in the same way as we define simplicial cohomology [111, Chapter 3]. This complex is isomorphic to the Morse complex [28, 7.4].

There are many other ways to prove this. A version of the argument using homology instead of cohomology can be found in [123, Chapter 3]. A completely different approach was given by Witten [256] and relates the Morse complex to the de Rham complex. □

Example 6.8 We calculate the Morse complex of a sphere with camel humps that is standing upright. There are two paths from the saddle point to the two humps, and there are two paths from the minimum to the saddle point. The latter two occur with opposite sign so they cancel out:

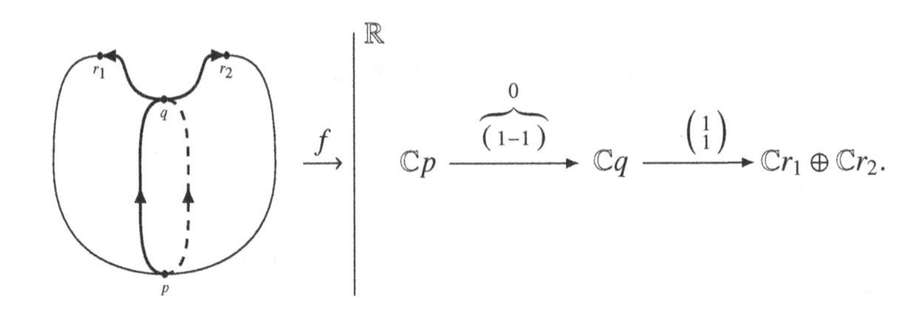

Observation 6.9 *We can recover the homology of a manifold by looking at a complex of critical points of a function on it. The differential of this complex counts flow trajectories.*

6.1.2 The Dynamics of Morse Theory

We are now going to reformulate Morse theory using symplectic geometry. For more background on symplectic geometry we refer to [52, 182]. Remember that a symplectic manifold is a $2n$-dimensional smooth real manifold $\mathbb{X}$ equipped with a closed nondegenerate 2-form ω:

$$\underbrace{d\omega = 0}_{\text{closed}} \quad \text{and} \quad \underbrace{\omega(X, -) = 0 \implies X = 0.}_{\text{nondegenerate}}$$

Given a smooth function h we can define a vector field X_h such that $dh = \iota_{X_h}\omega = \omega(X_h, -)$ and a flow $\phi_t^h \colon \mathbb{X} \to \mathbb{X}$ such that $\frac{d\phi_t^h(x)}{dt} = X_h(\phi_t^h(x))$. The function is usually called the *Hamiltonian* and the flow the *Hamiltonian flow*. This flow preserves the symplectic form and the Hamiltonian:

$$\frac{d}{dt}\phi_t^{h*}\omega = \mathcal{L}_{X_h}\omega = \iota_{X_h}d\omega + d\iota_{X_h}\omega = 0 + d^2h = 0,$$
$$\frac{d}{dt}\phi_t^{h*}h = \mathcal{L}_{X_h}h = dh(X_h) = \omega(X_h, X_h) = 0.$$

From a physical point of view the Hamiltonian represents the total energy of the system and the fact that h is invariant under the flow means that energy is conserved.

The standard example of a symplectic manifold is the cotangent space of a manifold $\mathsf{T}^*\mathbb{M}$ equipped with the 2-form

$$\omega = \sum_i dp_i \wedge dq^i,$$

where the q^i are local coordinates in $\mathbb{M}$ and the p_i are the coordinates in the cotangent space according to the basis $\{dq^i\}$. The q^i are called the position coordinates and the p_i the momentum coordinates. In this coordinate system the vector field X_h has the form

$$X_h = \sum_i \frac{\partial h}{\partial p_i} \cdot \frac{\partial}{\partial q^i} - \sum_i \frac{\partial h}{\partial q^i} \cdot \frac{\partial}{\partial p_i}.$$

If f is a function on a Riemannian manifold $(\mathbb{M}, g)$ we can pull it back to a function on $\mathsf{T}^*\mathbb{M}$ along the projection map. The Hamiltonian flow for this function is

$$X_f = -\sum_i \frac{\partial f}{\partial q^i} \cdot \frac{\partial}{\partial p_i}.$$

From a physical point of view this dynamical system is rather strange: the momentum of the particle will change but its position will not. This is because f contains a potential term (one involving the coordinates q^i) but no kinetic term (one involving the coordinates p_i). The flow ϕ_t^h moves the base space $\mathbb{M} \subset \mathsf{T}^*\mathbb{M}$ to another n-dimensional submanifold $\phi_t^h\mathbb{M} \subset \mathsf{T}^*\mathbb{M}$. The critical points of f are precisely the intersection points between $\phi_t^h\mathbb{M}$ and $\mathbb{M}$ because there the flow is stationary: $X_f(x) = 0$.

If we flow a Morse trajectory γ between two critical points, it will sweep out a surface with boundary on $\mathbb{M} \cup \phi_t^h\mathbb{M}$. This surface is very special: it is a holomorphic disk. To make sense of this statement we have to define a complex structure on $\mathsf{T}^*\mathbb{M}$. The tangent bundle $\mathsf{T}\mathbb{M}$ comes with the natural structure of a complex manifold. If the q^i are local coordinates for M we have local coordinates q^i, $\dot{q}^i$ for $\mathsf{T}\mathbb{M}$ according to the local basis $\frac{\partial}{\partial q^i}$. If we combine these as

$$z^i = q^i + i\dot{q}^i,$$

we get holomorphic coordinates for $\mathsf{T}\mathbb{M}$. If $\mathbb{M}$ has a metric g, there is a diffeomorphism between the tangent and cotangent bundles:

$$\phi_g \colon \mathsf{T}\mathbb{M} \to \mathsf{T}^*\mathbb{M} \colon X \mapsto g(X, -).$$

Under this diffeomorphism the coordinates p_i and $g_{ij}\dot{q}^j$ correspond. Here we used the Einstein summation convention and $g_{ij} := g(\frac{\partial}{\partial_i}, \frac{\partial}{\partial_j})$.

Lemma 6.10 *The map*

$$\psi\colon (-\infty,\infty) + [0,\epsilon]i \to \mathsf{T}^*\mathbb{M}\colon s + ti \to \phi_t^h \gamma(s)$$

is holomorphic.

Sketch of the proof We have to check the Cauchy–Riemann equations for ψ. The derivative in the s-direction is equal to $(\nabla f, 0)$ because it is the Morse flow for the q^i coordinates and does not change the p^i. The derivative in the t-direction only changes the p_i coordinates so it is $(0, -\frac{\partial f}{\partial q_i})$. If we rewrite this in $(q_i, \dot{q}_i)$ coordinates we get

$$\frac{\partial \psi}{\partial s} = (\nabla f, 0) \quad \text{and} \quad \frac{\partial \psi}{\partial t} = (0, -\nabla f) = -i\frac{\partial \psi}{\partial s}.$$

□

We can identify the strip $(-\infty,\infty)+[0,\epsilon]i$ holomorphically with the unit disk $\mathbb{D} = \{z \in \mathbb{k} \mid |z| \leq 1\}$ with the points $\{1,-1\}$ removed, so ψ can be seen as a holomorphic map

$$\psi\colon \mathbb{D} \setminus \{-1, 1\} \to \mathsf{T}^*\mathbb{M}$$

such that $\psi(e^{2\pi i u}) \in \mathbb{M}$ if $u \in [0, \frac{1}{2}]$ and $\psi(e^{2\pi i u}) \in \phi_\epsilon^h \mathbb{M}$ if $u \in [\frac{1}{2}, 1]$. We will call this a holomorphic 2-gon or digon.

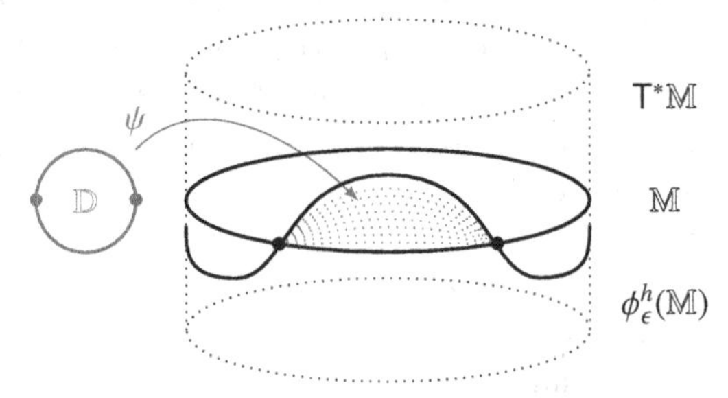

From this new point of view, the Morse cochain complex is spanned by intersection points between $\mathbb{L} := \phi_\epsilon^h \mathbb{M}$ and $\mathbb{M}$, while the differential counts the number of holomorphic 2-gons:

$$\mathsf{CM}^\bullet(\mathbb{M}, g, f) = \mathsf{CF}^\bullet(\mathbb{L}, \mathbb{M}) := \bigoplus_{p \in \mathbb{L}\cap\mathbb{M}} \mathbb{k}p \quad \text{and} \quad dp = \sum_{\substack{q \in \mathbb{L}\cap\mathbb{M} \\ |q| = |p|+1}} \sum_{p \overset{\psi}{\frown} q} \pm q.$$

What is the interpretation of the Morse index in this new formalism? First note that if $\mathbb{M}$ is an orientable Riemannian manifold we can associate to $\mathbb{M}$ a volume form

$$\Omega_{\mathbb{R}} = \sqrt{\det g_{ij}}\, dq^1 \wedge \cdots \wedge dq^n.$$

This volume form can be transported to a complex volume form on $\mathbb{TM}$,

$$\Omega_{\mathbb{C}} = \sqrt{\det g_{ij}}\, dz^1 \wedge \cdots \wedge dz^n.$$

If we look at $\Omega_{\mathbb{C}}|_{\mathsf{T}_p\mathbb{L}}$ then the image of this map will be a line $\mathbb{R}e^{i\theta_p} \subset \mathbb{C}$, where θ_p is an angle in $\mathbb{R}/\pi\mathbb{Z}$. In this particular case it is actually possible to lift this to a function $\theta^{\mathbb{L}}\colon \mathbb{L} \to \mathbb{R}$. This lift function is called a grading. Note that $\mathbb{M}$ itself comes with a grading $\theta^{\mathbb{M}}\colon \mathbb{M} \to \mathbb{R}\colon p \mapsto 0$ because $\Omega_{\mathbb{C}}|_{\mathsf{T}_p\mathbb{M}}$ is real valued.

If p is an intersection point between $\mathbb{L}$ and $\mathbb{M}$ then there is a unitary transformation $U\colon \mathsf{T}(\mathsf{T}^*_p\mathbb{M}) \to \mathsf{T}(\mathsf{T}^*_p\mathbb{M})$ that maps $\mathsf{T}_p\mathbb{L}$ to $\mathsf{T}_p\mathbb{M}$. Take $\Delta \in [0,\pi)$ such that $\det U = \pm e^{i\Delta}$. We define the degree of the intersection point p as

$$\deg p = \frac{\Delta - (\theta^{\mathbb{M}} - \theta^{\mathbb{L}})}{\pi}.$$

Note that because $U^*\Omega_p = \det(U)\Omega_p$ the phase $\theta^{\mathbb{M}} - \theta^{\mathbb{L}}$ must be in $\Delta + \pi\mathbb{Z}$, so $\deg p$ is an integer.

Example 6.11 Let $\mathbb{M}$ be the circle $\mathbb{R}/2\pi\mathbb{Z}$ with the standard metric. For the Hamiltonian we take $h = \cos q$, which has two critical points p_0, p_1 at $q = 0$ and $q = \pi$. The symplectic manifold $\mathsf{T}^*\mathbb{M}$ is a cylinder and the trajectories are vertical lines except for $q = 0, \pi$, where the trajectories are points.

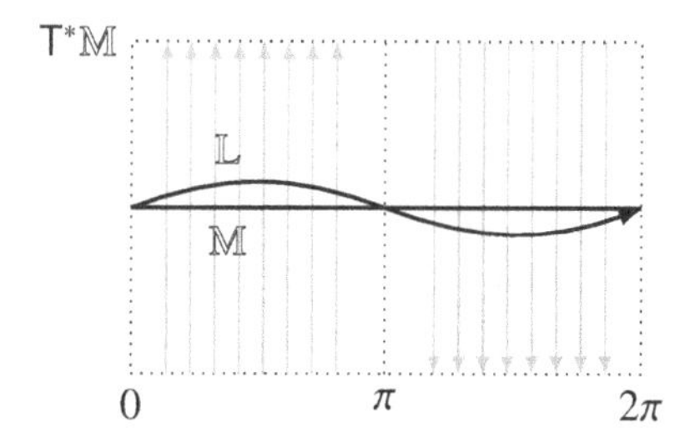

In the picture are two Lagrangian submanifolds, $\mathbb{M}$ and its flowed version $\mathbb{L} = \phi^h_\epsilon(\mathbb{M})$. For the function $\theta^{\mathbb{L}}$ we can take the angle that $\mathsf{T}_p\mathbb{L}$ makes at each point, so $\theta_{p_0} = \epsilon$ and $\theta_{p_1} = -\epsilon$.

The unitary matrices that map $\mathsf{T}_p\mathbb{L}$ to $\mathsf{T}_p\mathbb{M}$ are

$$U_0 = e^{i(\pi-\epsilon)} \quad \text{and} \quad U_1 = e^{i\epsilon},$$

so the degrees become

$$\deg p_0 = \frac{\pi - \epsilon - (0 - \epsilon)}{\pi} = 1 \quad \text{and} \quad \deg p_1 = \frac{\epsilon - (0 + \epsilon)}{\pi} = 0.$$

Observation 6.12 *We can recover the homology of a manifold by looking at a complex of intersection points between the cotangent section $\mathbb{M} \subset \mathsf{T}^*\mathbb{M}$ and a*

Hamiltonian deformation. The differential counts holomorphic strips between intersection points.

6.1.3 The Homology of the Loop Space

Let us now have a look at the other end of the dynamical spectrum, when the Hamiltonian consists of kinetic energy only:

$$h = g^{ij} p_i p_j.$$

This describes a particle freely moving on $\mathbb{M}$. From classical mechanics we know that such a particle follows a geodesic line on $\mathbb{M}$ with a constant speed. On $\mathbb{M} \subset \mathsf{T}^*\mathbb{M}$ we have $p_i = 0$ so the speed is zero and nothing happens: $\phi^h_\epsilon(\mathbb{M}) = \mathbb{M}$.

If we take a fiber of the cotangent bundle $\mathbb{L} = \mathsf{T}^*_x\mathbb{M}$ and identify it with $\mathsf{T}_x\mathbb{M}$ using the metric then the image of a point $(q^i, \dot{q}^i)$ under ϕ^h_t will lie on a geodesic line in the $\dot{q}^i$-direction at a distance $t\|\dot{q}^i\|$.

With this in mind, the intersection points between $\phi^h_\epsilon(\mathbb{L})$ and $\mathbb{L}$ represent closed geodesics that start and end at x. If we have such a simple closed geodesic $\gamma\colon [0,1] \to \mathbb{M}$ of length ℓ then for any $z \in \mathbb{Z}$ the point $\left(x, \frac{z\ell\dot{\gamma}(0)}{\epsilon\|\dot{\gamma}(0)\|}\right)$ will lie in $\phi^h_\epsilon(\mathbb{L}) \cap \mathbb{L}$. All intersection points arise in this way.

Consider the space of all loops in $\mathbb{M}$ with base point x:

$$\mathcal{L}_x(\mathbb{M}) := \left\{\gamma\colon \mathbb{S}^1 \to \mathbb{M} \mid \gamma(0) = x\right\}.$$

On this loop space we have a function $f\colon \mathcal{L}_x(\mathbb{M}) \to \mathbb{R}\colon \gamma \mapsto \int_{\mathbb{S}^1} \|\dot{\gamma}(t)\|\, dt$ that assigns to each loop its length. The geodesic loops are precisely the critical points of this function; it makes sense to construct a Morse-like complex that calculates the homology of $\mathcal{L}_x(\mathbb{M})$. Again it should be spanned by the intersection points between $\mathbb{L}$ and $\phi^h_t(\mathbb{L})$. But now these intersection points do not come from stationary trajectories but from cyclic ones. With each intersection point u there is a flow curve $\tilde{v}\colon [0,\epsilon] \to \mathsf{T}^*\mathbb{M}$ that starts and ends at u. Therefore, the differential should count strips

$$\psi\colon (-\infty, \infty) + [0,\epsilon]i \to \mathsf{T}^*\mathbb{M}\colon s + ti \to \phi^h_t\gamma(s)$$

for which the limits at $s = \pm\infty$ are precisely such flow curves. As in the Morse case, these strips should satisfy an appropriate version of the Cauchy–Riemann equations. Note that it is not easy to make this work because Morse theory on infinite-dimensional spaces becomes very technical.

Example 6.13 Let $\mathbb{M} = \mathbb{S}^1$; then $\mathsf{T}^*\mathbb{M}$ is a cylinder. The canonical coordinates are (q, p) where q represents the angle on the circle and p the height on the cylinder. The vertical line at the origin is $\mathbb{L} = \mathsf{T}_0\mathbb{S}^1$ and $\phi_t^h(\mathbb{L})$ is a spiral that wraps around the cylinder with the equation $q = tp$. The intersection points are given by $(0, 2\pi k/t)$ with $k \in \mathbb{Z}$.

The loop space has one connected component for each winding number and in each component there is precisely one critical point so there can be no trajectories between the critical points. Therefore, the differential on the complex spanned by the intersection points between $\phi_t^h(\mathbb{L})$ and $\mathbb{L}$ is zero.

Observation 6.14 *The homology of the loop space of a manifold can be computed by looking at intersection points between a cotangent fiber $\mathsf{T}_x^*\mathbb{M}$ and a Hamiltonian deformation. The differential counts holomorphic strips between intersection points.*

6.2 The Basic Fukaya Category

The discussion above opens several possible directions for generalizations. We can try to develop an intersection theory for Lagrangian submanifolds in any symplectic manifold and we can upgrade the differential to higher products by looking at holomorphic n-gons. This leads to the idea of a Fukaya category, but there are many different notions of Fukaya categories, depending on the type of symplectic manifold and what kind of Lagrangian submanifolds we want to include. In this section we will define the most basic one and work out the easiest example.

6.2.1 The Setting

To get a sensible theory we should of course impose restrictions on the complexity of the geometry of the symplectic manifolds and the Lagrangian submanifolds. We will therefore start with a setting that stays quite close to the case of cotangent bundles.

Definition 6.15 A symplectic manifold $(\mathbb{X}, \omega)$ is called *exact* if there is a 1-form θ such that $\omega = d\theta$. The 1-form θ is the *Liouville form* and the vector field X_θ such that $\omega(X_\theta, -) = \theta$ is called the *Liouville vector field.*

The pair $(\mathbb{X}, \theta)$ is called a *Liouville manifold with cylindrical ends* if there is a compact manifold with boundary $\mathbb{X}_0$ such that

- $\mathbb{X}$ is made from $\mathbb{X}_0$ by sticking cylinders on the boundary:

$$\mathbb{X} = \mathbb{X}_0 \cup_{\partial\mathbb{X}_0} \partial\mathbb{X}_0 \times [1, +\infty),$$

- X_θ points outwards on the cylinder:

$$X_\theta|_{\partial\mathbb{X}_0\times[1,+\infty)} = r\frac{\partial}{\partial r},$$

where r is the coordinate on $(1, +\infty)$.

The compact manifold $\mathbb{X}_0$ is called the *Liouville domain* [15].

Example 6.16 The standard example of a Liouville manifold with cylindrical ends is the cotangent bundle of a compact manifold $\mathbb{X} = \mathsf{T}^*\mathbb{M}$. The Liouville form is $\theta = p_i\, dq_i$ and the corresponding Liouville vector field is $X_\theta = p_i\frac{\partial}{\partial p_i}$. For $\mathbb{X}_0$ we can take the cotangent unit ball $\mathbb{X}_0 = \{x \in \mathbb{X} \mid ||p|| \leq 1\}$ and for the coordinate r we can take $||p||$ (note that to define the norm of p we need a metric on $\mathbb{M}$).

Definition 6.17 A *Lagrangian submanifold* is an oriented n-dimensional submanifold $\mathbb{L}$ for which $\omega(\mathsf{T}\mathbb{L}, -) = 0$. It is called *exact* if there is a function $f : \mathbb{L} \to \mathbb{R}$ such that $\theta|_{\mathbb{L}} = df$ and it has *conical ends* if on the cylinders $\partial\mathbb{X}_0 \times [1, +\infty)$ the Liouville vector field X_θ is tangent to $\mathbb{L}$. We say that two Lagrangian submanifolds $\mathbb{L}_1, \mathbb{L}_2$ intersect *transversally* if for each intersection point $p \in \mathbb{L}_1 \cap \mathbb{L}_2$ we have $\mathsf{T}_p\mathbb{X} = \mathsf{T}_p\mathbb{L}_1 \oplus \mathsf{T}_p\mathbb{L}_2$.

Example 6.18 The base of the cotangent bundle is a compact exact Lagrangian submanifold, while the cotangent fiber at any point $x \in \mathbb{M}$ is an example of an exact Lagrangian submanifold with conical ends. In both cases we have that the restriction of θ is zero, so it is the differential of a constant function. The base and the cotangent fiber intersect transversally at x.

Definition 6.19 Given a $2n$-dimensional symplectic manifold $(\mathbb{X}, \omega)$ we define an *almost-complex structure* as a map $J : \mathsf{T}\mathbb{X} \to \mathsf{T}\mathbb{X}$ with $J^2 = -1$. We say that J is *compatible with* ω if $g(-, -) := \omega(-, J-)$ is a Riemannian metric for $\mathbb{X}$. The triple $(\mathbb{X}, \omega, J)$ is called an *almost-Kähler manifold*.

Remark 6.20 If $\mathbb{X}$ is a complex manifold (i.e. there are charts $\phi_U : U \to \mathbb{C}^n$ such that $\phi_U \circ \phi_{U'}^{-1}$ are holomorphic), we can see $\mathbb{X}$ also as a real manifold and the map $J : \mathsf{T}\mathbb{X} \to \mathsf{T}\mathbb{X} : z \to iz$ as an almost-complex structure. The reverse is in general not true: not every almost-complex structure comes from an atlas of complex charts. If it does we drop the "almost" from our definitions and speak of *complex structures* and *Kähler manifolds*.

Remark 6.21 For a given pair $(\mathbb{X}, \omega)$ the space of compatible almost-complex structures is contractible. The same is not true for the space of all complex structures.

Definition 6.22 For any open subset $\mathbb{U} \subset \mathbb{C}$ a map $\phi\colon \mathbb{U} \to \mathbb{X}$ is *pseudo-holomorphic* if it satisfies the Cauchy–Riemann equation:

$$\frac{\partial \phi(u+iv)}{\partial u} + J\frac{\partial \phi(u+iv)}{\partial v} = 0.$$

Example 6.23 The complex manifold $\mathbb{C}^{*n}$ can be identified with the Liouville manifold $(\mathsf{T}^*(\mathbb{S}^1)^n, p_i\, dq_i)$ via $z_i = e^{p_i + iq_i}$; this gives a complex structure on $\mathbb{X} = \mathsf{T}^*(\mathbb{S}^1)^n$. So $(\mathbb{X}, \omega, J)$ is a Kähler manifold with

$$\begin{aligned}
\omega &= d\theta = dp_i \wedge dq_i = \frac{i}{2} d\ln z_i \wedge d\ln \bar{z}_i = \frac{i}{2}\frac{dz_i}{z_i} \wedge \frac{d\bar{z}_i}{\bar{z}_i},\\
J &= i\frac{\partial}{\partial z_i} \otimes_{\mathbb{C}} dz_i = \frac{\partial}{\partial q_i} \otimes dp_i - \frac{\partial}{\partial p_i} \otimes dq_i,\\
g &= dp_i \otimes dp_i + dq_i \otimes dq_i.
\end{aligned}$$

.

If $f \in \mathbb{C}[X_1^{\pm 1}, \ldots, X_n^{\pm 1}]$ is a function such that

$$\mathbb{X}_f := \{(z_1, \ldots, z_n) \in \mathbb{C}^{*n} \mid f(z_1, \ldots, z_n) = 0\}$$

is a smooth submanifold of $\mathbb{C}^{*n}$ then $(\mathbb{X}_f, \theta|_{\mathbb{X}_f})$ is a Liouville manifold with cylindrical ends with a compatible complex structure.

6.2.2 A First Attempt at a Definition

In this section $(\mathbb{X}, \theta)$ will be a Liouville manifold with cylindrical ends. We will write ω for the symplectic form $d\theta$ and assume that it comes with a compatible almost-complex structure J and induced metric g.

Definition 6.24 Given two transversal compact exact Lagrangian submanifolds $\mathbb{L}_1, \mathbb{L}_2 \subset \mathbb{X}$, we define the Floer complex as the $\mathbb{Z}_2$-graded vector space

$$\mathsf{CF}^{\pm}_{\Bbbk}(\mathbb{L}_1, \mathbb{L}_2) := \bigoplus_{p \in \mathbb{L}_1 \cap \mathbb{L}_2} \Bbbk p.$$

To define the degree of p, take positively oriented bases $v_1, \ldots, v_n$ and $w_1, \ldots, w_n$ for $\mathsf{T}_p\mathbb{L}_1$ and $\mathsf{T}_p\mathbb{L}_2$. We set $\deg p = 0$ if $v_1, \ldots, v_n, w_1, \ldots, w_n$ forms a positively oriented basis for $\mathsf{T}_p\mathbb{X}$, and $\deg p = 1$ if it forms a negatively oriented basis.

Definition 6.25 If $\mathbb{L}_0, \ldots, \mathbb{L}_k$ are transversally intersecting exact compact oriented Lagrangian submanifolds then we define $\mu\colon \mathsf{CF}^{\pm}_{\Bbbk}(\mathbb{L}_1, \mathbb{L}_0) \otimes \cdots \otimes \mathsf{CF}^{\pm}_{\Bbbk}(\mathbb{L}_k, \mathbb{L}_{k-1}) \to \mathsf{CF}^{\pm}_{\Bbbk}(\mathbb{L}_k, \mathbb{L}_0)$,

$$\mu(p_1, \ldots, p_k) := \sum_{p_0 \in \mathbb{L}_0 \cap \mathbb{L}_k} \sum_{\psi} \sigma_\psi \cdot p_0, \tag{6.1}$$

where we sum over all pseudo-holomorphic maps $\psi\colon \mathbb{D} \setminus \{e^{\frac{2r\pi i}{k+1}} \mid r \in \mathbb{Z}\} \to \mathbb{X}$ such that

- the boundary maps to the Lagrangians: $\psi(e^{it\frac{2\pi}{k+1}}) \in \mathbb{L}_i$ if $t \in (i, i+1)$,
- the limits at the roots of unity $\zeta^r = e^{\frac{2r\pi i}{k+1}}$ correspond to the intersection points: $\lim_{t\to\xi_r} \psi(z) = p_r$,
- the maps are considered up to reparametrizations of $\mathbb{D}$ fixing the ζ^r,
- the maps ψ cannot be deformed,
- each term gets a sign σ_ψ depending on the orientation of the Lagrangians and possibly some extra data.

Pictorially we can represent such a map as

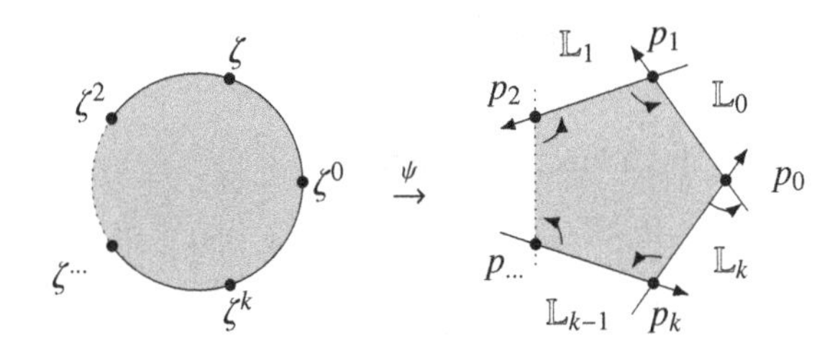

where we have added small arrows around the intersection points p_i to indicate the direction of the corresponding morphisms. We have also drawn the target with corners, to stress the fact that ψ does not need to be holomorphic at the ζ^i.

In the case where the Lagrangians are one-dimensional, σ_ψ has a factor of -1 for each odd degree p_i, $i > 0$ whose target Lagrangian runs against the positive direction on the disk.

We would like these products to be well defined and to satisfy the A_∞-relations, but in order for that we need to resolve some technical issues.

6.2.3 Moduli Spaces

First we want to make more precise what we mean by saying that the map ψ cannot be deformed. For each sequence of intersection points $p_0, \dots, p_k$ we can construct a moduli space $\mathcal{M}(p_0, \dots, p_k)$ classifying all maps satisfying the relevant boundary conditions up to reparametrization. This moduli space naturally splits as a disjoint union according to the homology class of ψ relative to the boundary $\cup_i \mathbb{L}_i$:

$$\mathcal{M}(p_0, \dots, p_k) = \bigsqcup_{u \in \mathsf{H}_2(\mathbb{X}, \cup_i \mathbb{L}_i)} \mathcal{M}_u.$$

Ideally, we want each of the $\mathcal{M}_u$ to be a smooth orientable manifold of the correct expected dimension; however, this is not easy to accomplish.

- To ensure smoothness we have to assume J is generic or even deform the Cauchy–Riemann equation by adding a time-dependent Hamiltonian flow and use a time-dependent almost-complex structure. (This is analogous to assuming that f is Morse–Smale when we constructed the Morse complex.)
- To ensure orientability, the Lagrangian submanifolds need to be orientable and equipped with a spin structure [218, II.11]. When $\mathbb{L}$ is a closed curve there are two spin structures, the trivial one and the nontrivial one, corresponding to the two local $\mathbb{Z}_2$-systems on a circle. The nontrivial one can be described by putting a marked point on the curve where the trivialization breaks down. Each marked point on the disk introduces an extra minus sign in σ_ψ.

In our further discussion we will assume that the $\mathcal{M}_u$ are oriented manifolds. When $\dim \mathcal{M}_u = 0$ this manifold will be a discrete set of points ψ, which each come with a sign σ_ψ representing the orientation. We will only sum over those ψ that sit in a zero-dimensional $\mathcal{M}_u$.

To make sure that the definition above is well defined, the total number of maps ψ that contribute to the sum needs to be finite. This is not immediately obvious: even if all the zero-dimensional $\mathcal{M}_u$ have a finite number of points, in total there might be an infinite number. Luckily the restrictions we imposed help us out.

Lemma 6.26 *If the symplectic manifold and the compact Lagrangian submanifolds are all exact, the sum in* (6.1) *is finite.*

Sketch of the proof This uses Gromov's compactness theorem [100], which states that for a given upper bound there are only a finite number of holomorphic maps with an area smaller than that upper bound. Because of the exactness of ω the area of a map ψ is given by an integral over the boundary, which we can decompose into a sum of integrals over each of the Lagrangian submanifolds:

$$\int_{\mathbb{D}} \psi^*\omega = \int_{\mathbb{D}} d\psi^*\theta = \int_{\partial\mathbb{D}} \psi^*\theta = \sum_i \int_{p_i}^{p_{i+1}} \theta|_{\mathbb{L}_i}.$$

Due to the exactness we can write each term in the sum as a difference,

$$\int_{p_i}^{p_{i+1}} \theta|_{\mathbb{L}_i} = \int_{p_i}^{p_{i+1}} df_i = f_i(p_{i+1}) - f_i(p_i),$$

where $f_i\colon \mathbb{L}_i \to \mathbb{R}$ is a primitive function for $\theta|_{\mathbb{L}_i}$. Because $\mathbb{L}_i$ is compact, the integral is bounded by the difference between the minimum and maximum of f_i and therefore the area of ψ is bounded. □

6.2.4 Bubbles

To ensure that the products satisfy the A_∞-relations, we want to mimic the case of the Morse differential in Theorem 6.7. There we glued two trajectories together and deformed them to a one-parameter family that broke down to another pair of trajectories. The two broken pairs then canceled each other out in the expression for d^2.

Here we can do something similar: we can glue two disks together at a common marked point, deform it and break it down into two other disks.

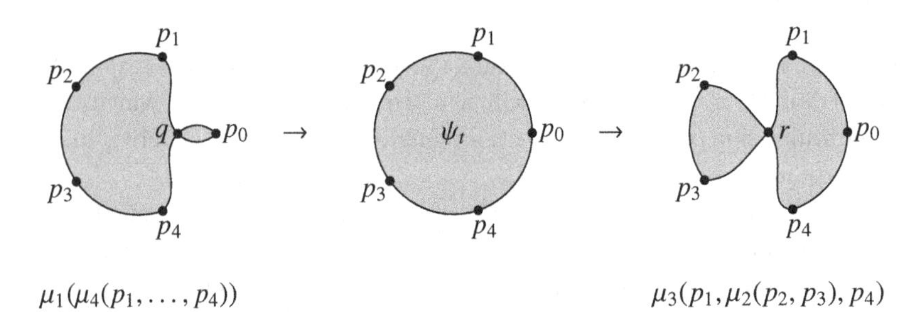

Both ends can be interpreted as a term in the A_∞-rule, which occur with opposite sign, so they cancel out.

At least that is what is supposed to happen in theory, but in reality there are two types of degenerations that do not correspond to terms in the A_∞-axioms. This phenomenon is called *bubbling*.

- *Disk bubbles* occur when an extra disk appears at one of the boundaries. This disk can be seen as an extra contribution from a nullary product μ_0. In that case, the structure is not an ordinary A_∞-structure but a curved one.
- *Sphere bubbles* occur when a sphere is cut off in the middle of the disk.

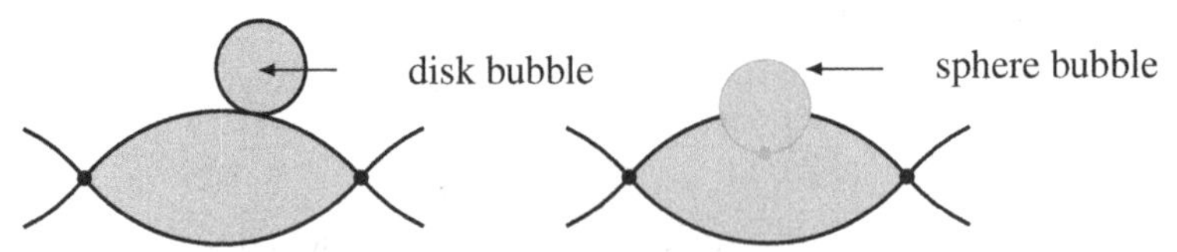

Bubbles are analogous to Morse trajectories that contain loops. In Morse theory these are impossible because trajectories flow upwards along the Morse function. In the exact setting we can say something similar.

Lemma 6.27 *The μ_k satisfy the A_∞-axioms.*

Sketch of the proof In the exact setting both types of bubbles give ψ with zero area. These can be used to make an infinite number of maps with an area bounded by a given constant, which contradicts Gromov's compactness theorem.

The full proof can be found in [218, Part II]. It involves constructing compactifications of the moduli spaces $\mathcal{M}_u(p_0, \ldots, p_k)$ whose boundary is made up of products of moduli spaces with fewer points. On these spaces the A_∞-products can be interpreted as a boundary operator d and the A_∞-relations then follow from the fact that $d^2 = 0$. □

6.2.5 Floer Cohomology

When all this is taken care of we arrive at the main theorem of Floer theory.

Theorem 6.28 (Floer [79]) *Fix a field $\Bbbk$ with characteristic 2 and a Liouville manifold with cylindrical ends and with a compatible almost-complex structure $(\mathbb{X}, \theta, J)$. Let $\mathbb{L}_1$, $\mathbb{L}_2$ be compact exact oriented Lagrangian submanifolds that intersect transversally and assume J is generic.*

(i) *The pair $(\mathsf{CF}^{\pm}_{\Bbbk}(\mathbb{L}_1, \mathbb{L}_2), \mu_1)$ is a $\mathbb{Z}_2$-graded complex.*

(ii) *If $J_t, t \in [0, 1]$ is a family of almost-complex structures then*

$$\mathsf{CF}^{\pm}_{\Bbbk}(\mathbb{L}_1, \mathbb{L}_2)_{J_0} \cong \mathsf{CF}^{\pm}_{\Bbbk}(\mathbb{L}_1, \mathbb{L}_2)_{J_1}$$

are quasi-isomorphic provided J_0, J_1 are generic.

(iii) *If ϕ^h_t is a Hamiltonian flow then*

$$\mathsf{CF}^{\pm}_{\Bbbk}(\mathbb{L}_1, \mathbb{L}_2) \cong \mathsf{CF}^{\pm}_{\Bbbk}(\phi^h_\epsilon \mathbb{L}_1, \mathbb{L}_2) \cong \mathsf{CF}^{\pm}_{\Bbbk}(\mathbb{L}_1, \phi^h_\epsilon \mathbb{L}_2)$$

are quasi-isomorphic provided the Lagrangian submanifolds appearing are transversal.

In other words, the Floer cohomology is an invariant that only depends on the Hamiltonian isotopy classes of the Lagrangian submanifolds and the symplectic structure.

Remark 6.29 As already mentioned above, the Lagrangians need to be equipped with spin structures for this theorem to work over a field with characteristic different from 2 [218, II.11].

6.2.6 Precategories

Now we would like to bundle all these data into a category. The objects should be exact compact Lagrangian submanifolds in $\mathbb{X}$, and the morphism spaces the Floer complexes. Unfortunately, life is not that simple: morphism spaces are only defined between transversally intersecting Lagrangians, so in particular there is no morphism space between an object and itself. We can take care of this problem with the help of the precategories from Section 3.4.4.

Definition 6.30 Assume $(\mathbb{X}, \theta, J)$ is a Liouville manifold with cylindrical ends and a compatible almost-complex structure. The *basic Fukaya precategory* ${}^{\mathsf{ba}}\underline{\mathbf{Fuk}}^{\pm}(\mathbb{X}, \theta, J)$ is the $\mathbb{Z}_2$-graded precategory with the following data:

- The objects are compact oriented exact Lagrangian submanifolds in $\mathbb{X}$ equipped with spin structures.
- The transversal sequences consist of n-tuples of Lagrangian submanifolds that are pairwise transversally intersecting with different intersection points and for which J is generic.
- For each $(\mathbb{L}_1, \mathbb{L}_2) \in {}^{\mathsf{ba}}\underline{\mathbf{Fuk}}^{\pm}(\mathbb{X}, \theta, J)^2_{\mathrm{tr}}$ we set

$$ {}^{\mathsf{ba}}\underline{\mathbf{Fuk}}^{\pm}(\mathbb{L}_1, \mathbb{L}_2) = \mathsf{CF}^{\pm}_{\Bbbk}(\mathbb{L}_1, \mathbb{L}_2). $$

- For each $(\mathbb{L}_0, \ldots, \mathbb{L}_k) \in {}^{\mathsf{ba}}\underline{\mathbf{Fuk}}^{\pm}(\mathbb{X}, \theta, J)^{k+1}_{\mathrm{tr}}$ we define products μ_k as above.

Lemma 6.31 *${}^{\mathsf{ba}}\underline{\mathbf{Fuk}}^{\pm}(\mathbb{X}, \theta, J)$ is a precategory with enough quasi-isomorphisms.*

Sketch of the proof Clearly, every subsequence of a transversal sequence is again transversal and because we are in a bubble-free situation the A_∞-axioms hold whenever they are defined.

To show that it has enough quasi-isomorphisms, assume that $\mathbb{L}$ is a compact Lagrangian submanifold. By [252, 6.1] we can find a neighborhood of $\mathbb{L} \subset \mathbb{X}$ that is isomorphic to $\mathsf{T}^*\mathbb{L}$. If we choose a Morse–Smale function f on $\mathbb{L}$ with one local minimum in each connected component, we can use this to flow $\mathbb{L}$ to a version $\phi^f_\epsilon\mathbb{L}$. The sum of these minima gives a morphism in ${}^{\mathsf{ba}}\underline{\mathbf{Fuk}}^{\pm}(\mathbb{L}, \phi^f_\epsilon\mathbb{L})$ with index 0, which will correspond to the identity in $\mathsf{H}^\bullet(\mathbb{L}, \Bbbk) \cong \mathsf{HCF}_{\Bbbk}(\mathbb{L}, \phi^f_\epsilon\mathbb{L})$ and hence it will act as a quasi-isomorphism.

Because f can be any Morse–Smale function on $\mathbb{L}$ we have many quasi-isomorphisms and given a set of other Lagrangians it will be possible to find an f such that $\mathbb{L}_+ = \phi^f_\epsilon\mathbb{L}$ and $\mathbb{L}_- = \phi^f_\epsilon\mathbb{L}$ are transversal to all of them and compatible with J. □

Definition 6.32 The *basic Fukaya category* of ${}^{\mathsf{ba}}\mathbf{Fuk}^{\pm}(\mathbb{X}, \theta, J)$ is defined as the categorification of the basic Fukaya precategory.

Theorem 6.33 *For two different almost-complex structures J_0, J_1, the Fukaya categories ${}^{\mathrm{ba}}\mathsf{Fuk}^{\pm}(\mathbb{X}, \theta, J_0)$ and ${}^{\mathrm{ba}}\mathsf{Fuk}^{\pm}(\mathbb{X}, \theta, J_1)$ are A_∞-quasi-isomorphic.*

Sketch of the proof Because the space of complex structures is contractible we can construct a path $\gamma\colon J_0 \to J_1$. In analogy to Theorem 6.28, this path induces quasi-isomorphisms between $\mathsf{CF}^{\pm}_{\Bbbk}(L_1, L_2)_{J_0}$ and $\mathsf{CF}^{\pm}_{\Bbbk}(L_1, L_2)_{J_1}$, which can be turned into an A_∞-functor between the precategories that induces a quasi-equivalence between the categories. □

Example 6.34 Take $\mathbb{X} = \mathsf{T}^*\mathbb{S}^1$ with $\theta = p\,dq$ and for the complex structure J we identify $\mathbb{X}$ with $\mathbb{C}/2\pi\mathbb{Z}$. If an embedding $\mathbb{L} \subset \mathbb{X}$ is exact, the tangent spaces can never be vertical. This implies that the embedding of each connected component of $\mathbb{L}$ looks like the graph of a 2π-periodic function $p\colon \mathbb{R} \to \mathbb{R}$.

Such a function gives an exact embedding if p is the derivative of a 2π-periodic primitive function f or, in other words, $\int_0^{2\pi} p\,dt = 0$. Note that two such embeddings will always intersect, so $\mathbb{L}$ can only have one connected component. This component can be oriented in two ways depending on whether we follow the embedded circle in the positive or negative direction. This gives a bijection

$$\mathrm{Ob}\,{}^{\mathrm{ba}}\mathsf{Fuk}^{\pm}(\mathbb{X}) \leftrightarrow \{df \mid f\colon \mathbb{S}^1 \to \mathbb{R}\} \times \{+, -\}.$$

Take two Lagrangian embeddings $\mathbb{L}_1$, $\mathbb{L}_2$ that intersect transversally. In that case, there will be $2r$ intersection points, which we name u_i, v_i depending on whether $\mathbb{L}_2$ crosses $\mathbb{L}_1$ from below or from above. If $\mathbb{L}_1$, $\mathbb{L}_2$ are oriented in the same way the u_i get $\mathbb{Z}_2$-degree 0 and the v_i degree 1. If they are oriented differently the degrees are reversed.

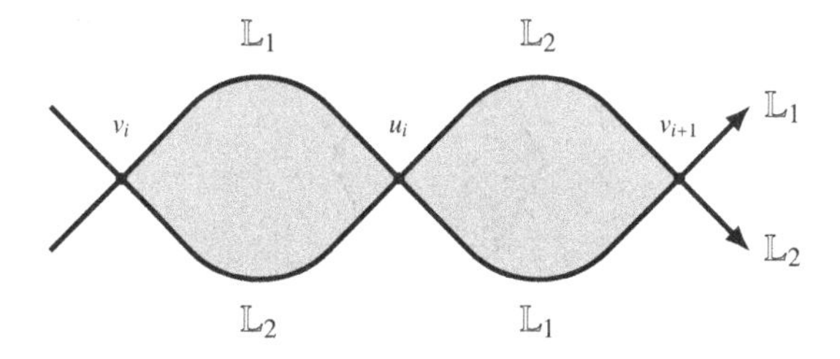

There will be precisely one holomorphic strip from a u to a v adjacent to it. The one next to it gets a plus sign and the one before it gets a minus sign:

$$\mathsf{CF}^{\pm}_{\Bbbk}(\mathbb{L}_1, \mathbb{L}_2) = \bigoplus_{i=1}^{r} \Bbbk u_i \oplus \Bbbk v_i \quad \text{with } \mu_1(u_i) = v_{i+1} - v_i.$$

The homology of this complex is spanned by $\iota = \sum u_i$ and $\xi = v_1$ (instead of v_1 we can choose any v_i as they are all equal in homology). If we go to the minimal model, for every transversal pair $(\mathbb{L}_1, \mathbb{L}_2)$ we get

$$\mathsf{H}^{\mathsf{ba}}\underline{\mathbf{Fuk}}^{\pm}(\mathbb{L}_1, \mathbb{L}_2) = \Bbbk\iota^{[12]} \oplus \Bbbk\xi^{[12]}.$$

To categorify this precategory, observe that every $\iota^{[12]}$ is a quasi-isomorphism. Indeed, if $u^{[13]}: \mathbb{L}_1 \to \mathbb{L}_3$ is a degree 0 intersection point then it sits in a unique triangle of degree 0 intersection points spanned by the $u^{[12]}$ that precedes it on $\mathbb{L}_1$ and the $u^{[23]}$ that follows it on $\mathbb{L}_2$, or vice versa depending on whether $\mathbb{L}_2$ lies above $u^{[13]}$ or below. Therefore $\iota^{[23]}\iota^{[12]} = \iota^{[13]}$.

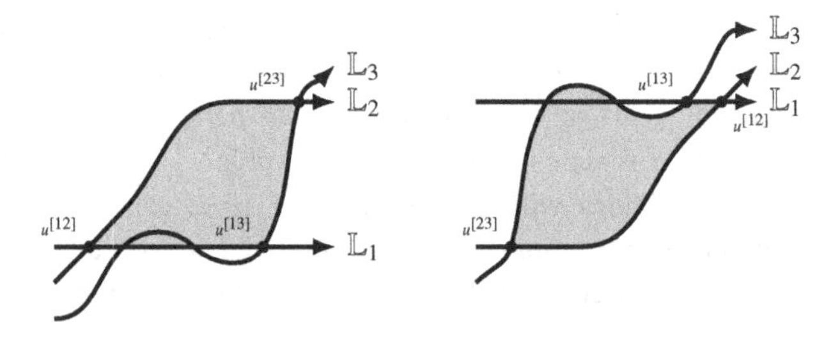

Similarly, we can deduce that $\xi^{[23]}\iota^{[12]} = \iota^{[23]}\xi^{[12]} = \xi^{[13]}$ and $\xi^{[23]}\xi^{[12]} = 0$.

After categorification, all objects with the same orientation become isomorphic and all the ι become units. After an A_∞-isomorphism we can turn them into strict units. This means that the only higher products that could be nonzero involve only ξ's. However, if we look at any n-gon bounded by Lagrangians going in the same direction around the cylinder, they cannot all go in the same direction around the polygon. This means that there are at least two points where the orientation changes direction. At least one of those two points will show up as an ι inside the product and therefore $\mu(\xi, \dots, \xi) = 0$.

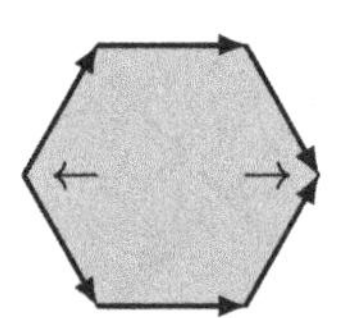

We can conclude that the Fukaya category is equivalent to an ordinary $\mathbb{Z}_2$-graded category with two objects $\{\mathbb{L}_+, \mathbb{L}_-\}$ and

$$^{\mathsf{ba}}\mathbf{Fuk}^{\pm}(\mathbb{L}_+, \mathbb{L}_+) = {}^{\mathsf{ba}}\mathbf{Fuk}^{\pm}(\mathbb{L}_-, \mathbb{L}_-) = \Bbbk[\xi]/(\xi^2) = \Lambda_1^{\pm},$$
$$^{\mathsf{ba}}\mathbf{Fuk}^{\pm}(\mathbb{L}_+, \mathbb{L}_-) = {}^{\mathsf{ba}}\mathbf{Fuk}^{\pm}(\mathbb{L}_-, \mathbb{L}_+) = \Bbbk[\xi]/(\xi^2)[1] = \Lambda_1^{\pm}[1].$$

If we take the derived version we get

$$\mathsf{D}^{\mathsf{ba}}\mathbf{Fuk}^{\pm}(\mathbb{X}, \theta, J) \overset{\infty}{=} \mathsf{D}\Lambda_1^{\pm}.$$

6.2.7 Upgrading the Grading

In our construction we only gave the category a $\mathbb{Z}_2$-grading but sometimes it is possible to construct a $\mathbb{Z}$-graded category. Ideally, a thorough treatment of gradings of Fukaya categories should include a discussion about the Maslov index [12, 173, 238], which is the generalization of the Morse index to Floer theory. We will however follow a more basic approach in line with Section 6.1.2.

Definition 6.35 A *Calabi–Yau manifold* $(\mathbb{X}, \Omega)$ is a Kähler manifold $(\mathbb{X}, \omega, J)$ equipped with a nowhere-vanishing complex volume form $\Omega \in \Omega^n_{\mathbb{C}}(\mathbb{X})$.

Analogously to the discussion in Section 6.1.2, we can use this volume form to define graded Lagrangians and put a grading on the Floer complex.

Definition 6.36 A Ω-*graded Lagrangian* is a pair $(\mathbb{L}, \theta)$ consisting of a Lagrangian submanifold $\mathbb{L}$ and a function $\theta\colon \mathbb{L} \to \mathbb{R}$, such that $e^{-i\theta}\Omega$ is a real volume form for $\mathbb{L}$. A Lagrangian is called *special* if it can be graded by a constant function.

Not every Lagrangian submanifold can be graded. For each point $x \in \mathbb{L}$ we have that $\theta(x)$ is determined up to a multiple of π. If we fix θ at x and take a curve $\gamma\colon x \to y$ we can continue θ along that curve to obtain a value for $\theta(y)$. For $\mathbb{L}$ to be gradable this continuation should be independent of the path. If $\mathbb{L}$ is connected and gradable then there is precisely one grading for every $k \in \mathbb{Z}$ because two gradings will differ globally by a multiple of π. A grading automatically gives an orientation to the Lagrangian: just take $e^{-i\theta}\Omega$ for the real volume form. Two gradings that differ by an odd multiple of π will have opposite orientations.

Definition 6.37 Suppose $(\mathbb{L}_1, \theta_1)$ and $(\mathbb{L}_2, \theta_2)$ are two graded Lagrangians that intersect transversally at x. Let $U\colon \mathsf{T}_x\mathbb{X} \to \mathsf{T}_x\mathbb{X}$ be a unitary map for the almost-complex structure J such that $U(\mathsf{T}_x\mathbb{L}_1) = \mathsf{T}_x\mathbb{L}_2$ and take $\Delta \in [0, \pi)$ such that $\det U = \pm e^{i\Delta}$. Then we define

$$\deg x = \frac{\Delta - (\theta_2(x) - \theta_1(x))}{\pi}$$

for $x \in \mathsf{CF}^\bullet_{\Bbbk}(\mathbb{L}_1, \mathbb{L}_2)$.

Lemma 6.38 *The degree of x is well defined and the product μ_k has degree $2 - k$.*

Sketch of the proof First of all $\deg x$ is an integer because $\Omega|_{\mathsf{T}_x\mathbb{L}_2} = \pm \det U \Omega|_{\mathsf{T}_x\mathbb{L}_2}$ so $\Delta - (\theta_2(x) - \theta_1(x))$ is a multiple of π. It is also well defined because U is determined up to products with unitary transformations that fix

$\mathsf{T}_x\mathbb{L}_1$ or $\mathsf{T}_x\mathbb{L}_2$. These are in fact orthogonal transformations of these subspaces, so they have determinant ± 1.

To find the degree of μ_k we can transform the domain of each pseudo-holomorphic map from a disk to a convex polygon with angles these Δ. The result then follows from the fact that a sum of the internal angles in a k-gon is $(k-2)\pi$. □

Definition 6.39 The *$\mathbb{Z}$-graded basic Fukaya precategory* ${}^{\mathsf{ba}}\underline{\mathsf{Fuk}}^\bullet(\mathbb{X}, \Omega)$ is the precategory whose transversal sequences are n-tuples of mutually transversally intersecting compact exact Ω-graded Lagrangian submanifolds with spin structures. The hom-spaces are the Ω-graded Floer complexes and the multiplications are the usual. The *$\mathbb{Z}$-graded Fukaya category* ${}^{\mathsf{ba}}\mathsf{Fuk}^\bullet(\mathbb{X}, \Omega)$ is its categorification.

The $\mathbb{Z}$-graded Fukaya precategory depends on Ω, but different choices for Ω can result in isomorphic categories. We say that $\Omega_0 \cong \Omega_1$ are *homotopic* if there is a path of complex volume forms between them.

Lemma 6.40 *Homotopic volume forms give rise to isomorphic Fukaya precategories.*

Sketch of the proof Let Ω_t, $t \in [0, 1]$ be a path of volume forms. At every point x we can measure how much the phase of $\Omega(x)$ shifts during the path:

$$\vartheta(x) = \int_0^1 d \arg \Omega_t(x).$$

If $\theta\colon \mathbb{L} \to \mathbb{R}$ is an Ω_0-grading then $\theta - \vartheta$ will be an Ω_1-grading and vice versa. Hence we get a bijection between Ω_0-graded Lagrangian submanifolds and Ω_1-graded Lagrangian submanifolds. The gradings of the Floer complexes induced by these gradings are the same because the term ϑ cancels out. □

Example 6.41 We can identify $\mathbb{X} = \mathsf{T}^*\mathbb{S}^1$ with $\mathbb{C}^*$ using the coordinate $z = e^{p+qi}$. A complex volume form is a 1-form that vanishes nowhere, e.g. $\Omega = z^r\, dz$. To grade the base Lagrangian $\mathbb{S}^1$ we have to choose $\theta(q) = -rq - q + k\pi$, but unless $r = -1$ this is not a function because it is not periodic.

This implies that the graded basic Fukaya category is empty unless $\Omega = dz/z$. In that case, it is equivalent to a category with objects $\mathbb{L}[k]$, $k \in \mathbb{Z}$, corresponding to the graded Lagrangians $(\mathbb{S}^1, k\pi)$ with ${}^{\mathsf{ba}}\mathsf{Fuk}^\bullet(\mathbb{L}[i], \mathbb{L}[j]) = \Bbbk[\xi]/(\xi^2)[j-i] = \Lambda_1^\bullet[j-i]$. On the derived level we can conclude

$$\mathtt{D}\,{}^{\mathsf{ba}}\mathsf{Fuk}^\bullet\left(\mathbb{C}^*, \frac{dz}{z}\right) \stackrel{\infty}{=} \mathtt{D}\Lambda_1^\bullet \stackrel{\infty}{=} \mathtt{D}\,\mathtt{mod}^\bullet\, \Bbbk[\![X]\!].$$

6.3 Variations

We will now construct some other Fukaya-type categories that are variations on the previous construction. Each of these variations comes in two flavors: a $\mathbb{Z}_2$-graded version and a $\mathbb{Z}$-graded version, depending on whether there is a complex volume form Ω around. To avoid repetition and notational cluttering, we will omit Ω but use the $\mathbb{Z}$-graded form of the category in the notation.

6.3.1 Immersions and Local Systems

It is possible to extend the scope of Floer cohomology in two ways: we could allow immersed Lagrangians and we could equip the Lagrangians with local systems.

Definition 6.42 An *immersed Lagrangian with a local system*, also called a *Lagrangian brane*, consists of a triple $(\mathbb{L}, \mathcal{L}, \phi)$, where

- $\mathbb{L}$ is an n-dimensional oriented compact manifold (with spin structure if $\operatorname{char} \Bbbk \neq 2$),
- $\mathcal{L}\colon \Pi_1(\mathbb{L}) \to \mathtt{vect}(\Bbbk)$ is a local system,
- $\phi\colon \mathbb{L} \to \mathbb{X}$ is an immersion such that for all $p \in \mathbb{L}$ we have that $\phi_* \mathsf{T}_p\mathbb{L}$ is Lagrangian.

Moreover, we impose that the immersion self-intersects transversally: at most two points of $\mathbb{L}$ can be mapped to the same point in $\mathbb{X}$ and if $\phi(p_1) = \phi(p_2) = x$ then $\mathsf{T}_x\mathbb{X} = \phi_*\mathsf{T}_{p_1}\mathbb{L} \oplus \phi_*\mathsf{T}_{p_2}\mathbb{L}$. Such points are called *double points*. We say that $(\mathbb{L}, \mathcal{L}, \phi)$ is *exact* if there is an $f\colon \mathbb{L} \to \mathbb{R}$ such that $\phi^*\theta = df$.

If we have exact transversally intersecting Lagrangian branes $\mathbb{L}_i{:=}(\mathbb{L}_i, \mathcal{L}_i, \phi_i)$ that do not intersect at double points, we can define the Floer complex as

$$\mathsf{CF}^\bullet_\Bbbk(\mathbb{L}_1, \mathbb{L}_2) := \bigoplus_{p \in \phi_1\mathbb{L}_1 \cap \phi_2\mathbb{L}_2} \mathtt{vect}(\mathcal{L}_1(\phi_1^{-1}p), \mathcal{L}_2(\phi_2^{-1}p))p,$$

and the product $\mu\colon \mathsf{CF}_\Bbbk(\mathbb{L}_1, \mathbb{L}_0) \otimes \cdots \otimes \mathsf{CF}_\Bbbk(\mathbb{L}_k, \mathbb{L}_{k-1}) \to \mathsf{CF}^\bullet_\Bbbk(\mathbb{L}_k, \mathbb{L}_0)$ becomes

$$\mu(\alpha_1 p_1, \ldots, \alpha_k p_k) := \sum_{p_0 \in \phi_0\mathbb{L}_0 \cap \phi_k\mathbb{L}_k} \sum_{\psi} \underbrace{\mathcal{L}_0(\gamma_0)\alpha_1\mathcal{L}_1(\gamma_1)\cdots\alpha_k\mathcal{L}_k(\gamma_k)}_{(*)} \epsilon_\psi \cdot p_0,$$

(the inner sum is over discs ψ with boundary points $p_0, p_1, \ldots, p_k$, as pictured under the summation sign)

where γ_r stands for the path from p_{r+1} to p_r along the boundary of $\operatorname{Im}\psi$ pulled back to $\mathbb{L}_r$:

$$\gamma_r\colon [0,1] \to \mathbb{L}_r\colon t \mapsto \phi_r^{-1}\psi\left(e^{\frac{2(r+1-t)\pi i}{k+1}}\right).$$

In other words, the elements in the Floer complex become morphisms between fibers at intersection points and the products measure the transport along the boundary of the immersed disk. The linear map $\mathcal{L}_i(\gamma_i)$ only depends on the homotopy class of the path and therefore the factor $(*)$ remains the same under deformations and degenerations of ψ. This is important for the cancelation of the terms in the A_∞-axioms.

In this generalized setting, the main theorem for Floer theory still holds provided we interpret $\phi^h_\epsilon(\mathbb{L})$ as $(\mathbb{L}, \mathcal{L}, \phi^h_\epsilon\phi)$. The old Floer theory can be recovered by treating an ordinary Lagrangian submanifold $\mathbb{L}$ as a brane with a trivial one-dimensional local system: $\mathcal{L}(p) = \mathbb{k}$ and $\mathcal{L}(\gamma) = 1$ for all points p and curves γ.

The idea of local systems can also be used to describe different spin structures on $\mathbb{L}$. Given one spin structure on $\mathbb{L}$, all others can be identified with $\mathsf{H}_1(\mathbb{L}, \mathbb{Z}_2)$. The latter space is equal to the space of local systems $\mathcal{S}$ such that $\mathcal{S}(p)$ is one-dimensional and $\mathcal{S}(\gamma) = \pm 1$ if γ is a loop. The Lagrangian brane $(\mathbb{L}, \mathcal{S}, \phi)$ can be interpreted as the same Lagrangian with a different spin structure.

Definition 6.43 The *exact Fukaya precategory* ${}^{\mathsf{ex}}\mathsf{Fuk}^\bullet(\mathbb{X})$ has as transversal sequences collections of exact Lagrangian branes that intersect transversally at different points and the products are defined as above. The *exact Fukaya category* ${}^{\mathsf{ex}}\mathsf{Fuk}^\bullet(\mathbb{X})$ is defined as its categorification.

Example 6.44 Let us return to the cotangent bundle of the circle $\mathsf{T}^*\mathbb{S}^1 = \mathbb{C}^*$ equipped with $\Omega = dz$.

- **Local systems.** To describe a local system $\mathcal{L}$ on $\mathbb{S}^1$ we have to pick a base point $b \in \mathbb{S}^1$ and choose a basis in $\mathcal{L}(b)$. This gives a matrix $M = \mathcal{L}(\ell)$ where ℓ is the positive loop around the circle with base point b. This local system corresponds to the representation $\rho_M : \pi_1(\mathbb{S}^1) \to \mathsf{GL}_n : \ell^k \mapsto M^k$.

  For every other point x there is a unique minimal path $x \leftarrow b$ (up to homotopy) in the positive direction that connects b to x. We can choose bases in $\mathcal{L}(x)$ such that $\mathcal{L}(x \leftarrow b) = \mathbb{1}$. Once these bases are fixed we have $\mathcal{L}(\gamma) = M^k$, where M is an invertible matrix and k is the number of times γ crosses b in the positive direction minus the number of times γ crosses the zero in the negative direction.

  Now consider two embedded exact positively oriented Lagrangian submanifolds $\mathbb{L}_1$, $\mathbb{L}_2$ and local systems $\mathcal{L}_1$, $\mathcal{L}_2$ given by matrices M_1, M_2.

The Lagrangians intersect at $2r$ points u_i, v_i. If we assume that the base points both lie between u_r and v_1 then the Floer complex becomes

$$\mathsf{CF}(\mathbb{L}_1, \mathbb{L}_2) = \bigoplus_{i=0}^{r} \mathrm{Mat}_{n_2 \times n_1}(\mathbb{k}) u_i \oplus \mathrm{Mat}_{n_2 \times n_1}(\mathbb{k}) v_i,$$

with $\mu_1(Au_i) = Av_{i+1} - Av_i$ if $i \neq r$ and $\mu_1(Au_r) = M_2 A M_1^{-1} v_1 - Av_r$. The homology of this complex can be interpreted as $\mathrm{Ext}^\bullet_{\pi_1(\mathbb{S}^1)}(\rho_{M_1}, \rho_{M_2})$. So if we allow local systems, the Fukaya category behaves like the category of finite-dimensional $\pi_1(\mathbb{S}^1)$-modules.

- **Immersions.** An immersion of $\mathbb{L}$ looks like $\phi\colon t \mapsto (kt, p(t))$, where $k \in \mathbb{Z}$ is the winding number and p is a 2π-periodic function. Such a function gives an exact immersion if p is the derivative of a 2π-periodic function primitive function f.

  Let us compare the base circle $\mathbb{L}_1 = \mathbb{S}^1 \subset \mathsf{T}^*\mathbb{S}^1$ and a two-dimensional local system determined by $M_1 \in \mathrm{Mat}_{2\times 2}(\mathbb{k})$ with an immersed Lagrangian that wraps around the cylinder twice and a one-dimensional local system with transport $M_2 \in \mathbb{k}$:

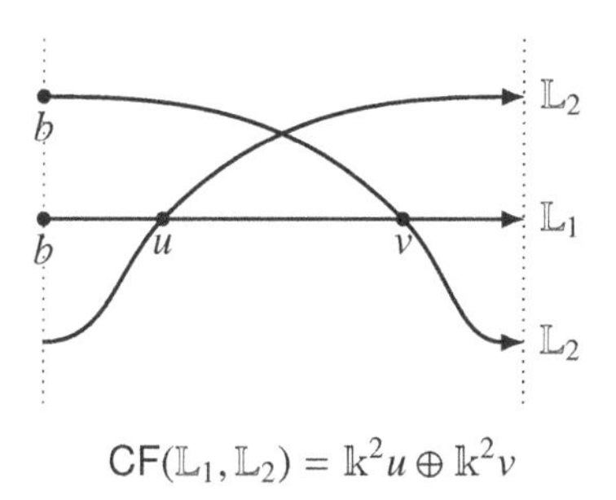

$$\mathsf{CF}(\mathbb{L}_1, \mathbb{L}_2) = \mathbb{k}^2 u \oplus \mathbb{k}^2 v$$

  with $\mu_1(Au) = M_1 Av - M_1^{-1} A M_2^{-1} v$, which is zero if $M_1^2 = M_2 \mathbb{1}_2$. If that is the case, there are two nonzero degree 0 morphisms. Further investigation shows that if $M_1 = \begin{pmatrix} 0 & M_2 \\ 1 & 0 \end{pmatrix}$ then the two objects are isomorphic. Intuitively this makes sense because then the local system on $\mathbb{L}_1$ is the pullback of the local system on $\mathbb{L}_2$ under the projection map $\mathbb{L}_2 \to \mathbb{L}_1$.

  This idea can be used to show that every immersed Lagrangian that wraps n times around the cylinder with a k-dimensional local system is isomorphic to an embedded Lagrangian with an $(n \times k)$-dimensional local system.

- **Multiple components.** We can also immerse multiple components. In this case, $\mathbb{L} = \mathbb{L}_1 \sqcup \cdots \sqcup \mathbb{L}_k$ and the restriction of the immersion and the local system to each component can be seen as a separate object in the

Fukaya category. It is not difficult to show that in the additive completion the big object is isomorphic to the direct sum of the smaller objects:

$$(\mathbb{L}, \phi, \mathcal{L}) \cong \bigoplus_i (\mathbb{L}_i, \phi|_{\mathbb{L}_i}, \mathcal{L}|_{\mathbb{L}_i}).$$

From this discussion it is clear that the objects in the Fukaya category are direct sums of graded circles with a local system. Each is equivalent to the shift of a finite-dimensional module of $\pi_1(\mathbb{S}^1)$ or its group ring $\Bbbk[X, X^{-1}]$. Just like for the polynomial ring, every object in the derived category $\mathtt{Dmod}^\bullet\, \Bbbk[X, X^{-1}]$ is of this form so we can conclude that

$$^{\mathsf{ex}}\mathtt{Fuk}^\bullet(\mathbb{X}) \overset{\infty}{\cong} \mathtt{Dmod}^\bullet\, \Bbbk[X, X^{-1}].$$

From this perspective it is natural to include immersions and local systems all at once. This version of the Fukaya category is better behaved because it is closed under direct sums, shifts and even cones. Therefore, if we talk about the Fukaya category we will always mean this version and not the basic version that we started with.

6.3.2 Wrapped Fukaya Categories

In the last example we saw that we could identify the Fukaya category of the cotangent bundle of a circle with the A_∞-category of finite-dimensional representations of $\pi_1(\mathbb{S}^1)$. From a representation-theoretical perspective it would also be desirable to include some nice infinite-dimensional modules as well, such as the finitely generated projectives. In order to do that, we should include some noncompact Lagrangian submanifolds in the construction of the Fukaya category. The obvious choice for these is the conical ones.

We start off with a Liouville manifold with cylindrical ends and a compatible almost-complex structure $(\mathbb{X}, \theta, J)$. On $\mathbb{X}$ we choose a Hamiltonian $h\colon \mathbb{X} \to \mathbb{R}$ that is zero on the Liouville domain $\mathbb{X}_0$ and quadratic in the outwards coordinate r on the tubes. This Hamiltonian generates a flow that we will call the *Reeb flow* and its flow lines $\gamma_x\colon [0,1] \to \mathbb{X}\colon t \mapsto \phi_t^h(x)$ are called *Reeb chords*. The Reeb flow preserves the coordinate r so the Reeb chords each cycle around in $\partial\mathbb{X}_0 \times \{r\}$ for a given r. As r gets bigger they will cycle around faster because h is quadratic in r.

Definition 6.45 If $\mathbb{L}_1$ and $\mathbb{L}_2$ are exact Lagrangian submanifolds such that $\phi_1^h\mathbb{L}_1$ and $\mathbb{L}_2$ intersect normally, we define the wrapped Floer complex as the vector space spanned by the Reeb chords that connect $\mathbb{L}_1$ with $\mathbb{L}_2$:

$$\mathsf{CW}(\mathbb{L}_1, \mathbb{L}_2; h) = \bigoplus_{\substack{\gamma(0)\in\mathbb{L}_1 \\ \gamma(1)\in\mathbb{L}_2}} \Bbbk\gamma.$$

As a vector space, this is isomorphic to $\mathsf{CF}^{\pm}(\phi_1^h\mathbb{L}_1, \mathbb{L}_2)$ because every Reeb chord corresponds to an intersection point. We can use this identification to give a $\mathbb{Z}_2$-grading to $\mathsf{CW}(\mathbb{L}_1, \mathbb{L}_2; h)$ or a $\mathbb{Z}$-grading if $\mathbb{X}$ is Calabi–Yau.

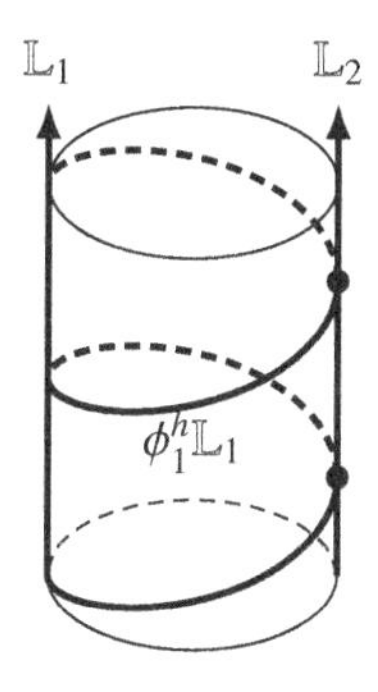

Intersection points

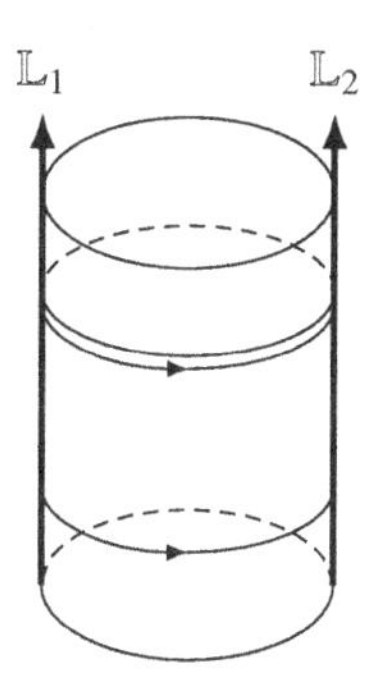

Reeb chords

To define the product we will again count immersed disks, but there is a matching problem. If a Reeb chord γ_1 corresponds to an intersection point p_1 between $\phi_1^h\mathbb{L}_0$ and $\mathbb{L}_1$, and a Reeb chord γ_2 to an intersection point p_2 between $\phi_1^h\mathbb{L}_1$ and $\mathbb{L}_2$, then $\mathbb{L}_1$ and $\phi_1^h\mathbb{L}_1$ do not match up. We can solve this by flowing p_1 to the intersection point $\phi_1^h p_1$ between $\phi_2^h\mathbb{L}_0$ and $\phi_1^h\mathbb{L}_1$. However, if we then count triangles with p_2 and $\phi_1^h p_1$ as corners, the third point will sit in $\phi_2^h\mathbb{L}_0 \cap \mathbb{L}_2$. Because $\phi_2^h\mathbb{L}_0 = \phi_1^{2h}\mathbb{L}_0$, this corresponds to a Reeb chord in $\mathsf{CW}(\mathbb{L}_0, \mathbb{L}_2; 2h)$.

Therefore, the product $\tilde{\mu}_k$ will be a map

$$\mathsf{CW}(\mathbb{L}_{k-1}, \mathbb{L}_k; h) \otimes \cdots \otimes \mathsf{CW}(\mathbb{L}_0, \mathbb{L}_1; h) \to \mathsf{CW}(\mathbb{L}_0, \mathbb{L}_k; kh).$$

Phrased in terms of Reeb chords it will count the following objects. Suppose that $\gamma_i \in \mathsf{CW}(\mathbb{L}_{i-1}, \mathbb{L}_i; h)$ are Reeb chords for $i \in \{1, \ldots, k\}$ and that $\gamma_0 \in \mathsf{CW}(\mathbb{L}_{i-1}, \mathbb{L}_i; kh)$. Define $\mathbb{D}_{k+1}$ to be the closed unit disk with $k + 1$ circle segments removed,

$$\mathbb{D}_{k+1} = \mathbb{D} \setminus \cup_i \left\{ e^{\frac{2\iota\pi i}{2k+2}} \mid t \in (2i + 1, 2i + 2) \right\},$$

and look at maps $\psi \colon \mathbb{D}_k \to \mathbb{X}$ up to reparametrization such that

- $\psi(e^{\frac{2\iota\pi i}{2k+2}}) \in \mathbb{L}_i$ if $i \in (2i, 2i + 1)$,
- $\lim_{u \to e^{\frac{2\iota\pi i}{2k+2}}} \psi(u) = \gamma_i(t - 2i)$ if $t \in (2i + 1, 2i + 2)$ and $i \in \{1, \ldots, k\}$,
- $\lim_{u \to e^{\frac{2\iota\pi i}{2k+2}}} \psi(u) = \gamma_0(1 - t)$ if if $t \in (0, 1)$,
- ψ satisfies a certain deformed Cauchy–Riemann equation.

Just like in the original case, the higher products should be sums of all these maps that cannot be deformed.

To get a genuine product we should compose $\tilde{\mu}_k$ with a map S_k : $\mathsf{CW}(\mathbb{L}_0, \mathbb{L}_k; kh) \to \mathsf{CW}(\mathbb{L}_0, \mathbb{L}_k; h)$. Because $\phi_1^h \mathbb{L}_0$ and $\phi_1^{kh} \mathbb{L}_0$ are Hamiltonian isotopic, these two complexes are quasi-isomorphic, so we can choose S_k to be a quasi-isomorphism. Choosing all these maps S_k for all possible Lagrangian submanifolds in a natural way such that the $\mu_k = S_k \tilde{\mu}_k$ satisfy the A_∞-relations is quite technical. For the details, we refer to [15, 6].

If $\mathbb{L}_1$ and $\mathbb{L}_2$ are compact Lagrangian submanifolds contained in the Liouville domain $\mathbb{X}_0$ the construction reduces to the original construction because in that case $h|_{\mathbb{X}_0} = 0$ and the flow is the identity. We can also introduce immersions and local systems in the same way as we did before.

Definition 6.46 ${}^{\mathtt{wr}}\underline{\mathtt{Fuk}}^\bullet(\mathbb{X}, h)$ is the precategory whose objects are the immersed exact conical Lagrangian submanifolds, including the compact ones, equipped with spin structures and local systems. The hom-spaces are the wrapped Floer complexes $\mathsf{CW}^\bullet(\mathbb{L}_0, \mathbb{L}_1; h)$ and the products the μ_k as above. The *wrapped Fukaya category* ${}^{\mathtt{wr}}\mathtt{Fuk}^\bullet(\mathbb{X})$ is its categorification.

Up to A_∞-quasi-equivalence ${}^{\mathtt{wr}}\mathtt{Fuk}^\bullet(\mathbb{X})$ is independent of the complex structure and the choice of h, and there is fully faithful embedding

$$ {}^{\mathtt{ex}}\mathtt{Fuk}^\bullet(\mathbb{X}) \subset {}^{\mathtt{wr}}\mathtt{Fuk}^\bullet(\mathbb{X}). $$

Example 6.47 If we return to the example of $\mathsf{T}^*\mathbb{S}^1$ then any cotangent fiber $\mathbb{L} = \mathsf{T}_x^*\mathbb{S}^1$ is an exact conical Lagrangian. Each Reeb chord γ is determined by its winding number $w(\gamma)$, so if we represent it by $X^{w(\gamma)}$ we get

$$ \mathsf{CW}(\mathbb{L}, \mathbb{L}; h) \cong \mathbb{k}[X, X^{-1}]. $$

For the standard $\Omega = dz/z$-grading $(\mathbb{L}, \frac{\pi}{2})$, the degree of all X^i is 0, so the only nontrivial product is μ_2. As the total winding number of an immersed disk is zero, the only possibility for a nontrivial product is $\mu(X^i, X^j) = X^{i+j}$. That this is indeed the case is shown by the following immersions, which we have drawn in the universal cover of the cylinder:

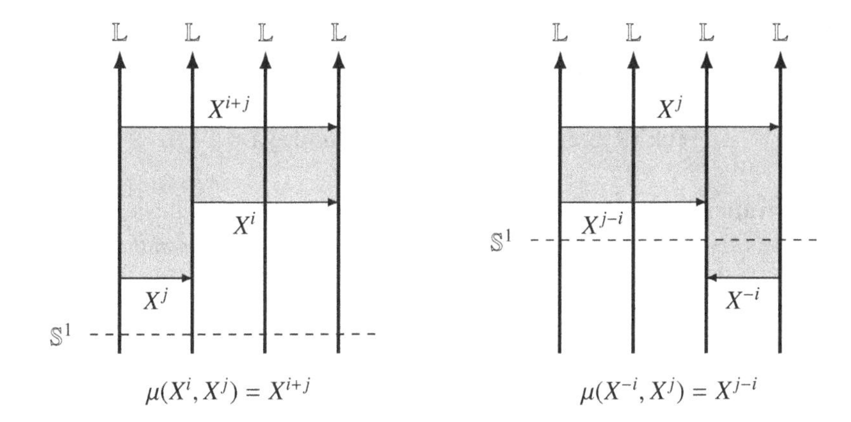

Other exact conical Lagrangians can be seen as two straight ends connected by a curve near the base circle that wraps around the cylinder a couple of times. If the two ends are at the same side then the object in the Fukaya category is a zero object because $\mathbb{1} = d\gamma$, where γ is the smallest Reeb chord that connects the two ends. If the two ends are at opposite sides, the object is isomorphic to $\mathsf{T}^*\mathbb{S}^1$ (with the correct orientation).

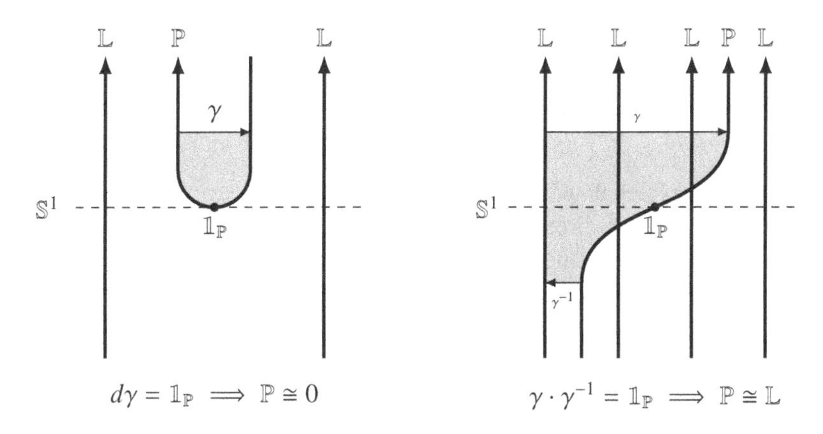

Because $\mathbb{L}$ is contractible, all local systems on $\mathbb{L}$ are trivial and all extra connected objects are isomorphic to (a shift of) a direct sum of cotangent fibers. The latter correspond to free rank $\Bbbk[X, X^{-1}]$-modules. Therefore we get

$$^{\mathtt{wr}}\mathtt{Fuk}^\bullet\left(\mathbb{X}, \frac{dz}{z}\right) \overset{\infty}{=} \mathtt{D}\,\mathtt{Mod}^\bullet\, \Bbbk[X, X^{-1}].$$

If we choose the grading $\Omega = z^k\, dz$ with $k \neq -1$ then we can still give the cotangent fiber a grading because it is contractible. The morphism

$X^i \in \mathsf{CW}(\mathbb{L}, \mathbb{L}) \cong \Bbbk[X, X^{-1}]$ will have degree $i(k+1)$. There are no compact objects because the ordinary Fukaya category is empty:

$$^{\mathrm{wr}}\mathrm{Fuk}^{\bullet}(\mathbb{X}, z^k dz) \overset{\infty}{=} \mathrm{D}\Bbbk[X, X^{-1}] \quad \text{with} \quad \deg X = k - 1.$$

Observation 6.48 *The relation between the wrapped Fukaya category and the ordinary Fukaya category is analogous to the relation between the derived category of finite-dimensional modules and the category of finitely generated modules.*

6.3.3 Partially Wrapped Fukaya Categories

Let $(\mathbb{X}, \theta)$ be a Liouville manifold with cylindrical ends. In general, the Reeb flow on the tubes will cause the endomorphism spaces of conical Lagrangians to be infinite-dimensional. In some situations this is desirable, but in other situations this can be a nuisance. Therefore we introduce a way to interrupt the flow.

Definition 6.49 Let $\mathbb{U}_0$ be a hypersurface in $\partial\mathbb{X}_0$ and set $\mathbb{U} := \mathbb{U}_0 \times [1, \infty)$. This forms a hypersurface in the cylindrical ends of $\mathbb{X}$ that we call a *stop* if the Reeb flow is everywhere normal to the surface.

Because the Reeb flow is everywhere normal to the surface, the Reeb chords can only flow in one direction through the hypersurface. This means that we can define the $\mathbb{U}$-degree of a Reeb chord as the number of times it crosses the $\mathbb{U}$.

Lemma 6.50 *For Lagrangian submanifolds disjoint from $\mathbb{U}$, the products in the wrapped Fukaya category can only decrease the $\mathbb{U}$-degree. In other words, if*

$$\mu_k(\gamma_1, \ldots, \gamma_k) = \sum \alpha_i \zeta_i$$

then the $\mathbb{U}$-degree of ζ_i is at most the sum of the $\mathbb{U}$-degrees of the γ_i.

Sketch of the proof If we look at an immersed disk $\psi\colon \mathbb{D} \to \mathbb{X}$ bounded by the γ_i and one ζ then $\psi^{-1}\mathbb{U}$ is a union of oriented curves. These curves can only enter through γ_i and leave through ζ_i because it is the only curve that is oriented oppositely on the boundary of $\mathbb{D}$.

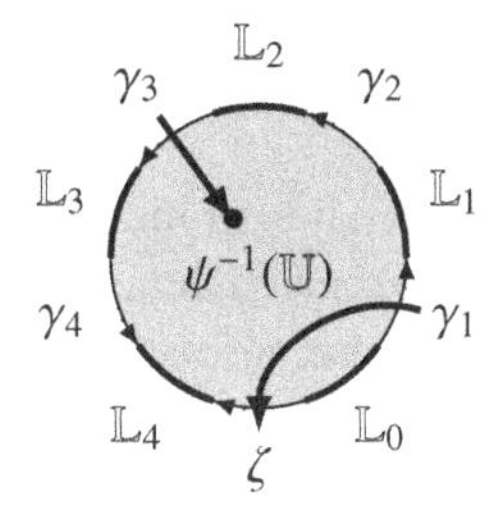

The curves can also stop inside $\mathbb{D}$ at a point of $\psi^{-1}(\mathbb{U}_0)$ if the holomorphic disk lies partly outside the tube. □

Definition 6.51 If σ is a set of stops that are possibly intersecting, we define the *partially wrapped Fukaya precategory* ${}^{\mathrm{wr}}\underline{\mathtt{Fuk}}^{\bullet}{}_{\sigma}(\mathbb{X})$ as the precategory whose objects are the objects in ${}^{\mathrm{wr}}\underline{\mathtt{Fuk}}^{\bullet}(\mathbb{X})$ that are disjoint from all stops $\mathbb{U} \in \sigma$ and the morphism spaces are those that have $\mathbb{U}$-degree 0 in ${}^{\mathrm{wr}}\underline{\mathtt{Fuk}}^{\bullet}(\mathbb{X})$ for all $\mathbb{U} \in \sigma$. The *partially wrapped Fukaya category* ${}^{\mathrm{wr}}\mathtt{Fuk}^{\bullet}{}_{\sigma}(\mathbb{X})$ is its categorification.

Example 6.52 Again look at the cotangent space of the circle. In that case, $\mathbb{X}_0$ is the disjoint union of two circles and $\mathbb{U}_0$ will be a choice of points on these circles. For each point there will be an outward-pointing ray on the cylinder. The rays divide the upper and lower tubes in different sectors, between which there are no Reeb chords.

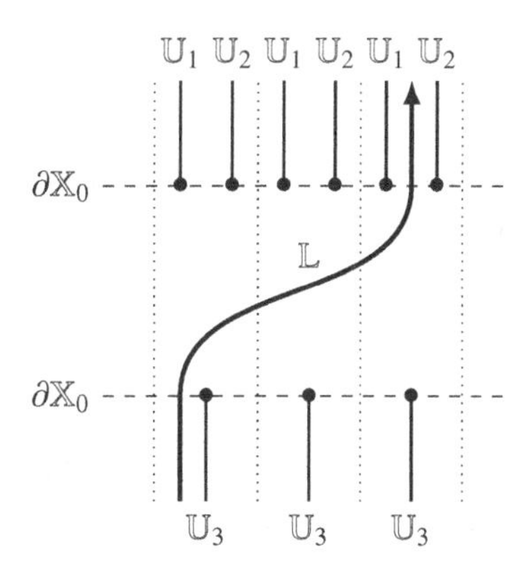

A conical Lagrangian submanifold is characterized by the sectors it starts and ends in and the number of times it wraps around the middle. All submanifolds of the same type give rise to isomorphic objects. Note that objects that start and end at the same side of the cylinder but in different sectors are nonzero objects in the partially wrapped Fukaya category. If we remove the stops these objects become zero in the fully wrapped category. Now let us focus on three special cases.

- If there is just one stop (assume it is in the lower end of the cylinder) then there is just one noncompact Lagrangian up to isotopy. The endomorphism ring of this Lagrangian is $\Bbbk[X]$ because all negative powers of X are killed by the stop. Therefore,

$$^{\mathtt{wr}}\mathtt{Fuk}^{\bullet}{}_{\sigma}(\mathbb{X}) \stackrel{\infty}{=} \mathtt{D\,Mod}^{\bullet}\,\Bbbk[X].$$

- If there are n stops at the same end then there are n different Lagrangians that go from the bottom to the top. Their endomorphism ring is the path algebra of the cyclic quiver. There is also, for every sector and every k, a Lagrangian that comes from the top, then moves k sectors to the right in $\mathbb{X}_0$ and then goes upwards again. Finally, there are also the compact Lagrangians. If we compare this to Sections 4.2.3 and 4.3.1 we can conclude that

$$^{\mathtt{wr}}\mathtt{Fuk}^{\bullet}{}_{\sigma}(\mathbb{X}) \stackrel{\infty}{=} \mathtt{D\,Mod}^{\bullet}\,\Bbbk[X] \star \mathbb{Z}_n.$$

- If there is one stop at each end, a comparison with Section 4.4.2 will show that

$$^{\mathtt{wr}}\mathtt{Fuk}^{\bullet}{}_{\sigma}(\mathbb{X}) \stackrel{\infty}{=} \mathtt{D\,Mod}^{\bullet}\,\circ\!\!\Longrightarrow\!\!\circ.$$

Given a Liouville manifold $\mathbb{X}$ and a set of stops σ, there is a natural A_∞-functor coming from the natural embedding of the precategories:

$$^{\mathtt{wr}}\mathtt{Fuk}^{\bullet}{}_{\sigma}(\mathbb{X}) \to {}^{\mathtt{wr}}\mathtt{Fuk}^{\bullet}(\mathbb{X}).$$

In each of the examples above, the functor is in fact a *localization*: a universal construction that inverts all morphisms whose cones lie in a given subcategory. In the first example we invert all positive powers of X. These are precisely the morphisms whose cone (cokernel) is a nilpotent $\Bbbk[X]$-module or, in other words, those that can be seen as finite-dimensional modules of $\Bbbk[[X]]$. We write this as

$$\mathtt{D\,Mod}^{\bullet}\,\Bbbk[X, X^{-1}] \stackrel{\infty}{=} \frac{\mathtt{D\,Mod}^{\bullet}\,\Bbbk[X]}{\mathtt{D\,mod}^{\bullet}\,\Bbbk[[X]]}.$$

The behavior of partially wrapped Fukaya categories and the connection between removing stops and localization are studied in [236] and [96].

6.3.4 Directed Fukaya Categories

Directed Fukaya categories are a way to define Fukaya categories for Landau–Ginzburg models. In 1998, Kontsevich proposed these models to study mirror symmetry for Fano varieties [142]. Their symplectic geometry is captured by the directed Fukaya category, which is also known as the Fukaya–Seidel category. The details of its construction were laid out by Seidel in his book [218]. A slightly different approach can be found in the work by Abouzaid [5], which establishes the connection with mirror symmetry for toric Fano varieties.

Definition 6.53 A *Landau–Ginzburg model* is a pair $(\mathbb{X}, W)$, where $\mathbb{X} = (\mathbb{X}, \theta, J)$ is an exact Kähler manifold and $W\colon \mathbb{X} \to \mathbb{C}$ a holomorphic map.

We could think of $\mathbb{X} = \mathbb{C}^n$ with $z_i = p_i + iq_i$ and $\theta = p_i\, dq^i$, but more fancy examples such as Liouville manifolds with cylindrical ends will also work.

Definition 6.54 A point $p \in \mathbb{X}$ is called *critical* if $\partial_{z_i} W(p) = 0$ for some local complex coordinates. The value $W(p)$ is called a *critical value*. A critical point is *nondegenerate* if the Hessian $\partial_{z_i}\partial_{z_j} W(p)$ is nondegenerate. If W has only a finite number of critical points $\{p_1, \ldots, p_k\}$, which are all nondegenerate, we will call $W\colon \mathbb{X} \to \mathbb{C}$ a *Lefschetz fibration*.

A Lefschetz fibration is the complex analog of a Morse function, but its behavior is rather different. Unlike in the real case, the Hessian can always be transformed into an identity matrix, so there is no notion of an index. Furthermore, if we look at the fibers of a Morse function on opposite sides of a critical value, these fibers will in general be different, but for a Lefschetz fibration all fibers of noncritical values will be isomorphic.

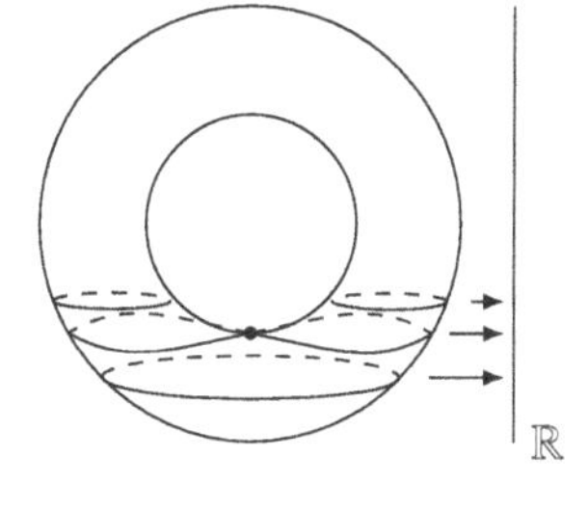

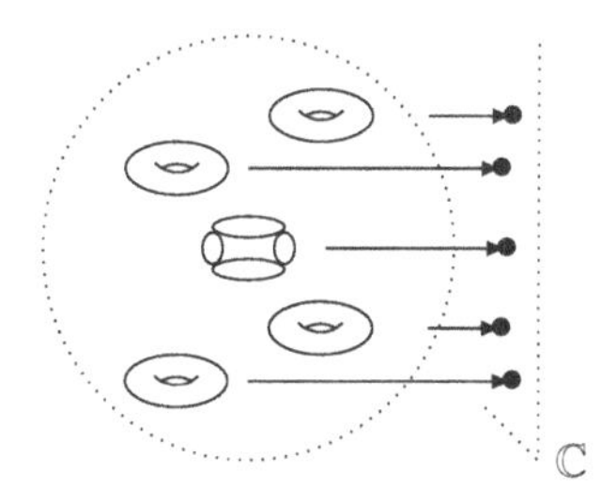

The reason for this is that we can connect any two noncritical values b, c by a path that avoids critical values. Take any such path $\gamma\colon [0, 1] \to \mathbb{C}$ with $\gamma(0) = b$ and $\gamma(1) = c$. Choose a smooth function $h := \mathbb{C} \to \mathbb{R}$ that is constant along γ and varies perpendicular to it. The function hW generates a Hamiltonian flow on $\mathbb{X}$ for which hW remains constant. This means that points that are projected by W onto the curve γ will stay projected onto the curve under the flow, and the flow will induce an isomorphism between $W^{-1}(b)$ and $W^{-1}(c)$.

If $c = W(p)$ is a critical value then $W^{-1}(c)$ will be singular in p. The flow construction from above breaks down because p will be a fixed point for the flow. This means that no point can reach it in finite time but points can flow towards p.

Definition 6.55 Let $\gamma\colon [0,1] \to \mathbb{C}$ be a smooth embedding with $\gamma(0) = b$ an ordinary complex number and $\gamma(1) = c$ a critical value coming from a critical point p. The *vanishing cycle* at $t \in [0,1]$ is the set of all points in the fiber $W^{-1}(\gamma(t))$ that flow to the critical point p. The *Lefschetz thimble* is the union of the vanishing cycle for all $t \in [0,1]$:

$$\mathbb{V}_{\gamma,p}(t) = \{x \in W^{-1}(\gamma(t)) \mid \lim_{u\to+\infty} \phi_u^{hW}(x) = p\} \quad \text{and} \quad \mathbb{U}_{\gamma,p} = \cup_{t\in[0,1]} \mathbb{V}_\gamma(t).$$

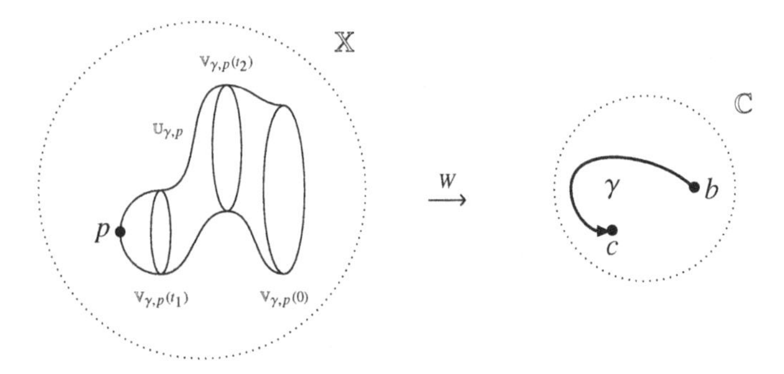

Example 6.56 Let $W\colon \mathbb{C}^n \to \mathbb{C}\colon (z_i) \mapsto \sum z_i^2$, $\gamma\colon [0,1] \to \mathbb{D}\colon t \mapsto (1-t)$ and $h\colon \mathbb{C} \to \mathbb{C}\colon z \mapsto \Im z$. If we set $z_i = q_i + ip_i$, we get $hW = \sum_i 2p_iq_i$ and hence the flow equations become

$$\dot{q}_i = -\frac{\partial hW}{\partial p_i} = -2q_i, \quad \dot{p}_i = \frac{\partial hW}{\partial q_i} = 2p_i,$$

which has as solutions $(q_i^0 e^{-2t}, p_i^0 e^{2t})$. The points that go to the critical point 0 are those with zero p_i^0. This means that

$$\mathbb{V}_{\gamma,0}(t) = \left\{(\vec{q},0) \mid \textstyle\sum_i q_i^2 = (1-t)\right\} \quad \text{and} \quad \mathbb{U}_{\gamma,0} = \left\{(\vec{q},0) \mid \textstyle\sum_i q_i^2 \le 1\right\}.$$

The vanishing cycle at $t = 0$ is a Lagrangian sphere inside the hypersurface $\sum z_i^2 = 1$, while the thimble is a real ball inside $\mathbb{C}^n$. If we take a different h, e.g. $h = g \cdot \Im z$ with $g\colon \mathbb{C} \to \mathbb{R}$, then the equations become

$$\dot{q}_i = -2q_i\left(g + p_i\frac{\partial gW}{\partial p_i}\right), \quad \dot{p}_i = 2p_i\left(g + q_i\frac{\partial gW}{\partial q_i}\right).$$

The exact solution of these equations is different but if $p_i = 0$ it will stay zero so the vanishing cycle and thimble stay the same.

Lemma 6.57 *The vanishing cycle and the Lefschetz thimble are exact Lagrangian submanifolds of $W^{-1}(\gamma(t))$ and $\mathbb{C}^n$. They only depend on γ and W, not on h.*

Sketch of the proof We can treat this problem locally and then it reduces to the calculation above. □

Assume now that all critical values lie inside a disk $\mathbb{D} \subset \mathbb{C}$ and choose a base point b on the boundary of the disk. For any curve from the base point to a critical value $\gamma\colon b \to c$ and a critical point $p \in W^{-1}(c)$, we can construct a vanishing cycle and a thimble. These will be different for different curves even if they flow towards the same critical point. To study the thimbles we would like to put them into a category.

Definition 6.58 The *directed Fukaya precategory* ${}^{\mathrm{dr}}\underline{\mathtt{Fuk}}^{\bullet}(\mathbb{X}, W)$ has as objects all the Lefschetz thimbles $\mathbb{U}_{\gamma,p}$, where $\gamma\colon [0,1] \to \mathbb{D}$ is a curve from the base point b to any critical value $c = W(p) \in \mathbb{D}$ such that the tangent vector of γ at b points inwards in the disk. On these thimbles we put orientations, spin structures and local systems.

A sequence of objects $(\mathbb{U}_{\gamma_1,p_1}, \ldots, \mathbb{U}_{\gamma_k,p_k})$ is transversal if the thimbles intersect transversally and the arguments $\dot{\gamma}_1(0), \ldots, \dot{\gamma}_k(0)$ increase.

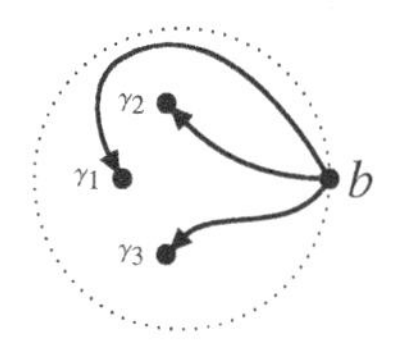

In other words, if we walk anticlockwise in a small circle around b then after entering $\mathbb{D}$ we first cross γ_1, then γ_2 and so on. The hom-spaces and products are defined in the usual way. The *directed Fukaya category* ${}^{\mathrm{dr}}\mathtt{Fuk}^{\bullet}(\mathbb{X}, W)$ is its categorification.

Remark 6.59 Although its definition depends on the boundary point (and the size of the disk), different choices will result in equivalent categories because we can find a flow that rotates the boundary and keeps the critical points fixed. This flow will induce an equivalence between the directed Fukaya categories for different base points.

Example 6.60 Take $\mathbb{X} = \mathbb{C}^*$ and let $W\colon \mathbb{C}^* \to \mathbb{C}\colon z \mapsto \frac{z-z^{-1}}{4}$. The critical points are $\pm i$ with critical values $\pm\frac{i}{2}$. If we take $b = 1$ then $W^{-1}(b) = \{p_+, p_-\}$ with $p_\pm = 2 \pm \sqrt{5}$. Any given curve $\gamma\colon 1 \to a$ can be lifted along W to two curves $\gamma^\pm := \gamma \pm \sqrt{4\gamma^2 + 1}$. The vanishing cycle at t will be $\{\gamma_+(t), \gamma_-(t)\}$ and the thimble will be the union of the two curves. Let us look at three examples:

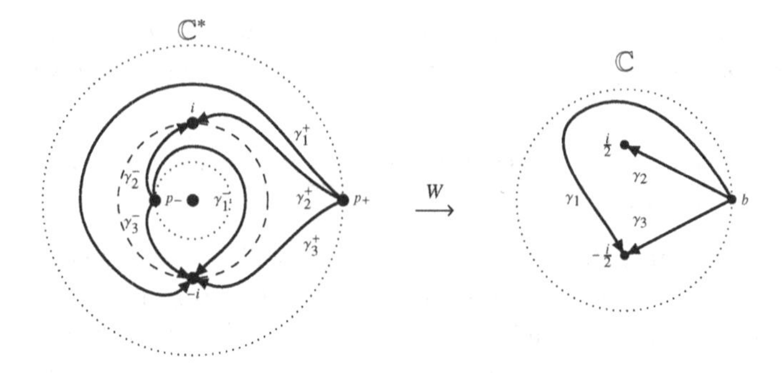

As there are no immersed disks, we have

$$
\begin{aligned}
{}^{\mathrm{dr}}\mathbf{Fuk}^{\bullet}(\mathbb{U}_1, \mathbb{U}_2) &= \mathbb{k}p_+ + \mathbb{k}p_-, \\
{}^{\mathrm{dr}}\mathbf{Fuk}^{\bullet}(\mathbb{U}_2, \mathbb{U}_3) &= \mathbb{k}p_+ + \mathbb{k}p_-, \\
{}^{\mathrm{dr}}\mathbf{Fuk}^{\bullet}(\mathbb{U}_1, \mathbb{U}_3) &= \mathbb{k}p_+ + \mathbb{k}p_- + \mathbb{k}c,
\end{aligned}
$$

where c represents the point $-i$.

If we unroll the cylinder we see a remarkable similarity with the partially wrapped Fukaya category of $\mathsf{T}^*\mathbb{S}^1$ with two stops, one at each side. If we place stops next to p_+ and p_- and prolong the thimbles to infinity on both ends, we get objects in the partially wrapped category whose morphism spaces are exactly the same.

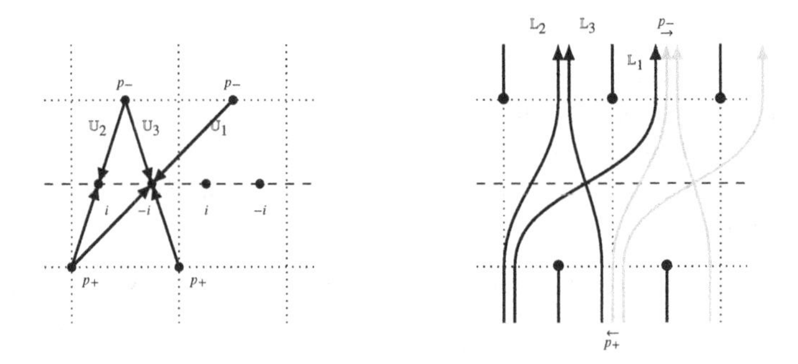

With some extra work this can be extended to a functor

$$
{}^{\mathrm{dr}}\mathbf{Fuk}^{\bullet}(\mathbb{C}^*, W) \to {}^{\mathrm{wr}}\mathbf{Fuk}^{\bullet}{}_{\sigma}(\mathbb{C}^*),
$$

which becomes an equivalence if we go to the derived categories. This example illustrates that, in many cases, partially wrapped and directed Fukaya categories can be seen as two sides of the same coin.

6.3.5 Relative Fukaya Categories

The settings we have so far considered are all examples where the symplectic manifold is not compact. This is because we have worked only with exact symplectic manifolds.

Lemma 6.61 *An exact symplectic manifold cannot be compact.*

Sketch of the proof If $(\mathbb{X}, \omega)$ is a $2n$-dimensional symplectic manifold then ω^n will be a volume form for $\mathbb{X}$, but if $\omega = d\theta$ then the cohomology class of ω and hence ω^n will be zero, which is impossible. □

Defining a Fukaya category for compact manifolds is quite difficult and therefore we will first construct an intermediate, which was introduced by Seidel in [217]. Let $\mathbb{X}$ be a compact Kähler manifold and let $\mathbb{D} = \cup_{i\in I}\mathbb{D}_i$ be a union of smooth transversally intersecting complex hypersurfaces such that $\mathbb{X} \setminus \mathbb{D}$ is Liouville with cylindrical ends. This means we can define the Fukaya category over $\mathbb{X} \setminus \mathbb{D}$ in the usual way.

What happens if we add in $\mathbb{D}$? Each object in ${}^{\mathrm{ex}}\mathtt{Fuk}^{\bullet}(\mathbb{X} \setminus \mathbb{D})$ can also be seen as a Lagrangian brane in $\mathbb{X}$ and the intersection points between such objects are the same in $\mathbb{X}$ as in $\mathbb{X} \setminus \mathbb{D}$. The difference lies, however, in the products: there could be additional pseudo-holomorphic disks in $\mathbb{X}$ that intersect divisors and there could even be an infinite number of them.

In order to keep track of them we consider the monoid G freely generated by the $\mathbb{D}_i$ and its formal power series ring $\mathbb{k}[[G]]$. Using the technology described in Section 3.4.3 it should be possible to deform the exact Fukaya category ${}^{\mathrm{ex}}\mathtt{Fuk}^{\bullet}(\mathbb{X} \setminus \mathbb{D})$ to a curved G-filtered A_∞-category ${}^{\mathrm{cu}}\mathtt{Fuk}^{\bullet}(\mathbb{X}, \mathbb{D})$ such that for each $g = g_1\mathbb{D}_1 + \cdots + g_k\mathbb{D}_k$, we have

$$\mu_{k,g}(p_1, \ldots, p_k) = \sum_{p_0} \sum_{\psi} \sigma^{\psi} p_0,$$

where ψ runs over all holomorphic maps whose intersection number with $\mathbb{D}_i$ is g_i. For this to work there are a number of issues that need to be addressed:

- For each intersection type there should only be a finite number of disks.
- There could be disk bubbles leading to a μ_0.
- There could be sphere bubbles, which could even be wholly contained in a divisor, resulting in negative intersection numbers.

To resolve this, one has to put restrictions on the type of divisors that are allowed and possibly to work over larger coefficient rings. For more details we refer to [217] and [224]. A construction in full generality is still work in progress but conceptually the idea is the following.

Definition 6.62 The *relative Fukaya category* ${}^{\mathrm{cu}}\mathtt{Fuk}^{\bullet}(\mathbb{X},\mathbb{D})$ is a curved deformation of the exact Fukaya category ${}^{\mathrm{ex}}\mathtt{Fuk}^{\bullet}(\mathbb{X}\setminus\mathbb{D})$ whose products count holomorphic disks sorted by intersection type with $\mathbb{D}$.

Remark 6.63 An important variation to this recipe is the following. Instead of viewing $\mathbb{D}$ as a union of components $\mathbb{D}_i$, we could see $\mathbb{D}$ as an $\mathbb{N}$-linear combination $\mathbb{D} = \sum a_i\mathbb{D}_i$. This determines a map $\mathbb{k}[[G]] \to \mathbb{k}[[\hbar]]\colon \hbar^{\mathbb{D}_i} \mapsto \hbar^{a_i}$ which can be used to view the relative Fukaya category as a deformation over the ring $\mathbb{k}[[\hbar]]$ instead of $\mathbb{k}[[G]]$.

Remark 6.64 There are different ways to turn this curved category into an uncurved one. The easiest is to take the curved twisted complexes and look at the subcategory of the objects with curvature $\widehat{\mu}_0 = \lambda\mathbb{1}$ with $\lambda \in \mathbb{k}[[G]]$ (or $\mathbb{k}[[\hbar]]$). If we forget about $\widehat{\mu}_0$, this is an uncurved filtered category, which we denote by ${}^{\lambda}\mathtt{Fuk}^{\pm}(\mathbb{X},\mathbb{D})$. If $\mathbb{X}$ is Calabi–Yau we can put a $\mathbb{Z}$-grading on the Fukaya category and then the degree of λ has to be 2, so λ can only be 0. In that case, we drop the prefix and write $\mathtt{Fuk}^{\bullet}(\mathbb{X},\mathbb{D})$. These categories are defined over $\mathbb{k}[[G]]$, which is not a field so we have to be careful when taking minimal models.

We can follow the strategy of Section 3.4.3 and tensor this category with $\mathbb{k}((G))$ to get an A_∞-category over a field and take the minimal model. We will denote this category by ${}^{\lambda}\mathtt{DFuk}^{\pm}(\mathbb{X},\mathbb{D}) \otimes \mathbb{k}((G))$ or $\mathtt{D}\,\mathtt{Fuk}^{\bullet}(\mathbb{X},\mathbb{D}) \otimes \mathbb{k}((G))$.

In some cases it is possible that the products $\bar{\mu}_k$ in ${}^{\mathrm{cu}}\mathtt{Fuk}^{\pm}(\mathbb{X},\mathbb{D})$ are not formal power series, but actual polynomials. Then we can specialize these polynomials to values in $\mathbb{C}$ by choosing a map $\rho\colon G \to \mathbb{C}$. If, additionally, $\bar{\mu}_k = 0$ for $k \gg 0$ we look at the twisted complexes for which $\rho(\hat{\mu}_0) = \lambda\mathbb{1}$ with $\lambda \in \mathbb{C}$. This results in uncurved A_∞-categories over $\mathbb{C}$.

Example 6.65 Let $\mathbb{X}$ be the Riemann sphere and take the divisor $\mathbb{D} = \{0\} + \{\infty\}$. In that case, $\mathbb{X} \setminus \mathbb{D} = \mathbb{C}^* = \mathsf{T}^*\mathbb{S}^1$ and we work over $\mathbb{k}[[\hbar]]$. The relative Fukaya category is a deformation of

$${}^{\mathrm{ex}}\mathtt{Fuk}^{\pm}(\mathbb{C}^*) \overset{\infty}{=} \mathtt{Dmod}^{\pm}\mathbb{C}[X^{\pm 1}].$$

Note that it can only be a $\mathbb{Z}_2$-graded deformation because the volume form dz/z does not extend to the Riemann sphere.

If $(\mathbb{L},\rho_M)$ is the base circle with a local system given by a parallel transport $M \in \mathbb{C}^*$, then there are two holomorphic disks that bound $\mathbb{S}^1$: the upper half of the sphere and the lower half. Each half goes through one of the points $\{0,\infty\}$ and the orientations of the boundaries are opposite, so their transport along the circle are inverses. If we pack all the μ_0-products together we get

$$\hat{\mu}_0|_{(\mathbb{L},\rho)} = (M\hbar + M^{-1}\hbar)\iota.$$

To determine the products between endomorphisms of $(\mathbb{L}, \rho_M)$ we look at a sequence of Lagrangians homotopic to the circle:

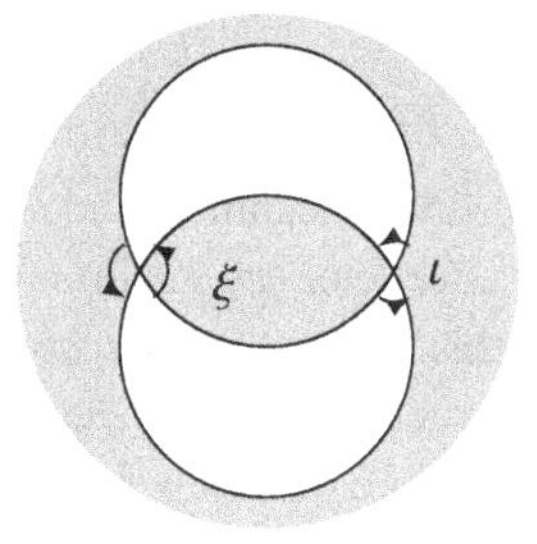

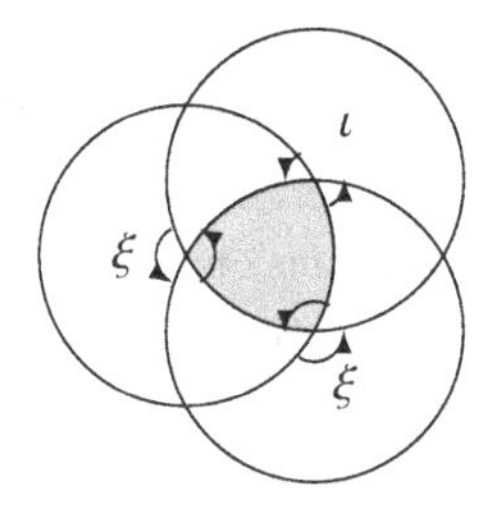

In this picture we stereographically projected the Riemann sphere onto the plane. For the unary product there are two disks contributing; for all other products there is only one because the other wraps multiple times around ∞, so it is not an immersion:

$$\hat{\mu}_1(\xi) = (M\hbar - M^{-1}\hbar)\iota,$$
$$\hat{\mu}_k(\xi, \ldots, \xi) = M\hbar\iota.$$

So unless $M = \pm 1$, the object $(\mathbb{L}, \rho_M)$ becomes a zero object. Let us denote the remaining two objects by $\mathbb{L}_\pm$. They are curved, with curvatures $\pm 2\hbar$, but we can consider them as objects in the uncurved categories ${}^{\pm 2\hbar}\mathbf{Fuk}^{\pm}(\mathbb{X}, \mathbb{D}) \otimes \Bbbk(\hbar)$. After an A_∞-isomorphism we can set the products $\hat{\mu}_{\geq 3}$ to zero and therefore we get that the endomorphism rings of these objects are

$$\frac{\Bbbk(\hbar)[\xi]}{(\xi^2 \pm \hbar)}.$$

6.3.6 The Full Fukaya Category

We arrive at the final stage of our journey: the full Fukaya category. For a given compact Kähler manifold $(\mathbb{X}, \omega, J)$, this category should include all Lagrangian branes. As we have already mentioned, the nonexactness implies that the symplectic area of the disks is not bounded above and hence we cannot apply Gromov's compactness theorem to show that there are only finitely many holomorphic polygons with boundaries on a given set of Lagrangian submanifolds. Therefore, the products can contain an infinite number of terms and might not be well defined.

Still, for a given upper bound the number of holomorphic polygons with an area below that bound will always be finite. So in general the product will be a series of terms corresponding to increasing area. This makes it possible to work in the filtered setting. Let $G \subset \mathbb{R}_{\geq 0}$ be the monoid generated by all areas that can occur as holomorphic disks. The ring $\Bbbk[\![G]\!]$ can be considered as a subring of the *Novikov field*, which consists of all power series of increasing real exponents:

$$\mathbb{A} = \left\{ \sum_{i=0}^{\infty} \lambda_i T^{r_i} \mid \lambda_i \in \Bbbk,\ r_i \in \mathbb{R} \text{ and } i > j \implies r_i > r_j \right\}.$$

Getting a well-defined category out of this is still very technical and requires some restrictions on the types of manifolds (e.g. acyclic or $\langle \pi_2(\mathbb{X}), \omega \rangle = 0$) and branes (e.g. monotone) we can include. If all this is taken into account we end up with the full Fukaya category. The construction of this category and its technical details are far beyond our scope. For more details we refer to the monographs [83, 84].

In the ideal situation, we end up with a G-filtered curved A_∞-category ${}^{\mathsf{cu}}\mathbf{Fuk}^{\pm}_{\mathbb{A}}(\mathbb{X})$ whose objects are compact Lagrangian branes with U_n-local systems. For any $\lambda \in \mathbb{A}$ the curved twisted objects with curvature λ form an ordinary A_∞-category ${}^{\lambda}\mathbf{Fuk}^{\pm}_{\mathbb{A}}(\mathbb{X})$ over the Novikov field.

Remark 6.66 The reason to work only with U_n-local systems instead of GL_n local systems is that the real part of the local system can be integrated in the position of the Lagrangian brane. For example, if we take $\mathsf{T}^*\mathbb{S}_1 = \mathbb{C}^*$ then the base circle $\{|z| = 1\}$ with a real local system can be interpreted as a shifted circle $\{|z| = r\}$ with a trivial local system. Note that the latter is not an exact Lagrangian, so it does not sit in ${}^{\mathsf{ex}}\mathbf{Fuk}^{\bullet}(\mathbb{C}^*)$ but it is an object in $\mathbf{Fuk}^{\bullet}_{\mathbb{A}}(\mathbb{C}^*)$.

Example 6.67 Let $\mathbb{T} = \mathbb{C}/(\mathbb{Z} + i\mathbb{Z})$ be the torus with the standard symplectic structure $z = q + ip$ and $\omega = a\, dp \wedge dq$. We consider three objects $\mathbb{L}_0, \mathbb{L}_1, \mathbb{L}_2$ in the Fukaya category, corresponding to the Lagrangian submanifolds with equations $q = 0$, $p = 0$ and $p + q = \frac{1}{2}$ respectively. There are three intersection points $p_1 = (0, 0) \in \mathbb{L}_0 \cap \mathbb{L}_1$, $p_2 = (\frac{1}{2}, 0) \in \mathbb{L}_1 \cap \mathbb{L}_2$ and $p_0 = (0, \frac{1}{2}) \in \mathbb{L}_0 \cap \mathbb{L}_2$.

To find the product $\mu(p_2, p_1)$ we need to count immersed triangles with p_0, p_1, p_2 as corners. If we go to the universal cover, all these triangles are isosceles right-angled triangles. If we fix p_1, such a triangle is fixed by the length of one of its sides and its orientation. The side is an odd number times one half, $\frac{2k+1}{2}$, and the area of the triangles is $\frac{a}{8}(2k+1)^2$. The resulting power series can be neatly interpreted in terms of the Jacobi theta function $\vartheta_2(z, t)$:

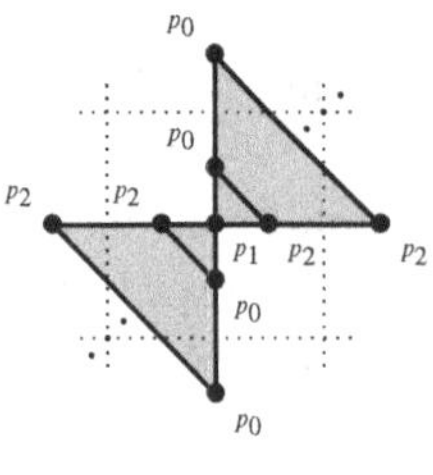

$$\hat{\mu}(p_2, p_1) = \sum_{k=0}^{+\infty} 2T^{\frac{a}{8}(2k+1)^2} p_0 = \vartheta_2(0, T^{\frac{a}{2}}) p_0.$$

6.3.7 Connections

Suppose $(\mathbb{X}, \omega, J)$ is a Kähler manifold with divisor $\mathbb{D}$ such that $\check{\mathbb{X}} := \mathbb{X} \setminus \mathbb{D}$ is Liouville, and let $W\colon \check{\mathbb{X}} \to \mathbb{C}$ be a holomorphic function. In the ideal situation, we can find a set of stops σ that blocks all Reeb chords between the different pieces of the conical ends of the thimbles. In this situation we get a nice diagram of A_∞-categories.

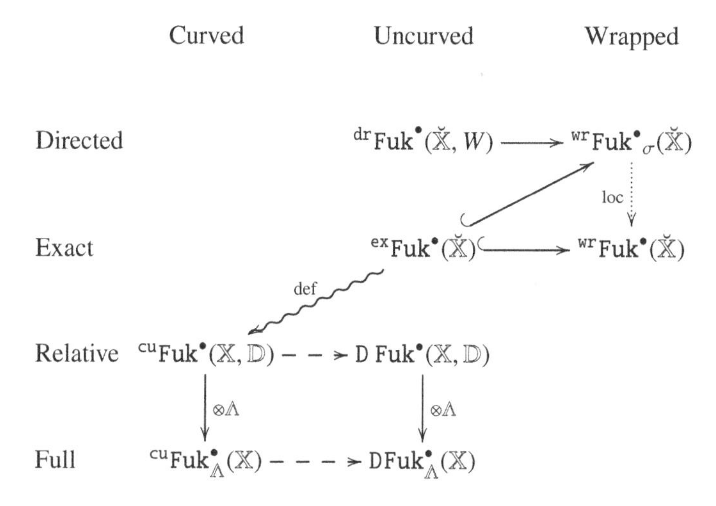

The snake arrow is a deformation, the dotted arrow a localization and the dashed arrows represent taking uncurved twisted complexes. The arrows pointing to the bottom row correspond to a change of base ring followed by an embedding.

6.4 Generators

We have seen many different possibilities for defining a Fukaya category. We can include or exclude different types of objects or allow or disallow certain morphisms. This leads to different but related categories. A part of this problem can be solved by going to the derived or split-derived categories. In this way we include all objects that should be there from a representation-theoretical point of view.

This approach also often makes it easier to describe the category because we can look for generators and try to view the category as the twisted completion of a small number of objects. Unlike for the whole category, it is then possible to describe all or most of the products of morphisms between these objects and in this way we get a fairly detailed description of the split-derived Fukaya category.

6.4.1 Cones in Fukaya Categories

To find out what it geometrically means to split-generate the Fukaya category we first need to look at the cone construction. Consider two Lagrangian submanifolds $\mathbb{L}_1$, $\mathbb{L}_2$ that intersect transversally at one point x. Around x we can choose canonical coordinates p_i, q_i such that $\mathbb{L}_1$ is given by the equations $p_i = 0$ and $\mathbb{L}_2$ by the equations $q_i = 0$.

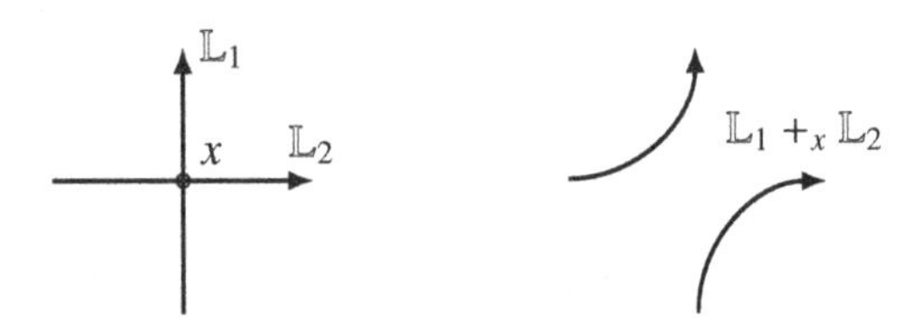

For any $\epsilon > 0$ the submanifold $\mathbb{L}_\epsilon$ given by the equations $p_i q_i + \epsilon = 0$ is a Lagrangian submanifold and for different ϵ there is a Hamiltonian isotopy connecting these submanifolds: ϕ_t^h with $f = \sum_i \ln p_i q_i$ in a neighborhood around $\mathbb{L}_\epsilon$. We can make a Lagrangian submanifold $\mathbb{L}_1 +_x \mathbb{L}_2$ that is equal to $\mathbb{L}_\epsilon$ in an open ball around x and equal to $\mathbb{L}_1 \cup \mathbb{L}_2$ outside a slightly larger open ball around x. In between the two balls we fudge the Lagrangian submanifold a bit to make the two pieces match up. This procedure works for any x but if we want the orientations to agree we also need that the $\mathbb{Z}_2$-degree of x is 1.

Theorem 6.68 *The object* $\mathbb{L}_1 +_x \mathbb{L}_2$ *is isomorphic to* $\mathrm{Cone}(x)$ *where* x *is considered as a degree* 1 *morphism in* $x\colon \mathbb{L}_1 \to \mathbb{L}_2$.

Sketch of the proof Take any Lagrangian submanifold $\mathbb{U}$ that transversally intersects $\mathbb{L}_1 +_x \mathbb{L}_2$ and does not enter the open ball around x where we performed the gluing. Because every intersection point lies either on $\mathbb{L}_1$ or on $\mathbb{L}_2$, the Floer complex splits as a direct sum

$$\mathsf{CF}(\mathbb{L}_1 +_x \mathbb{L}_2, \mathbb{U}) = \mathsf{CF}(\mathbb{L}_1, \mathbb{U}) \oplus \mathsf{CF}(\mathbb{L}_2, \mathbb{U}).$$

If we have a holomorphic n-gon for which one of the boundaries lies on $\mathbb{L}_1 +_x \mathbb{L}_2$ we can deform this by letting ϵ go to 0. In this way we get a holomorphic disk that has a boundary on $\mathbb{L}_1 \cup \mathbb{L}_2$. There are three possibilities: this boundary lies entirely on $\mathbb{L}_1$, entirely on $\mathbb{L}_2$ or it lies on both. In the first two cases the n-gon remains an n-gon, while in the last case it becomes an $(n+1)$-gon with x as an extra corner. Therefore, an $(n-1)$-ary product featuring $\mathbb{L}_1 +_x \mathbb{L}_2$ is the sum of the ordinary $(n-1)$-ary product and the n-products where one extra x is inserted. This is exactly how the product works in the cone construction. □

Remark 6.69 If $\mathbb{L}_1$ and $\mathbb{L}_2$ have more than one intersection point then $\mathbb{L}_1 +_x \mathbb{L}_2$ is not really an embedded Lagrangian but an immersed one. Another possibility is that $\mathbb{L}_1 +_x \mathbb{L}_2$ consists of different components. In that case, we should see each component as a direct summand of $\mathrm{Cone}(x)$.

Remark 6.70 There are several extensions to this construction.

- In the case where $\mathbb{L}_1$ and $\mathbb{L}_2$ come with local systems and the morphism is of the form αx, where $\alpha\colon \mathbb{L}_1(x) \to \mathbb{L}_2(x)$ is an invertible linear map, we can define a new local system on $\mathbb{L}_1 +_x \mathbb{L}_2$. Choose a base point on each component of $\mathbb{L}_1 +_x \mathbb{L}_2$ that under the degeneration to $\mathbb{L}_1 \cup \mathbb{L}_2$ does not go to x. Each loop through a base point on $\mathbb{L}_1 +_x \mathbb{L}_2$ degenerates into a broken path on $\mathbb{L}_1 \cup \mathbb{L}_2$ consisting of pieces in $\mathbb{L}_1$ and $\mathbb{L}_2$ glued at the ends in x. Use the local systems on $\mathbb{L}_1$ and $\mathbb{L}_2$ to get linear maps of the broken pieces and glue them together with α each time we go from $\mathbb{L}_1$ to $\mathbb{L}_2$ and with α^{-1} in the opposite direction.
- In the case where $\mathbb{L}_1$ and $\mathbb{L}_2$ intersect at multiple points $x_1, \dots, x_k$ we can interpret the cone of $\alpha_1 x_1 + \cdots + \alpha_k x_k$ as a new Lagrangian where we perform this construction at all points at once.

6.4.2 Generating Cotangent Bundles

Let us look at the case of the cylinder $\mathbb{X} = \mathsf{T}^*\mathbb{S}^1 = \mathbb{C}^*$ with the standard volume form dz. We had the following tower of inclusions:

$$^{\mathsf{ba}}\mathsf{Fuk}^\bullet\,\mathbb{X} \subset {}^{\mathsf{ex}}\mathsf{Fuk}^\bullet\,\mathbb{X} \subset {}^{\mathsf{wr}}\mathsf{Fuk}^\bullet\,\mathbb{X}.$$

The first had only one object $\mathbb{L}_1$ up to shift, with endomorphism ring $\mathbb{k}[\xi]/(\xi^2)$. In the second one, every indecomposable object corresponded up to shift with an object $\mathbb{L}_\rho$ consisting of a circle and a local system ρ. In the third one there are extra noncompact Lagrangians, such as the cotangent fiber $\mathbb{U} = \mathsf{T}^*_x\mathbb{S}^1$.

The latter has endomorphism ring $\mathbb{k}[X, X^{-1}] \cong \mathbb{k}[\pi_1(\mathbb{S}^1)]$ and it generates $^{\mathsf{wr}}\mathsf{Fuk}^\bullet\,\mathbb{X}$: if $\mathbb{L}$ is a noncompact Lagrangian it is isomorphic to a shift of $\mathbb{U}$, and if it is a circle with local system $\rho\colon \pi_1(\mathbb{S}^1) \to \mathsf{GL}_n$ then we can identify $\mathbb{L}$ with $\mathrm{Cone}(f)$, where $f = (1 + \rho(\ell))\mathbb{1}_n$ is seen as a degree 1 morphism from $\mathbb{U}^{\oplus n}$ to $\mathbb{U}[-1]^{\oplus n}$ and ℓ is the generator of the fundamental group of the circle.

Geometrically this can be seen as follows. If we take the cone over $\mathbb{1} + \rho(\ell)X$ we glue $\phi_1^h(\mathbb{U})$ and $\mathbb{U}$ together at two points. This results in three curves: one closed curve and two open curves, one at each end of the cylinder. The latter two are zero objects, while the closed curve is a compact Lagrangian with a local system. The transport around the circle is equal to $\rho(\ell)$ because at one point we connect the fibers with the identity matrix and at the other end with $\rho(\ell)$.

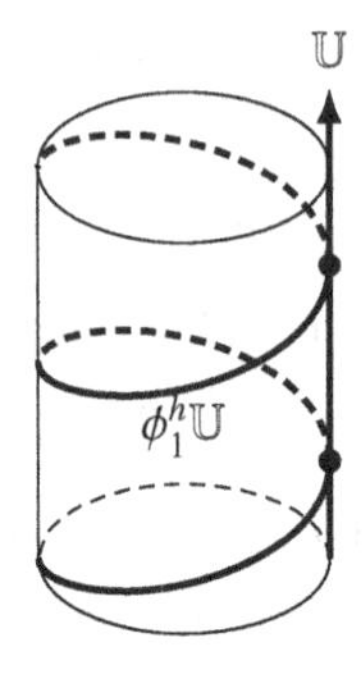

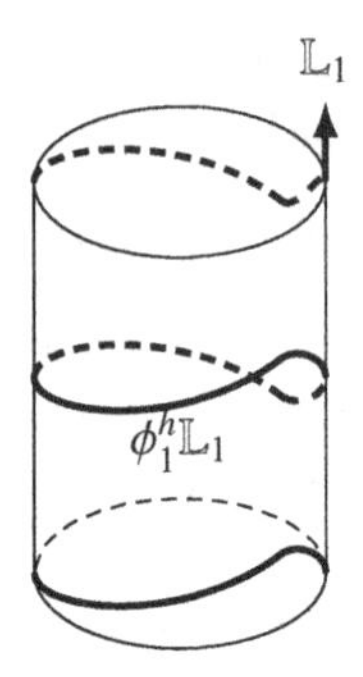

If we look at the exact Fukaya category, the situation is different. Because of the lack of these noncompact objects there is no finite set of generators. To generate all possible local systems we need a generator for each possible eigenvalue in the $\rho(\ell)$, so the split-derived Fukaya category is generated by all $\mathbb{L}_\lambda$, corresponding to a circle with a one-dimensional local system for which $\rho(\ell) = \lambda$. Out of these only the object $\mathbb{L}_1$ is contained in the basic Fukaya category and generates it:

$$\begin{array}{ccccc} \mathsf{D}^{\pi\,\mathsf{ba}}\mathsf{Fuk}^\bullet\,\mathbb{X} & \subset & \mathsf{D}^{\pi\,\mathsf{ex}}\mathsf{Fuk}^\bullet\,\mathbb{X} & \subset & \mathsf{D}^{\pi\,\mathsf{wr}}\mathsf{Fuk}^\bullet\,\mathbb{X} \\ \wr\| & & \wr\| & & \wr\| \\ \langle \mathbb{L}_1 \rangle & \subset & \langle \mathbb{L}_\lambda | \lambda \in \mathbb{k}^* \rangle & \subset & \langle \mathbb{U} \rangle. \end{array}$$

The generation result for the wrapped Fukaya category of the cylinder can be generalized to all cotangent bundles. This has been worked out in detail by Abouzaid in [7].

Theorem 6.71 (Abouzaid) *Let $\mathbb{M}$ be a compact connected Riemannian manifold. The wrapped Fukaya category of the cotangent bundle $\mathbb{X} = \mathsf{T}^*\mathbb{M}$ is generated by a cotangent fiber:*

$$\mathsf{D}^{\pi\,\mathsf{wr}}\mathsf{Fuk}^\bullet\,\mathsf{T}^*\mathbb{M} = \langle \mathsf{T}^*_x\mathbb{M} \rangle.$$

Sketch of the proof First of all note that all cotangent fibers are isomorphic. A geodesic between x and y induces a Reeb chord between $\mathsf{T}^*_x\mathbb{M}$ and $\mathsf{T}^*_y\mathbb{M}$, which is an invertible morphism because its inverse corresponds to the same geodesic traversed in the opposite direction.

Now take any Lagrangian submanifold $\mathbb{L}$ that intersects $\mathbb{M}$ transversally. Choose a function f on $\mathbb{M}$ that is only zero at the intersection points and look at the flow $\phi_t^f \mathbb{L}$. For $t \to \infty$ this flowed Lagrangian will become the union of cotangent fibers situated at the intersection points. This perspective allows us to see $\phi_t^f \mathbb{L}$ as a cone of a morphism between these cotangent fibers. This morphism is the sum of Reeb chords that, relative to the cotangent fibers, are isotopic to curves between these intersection points.

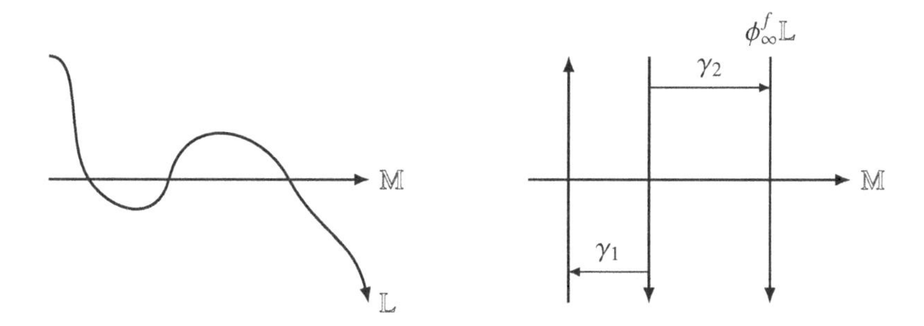

□

Corollary 6.72 (Abouzaid) *The wrapped Fukaya category of the cotangent bundle is split-derived equivalent to the cohomology ring of the loop space:*

$$\mathrm{D}^\pi\, {}^{\mathrm{wr}}\mathrm{Fuk}^\bullet\, \mathrm{T}^*\mathbb{M} \cong \mathrm{D}^\pi\, \mathrm{H}^\bullet(\mathcal{L}_x\mathbb{M}).$$

Sketch of the proof The Reeb chords from $\mathrm{T}_x\mathbb{M}$ to itself form the critical points of a Morse function on the loop space and therefore the ${}^{\mathrm{wr}}\mathrm{Fuk}^\bullet\, \mathbb{X}(\mathrm{T}_x\mathbb{M}, \mathrm{T}_x\mathbb{M})$ can be interpreted as a model for the cohomology of the loop space. □

6.4.3 Generating Directed Fukaya Categories

For directed Fukaya categories we can construct many sets of generators. Let $W\colon \mathbb{X} \to \mathbb{C}$ be a Lefschetz fibration with k different critical points p_i and corresponding critical values $c_i = W(p_i)$, which are all different. Choose k curves γ_i that start at the base point b and each go to a different critical value. Furthermore suppose that each γ_i starts in a different direction and they only intersect at b. We will call such a set an *exceptional collection*.

Theorem 6.73 (Seidel) *The set of thimbles of an exceptional collection split-generate the directed Fukaya category:*

$$\mathrm{D}^\pi\, {}^{\mathrm{dr}}\mathrm{Fuk}^\bullet(\mathbb{X}, W) = \langle \mathbb{U}_{\gamma_i} \mid i = 1, \ldots, k\rangle.$$

Sketch of the proof Every nonintersecting curve from b to a critical value can be extended to an exceptional collection. It can then be shown that each pair of exceptional collections can be turned into each other using a procedure called mutation. The mutation takes two curves next to each other at the base point and swaps their order by adding to one curve a loop around the other.

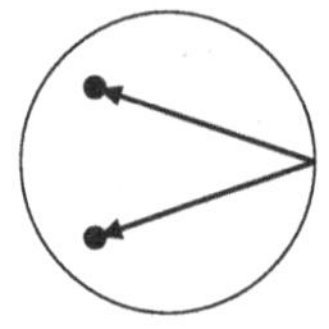
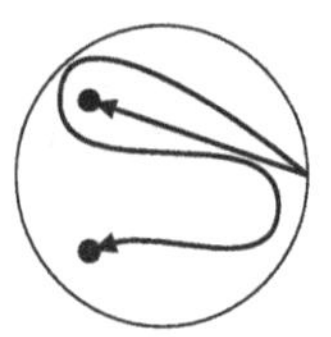

One can show that the new thimble in the mutated collection can be written as cones of the thimbles of the original one, and vice versa. Therefore, both collections generate the same category. As all thimbles occur in an exceptional collection, every exceptional collection generates all thimbles. □

Example 6.74 In Example 6.60 we studied the Lefschetz fibration $\pi\colon \mathbb{C}^* \to \mathbb{C}\colon z \mapsto \frac{z-z^{-1}}{4}$. This had two critical points with critical values $\pm\frac{i}{2}$. We considered three curves $\gamma_1, \gamma_2, \gamma_3$. Of these three curves γ_1, γ_2 form an exceptional collection and so do γ_2, γ_3. The former is a mutation of the latter. In both cases the endomorphism ring is the path algebra of the quiver $\circ \Longrightarrow \circ$. This implies that

$$\mathrm{D}^\pi\, {}^{\mathrm{dr}}\mathtt{Fuk}^\bullet(\mathbb{C}^*, W) \overset{\infty}{=} \mathrm{D}^\pi\,\mathtt{mod}^\bullet\, \circ \Longrightarrow \circ.$$

Example 6.75 Let $W\colon \mathbb{C} \to \mathbb{C}\colon Z \mapsto \frac{4}{5}Z^5 + Z$. The critical points of W are $\frac{\pm 1 \pm i}{2}$ with values $\frac{\pm 1 \pm i}{2}$. The fiber above the base point $1 \in \mathbb{C}$ consists of five points. In the picture below we have drawn six curves γ_i that connect the base point with these critical values. Each corresponds to a thimble $\mathbb{U}_i$ and a vanishing cycle $\mathbb{V}_i$. The vanishing cycle is a pair of points and the thimble a line joining them.

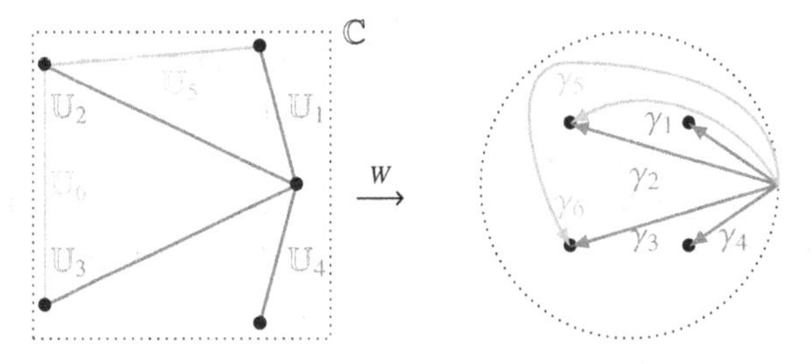

The quadruplet $\gamma_1, \ldots, \gamma_4$ form an exceptional sequence. In this case, all the vanishing cycles have the same point in common so there is a morphism between each pair of vertices. Further investigation shows that the endomorphism ring of the vanishing cycles is the path algebra of the linear quiver

If we mutate γ_2 around γ_1 it becomes γ_5. And if we mutate γ_3 first around γ_2 and then around γ_1 it becomes γ_5. If we substitute γ_2, γ_3 by γ_5, γ_6 the order of the vertices swaps because γ_6 and γ_5 start at a smaller angle than γ_1. Now only consecutive vanishing cycles intersect, so the endomorphism ring is the dual quiver algebra (in which all products of arrows are zero)

Both algebras generate the derived category of the linear quiver.

6.4.4 Generation Criteria

In general it is not so easy to determine whether a set of Lagrangian branes split-generates the Fukaya category. The geometric interpretation of the cone suggests that in order to find a set of split-generators, we need to look for a set of Lagrangians such that every Lagrangian can be Hamiltonianly isotoped to a limit that sits on this set.

Example 6.76 When $\mathbb{X}$ is a surface this can be accomplished by ensuring that the Lagrangians cut the surface into pieces with a trivial topology such as polygons or strips. Take for instance a compact Riemann surface Σ with a number of marked points $m_1, \dots, m_k$. The complement of the marked points $\Sigma^\times = \Sigma \setminus \{m_1, \dots, m_k\}$ can be given the structure of a Liouville manifold by cutting out a small disk around each marked point and gluing in an infinite cylinder. For each curve that connects two marked points we get a noncompact Lagrangian submanifold in ${}^{\mathrm{wr}}\mathbf{Fuk}^\bullet\, \Sigma^\times$, which we will call an arc. Below is an example of a two-punctured torus that is split into two squares by four arcs.

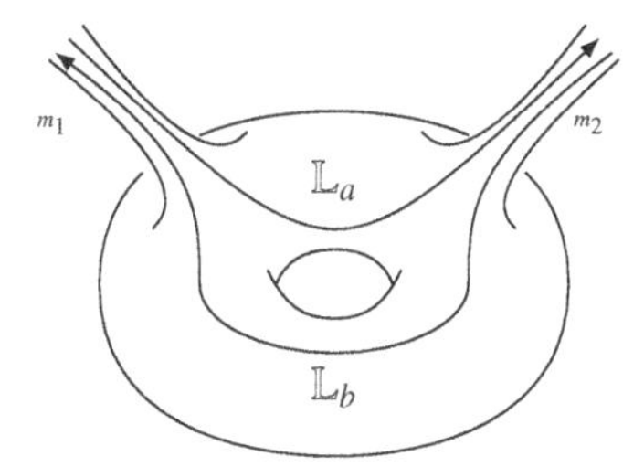

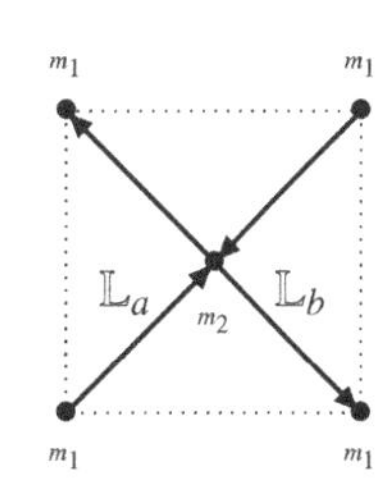

Each arc can be seen as an object in the wrapped Fukaya category and the Reeb chords between arcs are anticlockwise angles around the marked points. The cone of an angle glues two arcs together to a new arc. We can also glue more than two arcs together with a cone of a sum of angles. This enables us to make closed curves with local systems.

Theorem 6.77 *The wrapped Fukaya category* $\mathrm{D}^{\pi\,\mathrm{wr}}\mathbf{Fuk}^{\bullet}\,\Sigma^{\times}$ *is generated by each collection of arcs that divides the surface* Σ *into polygons.*

Sketch of the proof If the arcs divide the surface into polygons then we can approximate each curve by sequences of arcs that run along the boundary of the polygons through which the curve passes. Therefore we can make any curve using the cone construction.

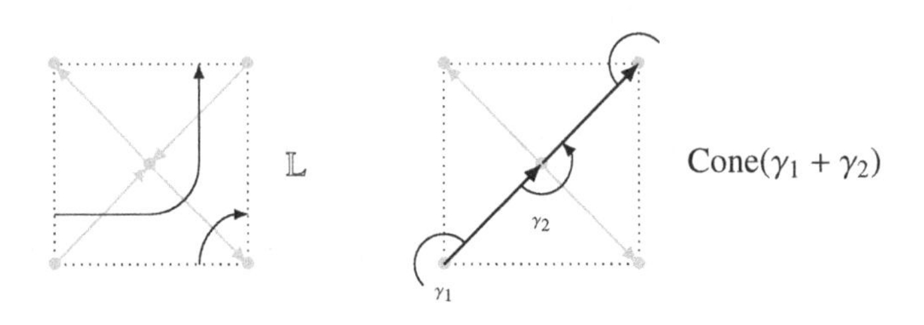

□

Remark 6.78 The idea of using arcs to generate Fukaya categories of surfaces will be explored in the third part of the book. It will enable us to define the Fukaya category combinatorially and avoid most of the technical details coming from symplectic geometry.

For higher-dimensional symplectic manifolds the situation is more difficult because Lagrangian branes do not cut these into disjoint pieces. Another approach to checking whether a set of branes generates the wrapped Fukaya category was worked out by Abouzaid in [6]. The idea is that in order to be split-generators, the morphisms between the branes must see enough of the Reeb flow. To measure this we have to compare two complexes.

Definition 6.79 Let $\{\mathbb{L}_i\}_{i\in I}$ be a collection of Lagrangian branes and L the endomorphism algebra of $\oplus_i\mathbb{L}_i$ in the Fukaya category. The *Hochschild homology* $\mathsf{HH}_\bullet(L)$ is the homology of the complex $\mathsf{CC}_\bullet := \oplus_{k=0}L\otimes L[1]^{\otimes k}$ with

$$d(\gamma_0\otimes\cdots\otimes\gamma_k) = \sum_{i>j}\pm\mu(\gamma_i,\ldots,\gamma_k,\ldots,\gamma_j)\otimes\gamma_{i+j+1}\otimes\cdots\otimes\gamma_{i-1}$$
$$+\sum_{i<j}\pm\gamma_0\otimes\cdots\otimes\gamma_i\otimes\mu(\gamma_{i+1},\ldots,\gamma_j)\otimes\gamma_{j+1}\otimes\cdots\otimes\gamma_k.$$

Remark 6.80 The Hochschild homology is an algebraic object and can be defined for any A_∞-algebra or category [253, 93]. For the polynomial ring $\mathbb{C}[X_1,\ldots,X_n]$ it is equal to the ring of polynomial differential forms

$$\mathbb{C}[X_1,\ldots,X_n]\langle dX_1,\ldots,dX_n\rangle.$$

This is know as the Hochschild–Kostant–Rosenberg theorem [160, 253] and it holds for all coordinate rings of smooth affine varieties (see Chapter 7).

Definition 6.81 Let $(\mathbb{X}, \theta)$ be a Liouville manifold with cylindrical ends. The *symplectic homology* is the homology of the complex $\mathsf{SC}_\bullet(\mathbb{X})$ spanned by Reeb flow cycles $\rho\colon \mathbb{S}^1 \to \mathbb{X}$ for a suitably generic time-dependent Hamiltonian with a differential that counts pseudo-holomorphic cylinders

$$d\rho_1 = \textstyle\sum_{\rho_2} \sum_\psi \sigma_\psi \rho_2 \quad \text{with} \quad \overset{\psi}{\to} \quad \rho_1 \quad \rho_2 \; .$$

Remark 6.82 The symplectic homology carries an A_∞-structure coming from counting holomorphic maps from the unit disk with a number of small disks cut out. For this structure it also has a unit up to homotopy. More information on symplectic homology can be found in [53, 78, 255, 182].

Definition 6.83 The *open–closed map* $\mathrm{OC}\colon \mathsf{CC}_\bullet(L) \to \mathsf{SC}^\bullet(\mathbb{X})$ counts signed holomorphic maps ψ of cylinders between a cycle of Reeb chords and a Reeb cycle: $\mathrm{OC}(\gamma_1 \otimes \cdots \otimes \gamma_k) = \sum_\rho \sum_\psi \sigma_\psi \rho$ with

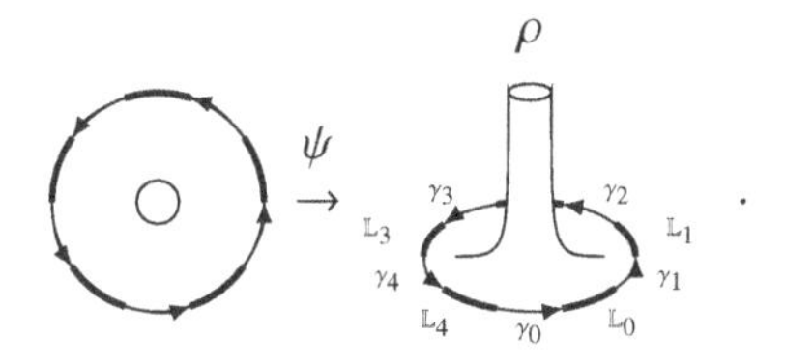

The map OC is a morphism of complexes and induces a map between the homologies

$$\underline{\mathrm{OC}}\colon \mathsf{HH}_\bullet(L) \to \mathsf{SH}_\bullet(\mathbb{X}).$$

Theorem 6.84 (Abouzaid's generation criterion) *The set $\{\mathbb{L}_i\}_{i\in I}$ split-generates the wrapped Fukaya category of $\mathbb{X}$ if the identity $1 \in \mathsf{SH}_\bullet(\mathbb{X})$ is contained in the image of $\underline{\mathrm{OC}}$.*

Sketch of the proof Given any brane $\mathbb{B}$ it is possible to use the criterion to find a combination of higher products $\mathbb{1}_\mathbb{B} = \sum_r \mu(\gamma_{r,1}, \ldots, \gamma_{r,k_r})$. Lemma 3.53 then tells us $\mathbb{B}$ is split-generated by $\mathbb{L}$. □

Example 6.85 The most basic example [6] of Abouzaid's criterion is when $\mathbb{X} = \mathsf{T}^*\mathbb{S}^1$. In that case, the cotangent fiber $\mathsf{T}_x\mathbb{M}$ is a generator. The Hochschild homology of $L = \mathbb{C}[X^{\pm 1}]$ is the ring of differential forms $\mathbb{C}[X^{\pm 1}]\langle dX\rangle$, where dX is shorthand for $1\otimes X - X\otimes 1$. In the symplectic homology each Reeb loop γ^i

that wraps i times around the cylinder gives rise to two elements, $\gamma^i[0]$, $\gamma^i[1]$, one of degree 0 and one of degree 1. (If the Hamiltonian is time independent, a closed Reeb orbit gives rise to a circle of different Reeb orbits, each starting at a different location in the orbit. After deforming the Hamiltonian, this circle symmetry breaks down and only two orbits remain.) If $i = 0$ these two elements correspond to the "constant loops" in the center where the flow is zero. This means that there is precisely one basis element for each degree and homology class. The open–closed map preserves both gradings and gives an isomorphism that maps X^i to $\gamma^i[0]$ and $X^i dX$ to $\gamma^i[1]$.

Remark 6.86 It is expected that there is also a version of this criterion for Fukaya categories of compact symplectic manifolds. In this setting the symplectic homology is substituted by quantum cohomology.

6.5 Exercises

Exercise 6.1 Let $\mathbb{T} = \mathbb{R}^2/2\pi\mathbb{Z}^2$. Determine the Morse cohomology of $\mathbb{T}$ using an appropriate Morse function, such as $(\theta_1, \theta_2) \mapsto \cos(\theta_1) + 2\sin(\theta_2)$.

Exercise 6.2 Let $\mathbb{M}$ be a manifold with a metric g. This defines a Hamiltonian $h = g^{ij} p_i p_j$ on the symplectic manifold $\mathrm{T}^*\mathbb{M}$

(i) $\mathbb{M} = \mathbb{R}^2/2\pi\mathbb{Z}^2$ with the standard metric coming from $\mathbb{R}^2$ and choose $x \in \mathbb{M}$. Show that $\mathrm{T}^*_x\mathbb{M}$ and $\phi^h_t \mathrm{T}^*_x\mathbb{M}$ intersect normally and describe the intersection locus.
(ii) $\mathbb{M}$ be the unit sphere with the standard metric coming from the embedding in $\mathbb{R}^3$ and choose $x \in \mathbb{M}$. Show that $\mathrm{T}^*_x\mathbb{M}$ and $\phi^h_t \mathrm{T}^*_x\mathbb{M}$ do not intersect normally.

Exercise 6.3 Find a θ that turns $\mathbb{C} \setminus \{0, 1\}$ into a Liouville manifold with three cylindrical ends at $0, 1, \infty$. Use this to make Liouville manifolds for punctured surfaces for all genera and number of punctures by cutting it into pairs of pants.

Exercise 6.4 Take an n-punctured surface of genus g with the structure of a Liouville manifold with cylindrical ends at the punctures and assume that $(g, n) \neq (0, 1), (0, 2)$.

(i) Show that if γ is a noncontractible simple closed curve, we can find a curve γ' that is exact and isotopic to γ. Show that the endomorphism ring in the exact Fukaya category of the object corresponding to this Lagrangian submanifold (with trivial spin structure) is $\mathbb{C}[\epsilon]/(\epsilon^2)$ with $\deg_{\mathbb{Z}_2} \epsilon = 1$.

(ii) Show that if γ is an open Lagrangian submanifold with two conical ends going to different punctures, the endomorphism ring in the wrapped Fukaya category of this object is $\mathbb{C}[X, Y]/(XY)$ with $\deg_{\mathbb{Z}_2} X = \deg_{\mathbb{Z}_2} Y = 1$.

Exercise 6.5 Consider the pair of pants as a Liouville manifold with cylindrical ends and place three stops, one on each cylindrical end.

(i) Show that a partially wrapped Fukaya category is hom-finite (i.e. the hom-spaces between objects are finite-dimensional).
(ii) Take three nonintersecting Lagrangian submanifolds with cylindrical ends that each end in a different pair of cylindrical ends. Describe the endomorphism ring of this triple. How many possibilities are there, depending on the positions of the stops?
(iii) Describe what happens with this endomorphism ring if the stops are removed.

Exercise 6.6 Let $\mathbb{X} = \mathsf{T}^*\mathbb{T}$, where $\mathbb{T}$ is the real torus. Choose position coordinates (q_1, q_2) on the torus and momentum coordinates p_1, p_2 for the cotangent directions and view $\mathbb{X}$ as a Liouville manifold with cylindrical ends.

(i) Let $x \in \mathbb{T} \subset \mathsf{T}^*\mathbb{T}$. Describe the Reeb chords from $\mathbb{L} = \mathsf{T}^*_p\mathbb{T}$ to itself. Explain why its endomorphism ring is $\mathbb{C}[\mathbb{Z}^2]$ concentrated in $\mathbb{Z}_2$-degree 0.
(ii) For each point $y \in \mathbb{T}$ and primitive vector $v \in \mathbb{Z}^2$, we can draw a line in $\mathbb{T}$ through y with normal v. Construct a stop such that every Reeb chord whose projection crosses that line in the v-direction gets a positive degree.
(iii) If $v_1, \ldots, v_k \in \mathbb{Z}^2$ are primitive vectors and $y_1, \ldots, y_k, x \in \mathbb{T}$ are different points, show that in the partially wrapped Fukaya category with stops for each pair (y_i, v_i) the endomorphism ring of $\mathbb{L}_x = \mathsf{T}^*_x\mathbb{T}$ is given by $\mathbb{C}[u \in \mathbb{Z}^2 \mid \langle v_i, u \rangle \leq 0]$.

Exercise 6.7 Consider the function $f\colon \mathbb{C}^3 \to \mathbb{C}\colon (X, Y) \mapsto XY - Z^3 + Z$.

(i) Determine the critical points and critical values of f.
(ii) Construct vanishing cycles and thimbles for the straight lines connecting $1 \in \mathbb{C}$ with the critical values. How do the vanishing cycles intersect?
(iii) Explain why the directed Fukaya category is equivalent to $\mathtt{Dmod}^{\pm} Q$ with $Q = \circ \longrightarrow \circ$.

Exercise 6.8 Let $W\colon \mathbb{C} \to \mathbb{C}\colon Z \mapsto \frac{4}{5}Z^5 + Z$ as in Example 6.75. For each possible orientation of the arrows in the linear quiver L_4, find generating sets for which the endomorphism ring is the path algebra of that quiver.

Exercise 6.9 Let $W\colon \mathbb{C}^* \to \mathbb{C}\colon Z \mapsto Z^2 + Z^{-1}$. Find a set of generators of the directed Fukaya category and describe its endomorphism ring.

Exercise 6.10 Consider $\mathsf{T}^*\mathbb{S}_1$ with k stops on one end and k stops on the other end. Let $\mathbb{L}_{i,0}$ be a Lagrangian that starts at one end between stops i and $i+1$ and ends between stops i and $i+1$ at the other end, while $\mathbb{L}_{i,1}$ starts between the same stops but ends one further along (between $i+1$ and $i+2$).

(i) Show that these generate the partially wrapped Fukaya category by approximating each curve by a cone over these curves.
(ii) Calculate the endomorphism ring of these Lagrangians.
(iii) Now consider $W\colon \mathbb{C}^* \to \mathbb{C}\colon Z \mapsto Z^k + Z^{-k}$. Explain why the directed Fukaya category of this Lefschetz fibration is equivalent to the partially wrapped Fukaya category we considered in this exercise.

7
The B-Side

In this chapter we will introduce the relevant categories for the B-model: the derived category of coherent sheaves and the category of singularities. Our discussion will be a bit different from the standard introduction to derived categories because we want to define it as an A_∞-category. To do this we need an enhancement of the category. We will do this using the sheaf cohomology of locally free sheaves. For more background about algebraic geometry we refer to [109, 74, 248]. For the classical approach to the theory of derived categories of (quasi-)coherent sheaves we refer to [186, 185, 157].

7.1 Varieties

7.1.1 Affine Rings and Varieties

The idea behind algebraic geometry is to see algebras as rings of functions over a space. In this way we get a kind of duality between rings and spaces, which can be made very precise for a special class of rings.

Definition 7.1 An *affine ring* is a finitely generated commutative algebra over $\mathbb{C}$ without nilpotent elements. We denote the category of these rings together with $\mathbb{C}$-algebra morphisms as $\mathtt{AffRing}$.

Let R be an affine ring:

$$R = \frac{\mathbb{C}[X_1, \dots, X_n]}{(f_1, \dots, f_k)}.$$

The set $\mathbb{V}(R) := \mathtt{AffRing}(R, \mathbb{C})$ of all one-dimensional representations $\rho \colon R \to \mathbb{C}$ can be seen as a subset of $\mathbb{C}^n$ by identifying ϕ with $(\rho(X_1), \dots, \rho(X_n))$:

$$\mathbb{V}(R) \cong \{x \in \mathbb{C}^n \mid f_1(x) = \cdots = f_k(x) = 0\}.$$

A ring morphism $\phi\colon R \to S$ gives rise to a continuous map $\mathbb{V}(\phi)\colon \mathbb{V}(S) \to \mathbb{V}(R)\colon \rho \to \rho \circ \phi$.

Definition 7.2 The category $\mathtt{AffVar}$ has as objects all the $\mathbb{V}(R)$ and as morphisms all the $\mathbb{V}(\phi)$. The objects are called *affine varieties* and the maps *affine maps*.

Example 7.3 If $R = \mathbb{C}[X_1, \ldots, X_n]$ then we call $\mathbb{V}(R)$ an *n-dimensional affine space* and denote it by $\mathbb{A}^n$. As a set it is equal to $\mathbb{C}^n$. The morphism $\phi\colon \mathbb{C}[X, Y] \to \mathbb{C}[X]\colon f(X, Y) \mapsto f(X, X^2)$ gives rise to an affine map $\mathbb{V}(\phi)\colon \mathbb{A}^1 \to \mathbb{A}^2\colon x \mapsto (x, x^2)$, which maps the affine line onto a parabola in $\mathbb{A}^2$.

The ring R itself can be seen as a ring of functions over $\mathbb{V}(R)$ by assigning to each element $r \in R$ the function $\mathrm{ev}_r\colon \mathbb{V}(R) \to \mathbb{C}\colon \rho \mapsto \rho(r)$. This suggests that we should be able to reconstruct R from $\mathbb{V}(R)$. Indeed, if $\mathbb{X}$ is an object in $\mathtt{AffVar}$ we can look at $\mathscr{O}(\mathbb{X}) := \mathtt{AffVar}(\mathbb{X}, \mathbb{A}^1)$, where $\mathbb{A}^1 = \mathbb{V}(\mathbb{C}[X])$ is identified with $\mathbb{C}$. The set $\mathscr{O}(\mathbb{X})$ has the structure of a ring coming from addition and multiplication in $\mathbb{C} = \mathbb{A}^1$. If $\phi\colon \mathbb{X} \to \mathbb{Y}$ is an affine map we can define a ring morphism

$$\mathscr{O}(\phi)\colon \mathscr{O}(\mathbb{Y}) \to \mathscr{O}(\mathbb{X})\colon f \mapsto f \circ \phi.$$

This turns $\mathscr{O}$ into a contravariant functor.

Theorem 7.4 *The functors* $\mathbb{V}$ *and* $\mathscr{O}$ *are antiequivalences:*

$$\mathtt{AffRing} \underset{\mathscr{O}}{\overset{\mathbb{V}}{\rightleftarrows}} \mathtt{AffVar}\,.$$

Sketch of the proof This is a categorical reformulation of Hilbert's Nullstellensatz, which implies that the map $R \to \mathscr{O}(\mathbb{V}(R))\colon r \mapsto \mathrm{ev}_r$ is an isomorphism. This relies crucially on the fact that R has no nilpotent elements, because for such elements ev_r would be the zero map. □

7.1.2 Modules as Sheaves

If R is an affine ring and $g \in R$ then the morphism $\iota\colon R \to R[g^{-1}]$ induces a map $\mathbb{V}(\iota)\colon \mathbb{V}(R[g^{-1}]) \to \mathbb{V}(R)$. This map is an embedding and its image is the set $\mathbb{U}_g$ containing all $\rho \in \mathbb{V}(R)$ for which $\rho(g) \neq 0$. These subsets are called *affine open subsets* and they generate a topology on $\mathbb{X} = \mathbb{V}(R)$, which is called the *Zariski topology*. Note that the intersection of two affine open subsets is again affine open ($\mathbb{U}_g \cap \mathbb{U}_h = \mathbb{U}_{gh}$) but the union of affine open subsets is not necessarily affine.

Definition 7.5 The category $\mathtt{Open}(\mathbb{X})$ has as objects all the *Zariski-open sets* of $\mathbb{X}$ and as morphisms the inclusions. The full subcategory of affine open subsets is denoted by $\mathtt{AffOpen}(\mathbb{X})$.

Remark 7.6 The Zariski topology is much coarser than the standard topology on $\mathbb{V}(R)$ seen as a subset of $\mathbb{C}^n$. In general it is not Hausdorff. In fact, if R has no zero divisors, every nonempty Zariski-open subset is dense.

If M is a module and $\rho\colon R \to \mathbb{C}$ is a point in $\mathbb{X} = \mathbb{V}(R)$ then the tensor product

$$M_\rho := M \otimes_\rho \mathbb{C} = \frac{M \otimes \mathbb{C}}{(mr \otimes 1 - m \otimes \rho(r) \mid r \in R)}$$

is a vector space over $\mathbb{C}$. If we vary ρ, this construction gives a "bundle" of vector spaces over $\mathbb{X}$, but unlike a vector bundle, M_ρ can have different dimensions for different $\rho \in \mathbb{X}$.

Every element $m \in M$ will define a vector $m \otimes 1 \in M_\rho$, so m can be viewed as a section of this bundle. If we restrict to an affine open subset $\mathbb{U}_g \subset \mathbb{X}$, we can interpret the elements of $M \otimes_R R[g^{-1}]$ as local sections with $m \otimes g^{-1}(\rho) = m \otimes \rho(g)^{-1} \in M_\rho$. If two affine open sets are contained in each other, $\iota\colon \mathbb{U}_g \to \mathbb{U}_h$, then $R[h^{-1}] \subset R[g^{-1}]$ and we have a restriction map in the opposite direction:

$$\mathscr{M}(\iota)\colon \mathscr{M}(\mathbb{U}_h) \to \mathscr{M}(\mathbb{U}_g)\colon M \otimes_R R[h^{-1}] \to M \otimes_R R[g^{-1}].$$

This tells us that $\mathscr{M}$ is a contravariant functor from $\mathtt{AffOpen}(\mathbb{X}) \to \mathtt{VECT}\,\mathbb{C}$ (or $\mathtt{MOD}\,R$ if you wish). Recall from Example 1.24 that a contravariant functor from $\mathtt{Open}(\mathbb{X})$ is called a presheaf. If we want to upgrade the construction $\mathscr{M}$ to a functor on $\mathtt{Open}(\mathbb{X})$, we need to work out the images of the nonaffine open sets. The idea for this is to glue sections together.

To make this precise let $\mathscr{S}\colon \mathtt{Open}(\mathbb{X}) \to \mathtt{VECT}\,\mathbb{C}$ be a presheaf, let $\mathbb{U} = \cup_{i \in I} \mathbb{U}_i$ be an open cover and write $\mathbb{U}_{ij}$ for the intersection $\mathbb{U}_i \cap \mathbb{U}_j$. From these data we can construct two maps

$$\mathscr{S}(\mathbb{U}) \xrightarrow{\iota} \prod_i \mathscr{S}(\mathbb{U}_i) \xrightarrow{\delta} \prod_{i<j} \mathscr{S}(\mathbb{U}_{ij})$$

with $\iota(x) = (\mathscr{S}(\mathbb{U}_i \to \mathbb{U})(x))_{i \in I}$ and $\delta(x_i)_{i \in I} = (\mathscr{S}(\mathbb{U}_{ij} \to \mathbb{U}_i)(x_i) - \mathscr{S}(\mathbb{U}_{ij} \to \mathbb{U}_j)(x_j))_{ij}$. The second map checks whether the local sections $(x_i)_{i \in I}$ all agree on overlaps. This is surely the case when the x_i all come from a section on $\mathbb{U}$, so $\delta\iota = 0$. It might however be the case that there are elements in $\operatorname{Ker}\delta$ that are not in $\operatorname{Im}\iota$, which indicates that some sections in $\mathscr{S}(\mathbb{U})$ are missing.

Definition 7.7 A presheaf $\mathscr{S}\colon \mathtt{Open}(\mathbb{X}) \to \mathtt{VECT}\,\mathbb{C}$ is a *sheaf* if for every open cover $\mathbb{U} = \cup_{i \in I} \mathbb{U}_i$ we have $\operatorname{Ker}\delta = \operatorname{Im}\iota$.

Lemma 7.8 *We can upgrade the functor* $\mathcal{M}\colon \texttt{AffOpen}(\mathbb{X}) \to \texttt{VECT}\,\mathbb{C}$ *to a sheaf* $\mathcal{M}\colon \texttt{Open}(\mathbb{X}) \to \texttt{VECT}\,\mathbb{C}$.

Sketch of the proof The idea is to define $\mathcal{M}(\mathbb{U})$ as $\operatorname{Ker}\delta$ for the cover of all affine open sets contained in $\mathbb{U}$. □

Example 7.9 Let us have a closer look at the affine plane $\mathbb{A}^2 = \mathbb{V}(R)$ with $R = \mathbb{C}[X, Y]$. It has an nonaffine open set $\mathbb{U} = \mathbb{U}_X \cup \mathbb{U}_Y = \mathbb{A}^1 \setminus \{(0,0)\}$. For $M = R$, the local sections $\mathcal{M}(\mathbb{U})$ are equivalent to pairs $(f(X,Y), g(X,Y)) \in \mathcal{M}(\mathbb{U}_X) \times \mathcal{M}(\mathbb{U}_Y) = R[X^{-1}] \times R[Y^{-1}]$ that agree in $\mathcal{M}(\mathbb{U}_{XY}) = R[X^{-1}, Y^{-1}]$. This means that (both) f (and g) cannot have negative powers of X (and Y). Therefore $\mathcal{M}(\mathbb{U}) \cong R$.

Definition 7.10 A sheaf of the form $\mathcal{M}\colon \texttt{Open}(\mathbb{X}) \to \texttt{VECT}\,\mathbb{C}$ that comes from some R-module M is called a *quasi-coherent sheaf*. If M is finitely generated then we call $\mathcal{M}$ a *coherent sheaf*.

Example 7.11 Consider the one-dimensional module $M = \mathbb{C}$, on which R acts via the representation $\sigma\colon R \to \mathbb{C}$. The vector space M_ρ will be one-dimensional if $\rho = \sigma$ and zero-dimensional otherwise. Likewise, the sections of M over $\mathbb{U}_g$ form one-dimensional space if $\sigma \in \mathbb{U}_g$ and a zero-dimensional space otherwise. This also holds for the nonaffine open sets. The coherent sheaf $\mathcal{M}$ is called a *skyscraper sheaf* on σ.

Example 7.12 If $M = R$ then the vector space M_ρ is one-dimensional for each point $\rho \in \mathbb{V}(R)$ and the space of sections on $\mathbb{U}_g$ is $\mathcal{O}(\mathbb{U}_g) = R[g^{-1}]$. The coherent sheaf $\mathcal{M}$ is called the *free sheaf of rank* 1 and represents the trivial rank 1 vector bundle over $\mathbb{V}(R)$.

Definition 7.13 The free sheaf of rank 1 can also be seen as a *sheaf of rings* because each $\mathcal{O}(\mathbb{U})$ has a ring structure. As such it is called the *structure sheaf* of $\mathbb{X}$ and denoted by $\mathcal{O}_\mathbb{X}$. If M is any R-module then the sheaf $\mathcal{M}$ is a *sheaf of* $\mathcal{O}_\mathbb{X}$*-modules*: for any open $\mathbb{U}$ the set $\mathcal{M}(\mathbb{U})$ will be an $\mathcal{O}_\mathbb{X}(\mathbb{U})$-module.

7.1.3 Gluing Varieties

Just like different patches of $\mathbb{R}^n$ can be glued together to a manifold we can try to glue different affine varieties to a bigger variety. We start with a space $\mathbb{X} = \mathbb{U}_1 \cup \cdots \cup \mathbb{U}_r$ such that each subset $\mathbb{U}_i$ is an affine variety with coordinate ring $R_i = \mathcal{O}(\mathbb{U}_i)$ and all intersections $\mathbb{U}_{ij} = \mathbb{U}_i \cap \mathbb{U}_j$ are affine open subsets of $\mathbb{U}_i$ and $\mathbb{U}_j$. The coordinate ring $R_{ij} := \mathcal{O}(\mathbb{U}_{ij})$ is equal to $R_i[g_{ij}^{-1}]$ and $R_j[g_{ji}^{-1}]$ for some functions $g_{ij} \in R_i$ and $g_{ji} \in R_j$ and we get restriction maps

$$R_i \to R_i[g_{ij}^{-1}] = R_{ij} = R_j[g_{ji}^{-1}] \leftarrow R_j.$$

Once these are fixed we can define coordinate rings $R_{i_1\dots i_k} = R_{i_1}[g_{i_1i_2}^{-1},\dots,g_{i_1i_k}^{-1}]$ for the higher intersections $\mathbb{U}_{i_1\dots i_k}$ and restriction morphisms $R_{i_1\dots i_k} \to R_{j_1\dots j_l}$ for all embeddings $\mathbb{U}_{j_1\dots j_l} \subset \mathbb{U}_{i_1\dots i_k}$.

The affine open sets of all the $\mathbb{U}_i$ together generate the Zariski topology on $\mathbb{X}$ and again we have a category $\mathtt{Open}(\mathbb{X})$. The space $\mathbb{X}$ comes with a natural sheaf of rings $\mathcal{O}_{\mathbb{X}}$ such that $\mathcal{O}_{\mathbb{X}}(\mathbb{U}_{i_1\dots i_k}) = R_{i_1\dots i_k}$, which is called the *structure sheaf*. For each open $\mathbb{U}_i$ there is a fully faithful embedding $\iota_i\colon \mathtt{Open}(\mathbb{U}_i) \subset \mathtt{Open}(\mathbb{X})$ and the restriction of $\mathcal{O}_{\mathbb{X}} \circ \iota_i$ will be the structure sheaf $\mathcal{O}_{\mathbb{U}_i}$.

Definition 7.14 The pair $(\mathbb{X}, \mathcal{O}_{\mathbb{X}})$ is called a *variety with affine cover* $\mathcal{U} = \{\mathbb{U}_1,\dots,\mathbb{U}_r\}$.[1]

Example 7.15 (Projective space) The projective space $\mathbb{P}^n$ consists of all one-dimensional subspaces of $\mathbb{C}^{n+1}$. We write $(x_0 : \cdots : x_n) \in \mathbb{P}^n$ for the subspace $\mathbb{C}(x_0,\dots,x_n) \subset \mathbb{C}^{n+1}$. We can cover $\mathbb{P}^n$ with $n+1$ open sets

$$\mathbb{U}_i = \{(x_0 : \cdots : x_n) \mid x_i \neq 0\} = \left\{\left(\frac{x_0}{x_i} : \cdots : 1 : \cdots : \frac{x_n}{x_i}\right)\right\} \cong \mathbb{C}^n.$$

The coordinate ring $\mathcal{O}(\mathbb{U}_i)$ can be identified with the polynomial ring $\mathbb{C}[\xi_{ki} \mid k \neq i]$, where ξ_{ki} represents the rational function $\frac{X_k}{X_i} \in \mathbb{C}(X_0,\dots,X_n)$. On the overlap we get embeddings

$$\mathbb{C}[\xi_{ki}|_{k\neq i}] \longrightarrow \mathbb{C}[\xi_{ki}|_{k\neq i}][\xi_{ji}^{-1}] = \mathbb{C}[\xi_{kl}|_{l=i,j}] = \mathbb{C}[\xi_{kj}|_{k\neq j}][\xi_{ij}^{-1}] \longleftarrow \mathbb{C}[\xi_{kj}|_{l\neq j}].$$

This is enough to calculate the structure sheaf on $\mathbb{P}^n$: in general, if $\mathbb{U} \in \mathtt{Open}(\mathbb{P}^n)$ we will have

$$\mathcal{O}(\mathbb{U}) := \left\{\tfrac{f}{g} \in \mathbb{C}(X_0,\dots,X_n) \mid \deg f = \deg g \text{ and } 0 \notin g(\mathbb{U})\right\}.$$

The first condition is needed for the value of f/g to be independent of the representative $(x_0 : \cdots : x_n)$ and the second for f/g to be defined in $\mathbb{U}$.

Example 7.16 (Projective varieties) More generally, if $R = \mathbb{C}[X_0,\dots,X_n]/(f_1,\dots,f_k)$ is a positively graded affine domain (i.e. $\deg X_i \geq 0$ and the f_i are homogeneous), we define

$$\mathbb{U}_i = \mathbb{V}(R[X_i^{-1}]_0)$$

and use the inclusions

$$R[X_i^{-1}]_0 \longrightarrow R[X_i^{-1}, X_j^{-1}]_0 \longleftarrow R[X_j^{-1}]_0$$

to glue the $\mathbb{U}_i$ together. The result is a space that we denote by $\mathbb{P}(R)$.

[1] Just like for manifolds where we ask that the space is Hausdorff, we need to impose some extra conditions on the cover for everything to work fine (e.g. separatedness) but we will not go into these details.

In the case $R = \mathbb{C}[X_0, \dots, X_n]$ for which all X_i have degree 1 we have that $\mathbb{P}(R)$ is equal to $\mathbb{P}^n$. For more general rings $R = \mathbb{C}[X_0, \dots, X_n]/(f_1, \dots, f_k)$ with $\deg X_i = 1$ we can identify $\mathbb{P}(R)$ with $\mathbb{X} := \{(x_0 : \cdots : x_n) \in \mathbb{P}^n \mid f_i(x_0, \dots, x_n) = 0\}$. Such varieties are called *projective varieties*.

If R is a polynomial ring but $X_0, \dots, X_k$ have degree 1 and the others degree 0 then $\mathbb{P}(R) = \mathbb{P}^k \times \mathbb{A}^{n-k}$. Adding relations will cut out subvarieties, which are called *quasi-projective varieties*. Take care: not all varieties are of this form.

Intuitively, a quasi-coherent sheaf on $(\mathbb{X}, \mathscr{O}_{\mathbb{X}})$ will be a sheaf that looks locally like a quasi-coherent sheaf over an affine open subset.

Definition 7.17 A sheaf of modules $\mathscr{M}$ over $\mathscr{O}_{\mathbb{X}}$ is called a (quasi)-coherent sheaf if and only if, for every affine open $\mathbb{U} \subset \mathbb{X}$, the restriction $\mathscr{M}|_{\mathsf{Open}(\mathbb{U})}$ is a quasi-coherent sheaf over $\mathscr{O}_{\mathbb{U}}$.

It is sufficient to check this condition for an affine cover, so to construct a quasi-coherent sheaf over $\mathbb{X} = \cup_i \mathbb{U}_i$ we need the following data:

- for each $\mathbb{U}_i$ an R_i-module M_i,
- for each $\mathbb{U}_{ij}$ an R_{ij}-module M_{ij} and isomorphisms of R_{ij}-modules

$$M_i \otimes_{R_i} R_{ij} \xrightarrow{\pi_{ij}} M_{ij} \xleftarrow{\pi_{ji}} M_j \otimes_{R_j} R_{ij}.$$

This gives us quasi-coherent sheaves $\mathscr{M}_i$ on $\mathbb{U}_i$, $\mathscr{M}_{ij}$ on $\mathbb{U}_{ij}$ and natural transformations $\pi_{ij} \colon \mathscr{M}_i|_{\mathsf{Open}\mathbb{U}_{ij}} \to \mathscr{M}_{ij}$. These transformations must satisfy a compatibility relation to make deeper intersections well defined.[2] Once all this is fixed, we extend this construction to a sheaf $\mathscr{M}$ on $\mathbb{X}$ by setting $\mathscr{M}(\mathbb{V}) = \operatorname{Ker} \delta$ with

$$\delta \colon \prod_i \mathscr{M}_i(\mathbb{V}_i) \to \prod_{i<j} \mathscr{M}_i(\mathbb{V}_{ij})$$
$$: (x_i) \mapsto \left(\pi_{ij}\mathscr{M}_i(\mathbb{V}_{ij} \to \mathbb{V}_i)(x_i) - \pi_{ji}\mathscr{M}_j(\mathbb{V}_{ij} \to \mathbb{V}_i)(x_j)\right)_{ij}$$

for every open $\mathbb{V} \subset \mathbb{X}$. In other words, the sections of $\mathbb{V}$ are gluings of sections on the $\mathbb{V}_i = \mathbb{V} \cap \mathbb{U}_i$. Because the sections of $\mathscr{O}_{\mathbb{X}}(\mathbb{V})$ are constructed in a similar way, one can show that $\mathscr{M}(\mathbb{V})$ is an $\mathscr{O}_{\mathbb{X}}(\mathbb{V})$-module and $\mathscr{M}$ is a quasi-coherent sheaf over $\mathbb{X}$.

Example 7.18 (Sheaves over projective varieties and graded modules) Consider a positively graded domain $R = \mathbb{C}[X_0, \dots, X_n]/(f_1, \dots, f_k)$ and $\mathbb{Z}$-graded R-module M. We can define a sheaf over $\mathbb{X} = \mathbb{P}(R)$ by setting

$$\mathscr{M}(\mathbb{U}_{X_i}) = (M \otimes_R R[X_i^{-1}])_0$$

[2] It is a nice exercise to write out the condition.

and using the standard maps

$$(M \otimes_R R[X_i^{-1}])_0 \longrightarrow (M \otimes_R R[X_i^{-1}, X_j^{-1}])_0 \longleftarrow (M \otimes_R R[X_j^{-1}])_0$$

for gluing. For $M = R$ we get a structure sheaf over $\mathbb{P}(R)$. If we set $M = R(k)$ where $R(k)_i = R_{i+k}$ then this new sheaf $\mathcal{O}(k)$ is not equal to the structure sheaf $\mathcal{O}$ for $k \neq 0$ because they have different sections: $\mathcal{O}(k)(\mathbb{U}_i) = R[X_i^{-1}]_k$ and $\mathcal{O}(k)(\mathbb{P}^n) = R_k$.

Different graded modules can give rise to the same sheaf. If M is a finite-dimensional R module then $X_i^n M = 0$ for n large enough, so $M \otimes_R R[X_i^{-1}] = 0$. This implies that $\mathcal{M}(\mathbb{U}_i) = 0$ for all X_i and hence $\mathcal{M}$ is the zero sheaf. Similarly, if M contains a finite-dimensional graded submodule N then M and M/N will give rise to the same sheaf.

Vice versa, if $\mathcal{M}$ is a sheaf over $\mathbb{P}(R)$ then we can cook up a corresponding graded module. First define $\mathcal{M}(k)$ as the sheaf for which

$$\mathcal{M}(k)(\mathbb{U}_i) = \mathcal{M}(\mathbb{U}_i) \otimes_{\mathcal{O}(\mathbb{U}_i)} \mathcal{O}(k)(\mathbb{U}_i)$$

for each affine open subset and then set

$$M = \bigoplus_{i \geq 0} \mathcal{M}(k)(\mathbb{X}).$$

One can check that the sheaf associated to M is isomorphic to the original $\mathcal{M}$ because they agree on all $\mathbb{U}_i$ and their intersections.

7.1.4 Sheaf Cohomology

Suppose we have a quasi-coherent sheaf $\mathcal{M}$ on a variety $(\mathbb{X}, \mathcal{O}_{\mathbb{X}})$ with an affine cover $\mathcal{U} = \{\mathbb{U}_1, \ldots, \mathbb{U}_r\}$ and let $M_{i_1 \ldots i_k} := \mathcal{M}(\mathbb{U}_{i_1} \cap \cdots \cap \mathbb{U}_{i_k})$. We have already seen that to construct global sections, we have to take the kernel of a certain map δ that measures the differences on overlaps. This map is actually the first map of a complex.

Definition 7.19 The Čech complex of $\mathcal{M}$ for the cover $\mathcal{U}$ is

$$\check{\mathsf{C}}^k(\mathcal{M}, \mathcal{U}) = \bigoplus_{i_1 < \cdots < i_k} M_{i_1 \ldots i_k} \quad \text{with } d_k = \sum_j \sum_{\{i_1 < \cdots < i_{k+1}\}} (-1)^j [M_{i_1 \ldots \widehat{i_j} \ldots i_k} \to M_{i_1 \ldots i_{k+1}}],$$

where the sum runs over all restriction maps.

The zeroth cohomology of this complex consists of all the global sections:

$$\mathsf{H}\check{\mathsf{C}}^0(\mathcal{M}, \mathcal{U}) = \mathcal{M}(\mathbb{X}).$$

This space does not depend on the specific cover; it is a property of the sheaf alone. This is a feature that is also true for the higher cohomology groups.

Lemma 7.20 *The cohomology of the Čech complex is independent of the affine cover.*

Sketch of the proof This is a consequence of a theorem by Serre which states that if an affine cover $\mathcal{U}'$ is a refinement of $\mathcal{U}$ then $\check{\mathsf{C}}(\mathcal{M},\mathcal{U}')^\bullet$ and $\check{\mathsf{C}}(\mathcal{M},\mathcal{U}')^\bullet$ are quasi-isomorphic. Furthermore, any two affine covers have a common refinement. □

Example 7.21 Let $\mathbb{X}$ be $\mathbb{P}^n = \mathbb{P}(R)$ with $R = \mathbb{C}[X_0,\dots,X_n]$ and $\mathcal{M} = \mathcal{O}(k)$. The space $M_{i_1\dots i_l}$ is in this case all functions in $R[X_{i_1}^{-1},\dots,X_{i_l}^{-1}]$ of degree k and the restriction maps are the standard embeddings. The complex $\check{\mathsf{C}}^\bullet(\mathcal{M},\mathcal{U})$ splits into a direct sum according to monomials $m = X_0^{s_0}\dots X_n^{s_n}$. For each monomial the complex has a one-dimensional vector space in each $M_{i_1\dots i_l}$ for which only the powers of the X_j with $j \in \{i_1,\dots,i_l\}$ can be negative.

Let us apply this to $\mathbb{P}^2$. The general complex looks as follows:

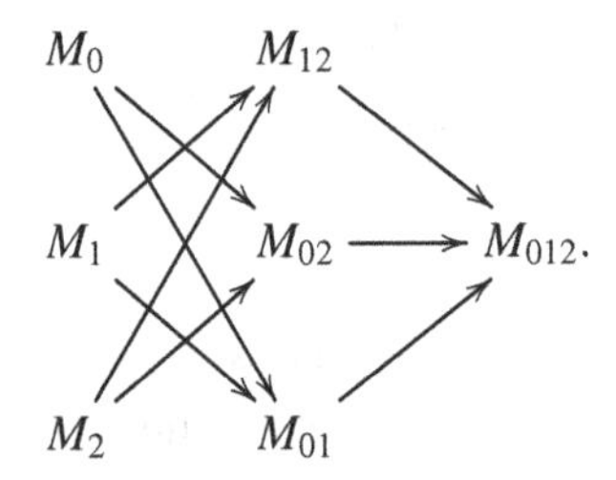

There are four possibilities for the monomial depending on how many of the X_i have a negative power. For each negative power X_i, we have to remove all the M without an i in its subscript.

$X_0X_1X_2$	X_0^{-1}	$X_0^{-1}X_1^{-1}$	$X_0^{-1}X_1^{-1}X_2^{-1}$
$\mathbb{C}$ $\mathbb{C}$ $\mathbb{C}$ $\mathbb{C} \to \mathbb{C}$ $\mathbb{C}$ $\mathbb{C}$	$\mathbb{C}$ 0 0 $\mathbb{C} \to \mathbb{C}$ 0 $\mathbb{C}$	0 0 0 0 $\mathbb{C}$ 0 $\mathbb{C}$	0 0 0 0 $\mathbb{C}$ 0 0
$H = \mathbb{C} \oplus 0 \oplus 0$	$H = 0 \oplus 0 \oplus 0$	$H = 0 \oplus 0 \oplus 0$	$H = 0 \oplus 0 \oplus \mathbb{C}$

From this we can conclude that

$$\mathsf{H}\check{\mathsf{C}}^i(\mathcal{O}(k),\mathcal{U}) = \begin{cases} R_k, & i = 0, \\ 0, & i = 1, \\ \langle X_0^{-r}X_1^{-s}X_2^{-t} \mid s,r,t \geq 1,\ k = -s-r-t\rangle, & i = 2. \end{cases}$$

Note that H^0 can only be nonzero if $k \geq 0$, while H^2 can only be nonzero if $k \leq -3$. For $i \neq 0, 2$ there is no cohomology at all.

Definition 7.22 A *locally free sheaf* over $\mathbb{X}$ is a quasi-coherent sheaf $\mathscr{F}$ such that there is an affine cover $\mathcal{U} = \{\mathbb{U}_i \mid i \in I\}$ for which all the $\mathscr{F}(\mathbb{U}_i)$ are free $\mathscr{O}(\mathbb{U})$ modules.

Locally free coherent sheaves over varieties are the algebraic equivalent of *vector bundles*. Indeed, if we consider an affine open $\mathbb{U}$ for which $\mathscr{F}(\mathbb{U})$ is free and finitely generated then $\mathscr{F}(\mathbb{U}) \cong \mathscr{O}(\mathbb{U})^r$ and above every point $\rho \in \mathbb{U}$ the fiber will be equal to $\mathscr{F}(\mathbb{U})_\rho = \mathscr{O}(\mathbb{U})^r_\rho = \mathbb{C}^r$. The number r is called the rank and if the rank is 1 we call $\mathscr{F}$ an *invertible sheaf*. The reason for this is that we can define a tensor product for sheaves such that for affine open subsets we have $\mathscr{M} \otimes \mathscr{N}(\mathbb{U}) := \mathscr{M}(\mathbb{U}) \otimes_{\mathscr{O}(\mathbb{U})} \mathscr{N}(\mathbb{U})$. For this operation the structure sheaf $\mathscr{O}$ is the identity element and the invertible sheaves are those that have an inverse: $\mathscr{M}^{-1}(\mathbb{U}) := \mathrm{Hom}_{\mathscr{O}(\mathbb{U})}(\mathscr{M}(\mathbb{U}), \mathscr{O}(\mathbb{U}))$.

Example 7.23 The sheaves $\mathscr{O}(k)$ over $\mathbb{X} = \mathbb{P}^n$ are locally free. Let $R = \mathbb{C}[X_0, \dots, X_n]$; then $\mathscr{O}(\mathbb{U}_i) = \mathbb{C}[\frac{X_0}{X_i}, \dots, \frac{X_n}{X_i}]$ while

$$\mathscr{O}(k)(\mathbb{U}_i) = R[X_i^{-1}]_k = X_i^k \mathbb{C}\left[\frac{X_0}{X_i}, \dots, \frac{X_n}{X_i}\right],$$

so the latter is a free $\mathscr{O}(\mathbb{U}_i)$-module of rank 1.

Example 7.24 If $\mathbb{X} = \mathbb{V}(R)$ is an affine variety then a coherent sheaf is locally free if and only if it is projective. This is the algebro-geometric version of the Serre–Swan theorem [222]. The proof can be found in [230, Tag 00NV]. We will illustrate it with an example of a locally free module that is not free.

Take for instance $\mathbb{V}(R)$ with $R = \mathbb{C}[X, Y]/(Y^2 - X^3 + X)$. This is an example of an affine elliptic curve embedded in $\mathbb{A}^2$. The module $F = RX + RY$ is not free but for $\mathbb{X} = \mathbb{U}_X \cup \mathbb{U}_{X^2-1}$ we see that

$$\begin{aligned} \mathscr{F}(\mathbb{U}_X) &= R[X^{-1}]X + R[X^{-1}]Y = R[X^{-1}], \\ \mathscr{F}(\mathbb{U}_{X^2-1}) &= R[(X^2-1)^{-1}]X + R[(X^2-1)^{-1}]Y \\ &= R[(X^2-1)^{-1}]Y \cong R[(X^2-1)^{-1}], \end{aligned}$$

so $\mathscr{F}$ is a locally free sheaf of rank 1. More generally, the maximal ideal $\mathfrak{p} \lhd R$ corresponding to a point on this curve is a locally free sheaf of rank 1 because we can find an open cover for which $\mathfrak{p}$ becomes principle. We will discuss this in more detail in the next section.

Example 7.25 (Sheaves from a divisor) Let $h(X, Y, Z)$ be a homogeneous function and suppose that $R = \mathbb{C}[X, Y, Z]/(h)$ defines a smooth curve

$\mathbb{X} = \mathbb{P}(R/(f))$ in $\mathbb{P}^2$. A divisor is a $\mathbb{Z}$-linear combination of points in $\mathbb{X}$: $D = a_1P_1 + \cdots + a_kP_k$. For each divisor we can define a sheaf $\mathscr{O}(D)$. Its local sections on the affine open $\mathbb{U}$ are the rational functions whose orders in P_i are at least a_i:

$$\mathscr{O}(D)(\mathbb{U}) = \left\{\tfrac{f}{g} \mid f, g \in O(\mathbb{U}),\ \forall P_i \in \mathbb{U} \text{ we have } \mathrm{ord}_{P_i} f - \mathrm{ord}_{P_i} g \geq a_i\right\}.$$

(Remember: if $\mathfrak{p} \lhd \mathscr{O}(\mathbb{U})$ is the maximal ideal corresponding to the point P then $\mathrm{ord}_{P_i} f$ is the largest exponent n such that $f \in \mathfrak{p}^n$.) Just like in the previous example, these sheaves are all locally free of rank 1 or line bundles. The degree of the line bundle $\mathscr{O}(D)$ is defined as $\sum a_i$.

Given a variety $(\mathbb{X}, \mathscr{O})$ and two quasi-coherent sheaves $\mathscr{M}$, $\mathscr{N}$ we define the hom-sheaf $\mathscr{H}\!om\,(\mathscr{M}, \mathscr{N})$ by

$$\mathscr{H}\!om\,(\mathscr{M}, \mathscr{N})(\mathbb{U}_i) := \mathrm{Hom}_{\mathscr{O}(\mathbb{U}_i)}(\mathscr{M}(\mathbb{U}_i), \mathscr{N}(\mathbb{U}_i)).$$

These hom-sheaves can be used to define a *category enriched in sheaves* $\mathscr{QC}oh\,\mathbb{X}$ whose objects are all the coherent sheaves and whose morphism spaces are the hom-sheaves. This category has full subcategories $\mathscr{C}oh\,\mathbb{X}$ and $\mathscr{V}ect\,\mathbb{X}$ which contain all coherent and all coherent locally free sheaves.

If we have an open cover $\mathcal{U}$ then we can turn $\mathscr{V}ect\,\mathbb{X}$ into a dg-category by applying the Čech complex construction. This category $\mathtt{Vect}^\bullet(\mathbb{X}; \mathcal{U})$ has the same objects as $\mathscr{V}ect\,\mathbb{X}$ but the hom-spaces are

$$\mathtt{Vect}^\bullet(\mathbb{X}; \mathcal{U})(\mathscr{M}, \mathscr{N}) := \check{\mathsf{C}}^\bullet(\mathscr{V}ect\,\mathbb{X}(\mathscr{M}, \mathscr{N}); \mathcal{U}).$$

For the product we use the cup product. If $f \in \check{\mathsf{C}}^k(\mathscr{V}ect\,\mathbb{X}(\mathscr{N}, \mathscr{P}); \mathcal{U})$ and $g \in \check{\mathsf{C}}^l(\mathscr{V}ect\,\mathbb{X}(\mathscr{M}, \mathscr{N}); \mathcal{U})$ we set

$$(f \smile g)_{i_0 \ldots i_{k+l}} = f_{i_0 \ldots i_k}|_{\mathbb{U}_{i_0 \ldots i_{k+l}}} \circ g_{i_k \ldots i_{k+l}}|_{\mathbb{U}_{i_0 \ldots i_{k+l}}}.$$

Lemma 7.26 *Different affine covers give rise to A_∞-quasi-isomorphic A_∞-categories:*

$$\mathtt{Vect}^\bullet(\mathbb{X}; \mathcal{U}) \overset{\infty}{=} \mathtt{Vect}^\bullet(\mathbb{X}; \mathcal{U}').$$

Sketch of the proof The proof is similar to the statement that the Čech cohomology does not depend on the affine cover: the quasi-isomorphisms coming from refinements are compatible with the cup product. □

Definition 7.27 The *category of perfect complexes of vector bundles* on $\mathbb{X}$ is the derived category of the dg-category of locally free sheaves on $\mathbb{X}$,

$$\mathtt{Perf}^\bullet\,\mathbb{X} := \mathtt{D}\,\mathtt{Vect}^\bullet(\mathbb{X}; \mathcal{U}).$$

Example 7.28 If $\mathbb{X} = \mathbb{V}(R)$ is an affine variety then the locally free sheaves correspond to the finitely generated projective modules of R. Moreover, we can use $\mathcal{U} = \{\mathbb{V}(R)\}$ as an affine cover of itself and then the Čech complex is concentrated in degree 0. If P, Q are two projective R-modules with corresponding sheaves $\mathscr{P}$, $\mathscr{Q}$, we have $\mathtt{Vect}^\bullet(\mathbb{X};\mathcal{U})(\mathscr{P},\mathscr{Q}) = \mathrm{Hom}_R(P,Q)$. This gives an equivalence of A_∞-categories between $\mathtt{Vect}^\bullet(\mathbb{X};\mathcal{U})$ and $\mathtt{Proj}^\bullet R = \mathtt{Split}\,\mathtt{Add}\,R$ (viewed as A_∞-categories concentrated in degree 0). After going to derived versions we get

$$\mathtt{Perf}^\bullet\,\mathbb{X} \stackrel{\infty}{=} \mathrm{D}\,\mathtt{Proj}^\bullet R = \mathrm{D}^\pi R.$$

So R is a split-generator of $\mathtt{Perf}^\bullet\,\mathbb{X}$.

Example 7.29 Let us have a closer look at $\mathbb{X} = \mathbb{P}^2$ and put $R = \mathbb{C}[X_0, X_1, X_2]$. Because they are locally free, all the $\mathscr{O}(k)$ are objects in $\mathscr{V}ect\,\mathbb{X}$ and we have

$$\mathscr{H}om\,(\mathscr{O}(k), \mathscr{O}(l)) = \mathscr{O}(l-k).$$

From our calculations in Example 7.21 we can deduce that

$$\mathtt{Perf}^\bullet\,\mathbb{X}(\mathscr{O}(k), \mathscr{O}(l)) = R_{l-k}$$

concentrated in degree 0 if $l \geq k$ and

$$\mathtt{Perf}^\bullet\,\mathbb{X}(\mathscr{O}(k), \mathscr{O}(l)) = 0$$

if $l < k < l-3$. If we look at the objects $\mathscr{O}$, $\mathscr{O}(1)$ and $\mathscr{O}(2)$ we see that these three objects have only degree 0 morphisms between them and

$$\mathrm{End}_{\mathtt{Perf}^\bullet\,\mathbb{X}}(\mathscr{O} \oplus \mathscr{O}(1) \oplus \mathscr{O}(2)) = \begin{pmatrix} \mathbb{C} & R_1 & R_2 \\ 0 & \mathbb{C} & R_1 \\ 0 & 0 & \mathbb{C} \end{pmatrix} =: B.$$

We can also write this algebra as a quiver

x_1, x_2, x_3 ; y_1, y_2, y_3

with relations $x_i y_j = x_j y_i$. A famous theorem by Beilinson [26] tells us that these three objects generate $\mathtt{Perf}^\bullet\,\mathbb{X}$. Therefore, this quiver with relations is called the *Beilinson quiver*:

$$\mathtt{Perf}^\bullet\,\mathbb{X} \stackrel{\infty}{=} \langle \mathscr{O}, \mathscr{O}(1), \mathscr{O}(2)\rangle \stackrel{\infty}{=} \mathrm{D}\,\mathrm{B}^\bullet.$$

7.1.5 Smoothness

If $\mathbb{X} = \mathbb{V}(R)$ is an affine variety we have seen that $\mathtt{Perf}^\bullet\,\mathbb{X} = \mathrm{D}\,\mathtt{Proj}^\bullet R$ is the minimal model of the dg-category of bounded complexes of projective R-modules $\mathtt{TwSplit}\,\mathtt{Add}\,R$. Because $\mathtt{Split}\,\mathtt{Add}\,R \subset \mathtt{Mod}^\bullet R$ we have $\mathtt{Perf}^\bullet\,\mathbb{X} \subset \mathrm{D}\,\mathtt{Mod}^\bullet R$.

Vice versa, we know that if M is any finitely generated R-module then we can construct a projective resolution for M. This gives us a complex of projective R-modules that is quasi-isomorphic to M. Unfortunately, such a complex is not always bounded, so it might not be in $\mathtt{TwSplit\,Add}\,R$.

Example 7.30 We consider two cubic curves.

- Let $R = \mathbb{C}[X, Y]/(Y^2 - X^3 - X^2)$ and S be the simple module $R/(X, Y)$. This module has a free resolution

$$R \xleftarrow{\;(Y\;X)\;} R \oplus R \xleftarrow{\begin{pmatrix} -X & -Y \\ Y & X^2+X \end{pmatrix}} R \oplus R \xleftarrow{\begin{pmatrix} -X^2-X & -Y \\ Y & X \end{pmatrix}} R \oplus R \longleftarrow \cdots$$

  which goes on indefinitely with the last two maps repeating. From this we can calculate that $\mathrm{Ext}^i_R(S, S) = \mathbb{C}^2$ for all $i > 0$ and hence S cannot have a bounded resolution.
- Let $R = \mathbb{C}[X, Y]/(Y^2 - X^3 + X)$ and S be the simple module $R/(X, Y)$. This module has a projective resolution

$$R \hookleftarrow RX + RY\ .$$

  Note that $RX + RY$ is projective but not free.

From a geometric point of view the difference between these two modules is that in the first case S corresponds to a singular point in $\mathbb{V}(R)$ and in the second case it is a smooth point.

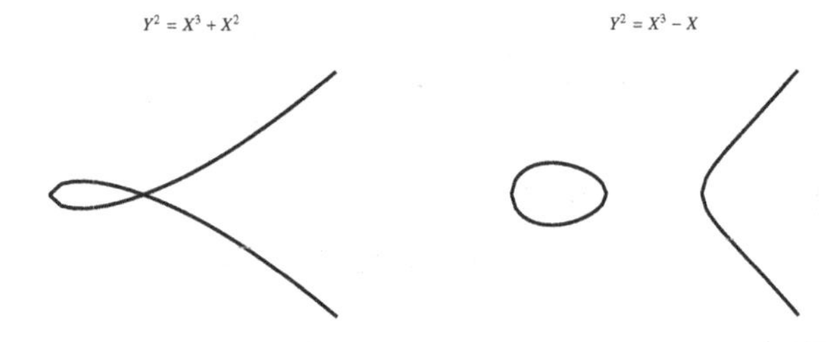

Intuitively, a smooth point $\rho \in \mathbb{V}(R)$ is a point whose surroundings locally look like an affine space. This can be made precise by looking at the completion $R_{\mathfrak{p}} = \varprojlim R/\mathfrak{p}^i$ with $\mathfrak{p} = \mathrm{Ker}\,\rho$. For the zero point in affine space $\rho \in \mathbb{A}^n = \mathbb{V}(\mathbb{C}[X_1, \ldots, X_n])$ we have $\mathfrak{p} = (X_1, \ldots, X_n)$, so this ring is the ring of formal power series $\mathbb{C}[\![X_1, \ldots, X_n]\!]$. In general, $R_{\mathfrak{p}}$ can always be written as $\mathbb{C}[\![X_1, \ldots, X_n]\!]/I$ where $I \in (X_1, \ldots, X_n)^2$.

Definition 7.31 A point $\rho \in \mathbb{V}(R)$ is *smooth of dimension n* if the completion $R_{\mathfrak{p}}$ with respect to $\mathfrak{p} = \mathrm{Ker}\,\rho$ is isomorphic to a formal power series ring $\mathbb{C}[\![X_1, \ldots, X_n]\!]$. A point in ρ is a general variety $(\mathbb{X}, \mathcal{O}_{\mathbb{X}})$ is smooth if it is smooth in an affine open neighborhood. A variety is *smooth* if and only if all its points are smooth.

Remark 7.32 If R is a domain then one can show that all smooth points have the same dimension. We will call this the dimension of $\mathbb{V}(R)$. A general variety is n-dimensional if it is covered by n-dimensional affine varieties.

Example 7.33 Let us return to the two elliptic curves.

- If $R = \mathbb{C}[X, Y]/(Y^2 - X^3 + X)$ then the formal completion around the zero point looks like $R_{\mathfrak{p}} = \mathbb{C}[[X, Y]]/(Y^2 - X^3 - X)$. In this ring we have

$$X = X^3 - Y^2 = (X^3 - Y^2)^3 - Y^2 = ((X^3 - Y^2)^3 - Y^2)^3 - Y^2 = \cdots = f(Y).$$

  Working this out, the terms containing powers of X get higher and higher degrees, so in the limit we get a formal power series $X = f(Y)$ and therefore $R_{\mathfrak{p}} = \mathbb{C}[[Y]]$.
- If $R = \mathbb{C}[X, Y]/(Y^2 - X^3 - X^2)$ then the formal completion around the zero point looks like $R_{\mathfrak{p}} = \mathbb{C}[[X, Y]]/(Y^2 - X^3 - X^2)$. This ring is not a formal power series ring because it has zero divisors: the expansions of $Y \pm X\sqrt{1 + X}$ into power series.

Theorem 7.34 (Local smoothness) *Let R be an affine domain, $\mathbb{X} = \mathbb{V}(R)$ and $\rho \in \mathbb{X}$ a point corresponding to the simple R-module S. The following statements are equivalent:*

(i) *the point ρ is smooth of dimension n,*
(ii) $\operatorname{Ext}^\bullet_R(S, S) \overset{\infty}{=} \Lambda^\bullet_n$,
(iii) *S has a finitely generated projective resolution of length n.*

Sketch of the proof These equivalences follow from the fact that the Ext-groups and resolutions can be calculated locally: $\operatorname{Ext}^\bullet_R(S, S) = \operatorname{Ext}_{R_\mathfrak{p}}(S, S) = \Lambda^\bullet_n$. For more information see [74, Chapter 19]. □

Statement (iii) can be rephrased in categorical terms: a point is smooth if and only if S is equivalent to an object in $\mathtt{Perf}^\bullet\,\mathbb{X} \subset \mathtt{D\,Mod}^\bullet R$. This local observation has a nice global version in terms of categories.

Theorem 7.35 (Global smoothness: affine case) *Let R be an affine domain and $\mathbb{X} = \mathbb{V}(R)$; then the following are equivalent:*

(i) *the variety $\mathbb{X}$ is smooth,*
(ii) *the embedding $\mathtt{Perf}^\bullet\,\mathbb{X} \subset \mathtt{D\,Mod}^\bullet R$ is an equivalence,*
(iii) *the hom-spaces in $\mathtt{D\,MOD}^\bullet R$ have bounded degree: $\mathtt{DMOD}^{\mathtt{i}} R(M, N) = 0$ if $|i| \gg 0$.*

Sketch of the proof It is a standard fact from algebraic geometry that $\mathbb{V}(R)$ is an n-dimensional smooth variety if and only if every R-module M has a projective resolution of at most length n [230, Tag 065U]. If M is finitely generated this resolution can also be chosen to be finitely generated and therefore every finitely generated module is contained in $\mathtt{Perf}^\bullet \mathbb{X}$.

For all modules this means that $\mathtt{MOD}^i R(M, N) = \mathrm{Ext}^i_R(M, N) = 0$ if $i > n$ or $i < 0$. If we allow twisted complexes these degrees can shift by a finite amount but as there are only finitely many summands in each twisted complex, the hom-spaces in $\mathtt{Tw}\,\mathtt{MOD}^\bullet R$ and $\mathtt{D}\,\mathtt{MOD}^\bullet R$ have bounded degrees. □

So far we have only considered affine varieties. If we want to extend these results to general varieties, we need to find suitable generalizations of $\mathtt{D}\,\mathtt{Mod}^\bullet R$ and $\mathtt{D}\,\mathtt{MOD}^\bullet R$. Recall from Chapter 2 that we constructed $\mathtt{Mod}^\bullet R$ as an A_∞-category by replacing each module by a free resolution. Likewise we should replace each coherent sheaf by a resolution of locally free sheaves.

Definition 7.36 A complex of sheaves on $\mathbb{X}$ is a diagram

$$\mathscr{P}^\bullet = \cdots \longrightarrow \mathscr{F}_{-1} \xrightarrow[d_{-1}]{} \mathscr{F}_0 \xrightarrow[d_0]{} \mathscr{F}_1 \xrightarrow[d_1]{} \mathscr{F}_2 \xrightarrow[d_2]{} \cdots,$$

where the $d_i \in \mathscr{H}\!om\,(\mathscr{F}_i, \mathscr{F}_{i+1})(\mathbb{X})$ and for each affine open $\mathbb{U} \subset \mathbb{X}$, the restriction $\mathscr{P}^\bullet(\mathbb{U})$ is a complex of $\mathscr{O}(\mathbb{U})$-modules. The complex $\mathscr{P}^\bullet$ is a locally free resolution of the sheaf $\mathscr{M}$, if for each affine open subset $\mathscr{P}^\bullet(\mathbb{U})$ is a projective resolution of $\mathscr{M}(\mathbb{U})$.

If we have resolutions $\mathscr{P}^\bullet$ and $\mathscr{Q}^\bullet$ for two sheaves $\mathscr{M}$, $\mathscr{N}$ then we can define the hom-sheaf between them by $\mathscr{H}\!om^\bullet(\mathscr{P}, \mathscr{Q})$, just as we did before. For each affine open $\mathbb{U}$ the space $\mathscr{H}\!om^\bullet(\mathscr{P}, \mathscr{Q})(\mathbb{U})$ is a complex with a differential coming from the differentials on $\mathscr{P}^\bullet$ and $\mathscr{Q}^\bullet$. If we take the Čech complex of the hom sheaf,

$$\check{\mathsf{C}}^\bullet\left(\mathscr{H}\!om^\bullet(\mathscr{P}, \mathscr{Q}); \mathcal{U}\right) := \bigoplus_{k,l} \bigoplus_{i_1,\ldots,i_k} \mathscr{H}\!om^l(\mathscr{P}, \mathscr{Q})(\mathbb{U}_{i_1\ldots i_k}).$$

On this space there are two differentials: the Čech differential d_C which increases the k-index and the hom-differential d_H which increases the l-index. These two differentials anticommute and therefore the total differential $d_T = d_C + d_H$ is a differential for the $k+l$-grading: $d_T^2 = d_C^2 + d_C d_H + d_H d_C + d_H^2 = 0$. Different choices of locally free resolutions and affine covers give rise to quasi-isomorphic complexes and the degree 0 components are the global sections of the hom-sheaf between $\mathscr{M}$ and $\mathscr{N}$:

$$\mathsf{H}^0\check{\mathsf{C}}\left(\mathscr{H}\!om^\bullet(\mathscr{P}, \mathscr{Q}); \mathcal{U}\right) = \mathscr{H}\!om\,(\mathscr{M}, \mathscr{N})(\mathbb{X}).$$

Definition 7.37 The category with as objects the locally free resolutions of (quasi-)coherent sheaves on $\mathbb{X}$ and as hom-spaces these Čech complexes is a dg-category with respect to these total differentials. We denote its minimal model by $\mathtt{Coh}^\bullet\,\mathbb{X}$ ($\mathtt{QCoh}^\bullet\,\mathbb{X}$) and its derived version by $\mathtt{D\,Coh}^\bullet\,\mathbb{X}$ ($\mathtt{D\,QCoh}^\bullet\,\mathbb{X}$). The latter is known as the *bounded derived category of (quasi-)coherent sheaves* on $\mathbb{X}$.

Because every object in $\mathtt{Vect}^\bullet\,\mathbb{X}$ is its own resolution, we have an embedding of $\mathtt{Vect}^\bullet\,\mathbb{X}$ into $\mathtt{Coh}^\bullet\,\mathbb{X} \subset \mathtt{QCoh}^\bullet\,\mathbb{X}$. By going to the derived categories we get the following towers of A_∞-categories:

$$\begin{array}{llllll} \text{Affine:} & \mathtt{Perf}^\bullet\,\mathbb{X} & \subset & \mathtt{D\,Mod}^\bullet R & \subset & \mathtt{D\,MOD}^\bullet R, \\ \text{General:} & \mathtt{Perf}^\bullet\,\mathbb{X} & \subset & \mathtt{D\,Coh}^\bullet\,\mathbb{X} & \subset & \mathtt{D\,QCoh}^\bullet\,\mathbb{X}. \end{array}$$

This translation enables us to upgrade the previous theorem to the general case.

Theorem 7.38 (Global smoothness: general case) *Let $\mathbb{X}$ be a variety; then the following are equivalent:*

(i) *the variety $\mathbb{X}$ is smooth,*
(ii) *the embedding $\mathtt{Perf}^\bullet\,\mathbb{X} \subset \mathtt{D\,Coh}^\bullet\,\mathbb{X}$ is an equivalence,*
(iii) *the hom-spaces in $\mathtt{D\,QCoh}^\bullet\,\mathbb{X}$ have bounded degree.*

Remark 7.39 If additionally $\mathbb{X}$ is a projective variety then the hom-spaces in $\mathtt{D\,Coh}^\bullet\,\mathbb{X}$ are not only bounded in degree but also finite-dimensional. In other words, $\mathtt{D\,Coh}^\bullet\,\mathbb{X}$ is a hom-finite A_∞-category.

Smooth varieties automatically come with a nice set of locally free sheaves that are analogs from differential geometry.

Definition 7.40 Let $(\mathbb{X}, \mathscr{O})$ be a variety.

- The *tangent sheaf* $\mathscr{T}$ is defined as the sheaf of derivations on $\mathbb{X}$,

$$\mathscr{T}(\mathbb{U}) := \mathrm{Der}(\mathscr{O}(\mathbb{U})) = \{\delta\colon \mathscr{O}(\mathbb{U}) \to \mathscr{O}(\mathbb{U}) \mid \delta(uv) = \delta(u)v + u\delta(v)\}.$$

- The *cotangent sheaf* $\mathscr{T}^*$ is defined as its dual,

$$\mathscr{T}^*(\mathbb{U}) := \mathrm{Hom}_{\mathscr{O}(\mathbb{U})}(\mathscr{T}(\mathbb{U}), \mathscr{O}(\mathbb{U})).$$

  Each expression of the form fdg with $f, g \in \mathscr{O}(\mathbb{U})$ can be interpreted as an element in $\mathscr{T}^*(\mathbb{U})$ by the identification $fdg(\delta) = f\delta(g)$.
- The *sheaf of differential k-forms* $\Omega^\bullet$ is the exterior algebra of the cotangent sheaf over the structure sheaf,

$$\Omega^\bullet(\mathbb{U}) := \bigwedge \mathscr{T}^*(\mathbb{U})^\bullet = \frac{\bigoplus_{n=0}^{\infty} \mathscr{T}^*(\mathbb{U})^{\otimes^n_{\mathscr{O}(\mathbb{U})}}}{\langle u \otimes v - v \otimes u \mid u, v \in \mathscr{T}^*(\mathbb{U})\rangle}.$$

- The *canonical sheaf* is the sheaf of top differential forms

$$\mathscr{K}(\mathbb{U}) := \Omega^n(\mathbb{U}) \quad \text{with } n = \dim \mathbb{X}.$$

Theorem 7.41 *If $(\mathbb{X}, \mathscr{O})$ is smooth then $\mathscr{T}$, $\mathscr{T}^*$ and Ω^k are locally free sheaves and $\mathscr{K}$ is a line bundle.*

Sketch of the proof See [230, Tag 05D4]. □

Example 7.42 If $\mathbb{X} = \mathbb{P}^1$ then we can write a derivation locally on $\mathbb{U}_0$ as $f(\xi)\frac{d}{d\xi}$ with $\xi = \frac{X}{Y}$. On $\mathbb{U}_1$ this derivation becomes $\frac{d\eta}{d\xi}\frac{d}{d\eta} = -f(\eta^{-1})\eta^2\frac{d}{d\eta}$, so if f has degree at most 2 then the derivation is globally defined. Therefore, $\mathscr{T}(\mathbb{X}) = \mathbb{C}^3$ and we get $\mathscr{T} = \mathscr{O}(2)$, while $\mathscr{T}^* = \Omega^1 = \mathscr{K} = \mathscr{O}(-2)$ because $d\xi = -\eta^{-2}d\eta$. In general, one can show that for $\mathbb{P}^n$, the canonical bundle is equal to $\mathscr{K} = \mathscr{O}(-n-1)$ because

$$d\xi_{10} \wedge \cdots \wedge \xi_{n0} = \pm\xi_{01}^{-n-1} d\xi_{01} \wedge \cdots \wedge \xi_{n1}.$$

7.2 Other Geometrical Objects

In this section we will consider some other objects that appear in algebraic geometry. Just like for varieties, we can construct affine and nonaffine versions. The latter are obtained by gluing affine copies together but we will not work this out in detail.

7.2.1 Support

If $\mathbb{X}$ is an affine variety and I is an ideal in $R = \mathscr{O}(\mathbb{X})$ such that R/I is an affine domain then we have a natural embedding of $\mathbb{Y} = \mathbb{V}(R/I)$ as a closed subset of $\mathbb{X}$. If M is an R/I-module then we can also consider it as an R-module. This gives us an embedding $\mathsf{Coh}\,\mathbb{Y} \subset \mathsf{Coh}\,\mathbb{X}$ as the subcategory of all sheaves over $\mathbb{X}$ supported on $\mathbb{Y}$.

As ordinary categories this is a fully faithful embedding, but not if we add in the extra A_∞-structure. Indeed, if $\mathbb{Y} = \{p\}$ is just one point in $\mathbb{X} = \mathbb{A}^1$ then the A_∞-endomorphism ring of the simple module S_p as an $\mathscr{O}(\mathbb{Y})$-module is $\mathrm{Ext}^\bullet_{\mathbb{C}}(S_p, S_p) = \mathbb{C}$, while as an $\mathscr{O}(\mathbb{X})$-module it is $\mathrm{Ext}^\bullet_{\mathbb{C}[X]}(S_p, S_p) = \mathbb{C}\langle\xi\rangle/(\xi^2)$.

Definition 7.43 The *derived category of coherent sheaves on $\mathbb{X}$ supported on $\mathbb{Y}$*, written as $\mathsf{D\,Coh}^\bullet{}_{\mathbb{Y}}(\mathbb{X})$, is the sub-$A_\infty$-category of $\mathsf{D\,Coh}^\bullet(\mathbb{X})$ generated by the sheaves on $\mathbb{Y}$ viewed as sheaves on $\mathbb{X}$.

Example 7.44 If $\mathbb{Y} = \{p\}$ is an n-dimensional smooth point in a variety $\mathbb{X}$ then $\mathtt{D\,Coh}^\bullet{}_{\mathbb{Y}}(\mathbb{X})$ is generated by the skyscraper sheaf $\mathscr{S}_p$. This means that $\mathtt{D\,Coh}^\bullet{}_{\mathbb{Y}}(\mathbb{X})$ contains the self-extensions of $\mathscr{S}_p$, which are sheaves in $\mathtt{Coh}\,\mathbb{X}$ but not contained in $\mathtt{Coh}\,\mathbb{Y} \subset \mathtt{Coh}\,\mathbb{X}$.

Because the endomorphism ring of $\mathscr{S}_p$ in $\mathtt{D\,Coh}^\bullet(\mathbb{X})$ is the exterior algebra $\Lambda_n^\bullet$, we have $\mathtt{D\,Coh}^\bullet{}_{\mathbb{Y}}(\mathbb{X}) \overset{\infty}{=} \mathtt{D}\Lambda_n^\bullet$. This category is equivalent to $\mathtt{D\,mod}^\bullet\,\mathbb{C}[[X_1, \dots, X_n]]$, so it only sees an infinitesimal neighborhood of $\mathbb{Y}$ in $\mathbb{X}$.

Remark 7.45 In general, if $\mathbb{Y}$ is singular but $\mathbb{X}$ is smooth then $\mathtt{D\,Coh}^\bullet{}_{\mathbb{Y}}(\mathbb{X})$ will behave much more nicely than $\mathtt{D\,Coh}^\bullet(\mathbb{Y})$.

7.2.2 Schemes

Schemes are a generalization of varieties where the coordinate ring is allowed to be any commutative ring R. The space associated to R is denoted by $\mathsf{Spec}(R)$ and it consists of all prime ideals of R. The affine open sets of $\mathsf{Spec}(R)$ are of the form $\mathbb{U}_f = \{\mathfrak{p} \in \mathsf{Spec}(R) \mid f \notin \mathfrak{p}\}$. Just like with ordinary varieties we can turn R into a sheaf of rings $\mathscr{O}$ over $\mathsf{Spec}(R)$ by putting $\mathscr{O}(\mathbb{U}_f) = R[f^{-1}]$. The scheme corresponding to R consists of the ringed space $\mathbb{X} = (\mathsf{Spec}(R), \mathscr{O})$ and we can define (quasi)-coherent sheaves in the usual way.

If R is an affine ring then we can embed $\mathbb{V}(R)$ into $\mathsf{Spec}(R)$ by mapping the representation $\rho\colon R \to \mathbb{C}$ to the maximal ideal $\operatorname{Ker}\rho$. Although $\mathsf{Spec}(R)$ usually has more points than $\mathbb{V}(R)$, these topological spaces look very similar. In particular, the associated categories of open subsets are equivalent: $\mathtt{Open}\mathbb{V}(R) \cong \mathtt{Open}\,\mathsf{Spec}(R)$. Therefore, the scheme versions of $\mathtt{Coh}^\bullet\,\mathbb{X}$, $\mathtt{QCoh}^\bullet\,\mathbb{X}$ and $\mathtt{Perf}^\bullet\,\mathbb{X}$ are the same as the variety versions and equivalent to $\mathtt{MOD}^\bullet\,R$, $\mathtt{Mod}^\bullet\,R$ and $\mathtt{D\,Proj}^\bullet\,R$ respectively.

Example 7.46 If $R = \mathbb{C}[X]$ then $\mathsf{Spec}(R)$ has one extra point in comparison to $\mathbb{V}(R)$. This point corresponds to the prime ideal (0). Because $f \notin (0)$, this point is contained in all the $\mathbb{U}_f$ and hence it is in the closure of any point. This point is called the generic point of $\mathsf{Spec}(R)$.

Example 7.47 If $R = \mathbb{C}[X]/(X^n)$ then $\mathsf{Spec}(R)$ has only one element: the prime ideal $\mathfrak{p} = (X)$. For $n = 1$ this scheme is the same as the affine variety consisting of one point. If $n > 1$ the scheme still has one point but the ring of functions around this point is larger: R instead of $\mathbb{C}$. Therefore, it is called a *fat point*. Its derived category is $\mathtt{D\,Coh}^\bullet\,\mathbb{X} = \mathtt{D\,Mod}^\bullet\,\mathbb{C}[X]/(X^n)$, which is not equal to $\mathtt{Perf}^\bullet\,\mathbb{C}[X]/(X^n)$ so the fat point is a singular scheme.

For more information about schemes we refer to [109, 184, 72].

7.2.3 Noncommutative Varieties

We can also try to make the jump to noncommutative rings. Ideally we would like to treat $\mathtt{Mod}^\bullet R$ or $\mathtt{MOD}^\bullet R$ as categories of sheaves over some topological space. In general, this is quite hard to do because there are different possibilities for the correct notion of a point (e.g. prime ideal, maximal ideal, simple representation, one-dimensional representation). Each of these possibilities has its own drawbacks and in general it is better to forget about the space altogether and just study the categories $\mathtt{Mod}^\bullet R$ or $\mathtt{MOD}^\bullet R$ and their derived versions $\mathtt{D}\,\mathtt{Mod}^\bullet R$ or $\mathtt{D}\,\mathtt{MOD}^\bullet R$.

Example 7.48 In some cases projective varieties can be viewed as affine noncommutative varieties. From Section 4.4.2 we know that the derived category $\mathtt{D}\,\mathtt{Coh}^\bullet \mathbb{P}^1$ is equivalent to $\mathtt{D}\,\mathtt{mod}^\bullet Q = \mathtt{D}\,\mathtt{Mod}^\bullet Q$, where

$$Q = \circ \Longrightarrow \circ\,.$$

The reason for this is that we found a split-generator for $\mathtt{D}\,\mathtt{Coh}^\bullet \mathbb{P}^1$ whose endomorphism ring is the path algebra of Q. Similarly, if we look at Example 7.29 we can conclude that $\mathtt{D}\,\mathtt{Coh}^\bullet \mathbb{P}^2 = \mathtt{Perf}^\bullet \mathbb{P}^2 = \mathtt{D}\,\mathtt{Mod}^\bullet B$, where B is the path algebra with relations coming from the Beilinson quiver.

If we view a noncommutative algebra A as a ring of functions over a noncommutative variety and its modules as quasi-coherent sheaves, we can transfer geometrical notions to the noncommutative setting. Just like for a commutative variety there will be three categories of interest: $\mathtt{D}\,\mathtt{MOD}^\bullet A$ takes the place of $\mathtt{D}\,\mathtt{QCoh}^\bullet \mathbb{X}$, the finitely generated modules $\mathtt{D}\,\mathtt{Mod}^\bullet A$ that of $\mathtt{D}\,\mathtt{Coh}^\bullet \mathbb{X}$ and $\mathtt{Perf}^\bullet A := \mathtt{D}\,\mathtt{Proj}^\bullet A$ assumes the role of $\mathtt{Perf}^\bullet \mathbb{X}$. However, if A is not Noetherian then $\mathtt{D}\,\mathtt{Mod}^\bullet A$ is not well behaved because then $\mathtt{Mod}\,A$ is not closed under taking kernels. Therefore we will now assume that A is Noetherian.

Definition 7.49 The (right) *global dimension* of an algebra A is the maximal number n such that $\mathrm{Ext}^n_A(M, N) \neq 0$ for some modules $M, N \in \mathtt{MOD}^\bullet A$. If there is no such n then we call the global dimension of A infinite.

Example 7.50 We state the global dimensions of some important algebras.

- The global dimension of a (skew) field is 0 because all modules are free.
- The global dimension of the coordinate ring of a smooth n-dimensional affine variety is n.
- The global dimension of a path algebra $\mathbb{C}Q$ is 1 because every module M has a resolution of the form

$$\bigoplus_{v\in Q_0} Mv \otimes v\mathbb{C}Q \xleftarrow{\;a\otimes t(a)-h(a)\otimes a\;} \bigoplus_{a\in Q_1} Mh(a) \otimes t(a)\mathbb{C}Q\,.$$

- The global dimension of the path algebra with relations coming from the Beilinson quiver is 2.
- Take care: the global dimension is not preserved under derived equivalence. Both

$$\circ \longrightarrow \circ \longrightarrow \circ \quad \text{and} \quad \circ \xrightarrow{\alpha} \circ \xrightarrow{\beta} \circ/(\beta\alpha)$$

  have the same derived category but the global dimension of the former is 1 and of the latter is 2.
- Having finite global dimension is not a very strong condition. Every path algebra with relations coming from a quiver without oriented cycles has finite global dimension.

Theorem 7.51 (Smoothness for Noetherian algebras) *If A is Noetherian and the global dimension is n then*

(i) $\mathtt{Perf}^\bullet A \subset \mathtt{D\,Mod}^\bullet A$ *is an equivalence,*
(ii) *the hom-spaces in* $\mathtt{D\,MOD}^\bullet A$ *are bounded in degree.*

Sketch of the proof If the global dimension is n, every module has a projective resolution of length at most n. Because A is Noetherian this resolution is finitely generated and hence in $\mathtt{Perf}^\bullet A$. For the second part, we know that the hom-spaces between modules are bounded in degree so if we take hom-spaces between twisted complexes of modules they will also be bounded. □

Remark 7.52 Because the translation of geometric ideas to the noncommutative setting is often imperfect, there are several different approaches to noncommutative algebraic geometry depending on what aspect is important [152, 145, 226] and there is much interplay with the field of representation theory of finite-dimensional algebras [97, 108].

7.2.4 Stacks

Suppose that $\mathbb{X}$ is an affine variety with an action of a finite group G. We would like to construct a new variety whose points correspond to the orbits in $\mathbb{X}$. In order to do that we need to find the coordinate ring of this new variety. The action of G on the original variety induces an action on its coordinate ring $R = \mathcal{O}(\mathbb{X})$:

$$\forall\, g \in G,\ \forall\, f \in R,\ \forall\, x \in \mathbb{X},\ \text{we have}\ \ g \cdot f(x) := f(g^{-1} \cdot x).$$

The functions on the orbit set can be seen as functions on $\mathbb{X}$ that are invariant under the group action. These form a ring

$$R^G := \{f \in R \mid \forall\, g \in G,\ g \cdot f = f\},$$

and we define the quotient variety as $\mathbb{X}/\!\!/G := \mathbb{V}(R^G)$. The natural embedding $R^G \to R$ gives rise to a projection map $\pi\colon \mathbb{X} \to \mathbb{X}/\!\!/G$ and because G is finite the fibers of π are precisely the orbits.

Definition 7.53 The variety $\mathbb{X}/\!\!/G := \mathbb{V}(R^G)$ is called the *categorical quotient.*

Example 7.54 Let $\mathbb{X} = \mathbb{A}^1$, $G = \mathbb{Z}_n$ and the action be multiplication by the nth roots of unity. In this case, $R = \mathbb{C}[X]$ and the functions that are invariant are the polynomial functions in X^n. Therefore, $\mathscr{O}(\mathbb{X}/\!\!/G) = \mathbb{C}[X^n]$ and $\mathbb{X}/\!\!/G \cong \mathbb{C}$.

The example above shows that the ring of invariants does not completely capture the quotient construction because in $\mathbb{X}/\!\!/G \cong \mathbb{C}$ all points look alike but in the quotient construction there is one special point: the zero orbit. This orbit has only one point while the other orbits have $\#G$ points. To get around this problem, we can construct a new ring.

Definition 7.55 If $\mathbb{X} = \mathbb{V}(R)$ is an affine variety and G a finite group acting on $\mathbb{X}$ then we define the *smash product* $R \star G$ as the ring whose elements are $\mathscr{O}(\mathbb{X})$-linear combinations of elements in G,

$$R \star G := \{\textstyle\sum_i f_i g_i \mid f_i \in R,\ g_i \in G\},$$

and whose product satisfies

$$f_1 g_1 \cdot f_2 g_2 = (f_1(g_1 \cdot f_2))g_1 g_2.$$

If G acts faithfully then one can show that the center of the smash product $R \star G$ is equal to the ring of invariants R^G. The ring $R \star G$ is a noncommutative $\mathbb{C}$-algebra, so we can consider it as a noncommutative variety.

Definition 7.56 The *quotient stack* $[\mathbb{X}/G]$ is the noncommutative variety corresponding to $R \star G$ so, by definition,

$$\begin{aligned} \mathtt{Perf}^\bullet[\mathbb{X}/G] &:= \mathtt{Perf}^\bullet R \star G, \\ \mathtt{D\,Coh}^\bullet[\mathbb{X}/G] &:= \mathtt{D\,Mod}^\bullet R \star G, \\ \mathtt{D\,QCoh}^\bullet[\mathbb{X}/G] &:= \mathtt{D\,MOD}^\bullet R \star G. \end{aligned}$$

Example 7.57 Let $\mathbb{X} = \mathbb{C}$, $G = \mathbb{Z}/n\mathbb{Z}$ and the action be multiplication by the nth roots of unity. In this case, Theorem 4.29 tells us that the quotient stack corresponds to the path algebra of the cyclic quiver with n vertices:

$\mathscr{O}(\mathbb{X}) \star G \cong \mathbb{C}Q$ with $Q =$ 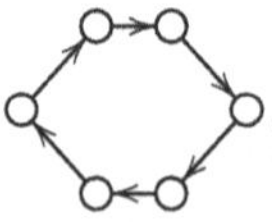 .

Therefore, $\mathtt{D\,Coh}^\bullet[\mathbb{X}/G] \cong \mathtt{D\,Mod}^\bullet Q$, while $\mathtt{D\,Coh}^\bullet \mathbb{X}/\!\!/G \cong \mathtt{D\,Mod}^\bullet \mathbb{C}[X]$, so the quotient stack is indeed something different from the quotient variety.

In general, the quotient stack behaves better from a geometrical point of view than the categorical quotient.

Example 7.58 If $G = \mathbb{Z}/2\mathbb{Z} \cong \{1, -1\} \cong \langle g \mid g^2 \rangle$ acts diagonally on $\mathbb{A}^2$ by multiplication then the ring of invariants $\mathbb{C}[X, Y]^G$ is generated by $\xi = X^2$, $\eta = Y^2$, $\zeta = XY$. This shows that $\mathbb{C}[X, Y]^G \cong \mathbb{C}[\xi, \eta, \zeta]/(\xi\eta - \zeta^2)$, so $\mathbb{X}/\!\!/G$ is in fact a cone, which is not a smooth variety. On the other hand, the coherent sheaves on the stack $[\mathbb{A}^2/G]$ can be seen as finitely generated A-modules where

$$A = \mathbb{C}[X, Y] \star G = \frac{\mathbb{C}\langle X, Y, g\rangle}{\langle XY - YX, gX + Xg, gY + Yg, g^2\rangle}.$$

This algebra is isomorphic to the path algebra of the quiver with relations

$$\circ \underset{x_2, y_2}{\overset{x_1, y_1}{\rightleftarrows}} \circ \qquad \mathcal{R} = \langle x_1y_2 - y_1x_2, x_2y_1 - y_2x_1\rangle.$$

Every A-module M has a resolution of the form

$$M \otimes_{\mathbb{C}G} A \xleftarrow{(X\ Y)} M \otimes_{\mathbb{C}G} A^{\oplus 2} \xleftarrow{\binom{-Y}{X}} M \otimes_{\mathbb{C}G}.$$

So A has global dimension 2. Because A is Noetherian we have

$$\mathtt{D\,Coh}^\bullet[\mathbb{A}^2/G] \cong \mathtt{Perf}^\bullet[\mathbb{A}^2/G].$$

Remark 7.59 One can also define a coherent sheaf on $[\mathbb{X}/G]$ as a G-equivariant sheaf on $\mathbb{X}$. This method allows us to define stacky quotients for nonaffine varieties. It is even possible to expand the notion of a stack beyond quotients of group actions. For a brief introduction to stacks we refer to [76, 196].

7.2.5 Landau–Ginzburg Models

Matrix factorizations were first introduced by Eisenbud in [73]. Later, this technique was picked up by physicists for their study of certain physical models called *Landau–Ginzburg models.* The latter consist of pairs $(\mathbb{X}, W)$ where $\mathbb{X}$ is a variety and $W: \mathbb{X} \to \mathbb{C}$ a holomorphic function. If $\mathbb{X}$ is affine, we can think of W as an element in $\mathcal{O}(\mathbb{X})$; this element is often called the *potential*.

Definition 7.60 Let (R, W) be a pair (R, W) consisting of an affine domain R and an element $W \in R$.

- A *matrix factorization* of (R, W) is a diagram

$$P = P_0 \underset{d_1}{\overset{d_0}{\rightleftarrows}} P_1 \quad \text{with } d_0d_1 = W\mathbb{1}_{P_1} \text{ and } d_1d_0 = W\mathbb{1}_{P_0},$$

where P_0, P_1 are finitely generated projective modules (i.e. vector bundles over $\mathbb{V}(R)$).

- The space of morphisms between two matrix factorizations P, Q is the direct sum $\bigoplus_{i,j\in\mathbb{Z}_2} \mathrm{Hom}_R(P_i, Q_j)$ with differential

$$d\begin{pmatrix} f_{00} & f_{01} \\ f_{10} & f_{11} \end{pmatrix} = \begin{pmatrix} 0 & d_1 \\ d_0 & 0 \end{pmatrix}_Q \begin{pmatrix} f_{00} & f_{01} \\ f_{10} & f_{11} \end{pmatrix} - \begin{pmatrix} f_{00} & -f_{01} \\ -f_{10} & f_{11} \end{pmatrix} \begin{pmatrix} 0 & d_1 \\ d_0 & 0 \end{pmatrix}_P .$$

- The *category of matrix factorizations* $\mathtt{MF}^{\pm}(R, W)$ is the minimal model of the dg-category with objects and morphisms as defined above.

Remark 7.61 If R is a polynomial ring then the *Quillen–Suslin theorem* [206] tells us that every projective R-module is free. With Example 3.65 in mind, we can interpret the P_i as $R^{\oplus k}$ and the d_i as matrices with coefficients in R. In other words, $(P_0 \oplus P_1[1], \begin{pmatrix} 0 & d_1 \\ d_0 & 0 \end{pmatrix})$ is a twisted complex for the curved algebra R_W:

$$\mathtt{MF}^{\pm}(R, W) = \mathsf{H}^0\mathtt{Tw}R_W = {}^0\mathtt{D}R_W.$$

Example 7.62 Consider the ring $R = \mathbb{C}[X, Y]$ and the polynomial $W = XY$. There is a natural matrix factorization:

$$S := R \underset{Y}{\overset{X}{\rightleftarrows}} R .$$

We have that $\mathrm{End}_{\mathrm{MF}}(S, S) = \mathbb{C}\mathbb{1}$. In fact, one can show that $\mathtt{MF}^{\pm}(R, W)$ is generated by S and therefore it is equivalent to $\mathtt{D}\mathbb{C}^{\pm}$, which is the $\mathbb{Z}_2$-graded derived category of a point, or the category of matrix factorizations $\mathtt{MF}^{\pm}(\mathbb{C}, 0)$.

This observation is not a coincidence. In general, $\mathtt{MF}^{\pm}(R, W)$ can be used to describe the singular locus of $W^{-1}(0) = \mathbb{V}(R/(W))$. For each $p \in W^{-1}(0) \subset \mathbb{C}^n$ we can construct a matrix factorization. Assume $R = \mathbb{C}[X_1, \ldots, X_n]$ with $p = 0 \in W^{-1}(0)$ and write $W = \sum X_r W_r$ for some polynomials W_r. Let $\Omega^{\pm} = R \otimes \Lambda_n^{\bullet}$ the space of differential forms viewed as a $\mathbb{Z}_2$-graded R-module. We equip $\Omega^{\pm}$ with an R-linear map

$$d\omega = \sum_r X_r \iota_{\partial_{X_r}} \omega + \sum_r W_r \, dX_r \wedge \omega.$$

Note that the first part of d has $\mathbb{Z}$-degree -1 because we fill in ∂_{X_r} into ω, while the second part has degree $+1$, because we wedge ω with a 1-form. Therefore, the total differential is only homogeneous for the $\mathbb{Z}_2$-grading. We can calculate that $d^2\omega = W\omega$ (see also Example 12.13) so the pair $M_p = (\Omega^{\pm}, d)$ is a matrix factorization of W and it has two important properties:

- If we choose a different split $W = \sum_r X_r W_r'$, the new matrix factorization is quasi-isomorphic to the old one.

- If one of the W_r has a constant term then $(\Omega^{\pm}, d)$ is quasi-equivalent to zero.

This implies that M_p only depends on p and is only nonzero if p is in the singular locus of $W^{-1}(0)$. We denote this space by

$$\mathtt{Sing}(W) = \{p \in W^{-1}(0) \mid (dW)_p = 0\}.$$

Example 7.63 Consider the ring $R = \mathbb{C}[X, Y]$ and the polynomial $W = XY(X - Y)$. There are three obvious matrix factorizations:

$$S_X := R \underset{Y(X-Y)}{\overset{X}{\rightleftarrows}} R\,, \quad S_Y := R \underset{X(X-Y)}{\overset{Y}{\rightleftarrows}} R\,, \quad S_{X-Y} := R \underset{XY}{\overset{X-Y}{\rightleftarrows}} R\,.$$

For each $Z \in \{X, Y, X - Y\}$ we have $\mathrm{End}_{\mathrm{MF}}(S_Z, S_Z) = \frac{\mathbb{C}[U]}{(U^2)}$, where U is one of the factors in the double product W/Z. Either of the factors will do because they represent the same cohomology class, possibly up to a sign.

On the other hand, $\mathtt{Sing}(W)$ consists of one point $p = 0$. By splitting W as $X(XY) + Y(-XY)$ we get a matrix factorization of the form

$$\begin{aligned} M_p &\cong \left(R \oplus R\,dX \wedge dY \oplus R\,dX \oplus R\,dY, \begin{pmatrix} 0 & 0 & X & Y \\ 0 & 0 & XY & XY \\ XY & -Y & 0 & 0 \\ -XY & X & 0 & 0 \end{pmatrix}\right) \\ &\cong \left(R \oplus R \oplus R[1] \oplus R[1], \begin{pmatrix} 0 & 0 & X & Y \\ 0 & 0 & 0 & X(X-Y) \\ Y(X-Y) & -Y & 0 & 0 \\ 0 & X & 0 & 0 \end{pmatrix}\right) \\ &\cong \mathrm{Cone}\left(S_X \xrightarrow{Y} S_X[1]\right). \end{aligned}$$

So M_p is generated by S_X. On the other hand, if $\phi\colon M_p \to M_p$ is the degree 1 morphism coming from the composition $M_p \to S_X \cong S_X[1] \to M_p$ then $\mathrm{Cone}(\phi) \cong S_X \oplus S_X[1]$. So M_p split-generates S_X. Similar reasoning shows that it also split-generates S_Y, S_Z and even the whole category.

These examples generalize to three important theorems:

Theorem 7.64 (Knörrer [140]) *If $W \in R = \mathbb{C}[X_1, \ldots, X_n]$ then*

$$\mathtt{MF}^{\pm}(R, W) \overset{\infty}{=} \mathtt{MF}^{\pm}(R[U, V], W + UV).$$

Theorem 7.65 (Dyckerhoff [69]) *If $W \in R = \mathbb{C}[X_1, \ldots, X_n]$ has just one isolated singular point, $\mathtt{Sing}(W) = \{p\}$, then $\mathtt{MF}^{\pm}(R, W)$ is split-generated by M_p.*

Theorem 7.66 (Buchweitz [46]) *The category $\mathtt{MF}^0(R, W)$ is equivalent to the category of singularities of $\mathbb{X} = W^{-1}(0)$, which is the quotient*

$$\mathtt{Sing}\,\mathbb{X} := \frac{\mathtt{D^bCoh}\,\mathbb{X}}{\mathtt{Perf}\,\mathbb{X}}.$$

This equivalence sends a matrix factorization P to the $R/(W)$-module $P_0/\operatorname{Im} d_1$ viewed as a sheaf over $\mathbb{X} = \mathbb{V}(R/(W))$.

Remark 7.67 Note that this theorem is formulated in terms of triangulated categories instead of A_∞-categories. To make sense of this equivalence as A_∞-categories one can use the Drinfeld quotient [67] and the category of matrix factorizations has to be interpreted as a 2-periodic $\mathbb{Z}$-graded category instead of a $\mathbb{Z}_2$-graded category.

Remark 7.68 $\mathtt{Sing}\,\mathbb{X}$ is called the category of singularities because the quotient construction kills all the sheaves that are supported on the smooth locus of $\mathbb{X}$. Those are sheaves with a finite locally free resolution and hence they are objects in $\mathtt{Perf}^\bullet\,\mathbb{X}$. If $\mathbb{X}$ is a smooth variety then $\mathtt{Perf}^\bullet\,\mathbb{X} = \mathtt{D}\,\mathtt{Coh}^\bullet\,\mathbb{X}$ so the quotient is zero.

The downside of matrix factorizations is that we are restricted to the $\mathbb{Z}_2$-graded setting. This can be circumvented by working with graded algebras.

Definition 7.69 A *graded Landau–Ginzburg model* is a pair (R, W) consisting of a graded ring R and a homogeneous element $W \in R$ of degree w.

- A *graded matrix factorization* of (R, W) consists of a diagram

$$P = P^0 \underset{d_1}{\overset{d_0}{\rightleftarrows}} P^1$$

  where P_0 and P_1 are projective graded R-modules and d_0, d_1 are maps of degree 0 and w such that $d_1 d_0 = W\mathbb{1}_{P^0}$ and $d_0 d_1 = W\mathbb{1}_{P^1}$.
- The homological shift of a graded matrix factorization P is

$$P[1] = P^1 \underset{-d_0}{\overset{-d_1}{\rightleftarrows}} P^0(w)\ .$$

  Note that $P[2]$ coincides with $P(w)$ where (w) is the shift for the R-degree.
- If P, Q are two graded matrix factorizations we define

$$\operatorname{Hom}^\bullet(P, Q) = \bigoplus_i \operatorname{grHom}_R(P^0, Q[i]^0) \oplus \operatorname{grHom}_R(P^1, Q[i]^1),$$

  where $\operatorname{grHom}_R(X, Y)$ stands for the degree-preserving R-module morphisms. On this space the standard differential has degree 1 for the homological grading.
- The *category graded of matrix factorizations* $\mathtt{grMF}^\bullet(R, W)$ is the minimal model of the dg-category with objects and morphisms as defined above.

Example 7.70 Let us return to $R = \mathbb{C}[X, Y]$ and $W = XY(X - Y)$. Up to homological shift the ordinary matrix factorization that factors W into X and $Y(X - X)$ can be seen as a graded matrix factorization in three ways:

$$R \underset{Y(X-Y)}{\overset{X}{\rightleftarrows}} R(1)\,, \quad R(1) \underset{Y(X-Y)}{\overset{X}{\rightleftarrows}} R(2)\,, \quad R(2) \underset{Y(X-Y)}{\overset{X}{\rightleftarrows}} R(3)\,.$$

For more details on graded matrix factorizations we refer to the work of Orlov [200], whose main result will also be discussed in the next section. Landau–Ginzburg models can also be developed in the nonaffine [201, 225] or stacky [216] settings and there is a vast literature on stable representation categories, which is the noncommutative analog of the category of singularities [46, 108].

7.3 Equivalences

We have seen that B-models can come in many different flavors, and different B-models often give rise to equivalent or related categories. In this section we will study this phenomenon in more detail.

7.3.1 Reconstructing Varieties

If A is an algebra then every $r \in Z(A)$ gives rise to a natural transformation $\nu_r \colon \mathbb{1}_{\mathtt{MOD}A} \to \mathbb{1}_{\mathtt{MOD}A}$ that assigns to each module M the element in $\mathrm{Hom}_R(M, M)$ that corresponds to multiplication by r. This is indeed a natural transformation because the action of r commutes with everything. The reverse also holds: every natural transformation of $\mathbb{1}_{\mathtt{Mod}\,A}$ to itself is of the form ν_r with $r = \nu_A(1)$, because if M is any module and $\phi_m \colon A \to M$ is the module morphism with $\phi_m(1) = m$ then

$$\nu_M(m) = \nu_M(\phi_m(1)) = \phi_m(\nu_A(1)) = \phi_m(r) = rm.$$

The same reasoning can be applied to other module categories that contain A itself such as $\mathtt{Mod}\,A$, $\mathtt{D}\,\mathtt{Mod}^\bullet A$, $\mathtt{Perf}^\bullet A$. For an affine variety this translates to

$$\mathrm{End}(\mathbb{1}_{\mathtt{D}\,\mathtt{Coh}^\bullet \mathbb{X}}) \cong \mathrm{End}(\mathbb{1}_{\mathtt{D}\,\mathtt{Mod}^\bullet \mathbb{V}(R)}) \cong R.$$

In other words, we can recover a commutative affine variety from its derived category as $\mathbb{V}(\mathrm{End}(\mathbb{1}_{\mathtt{D}\,\mathtt{Coh}^\bullet \mathbb{X}}))$.

For quasi-projective varieties this approach fails because in general these have very few global functions. Instead, we would like to reconstruct these as $\mathbb{X} = \mathbb{P}(R)$ of some graded ring R. Unlike the coordinate ring of an affine variety, this graded ring R is not uniquely determined by $\mathbb{X}$. Instead, there is a line bundle $\mathscr{L} = \mathscr{O}(1)$ such that $R = \oplus_i \mathscr{L}^{\otimes i}(\mathbb{X})$.

Definition 7.71 A line bundle $\mathscr{L}$ on $\mathbb{X}$ is called *very ample* if $\mathbb{X} = \mathbb{P}(R_{\mathscr{L}})$ with $R_{\mathscr{L}} := \bigoplus_i \mathscr{L}^{\otimes i}(\mathbb{X})$, and $\mathscr{L}$ is called *ample* if $\mathscr{L}^{\otimes k}$ is very ample for some $k \geq 1$.

Example 7.72 In general, a projective variety can have many ample line bundles. For the variety $\mathbb{P}^1$, all $\mathscr{O}(i)$ with $i > 0$ are ample. Take for instance $\mathscr{L} = \mathscr{O}(2)$; then

$$R_{\mathscr{L}} = \bigoplus_{i \geq 0} \mathscr{O}(2i) = \mathbb{C}[X^2, Y^2, XY] \cong \frac{\mathbb{C}[U, V, W]}{(UV - W^2)}.$$

By the last description, the variety $\mathbb{P}(R_{\mathscr{L}})$ can be interpreted as the conic $UV - W^2 = 0$ inside $\mathbb{P}^2$. This variety is isomorphic to $\mathbb{P}^1$ via the Veronese embedding $(x : y) \mapsto (x^2 : y^2 : xy)$.

From a categorical point of view we can recover $R_{\mathscr{L}}$ as a space of natural transformations. Any line bundle defines an endofunctor $- \otimes \mathscr{L}$ on $\mathscr{V}ect(\mathbb{X})$ by

$$\mathscr{M} \otimes \mathscr{L}(\mathbb{U}) := \mathscr{M}(\mathbb{U}) \otimes_{\mathscr{O}(\mathbb{U})} \mathscr{L}(\mathbb{U}).$$

This functor induces a dg-functor on $\mathtt{Vect}^\bullet(\mathbb{X})$ and an A_∞-functor $\mathcal{T}$ on its derived version $\mathtt{D}\,\mathtt{Vect}^\bullet\mathbb{X} = \mathtt{Perf}^\bullet\mathbb{X}$ (this functor is usually denoted in the derived context by $\otimes_L \mathscr{L}$). The global sections of $\mathscr{L}$ will correspond to the natural transformations between the identity functor and $\mathcal{T}$:

$$\mathscr{L}(\mathbb{X}) = \mathsf{Fun}(\mathbb{1}, \mathcal{T}).$$

(Here Fun stands for the category of functors from $\mathtt{Perf}^\bullet\mathbb{X}$ to itself with natural transformations as morphisms.) Because $\mathcal{T}$ is invertible, with $\mathcal{T}^{-1} = \otimes \mathscr{L}^{-1}$ and $\mathscr{L}^{-1} = \mathscr{H}om(\mathscr{L}, \mathscr{O}_{\mathbb{X}})$, we can identify

$$\mathsf{Fun}(\mathcal{T}^i, \mathcal{T}^j) = \mathsf{Fun}(\mathbb{1}, \mathcal{T}^{i-j}) = \mathsf{Fun}(\mathbb{1}, - \otimes \mathscr{L}^{i-j})$$

and therefore

$$R_{\mathcal{T}} := \bigoplus_{i \geq 0} \mathsf{Fun}(\mathbb{1}, \mathcal{T}^i)$$

acquires a ring structure from composition, which is isomorphic to $R_{\mathscr{L}}$. So reconstructing the variety amounts to finding a good $\mathcal{T}$.

As we saw in Definition 7.40, there is one intrinsically defined line bundle associated to a variety: the canonical line bundle. But unfortunately it is not necessarily an ample bundle, so it might not suit our needs. Therefore we make a distinction between the following types of varieties.

Definition 7.73 A smooth projective variety $\mathbb{X}$ is called

- *Fano* if the anticanonical bundle $\mathscr{K}^{-1}$ is ample,
- *Calabi–Yau* if the canonical bundle is trivial $\mathscr{K} = \mathscr{O}_{\mathbb{X}}$,
- *anti-Fano* if the canonical bundle $\mathscr{K}$ is ample.

Remark 7.74 These three possibilities are not exhaustive, but represent three basic cases roughly corresponding to manifolds with positive, zero or negative curvature respectively.

Example 7.75 Projective space $\mathbb{P}^n$ is a Fano variety because $\mathscr{K}^{-1} = \mathscr{O}(n+1)$ is ample. If f is a degree k homogeneous function in $R = \mathbb{C}[X_0, \dots, X_n]$ then the adjunction formula [109] tell us that if $\mathbb{P}(R/(f))$ is smooth, it will be Fano if $k < n + 1$, Calabi–Yau if $k = n + 1$ and anti-Fano if $k > n + 1$.

If we restrict to $n = 2$ then this illustrates the previous remark because $\mathbb{P}(R/(f))$ will be Fano if $k = 1, 2$, which means that it is a projective line or quadric and hence isomorphic to the Riemann sphere. If $k = 3$ the variety is Calabi–Yau and we get an elliptic curve, which is a torus and hence flat. The anti-Fano case corresponds to curves with higher genus, which have negative curvature.

All this indicates that we might be able to reconstruct a smooth projective variety from its derived category $\mathtt{D}\,\mathtt{Coh}^\bullet\,\mathbb{X} = \mathtt{Perf}^\bullet\,\mathbb{X}$ if it is Fano or of general type because then we can try to find the functor $- \otimes \mathscr{K}$.

Theorem 7.76 (Bondal–Orlov) *If $\mathbb{X}$ is Fano or anti-Fano then we can reconstruct $\mathbb{X}$ from $\mathtt{D}\,\mathtt{Coh}^\bullet\,\mathbb{X}$.*

Sketch of the proof The main tool of the proof is the *Serre functor*. If C is a $\mathbb{k}$-linear category then $\mathcal{F}\colon \mathsf{C} \to \mathsf{C}$ is called a *Serre functor* if there are natural isomorphisms

$$\nu_{X,Y}\colon \mathsf{C}(X, Y) \to \mathsf{C}(\mathcal{F}(Y), X)^\vee,$$

where $\vee$ is the $\mathbb{k}$-linear dual. If a category has a Serre functor it is uniquely determined up to a natural isomorphism, so it is an intrinsic object. The derived category of an n-dimensional smooth projective variety $\mathbb{X}$ has a Serre functor that is equal to $- \otimes \mathscr{K}[n]$. If $\mathscr{K}$ is ample or antiample, Bondal and Orlov [32] use this Serre functor to identify skyscraper sheaves and line bundles on $\mathbb{X}$ inside the derived category. □

7.3.2 Fourier–Mukai Transforms

If they are not Fano or of general type then nonisomorphic varieties can have equivalent derived categories. To find examples of this phenomenon we will introduce a large class of functors between derived categories of coherent sheaves and then examine when such functors are equivalences.

The inspiration for this construction comes from Fourier analysis. If $f: \mathbb{R} \to \mathbb{C}$ is a function then we can define its Fourier transform as

$$\mathcal{F}(f): \mathbb{R} \to \mathbb{C}: y \mapsto \int_{\mathbb{R}} f(x)e^{-ixy}\, dx.$$

Although this operation turns a function on $\mathbb{R}$ to a function on $\mathbb{R}$ it is better to consider it as a map from functions on one space $\mathbb{X} = \mathbb{R}$ (position space) to another space $\mathbb{Y}$ (momentum space), which accidentally happens to be isomorphic to $\mathbb{R}$. This becomes clear if we change $\mathbb{X}$ to the circle $\mathbb{S}_1$. In that case, $\mathbb{Y} = \mathbb{Z}$ and the Fourier transform turns a periodic function into its Fourier series.

How does the Fourier transform actually work? It starts with a function on $\mathbb{X}$, expands it into a function on $\mathbb{X} \times \mathbb{Y}$ by pulling it back along $\pi_{\mathbb{X}}: \mathbb{X} \times \mathbb{Y} \to \mathbb{X}$, then multiplies it by a function on $\mathbb{X} \times \mathbb{Y}$ and finally it integrates out the $\mathbb{X}$ component. We will now try to mimic this procedure for sheaves instead of functions.

Definition 7.77 Suppose that $\mathbb{X}$ and $\mathbb{Y}$ are affine varieties; then we define their *Cartesian product* as $\mathbb{X}\times\mathbb{Y} := \mathbb{V}(\mathcal{O}_{\mathbb{X}} \otimes_{\mathbb{C}} \mathcal{O}_{\mathbb{Y}})$. If $\mathbb{X}$ and $\mathbb{Y}$ are varieties with covers $\mathcal{U} = \{\mathbb{U}_i : i \in I\}$ and $\mathcal{V} = \{\mathbb{V}_j : j \in J\}$, we define the product $\mathbb{X} \times \mathbb{Y}$ as the variety that is covered by the $\mathbb{U}_i \times \mathbb{V}_j$ with the obvious gluing maps.

Remark 7.78 Note that, as a set, $\mathbb{X} \times \mathbb{Y}$ is the product of $\mathbb{X}$ and $\mathbb{Y}$ but not as a topological space (for the Zariski topology). Furthermore, one can also show that if $\mathbb{X}$ and $\mathbb{Y}$ are smooth then $\mathbb{X} \times \mathbb{Y}$ is also smooth.

Now suppose that $\mathbb{X}$ and $\mathbb{Y}$ are smooth varieties and $\mathcal{P}^\bullet$ is an object in $\mathtt{Perf}^\bullet(\mathbb{X} \times \mathbb{Y})$; then we construct a functor that consists of three parts:

(i) *Lifting to the product.* If $\mathcal{M}$ is a sheaf on $\mathbb{X}$ we define $\pi_{\mathbb{X}}^*\mathcal{M}$ by

$$\pi_{\mathbb{X}}^*\mathcal{M}(\mathbb{U}_i \times \mathbb{V}_j) = \mathcal{M}(\mathbb{U}_i) \otimes_{\mathbb{C}} \mathcal{O}(\mathbb{V}_i).$$

Note that if $\mathcal{M}$ is locally free then $\pi_{\mathbb{X}}^*\mathcal{M}$, so this gives a functor $\pi_{\mathbb{X}}^*: \mathcal{V}ect(\mathbb{X}) \to \mathcal{V}ect(\mathbb{X} \times \mathbb{Y})$ and hence also an A_∞-functor $\mathtt{D}\pi_{\mathbb{X}}^*: \mathtt{Perf}^\bullet(\mathbb{X}) \to \mathtt{Perf}^\bullet(\mathbb{X} \times \mathbb{Y})$.

(ii) *Tensoring with* $\mathscr{P}^\bullet$. For any finitely generated complex of locally free sheaves $\mathscr{P}^\bullet$ over a variety $\mathbb{V}$ (in our case $\mathbb{V} = \mathbb{X} \times \mathbb{Y}$) we can construct a sheafy functor

$$-\otimes \mathscr{P}^\bullet \colon \mathscr{V}ect^\bullet \mathbb{V} \to \mathscr{V}ect^\bullet \mathbb{V} \colon \mathscr{M}^\bullet \mapsto (\mathscr{M} \otimes \mathscr{P})^\bullet$$

with $(\mathscr{M} \otimes \mathscr{P})^i(\mathbb{U}) := \bigoplus_{k+l=i} \mathscr{M}^k(\mathbb{U}) \otimes \mathscr{P}^l(\mathbb{U})$ and $d_{\mathscr{M}\otimes\mathscr{P}} = d_{\mathscr{M}} \otimes \mathbb{1}_{\mathscr{P}} \pm \mathbb{1}_{\mathscr{M}} \otimes d_{\mathscr{P}}$, which can be turned into an A_∞-functor by taking Čech cohomology:

$$\mathtt{D}(-\otimes \mathscr{P}) \colon \mathtt{Perf}^\bullet \mathbb{V} \to \mathtt{Perf}^\bullet \mathbb{V}.$$

(iii) *Integrating out* $\mathbb{X}$. If $\mathscr{N}$ is a sheaf on $\mathbb{X} \times \mathbb{Y}$ we would like to set

$$\pi_{\mathbb{Y}*}\mathscr{N}(\mathbb{V}_j) = \mathscr{N}(\mathbb{X} \times \mathbb{V}_j).$$

But this operation is not very nice from an A_∞ point of view. Indeed, $\mathscr{N}(\mathbb{X}\times\mathbb{V}_j)$ is actually the zeroth Čech cohomology of the sheaf $\mathscr{N}(-\times \mathbb{V}_j) \colon \mathbb{U}_i \mapsto \mathscr{N}(\mathbb{U}_i \times \mathbb{V}_j)$ so this procedure takes the homology of a complex, which is bad. Instead it is better to keep the complex:

$$\mathtt{D}\pi_{\mathbb{Y}*}\mathscr{N}(\mathbb{V}_j) = \check{\mathtt{C}}(\mathscr{N}(-\times \mathbb{V}_j); \mathcal{U}).$$

This turns a locally free sheaf into a complex of locally free sheaves. Unfortunately, these are not necessarily finitely generated so we get a functor

$$\mathtt{D}\pi_{\mathbb{Y}*} \colon \mathtt{Perf}^\bullet \mathbb{X} \times \mathbb{Y} \to \mathtt{D}\,\mathtt{VECT}^\bullet \mathbb{Y}.$$

Remark 7.79 Because the $\mathscr{N}(\mathbb{U}_i \times \mathbb{V}_i)$ are in general not finitely generated as $\mathscr{O}(\mathbb{V}_i)$-modules, these Čech complexes are in $\mathtt{D}\,\mathtt{VECT}^\bullet\,\mathbb{Y}$ instead of $\mathtt{Perf}^\bullet\,\mathbb{Y}$. However, if $\mathbb{X}$ is a smooth projective variety, the cohomology of these complexes has finite rank and hence $\mathtt{D}\pi_{\mathbb{Y}*}$ can be seen as a functor to $\mathtt{Perf}^\bullet(\mathbb{Y})$.

Definition 7.80 Let $\mathbb{X}$ and $\mathbb{Y}$ be two smooth projective varieties. A Fourier–Mukai kernel is an object $\mathscr{F} \in \mathtt{D}\,\mathtt{Coh}^\bullet(\mathbb{X} \times \mathbb{Y})$. Represent $\mathscr{F}$ as a complex of locally free sheaves on $\mathbb{X} \times \mathbb{Y}$: $\mathscr{P}^\bullet \in \mathscr{V}ect^\bullet(\mathbb{X} \times \mathbb{Y})$. The *Fourier–Mukai transform* is the functor

$$\mathcal{F}_{\mathscr{F}} := \mathtt{D}\pi_{\mathbb{Y}*} \circ \mathtt{D}(-\otimes \mathscr{P}^\bullet) \circ \mathtt{D}\pi_{\mathbb{X}}^* \colon \mathtt{Perf}^\bullet\,\mathbb{X} \to \mathtt{Perf}^\bullet\,\mathbb{Y}.$$

Example 7.81 The notion of a Fourier–Mukai transform is very versatile and can be used to describe some well-known functors.

- If $\mathbb{Y}$ is a point then we can consider the structure sheaf $\mathscr{O}_{\mathbb{X}}$ as a sheaf on $\mathbb{X} \times \mathbb{Y}$. The corresponding Fourier–Mukai transform maps a sheaf $\mathscr{S}$ over $\mathbb{X}$ to its Čech cohomology viewed as an object in $\mathtt{Perf}^\bullet\,\mathbb{Y} \overset{\infty}{=} \mathtt{vect}^\bullet\,\mathbb{C}$.

- Let $\mathbb{X} = \mathbb{Y}$ and consider the structure sheaf of the diagonal

$$\mathcal{O}_\Delta(\mathbb{U}_i \times \mathbb{U}_j) = \mathcal{O}_\mathbb{X}(\mathbb{U}_i \cap \mathbb{U}_j).$$

  The Fourier–Mukai transform with kernel $\mathcal{O}_\Delta$ is isomorphic to the identity functor. This is similar to the formula

$$\int_\mathbb{R} f(y)\delta(x-y)\,dy = f(x),$$

  where δ is the Dirac delta function.
- If $\mathcal{S}$ is an invertible sheaf on $\mathbb{X}$ we can turn this into a sheaf on $\mathbb{X} \times \mathbb{X}$ concentrated on the diagonal:

$$\mathcal{S}_\Delta(\mathbb{U}_i \times \mathbb{U}_j) = \mathcal{S}_\mathbb{X}(\mathbb{U}_i \cap \mathbb{U}_j).$$

  The Fourier–Mukai transform with kernel $\mathcal{S}_\Delta$ corresponds to tensoring with $\mathcal{S}$.
- If $\phi\colon \mathbb{X} \to \mathbb{Y}$ is a morphism of varieties then we can construct the structure sheaf of the graph of the map

$$\mathcal{O}_\Gamma(\mathbb{U}) = \frac{\mathcal{O}_{\mathbb{X}\times\mathbb{Y}}(\mathbb{U})}{(f \mid f(x,\phi(x)) = 0)}.$$

  The Fourier–Mukai transform with kernel $\mathcal{O}_\Gamma$ corresponds to tensoring with the derived pushforward of ϕ.

In all these examples, skyscraper sheaves of points are sent to skyscraper sheaves of points. But there are also more intricate examples.

Example 7.82 If we look at the elliptic curve $\mathbb{E} = \mathbb{P}(R)$ with

$$R = \mathbb{C}[X,Y,Z]/(Y^2Z - X^3 + XZ^2),$$

we have a line bundle $\mathcal{L}_P{:=}\mathcal{O}(P-P_\infty)$ for each point $P \in \mathbb{E}$ (cf. Example 7.25). To distinguish between the two copies of the elliptic curve, one parametrizing the points and one parametrizing line bundles, we denote the latter by $\mathbb{E}^\vee$.

We can group all these line bundles together into a big line bundle $\mathcal{T}$ on $\mathbb{E}^\vee \times \mathbb{E}$ whose fiber at (P,Q) is the fiber of $\mathcal{L}_P$ at Q. This bundle is called the Poincaré bundle and we can use it to define a Fourier–Mukai transform from $\mathsf{D\,Coh}^\bullet\,\mathbb{E}^\vee$ to $\mathsf{D\,Coh}^\bullet\,\mathbb{E}$. This Fourier–Mukai transform will map the skyscraper sheaf at P to the line bundle $\mathcal{L}_P$. One can show that this functor is in fact an equivalence.

In [183] Mukai used a higher-dimensional analog of this construction to find an equivalence between an abelian variety $\mathbb{A}$ and its dual $\mathbb{A}^\vee$. In this way he obtained the first example of a derived equivalence between different varieties.

We end this section with a nice theorem by Orlov, which tells us that if one restricts to smooth projective varieties, the Fourier–Mukai construction is sufficient to describe all relevant derived equivalences between varieties.

Theorem 7.83 (Orlov [198]) *Every exact functor between the derived categories of smooth projective varieties is isomorphic to a Fourier–Mukai transform.*

For more information on Fourier–Mukai transforms we refer to the book by Huybrechts on the subject [124].

7.3.3 Moduli Spaces

In the last example we used a space $\mathbb{E}^\vee$ that parametrized certain objects in $\mathtt{DCoh}^\bullet\,\mathbb{E}$ (degree 0 line bundles). This space allowed us to transform each object in $\mathtt{DCoh}^\bullet\,\mathbb{E}$ into an object in $\mathtt{DCoh}^\bullet\,\mathbb{E}^\vee$. Of course, this idea is not restricted to derived categories of coherent sheaves; we could also look at other categories.

Let $A = \mathbb{C}Q/I$ be the path algebra of a quiver with relations and fix a dimension vector $\alpha\colon Q_0 \to \mathbb{N}$. We define the space

$$\mathsf{Rep}(Q,\alpha) = \bigoplus_{a\in Q_1} \mathrm{Mat}_{\alpha_{h(a)}\times\alpha_{t(a)}}(\mathbb{C}),$$

which can be seen as a big affine space with coordinate ring

$$R = \mathbb{C}[X^a_{ij} \mid a \in Q_0,\ 1 \le i \le \alpha_{h(a)},\ 1 \le j \le \alpha_{t(a)}].$$

Each point in this space is represented by an $\#Q_1$-tuple of matrices $(\rho_a)_{a\in Q_1}$ and corresponds to a representation $\rho\colon \mathbb{C}Q \to \mathrm{Mat}_{n\times n}(\mathbb{C})$ with $n = \sum_v \alpha_v$, which assigns to each a a block matrix with ρ_a in the appropriate block. We can group all these representations into a vector bundle $\mathsf{Rep}(Q,\alpha) \times \mathbb{C}^n$ on $\mathsf{Rep}(Q,\alpha)$. On the algebraic side this corresponds to the free R-module $T = R \otimes \mathbb{C}^n$ with a module morphism $\rho\colon \mathbb{C}Q \to \mathrm{End}_R(T)$.

Many points in $\mathsf{Rep}(Q,\alpha)$ correspond to isomorphic representations. In order to take care of this we can look at the action of the base change group

$$\mathsf{GL}_\alpha = \prod_{v\in Q_0} \mathsf{GL}_{\alpha_v}(\mathbb{C}),$$

which acts on $\mathsf{Rep}(Q,\alpha)$ by

$$(g \cdot \rho)_a := (g_{h(a)}\rho_a g^{-1}_{t(a)}).$$

The orbits of this action are precisely the isomorphism classes. It is tempting to take the quotient of this group action by considering the variety corresponding to the ring of invariants:

$$\mathsf{Rep}(Q,\alpha)/\!\!/\mathsf{GL}_\alpha := \mathbb{V}(R^{\mathsf{GL}_\alpha}).$$

Unlike in the case where the group was finite, this quotient does not parametrize all orbits. If the group is infinite it is possible that one orbit sits in the closure of another. When this happens every invariant function must have the same value on both orbits, so the two orbits will correspond to the same point in the quotient.

Example 7.84 If $(Q,\alpha) = ① \Leftarrow ①$ then $\mathsf{Rep}(Q,\alpha) = \mathbb{C}^2$ and $\mathsf{GL}_\alpha = \mathbb{C}^{*2}$ with $(g,h)\cdot(x,y) = (gxh^{-1}, gyh^{-1})$. The orbits of this action are the punctured lines $\mathbb{C}(x,y)\backslash\{(0,0)\}$ and the zero orbit $\{(0,0)\}$, which is contained in the closure of all orbits. Therefore, the categorical quotient only has one point. Indeed, the coordinate ring is $R = \mathbb{C}[X,Y]$ with action $(g,h)\cdot X = gXh^{-1}$, $(g,h)\cdot Y = gYh^{-1}$, so $R^{\mathsf{GL}_\alpha} = \mathbb{C}$ and $\mathsf{Rep}(Q,\alpha)/\!\!/\mathsf{GL}_\alpha = \{p\}$.

We can remedy this by throwing away the bad orbit and then taking the quotient afterwards. In this way we end up with $\mathbb{C}^2 \setminus \{(0,0)\}/\mathsf{GL}_\alpha = \mathbb{P}^1$, which is a projective variety that classifies all orbits except the zero. We can recover this quotient as $\mathbb{P}(R)$ where R gets a grading from the action: $f \in R$ has degree n if $(g,h)\cdot f = g^n h^{-n} f$.

We can generalize this example to the notion of a *GIT-quotient*.[3] For each vector $\theta\colon Q_0 \to \mathbb{Z}$ we can construct a multiplicative character

$$\chi_\theta\colon \mathsf{GL}_\alpha \to \mathbb{C}^*: g \mapsto \prod_{v\in Q_0} \det g_v^{\theta_v}.$$

We say that $f \in R$ is a *θ-semi-invariant* if $g\cdot f = \chi_\theta(g)f$. The *ring of semi-invariants* is the direct sum of all semi-invariants of powers of χ_θ:

$$R_\theta := \bigoplus_{n=0}^{\infty} \{f \in R \mid \forall\, g \in \mathsf{GL}_\alpha,\ g\cdot f = \chi_\theta(g)^n f\}.$$

This is a graded ring whose degree 0 part is the ring of invariants R^{GL_α}. The variety $\mathbb{P}(R_\theta)$ is covered by open pieces $\mathbb{U}_f = \mathbb{V}(R_\theta[f^{-1}]_0)$ where $f \in (R_\theta)_n$. The elements of $R_\theta[f^{-1}]_0$ can be seen as invariant functions over $\{\rho \in \mathsf{Rep}(Q,\alpha) \mid f(\rho) \neq 0\}$, so $\mathbb{U}_f = \{\rho \in \mathsf{Rep}(Q,\alpha) \mid f(\rho) \neq 0\}/\!\!/\mathsf{GL}_\alpha$.

[3] GIT is the abbreviation for geometric invariant theory.

Definition 7.85 We say that a representation ρ is *θ-semistable* if there is an $f \in (R_\theta)_{>0}$ for which $f(\rho) \neq 0$. The moduli-space of θ-semistable representations is

$$\mathbb{M}_\theta(Q, \alpha) = \mathbb{P}(R_\theta)$$

and it can be interpreted as

$$\mathsf{Rep}^{\theta\text{-ss}}(Q, \alpha) /\!\!/ \mathsf{GL}_\alpha := \{\rho \in \mathsf{Rep}(Q, \alpha) \mid \rho \text{ is } \theta\text{-semistable}\} /\!\!/ \mathsf{GL}_\alpha.$$

In general it is still possible that two nonisomorphic θ-semistable representations correspond to the same point in $\mathbb{M}_\theta(Q, \alpha)$. This happens when there are orbits of different sizes, because in that case a small orbit can lie in the closure of a big orbit. The biggest orbits are those of representations with the smallest stabilizer possible: the scalar matrices $\mathbb{C}^* \subset \mathsf{GL}_\alpha$. Therefore we call a semistable representation *θ-stable* if its stabilizer is $\mathbb{C}^*$.

The idea of (semi)stability is defined using the geometry of the group action of GL_α on the variety $\mathsf{Rep}(Q, \alpha)$, but it also has a nice interpretation in terms of the representation theory of the quiver.

Theorem 7.86 (King) *A representation $\rho \in \mathsf{Rep}(Q, \alpha)$ is θ-stable (θ-semistable) if and only if $\theta \cdot \alpha = 0$ and for all proper subrepresentations we have $\theta \cdot \beta > 0$ ($\theta \cdot \beta \geq 0$), where β is the dimension vector of the subrepresentation.*

Sketch of the proof We indicate only the main idea of this connection; for the full details we refer to [138]. First of all, note that the existence of a nonzero $n\theta$-semi-invariant implies $\theta \cdot \alpha = 0$ because the action of a scalar matrix $\lambda \mathbb{1}$ on ρ is trivial:

$$f(\rho) = f(\lambda \mathbb{1} \cdot \rho) = \chi_{n\theta}(\lambda \mathbb{1}) f(\rho) = \lambda^{n\theta \cdot \alpha} f(\rho).$$

If a representation ρ has a subrepresentation with dimension vector β then we can bring the representation into block upper-triangular form. If we conjugate this with a matrix of the form $g_t = \begin{pmatrix} t\mathbb{1}_\beta & 0 \\ 0 & \mathbb{1}_{\alpha-\beta} \end{pmatrix}$ then the limit $\lim_{t \to 0} g_t \cdot \rho$ exists because of this upper-triangular form. However, if f is an $n\theta$-semi-invariant then $f(g_t \cdot \rho) = t^{n\theta \cdot \beta} f(\rho)$, so if $\theta \cdot \beta < 0$ then the limit for $t \to 0$ does not exist and we get a contradiction. □

Definition 7.87 The stability condition θ is called *generic* for α if $\theta \cdot \alpha = 0$ and $\theta \cdot \beta \neq 0$ for every nonzero dimension vector $\beta < \alpha$. This implies that the notions θ-semistable and θ-stable coincide.

Theorem 7.88 (King [138]) *If θ is generic then $\mathbb{M}_\theta(Q, \alpha)$ is a smooth variety whose points correspond to the isomorphism classes of θ-stable α-dimensional representations of Q.*

Sketch of the proof The space $\mathsf{Rep}^{\theta\text{-ss}}(Q, \alpha)$ is an open subset of the smooth space $\mathsf{Rep}(Q, \alpha)$ and hence smooth itself. The fibers of the projection

$$\pi\colon \mathsf{Rep}^{\theta\text{-ss}}(Q, \alpha) \to \mathbb{M}^\theta(Q, \alpha)$$

are all isomorphic to the smooth variety $\mathsf{GL}_\alpha/\mathbb{C}^*$ and therefore the target $\mathbb{M}^\theta(Q, \alpha)$ must also be smooth. □

Example 7.89 If we return to the Kronecker quiver Q with $\alpha = (1, 1)$, we see that there are different possibilities depending on $\theta = (\theta_1, \theta_2)$:

- If $\theta_1 + \theta_2 \neq 0$ then there are no θ-semi-invariants, so $R_\theta = \mathbb{C}$ and $\mathbb{P}(R_\theta) = \emptyset$. There are no θ-semistable representations because $\alpha\cdot\theta = 0$, so $\mathbb{M}^\theta(Q, \alpha) = \emptyset$.
- If $\theta_1 = -\theta_2 = 0$ then the constant functions are θ-semi-invariants, so $R_\theta = \mathbb{C} \oplus \mathbb{C} \oplus \cdots$ and $\mathbb{P}(R_\theta) = \{p\}$. Note that θ is not generic. In this case, all representations are semistable but none of them is stable. Every orbit has the zero orbit in its closure so $\mathbb{M}^\theta(Q, \alpha)$ contains just one point.
- If $\theta_1 = -\theta_2 = k > 0$ then the degree k functions are θ-semi-invariants, so $R_\theta = \mathbb{C} \oplus R_k \oplus R_{2k} \oplus \cdots$ and $\mathbb{P}(R_\theta) = \mathbb{P}^1$. In this case, all representations are stable except the zero orbit because the latter has a subrepresentation with dimension vector $(0, 1)$.
- If $\theta_1 = -\theta_2 = k < 0$ then there are no θ-semi-invariants, so $R_\theta = \mathbb{C}$ and $\mathbb{P}(R_\theta) = \emptyset$. In this case, no representation is stable because they all have a subrepresentation with dimension vector $(1, 0)$.

If θ is generic we can parametrize all θ-stable representations up to isomorphism. Each of these representations can be seen as a vector space $\mathbb{C}^n$ with an action of $\mathbb{C}Q$. The natural question is whether we can bundle all these representations into a vector bundle over $\mathbb{M}^\theta(Q, \alpha)$? The naive idea would be to look at the bundle $\mathsf{Rep}(Q, \alpha) \times \mathbb{C}^n \to \mathsf{Rep}(Q, \alpha)$, restrict to the stable representations and then quotient out GL_α. Unfortunately, this does not work because every point (ρ, x) has $(\rho, \lambda x)$ in its orbit (recall that the scalar $\lambda \in \mathsf{GL}_\alpha$ acts trivially on ρ). This means that $(\rho, 0)$ sits in the closure of the orbit of all (ρ, x) and the quotient is just $\mathbb{M}^\theta(Q, \alpha)$ again.

To avoid this we assume that there is a morphism $\sigma\colon \mathsf{GL}_\alpha \to \mathbb{C}^*$ such that $\sigma(\lambda) = \lambda$ for all $\lambda \in \mathbb{C}^* \subset \mathsf{GL}_\alpha$ and we tweak the GL_α action to

$$g(\rho, x) := \left(g \cdot \rho, \sigma(g)^{-1} g x\right).$$

In this way the scalars act trivially on x and we can take the quotient of $\mathsf{Rep}^{\theta\text{-ss}}\theta(Q, \alpha) \times \mathbb{C}^n$ for this new action. This quotient is called the *universal bundle* and it can be seen as a sheaf of $\mathbb{C}Q$-modules over $\mathbb{M}^\theta(Q, \alpha)$.

Analogously to the commutative case, where the Fourier–Mukai kernel can be seen as a sheaf over $\mathbb{X}$ of sheaves over $\mathbb{Y}$, the universal bundle is a sheaf over $\mathbb{M}^\theta(Q,\alpha)$ of sheaves over the noncommutative variety associated to the noncommutative algebra $\mathbb{C}Q$. This allows us to define a noncommutative Fourier–Mukai transform $\mathcal{F}\colon \mathtt{D\,Coh}^\bullet\,\mathbb{M}^\theta(Q,\alpha) \to \mathtt{D\,Mod}^\bullet\,\mathbb{C}Q$.

Example 7.90 If we return to the Kronecker quiver Q with $\alpha = (1,1)$ and $\theta = (1,-1)$ we can use $\sigma(g,h) = h$. The GL_α-action on $T = \mathbb{C}[X,Y] \otimes \mathbb{C}^2 = \mathbb{C}[X,Y]e_1 \oplus \mathbb{C}[X,Y]e_2$ becomes $(g,h)e_1 = h^{-1}ge_1$ and $(g,h)e_2 = h^{-1}he_2 = e_2$. Therefore, the tautological bundle is $\mathcal{O}(1) \oplus \mathcal{O}$ over $\mathbb{M}^\theta(Q,\alpha) = \mathbb{P}^1$.

To find the image of $\mathcal{O}(k) \in \mathtt{Perf}^\bullet\,\mathbb{P}^1$ under the Fourier–Mukai transform, we tensor it with $\mathcal{T}_\theta$. In this way we get $\mathcal{O}(k+1) \oplus \mathcal{O}(k)$. The Čech complex of this bundle is an infinite-dimensional graded representation of $\mathbb{C}Q$:

$$\begin{array}{ccc}
\mathbb{C}[X] \oplus \mathbb{C}[X^{-1}] & \xrightarrow{(1\ X^k)} & \mathbb{C}[X,X^{-1}] \\
\left(\begin{smallmatrix} X & 0 \\ 0 & 1 \end{smallmatrix}\right)\Big\Downarrow \left(\begin{smallmatrix} 1 & 0 \\ 0 & X^{-1} \end{smallmatrix}\right) & & X \Big\Downarrow 1 \\
\mathbb{C}[X] \oplus \mathbb{C}[X^{-1}] & \xrightarrow{(1\ X^{k+1})} & \mathbb{C}[X,X^{-1}].
\end{array}$$

The horizontal arrows are the differentials of the Čech complex and the vertical arrows denote the action of the quiver arrows. Now we take the cohomology of this complex. If $k \geq -1$ the cohomology is concentrated in degree 0 and we get the representation

$$\mathbb{C}^{k+2} \overset{\left(\begin{smallmatrix} \mathbb{1}_k \\ 0 \end{smallmatrix}\right)}{\underset{\left(\begin{smallmatrix} 0 \\ \mathbb{1}_k \end{smallmatrix}\right)}{\Longleftarrow}} \mathbb{C}^{k+1}\ .$$

If $k < -1$ the cohomology is concentrated in degree 1 and we get the representation

$$\mathbb{C}^{-k-1} \overset{(\mathbb{1}_k\ 0)}{\underset{(0\ \mathbb{1}_k)}{\Longleftarrow}} \mathbb{C}^{-k}\ .$$

Remark 7.91 The construction of moduli spaces also works for path algebras with relations $A = \mathbb{C}Q/I$. In that case, we have to look at the subvariety

$$\mathsf{Rep}(A,\alpha) = \{\rho \in \mathsf{Rep}(Q,\alpha) \mid \rho(I) = 0\}$$

or some irreducible component. Its coordinate ring R_A is a quotient of R by a GL_α-closed ideal $\mathtt{i}$. If this component is smooth and θ is generic, then $\mathbb{M}^\theta(Q,\alpha) = \mathbb{P}(R_\theta/\mathtt{i} \cap R_\theta)$ is a smooth variety that comes with a locally free sheaf whose fibers are θ-stable A-modules. This can be used to construct a Fourier–Mukai functor but it is not easy to determine when this functor is an equivalence.

7.3.4 Tilting

The noncommutative version of the Fourier–Mukai transform is closely related to another method to construct equivalences: tilting.

Definition 7.92 Let $\mathtt{C}^\bullet$ be a minimal split-closed A_∞-category. A *tilting object* is a split-generator such that $\mathtt{C}^\bullet(T,T)$ is concentrated in degree 0.

If T is a tilting object then $A = \mathtt{C}^\bullet(T,T)$ is an ordinary algebra as all higher products are zero for degree reasons. Furthermore,

$$\mathtt{C}^\bullet \overset{\infty}{=} \mathtt{D}^\pi\, \mathtt{A}^\bullet \overset{\infty}{=} \mathtt{Perf}^\bullet A,$$

so all objects are isomorphic to finite complexes of finitely generated projective A-modules.

If $\mathtt{C}^\bullet$ is the minimal model of a dg-category $\mathtt{dgC}^\bullet$ and $A = \mathtt{dgC}^\bullet(T,T)$, we can construct these complexes explicitly. For any M in $\mathtt{dgC}^\bullet$ the complex $\mathtt{C}^\bullet(M,T)$ has the structure of a right A-module and $d(ma) = (dm)a \pm m(da) = (dm)a$, so it is in fact a complex of right A-modules. The map

$$\mathrm{Hom}(-,T)\colon\ \mathtt{dgC}^\bullet \to \mathtt{Tw}\,\mathtt{MOD}^\bullet A\colon T \to \mathtt{dgC}^\bullet(M,T)$$

is a dg-functor which, after going to the minimal models, gives an A_∞-functor $\mathtt{D}\,\mathrm{Hom}(-,T)\colon\ \mathtt{C}^\bullet \to \mathtt{D}\,\mathtt{MOD}^\bullet A$. Because the image of T is A as a right module over itself and T is a split-generator, the functor $\mathtt{D}^\pi\,\mathrm{Hom}(-,T)$ will embed $\mathtt{C}^\bullet$ into $\mathtt{D}\,\mathtt{MOD}^\bullet A$ as a full subcategory that is equivalent to $\mathtt{Perf}^\bullet A$.

In the special case where $\mathtt{dgC}^\bullet = \mathtt{Vect}^\bullet\mathbb{X}$ and $\mathscr{T}$ is a vector bundle (i.e. a *tilting bundle*), we get a functor from $\mathtt{Perf}^\bullet\mathbb{X}$ to $\mathtt{Perf}^\bullet A$ with

$$\mathtt{A} = \mathscr{H}\!om\,(\mathscr{T},\mathscr{T})(\mathbb{X}).$$

It takes a complex of vector bundles, then it constructs the hom-sheaf with $\mathscr{T}$ and it takes the Čech cohomology. But taking the hom-sheaf can be interpreted as tensoring over A with the dual sheaf $\mathscr{T}^\vee(\mathbb{U}) := \mathrm{Hom}(\mathscr{T}(\mathbb{U}),\mathbb{C})$, which is a sheaf of left A-modules. Therefore, the tilting functor is in fact a noncommutative Fourier–Mukai transform with kernel $\mathscr{T}^\vee(\mathbb{U})$.

Example 7.93 The projective line $\mathbb{P}^1$ has a tilting bundle $\mathscr{O} \oplus \mathscr{O}(1)$ and more generally $\mathbb{P}^k$ has a tilting bundle $\mathscr{O} \oplus \cdots \oplus \mathscr{O}(k)$. The corresponding endomorphism algebra is the path algebra of a quiver

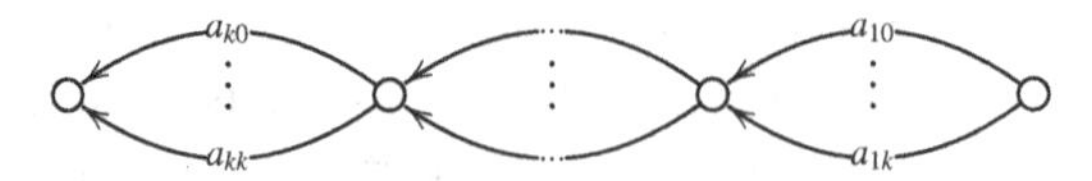

with relations $a_{i+1u}a_{iv} = a_{i+1v}a_{iu}$.

Example 7.94 Different path algebras with relations can also be related via tilting. The easiest example is

$$\circ \longrightarrow \circ \longrightarrow \circ \quad \text{and} \quad \circ \longrightarrow \circ \longleftarrow \circ,$$

where the latter is the endomorphism ring of $T = S_2 \oplus S_{23} \oplus P_2$ (using the notation of Section 4.2.1). For more information on tilting for finite-dimensional algebras we refer to [10].

Remark 7.95 It is quite rare for a derived category of coherent sheaves of a projective variety to have a tilting bundle. In dimension 1, only the projective line has one, but all other curves do not. In dimension 2, rational surfaces have tilting bundles [120]. In general, one can show that if a variety admits a tilting bundle then its Grothendieck group (see Chapter 11) is free of finite rank. In particular, this excludes all projective Calabi–Yau varieties.

7.3.5 Matrix Factorizations and Projective Varieties

Consider the situation where $R = \mathbb{C}[X_0, \ldots, X_n]$ and $f \in R$ is a homogeneous polynomial of degree k. In this case, we can consider two categories:

- the derived category of coherent sheaves $\mathtt{D}\,\mathtt{Coh}^\bullet\,\mathbb{X}$ with $\mathbb{X} = \mathbb{P}(R/(f))$,
- the category of graded matrix factorization $\mathtt{grMF}^\bullet(R, f)$.

In general, these two categories are not equivalent but they are closely related.

Example 7.96 We will examine the case $n = 1$ with three functions $f_1 = X$, $f_2 = XY$ and $f_3 = XY(X - Y)$. The variety $\mathbb{X} = \mathbb{P}(R/(f_k))$ consists of k points in $\mathbb{P}^1$ and $\mathbb{X} \subset \{(0 : 1), (1 : 0), (1 : 1)\}$. This implies that $\mathbb{X} \cong \mathbb{V}(\mathbb{C}^k)$ and hence $\mathtt{D}^{\mathrm{b}}\mathtt{Coh}\,\mathbb{X} \cong \mathtt{D}^{\mathrm{b}}\mathtt{mod}\mathbb{C}^k \cong \mathtt{D}^{\mathrm{b}}\mathtt{mod}Q$, where Q is a quiver with k vertices and no arrows.

- The category of graded matrix factorizations for f_1 is trivial because $\mathbb{V}(R/(f_1))$ is an affine line, which has no singularities.
- For f_2 the category is generated by the objects

$$S_X = R \underset{Y}{\overset{X}{\rightleftarrows}} R(1) \quad \text{and} \quad S_Y = R \underset{X}{\overset{Y}{\rightleftarrows}} R(1)\,.$$

  The endomorphism rings of S_X and S_Y are equal to $\mathbb{C}$ and there are no morphisms between them. Therefore, $\mathtt{grMF}^\bullet(R, f_2) \overset{\infty}{=} \mathtt{D}\,\mathtt{mod}^\bullet\,Q$, where Q is a quiver with two vertices and no arrows.

- For f_3 the category is generated by $S_X(i)$, $S_Y(i)$ and $S_{X-Y}(i)$. If we look at the endomorphism algebra of $S_X \oplus S_Y \oplus S_{X-Y}$, we see that it is equal to $\mathbb{C}^3$, so these three objects generate a subcategory of $\mathtt{grMF}^\bullet(R, f_2)$ that is equivalent to $\mathtt{D\,mod}^\bullet\,\mathbb{C}^3 \overset{\infty}{=} \mathtt{D\,Coh}^\bullet\,\mathbb{X}$. This subcategory is however not equivalent to $\mathtt{grMF}^\bullet(R, f_2)$.

The different behavior depending on the degree of the function generalizes to all hypersurfaces of projective spaces.

Theorem 7.97 (Orlov) *Let $\mathbb{X} = \mathbb{P}(R/(f))$ be a smooth subvariety of $\mathbb{P}^n = \mathbb{P}(R)$; then depending on $k = \deg f$ we have the following functors:*

(i) *If $k < n + 1$ there is a fully faithful embedding $\mathcal{F}\colon \mathtt{grMF}^0(R, f) \to \mathtt{D^bCoh}\,\mathbb{X}$.*
(ii) *If $k = n + 1$ there is an equivalence $\mathcal{F}\colon \mathtt{D^bCoh}\,\mathbb{X} \to \mathtt{grMF}^0(R, f)$.*
(iii) *If $k > n + 1$ there is a fully faithful embedding $\mathcal{F}\colon \mathtt{D^bCoh}\,\mathbb{X} \to \mathtt{grMF}^0(R, f)$.*

Sketch of the proof The two categories involved can be constructed as quotients of a larger category:

$$\mathtt{D^bCoh}\,\mathbb{X} \cong \frac{\mathtt{D^bgrMod}\,R}{\langle \mathtt{grmod}\,R\rangle} \quad \text{and} \quad \mathtt{grMF}^0(R, W) \cong \frac{\mathtt{D^bgrMod}\,R}{\langle \mathtt{grProj}\,R\rangle},$$

where the respective denominators are the subcategories generated by the finite-dimensional graded modules and the projective graded modules. In [200] the result is then obtained by looking at how these denominators fit inside $\mathtt{D^bgrMod}\,R$.

If $k = n + 1$ there is also a nice geometrical approach to this theorem due to Segal and Shipman [216, 225]. By reinterpreting the grading on R as a group action, one can see $\mathtt{grMF}(R, W)$ as a category of matrix factorizations on a stack $\mathtt{MF}([\mathbb{A}^{n+1}/\mathbb{Z}_{n+1}], W)$. This stack is derived equivalent to the total space of the canonical bundle of $\mathbb{Y} = |\mathcal{K}_{\mathbb{P}^n}|$ and therefore we get an equivalence with $\mathtt{MF}(\mathbb{Y}, W)$. On $\mathbb{Y}$ the zero locus of W consists of two pieces: the fibers that sit on top of the zero locus of W in $\mathbb{P}^n \subset |\mathcal{K}_{\mathbb{P}^n}| = \mathbb{Y}$ and $\mathbb{P}^n$ itself. The singular locus is therefore equal to $\mathbb{X} \subset \mathbb{P}^n \subset \mathcal{K}_{\mathbb{P}^n}$. Locally it is a normal intersection and therefore one can use Knörrer periodicity to obtain an equivalence between $\mathtt{MF}(\mathbb{Y}, W)$ and $\mathtt{MF}(\mathbb{X}, 0) \cong \mathtt{D^bCoh}\,\mathbb{X}$. □

7.4 Exercises

Exercise 7.1 Let $\mathbb{X}_1, \mathbb{X}_2$ be two affine varieties.

(i) Show that if $\mathbb{X}_1, \mathbb{X}_2$ have the same finite number of points then they are isomorphic affine varieties.
(ii) Show that if $\mathsf{Coh}\,\mathbb{X}_1 \cong \mathsf{Coh}\,\mathbb{X}_2$ then they are isomorphic affine varieties (use Morita equivalence).
(iii) Show that for any variety $\mathbb{X}$ we have $\mathscr{O}(\mathbb{X}) \cong \mathrm{End}\,\mathbb{1}_{\mathsf{Perf}^\bullet\,\mathbb{X}} \cong \mathrm{End}\,\mathbb{1}_{\mathsf{D\,Coh}^\bullet\,\mathbb{X}}$.

Exercise 7.2 Let $\mathbb{X}_1, \mathbb{X}_2$ be two varieties with affine covers $(\mathbb{U}_{1,i})_i$, $(\mathbb{U}_{2,j})_j$ and look at the product $\mathbb{X}_1 \times \mathbb{X}_2$ as the variety with open cover $(\mathbb{U}_{1,i} \times \mathbb{U}_{2,j})_{ij}$.

(i) Show that if $\mathbb{X}_1 = \mathbb{P}^n$ and $\mathbb{X}_2 = \mathbb{P}^m$ then we can see $\mathbb{X}_1 \times \mathbb{X}_2$ as $\mathbb{P}(R)$ where R is the subring of $\mathbb{C}[X_0, \dots, X_n, Y_0, \dots, Y_m]$ generated by the monomials $X_i Y_j$.
(ii) Let $R = \mathbb{C}[X_1, \dots, X_4]/(X_1^2 + \cdots + X_4^2)$ with $\deg X_i = 1$. Show that $\mathbb{P}(R)$ is isomorphic to $\mathbb{P}^1 \times \mathbb{P}^1$.
(iii) If $\mathscr{M}_1, \mathscr{M}_2$ are sheaves over $\mathbb{X}_1, \mathbb{X}_2$ then we can define $\mathscr{M}_1 \otimes \mathscr{M}_2$ over $\mathbb{X}_1 \times \mathbb{X}_2$ by demanding that

$$\mathscr{M}_1 \otimes \mathscr{M}_2(\mathbb{U}_{1,i} \times \mathbb{U}_{2,j}) = \mathscr{M}_1(\mathbb{U}_{1,i}) \otimes \mathscr{M}_2(\mathbb{U}_{2,j}).$$

Show that if $\mathscr{M}_1, \mathscr{M}_2$ are locally free then $\mathscr{M}_1 \otimes \mathscr{M}_2$ is also locally free.

Exercise 7.3 Let $\mathbb{E} = \mathbb{P}(R)$ with $R = \mathbb{C}[X, Y, Z]/(Y^2 Z - (X - \lambda_1 Z)(X - \lambda_2 Z)(X - \lambda_3 Z))$.

(i) Show that $\mathbb{E}$ is smooth if and only if the λ_i are all different. Assume for the rest of the exercise that we are in this case.
(ii) Show that the sheaf cohomology of the structure sheaf is $\mathbb{C} \oplus \mathbb{C}$ concentrated in degrees 0, 1.
(iii) Let $\mathbb{S}_p$ be the skyscraper sheaf at a point $p = (x : y : z) \in \mathbb{E}$ and construct a locally free resolution for it.
(iv) Use this to compute the endomorphism algebra of $\mathscr{O} \oplus \mathscr{S}_p$ inside $\mathsf{D\,Coh}^\bullet\,\mathbb{E}$. Show that as an algebra it is isomorphic to

$$\mathtt{E} = \frac{\mathbb{C}Q}{\langle ba, c^2 \rangle} \quad \text{with } Q = c \circlearrowright \circ \underset{b}{\overset{a}{\rightleftarrows}} \circ .$$

(v) One can show that $\mathscr{O}$ and $\mathbb{S}_p$ split-generate the derived category. Explain why this implies that the A_∞-structure on $\mathtt{E}$ must be nontrivial.

Exercise 7.4 Let $R = \mathbb{C}[X_0, \dots, X_n]/(f)$ where f is an irreducible homogeneous function of degree d.

(i) Show that $\mathbb{V}(R)$ is smooth if and only if f is linear and hence $\mathbb{V}(R) \cong \mathbb{A}^n$.

(ii) Show that $\mathbb{P}(R)$ is smooth for generic f if $d > 1$ (i.e. smoothness is a dense open condition of the set of all degree d homogeneous polynomials $\mathbb{C}[X_0, \ldots, X_n]_d$).
(iii) Show that if $\mathbb{P}(R)$ is smooth then $\mathscr{K} \cong \mathscr{O}(d - n - 1)$, so if $d = n + 1$ then $\mathbb{P}(R)$ is Calabi–Yau.

Exercise 7.5 Let $\mathbb{X} = \mathbb{P}(R)$ with $R = \mathbb{C}[X, Y, U, V]/(XY - UV)$ and all variables have degree 1.

(i) Determine the endomorphism ring of $\mathscr{O}(0) \oplus \mathscr{O}(1) \oplus \mathscr{O}(2) \oplus \mathscr{O}(3)$ in $\mathtt{Perf}^\bullet \mathbb{X}$ and draw its quiver.
(ii) Identify $\mathbb{X}$ with $\mathbb{P}^1 \times \mathbb{P}^1$. For each $(i, j) \in \mathbb{Z}^2$ we define $\mathscr{O}(i, j) = \mathscr{O}_{\mathbb{P}^1}(i) \otimes \mathscr{O}_{\mathbb{P}^1}(j)$. Determine the endomorphism ring of $\mathscr{O}(0, 0) \oplus \mathscr{O}(1, 0) \oplus \mathscr{O}(0, 1) \oplus \mathscr{O}(1, 1)$ and draw its quiver.

Exercise 7.6 Let $f(X) = (X - \lambda_1)^{m_1} \cdots (X - \lambda_k)^{m_k} \in \mathbb{C}[X]$ and denote the exponent vector by $\vec{m}$. For every sequence $\vec{r} = (r_i)$ with $0 \le r_i \le m_i$ we have a matrix factorization $M_{\vec{r}} \in \mathtt{MF}^\pm(\mathbb{C}[X], f)$ that splits f as

$$(X - \lambda_1)^{r_1} \cdots (X - \lambda_k)^{r_k} \cdot (X - \lambda_1)^{m_1 - r_1} \cdots (X - \lambda_k)^{m_k - r_k}.$$

(i) Show that $M_{\vec{r}}$ is the zero object if and only if $\vec{r} = 0$ or $\vec{r} = \vec{m}$.
(ii) Show that

$$M_{\vec{r}} \cong M_{(r_1, 0, \ldots, 0)} \oplus \cdots \oplus M_{(0, 0, \ldots, r_k)}.$$

(iii) Describe the endomorphism ring of $M_{\vec{r}}$.
(iv) Describe the category of graded matrix factorizations $\mathtt{grMF}^\bullet(\mathbb{C}[X], X^n)$.

Exercise 7.7 Consider the category $\mathtt{D\,Coh}^\bullet \mathbb{P}^1$.

(i) Describe the Fourier–Mukai transform with kernel the skyscraper sheaf of the point $(p_1, p_2) \in \mathbb{P}^1 \times \mathbb{P}^1$.
(ii) Describe the Fourier–Mukai transform with kernel the line bundle $\mathscr{O}(i, j) = \mathscr{O}(i) \otimes \mathbb{S}(j)$.
(iii) Find a kernel such that the Fourier–Mukai transform corresponds to the (derived) pullback of the automorphism $\phi\colon \mathbb{P}^1 \to \mathbb{P}^1 : (x : y) \to (y : x)$.

Exercise 7.8 Bondal and Orlov's theorem states that we can recover a smooth Fano variety or variety of general type from its derived category of coherent sheaves by looking at point-like objects. These are objects whose endomorphism ring in the derived category are exterior algebras Λ_n. This idea does not work in the Calabi–Yau case.

(i) Show that for an elliptic curve $\mathbb{E} = \mathbb{P}(R)$ with $R = \mathbb{C}[X, Y, Z]/(Y^2 Z - (X - \lambda_1 Z)(X - \lambda_2 Z)(X - \lambda_3 Z))$ the line bundles are also point-like objects.

(ii) Give an example of a Fourier–Mukai transform that is an equivalence and transforms skyscraper sheaves into line bundles.

Although it is not possible in general to recover a Calabi–Yau variety from a derived category, in the case of elliptic curves one can still show that $\mathtt{D\,Coh}^\bullet\,\mathbb{E}_1 \cong \mathtt{D\,Coh}^\bullet\,\mathbb{E}_2$ implies that $\mathbb{E}_1 \cong \mathbb{E}_2$ because one can reconstruct $\mathbb{E}$ as the moduli space of objects with a fixed class in the numerical Grothendieck group (see [205]).

Exercise 7.9 The blow-up of $\mathbb{A}^2$ in the zero point $(0,0)$ is the variety $\tilde{\mathbb{A}}^2 := \mathbb{P}(R)$ with

$$R = \mathbb{C}[x, y, X, Y]/(Xy - xY),$$

where x, y have degree 0 and X, Y degree 1.

(i) Find an affine cover for $\tilde{\mathbb{A}}^2$.
(ii) Let $Q = \circ \rightrightarrows \circ$. Show that the moduli space for $\alpha = (1, 1)$ and $\theta = (-1, 1)$ is $\tilde{\mathbb{A}}^2$.
(iii) Show that the corresponding tautological bundle $\tilde{\mathbb{A}}^2$ is a tilting bundle for $\tilde{\mathbb{A}}^2$. What is its endomorphism ring?

Exercise 7.10 Consider the space $\mathbb{P}^1$ with an action of $G = \{1, -1\}$,

$$g \cdot (x, y) := (x : gy).$$

The weighted projective line $\mathbb{P}^1_{2,2}$ is defined as the stacky quotient $[\mathbb{P}^1/G]$.

(i) Define the notion of a G-equivariant sheaf on $\mathbb{P}^1_{2,2}$ and classify the G-equivariant line bundles. We set $\mathtt{D\,Coh}^\bullet[\mathbb{P}^1/G]$ as the derived category of G-equivariant sheaves.
(ii) Interpret $\mathtt{D\,Coh}^\bullet[\mathbb{P}^1/G]$ as the derived category of sheaves of modules over a sheaf of algebras over $\mathbb{P}^1$.
(iii) Find a tilting object $\mathtt{D\,Coh}^\bullet[\mathbb{P}^1/G]$ and calculate its endomorphism ring.

8
Mirror Symmetry

Now that we have defined A_∞-categories associated to the A-model and the B-model, a natural question arises:

Is it possible to find A-models and B-models that describe the same underlying A_∞-category?

This phenomenon goes under the name of homological mirror symmetry. A pair of an A-model and a B-model that give rise to the same category is called a *mirror pair*. In this chapter we will discuss how to find possible mirror pairs, how to prove that they are indeed mirror pairs and how they are related geometrically.

8.1 The Complex Torus

The starting point of our investigation is an example that we have already studied in Chapter 5.

Example 8.1 Let us look at the cylinder $\mathbb{X} = \mathsf{T}^*\mathbb{S}^1 \cong \mathbb{C}^*$ with a volume form $\Omega = dz/z$ for the grading. This can be seen as a Liouville manifold with cylindrical ends.

- The corresponding wrapped Fukaya category ${}^{\mathtt{wr}}\mathtt{Fuk}^\bullet(\mathbb{X}, \Omega)$ is generated by a cotangent fiber, whose endomorphism ring is $R = \mathbb{C}[X, X^{-1}]$ concentrated in degree 0. This implies that $\mathtt{D}\,{}^{\mathtt{wr}}\mathtt{Fuk}^\bullet(\mathbb{X}, \Omega) \overset{\infty}{=} \mathtt{D}R \overset{\infty}{=} \mathtt{D}\,\mathtt{Mod}^\bullet R$. The latter is isomorphic to $\mathtt{D}\,\mathtt{Coh}^\bullet\,\mathbb{Y}$, where $\mathbb{Y} = \mathbb{V}(R) \cong \mathbb{C}^*$. Therefore we can consider $(\mathsf{T}^*\mathbb{S}^1, \mathbb{C}^*)$ as a mirror pair:

$$\mathtt{D}\,{}^{\mathtt{wr}}\mathtt{Fuk}^\bullet(\mathsf{T}^*\mathbb{S}^1, \Omega) \overset{\infty}{=} \mathtt{D}\,\mathtt{Coh}^\bullet\,\mathbb{C}^*.$$

- We can also look at the exact Fukaya category of $\mathsf{T}^*\mathbb{S}^1$. In that case, the objects are exact compact Lagrangians with a local system. Up to isotopy there is only one such Lagrangian, the circle itself, while the local system can be seen as a finite-dimensional representation of $\pi_1(\mathbb{S}^1) = \mathbb{Z}$ or its group ring $\mathbb{C}[\pi_1(\mathbb{S}^1)] \cong \mathbb{C}[X, X^{-1}] \cong R$. This gives an equivalence $\mathtt{D\,mod}^\bullet R \cong \mathtt{D\,Fuk}^\bullet \mathbb{X}$. From the B-side perspective, the finite-dimensional representations of R are those generated by the skyscraper sheaves $\mathscr{S}_p$ on $\mathbb{Y} = \mathbb{C}^*$; these are also called the *torsion sheaves*:

$$\mathtt{D}^{\,\mathtt{ex}}\mathtt{Fuk}^\bullet(\mathsf{T}^*\mathbb{S}^1, \Omega) \overset{\infty}{=} \mathtt{D\,tor}^\bullet \mathbb{C}^* := \langle \mathscr{S}_p \mid p \in \mathbb{C}^* \rangle \subset \mathtt{D\,Coh}^\bullet \mathbb{C}^*.$$

- Finally, we can look at the $\mathbb{Z}_2$-graded wrapped Fukaya category. This is again generated by a cotangent fiber, so it is equivalent to $\mathtt{D}^\pm R$. On the B-side we can realize this category as a category of matrix factorizations $\mathtt{MF}^\pm(\mathbb{C}^3, (XY - 1)Z)$. If we write F for the free rank 1 module of $\mathbb{C}[X, Y, Z]$, the latter is generated by the object $(F \oplus F, \left(\begin{smallmatrix} 0 & XY-1 \\ Z & 0 \end{smallmatrix}\right))$, whose endomorphism ring is $\mathbb{C}[X, Y, Z]/(XY - 1, Z) \cong \mathbb{C}[X, X^{-1}] = R$. Geometrically, this B-model corresponds to the singular locus of a hypersurface in $\mathbb{C}^3$ consisting of a quadric and a plane. Their intersection is a conic section and it is isomorphic to the affine variety $\mathbb{C}^*$:

$$\mathtt{D}^{\mathtt{wr}}\mathtt{Fuk}^\pm(\mathsf{T}^*\mathbb{S}^1) \overset{\infty}{=} \mathtt{MF}^\pm(\mathbb{C}^3, (XY - 1)Z).$$

So the fact that $\mathbb{C}^*$ forms a mirror pair with $\mathsf{T}^*\mathbb{S}^1$ manifests itself in three different equivalences of A_∞-categories.

The previous example generalizes to higher dimensions: if we put $\mathbb{X} = (\mathsf{T}^*\mathbb{S}^1)^n = \mathsf{T}^*(\mathbb{S}^1)^n$ and $\Omega = \frac{dz_1}{z_1} \wedge \cdots \wedge \frac{dz_n}{z_n}$, then the wrapped Fukaya category will still be split-generated by a cotangent fiber, but now its endomorphism ring will be $\mathbb{C}[X_1^{\pm 1}, \ldots, X_n^{\pm 1}]$ and the corresponding affine variety is $\mathbb{C}^{*n}$. Therefore, $(\mathsf{T}^*(\mathbb{S}^1)^n, \mathbb{C}^{*n})$ are mirror pairs.

The base space $\mathbb{L} = (\mathbb{S}^1)^n$ is an exact Lagrangian torus inside $\mathbb{X}$. From Chapter 5 we know that the endomorphism ring of $\mathbb{L}$ equipped with a trivial local system is equal to the cohomology ring of the torus, which is the exterior algebra in n variables Λ_n. Putting a nontrivial local system on $\mathbb{L}$ amounts to choosing a representation of $\pi_1(\mathbb{L}) = \mathbb{Z}^n$. The group ring of this fundamental group is $\mathbb{C}[X_1^{\pm 1}, \ldots, X_n^{\pm 1}]$, so each finite-dimensional module of this ring gives an object in the exact Fukaya category. From the B-side perspective the simple (one-dimensional) modules correspond to the skyscraper sheaves on $\mathbb{V}(\mathbb{C}[X_1^{\pm 1}, \ldots, X_n^{\pm 1}]) \cong \mathbb{C}^{*n}$. In general, the sheaves that correspond to finite-dimensional modules of the coordinate ring are also called the torsion sheaves. Finally, the ungraded version can be extended as well and we get the following table of equivalences.

A-model	Algebra	B-model
$\mathtt{D}^{\mathtt{wr}}\mathtt{Fuk}^\bullet(\mathsf{T}^*(\mathbb{S}^1)^n, \Omega)$	$\mathtt{D\,Mod}^\bullet\,\mathbb{C}[X_1^{\pm 1}, \ldots, X_n^{\pm 1}]$	$\mathtt{D\,Coh}^\bullet\,\mathbb{C}^{*n}$
$\mathtt{D}^{\mathtt{ex}}\mathtt{Fuk}^\bullet(\mathsf{T}^*(\mathbb{S}^1)^n, \Omega)$	$\mathtt{D\,mod}^\bullet\,\mathbb{C}[X_1^{\pm 1}, \ldots, X_n^{\pm 1}]$	$\mathtt{D\,tor}^\bullet\,\mathbb{C}^{*n}$
$\mathtt{D}^{\mathtt{wr}}\mathtt{Fuk}^\pm(\mathsf{T}^*(\mathbb{S}^1)^n)$	$\mathtt{D}^\pm\mathbb{C}[X_1^{\pm 1}, \ldots, X_n^{\pm 1}]$	$\mathtt{MF}^\pm(\mathbb{C}^{n+2}, (X_0 \cdots X_n - 1)Z)$

In the example above we see that $\mathbb{X} = \mathsf{T}^*(\mathbb{S}^1)^n \cong \mathbb{C}^{*n}$ is mirror to $\mathbb{Y} = \mathbb{C}^{*n}$, which is isomorphic to $\mathbb{X}$ itself. This point of view is a bit deceptive because in reality $\mathbb{X}$ and $\mathbb{Y}$ should not be considered the same, but dual.

The variety $\mathbb{C}^{*n}$ is a group for the multiplication and as such it is also called the *n-dimensional complex torus* and denoted by $\mathbb{T}$. To this group we can associate two other groups: the *group of characters* $M := \mathtt{Grp}(\mathbb{T}, \mathbb{C}^*)$ and the *group of one-parameter subgroups* $N := \mathtt{Grp}(\mathbb{C}^*, \mathbb{T})$. Both groups are isomorphic to $\mathbb{Z}^n$: for each $u, v \in \mathbb{Z}^n$ we can define

$$\chi_u : \mathbb{T} \to \mathbb{C} : (x_1, \ldots, x_n) \mapsto x_1^{u_1} \ldots x_n^{u_n} \in M,$$
$$\lambda_v : \mathbb{C}^* \to \mathbb{T} : t \mapsto (t_1^{v_1}, \ldots, t_n^{v_n}) \in N.$$

These groups are dual via the natural pairing $\langle u, v\rangle = u_1 v_1 + \cdots + u_n v_n$, which satisfies $\chi_u \circ \lambda_v(t) = t^{\langle u,v\rangle}$. We can recover $\mathbb{T}$ as $\mathtt{Grp}(M, \mathbb{C}^*)$ or $\mathbb{V}(\mathbb{C}[M])$, so it makes sense to define the *dual torus* $\mathbb{T}^\vee = \mathtt{Grp}(N, \mathbb{C}^*) = \mathbb{V}(\mathbb{C}[N])$.

Remember that the points of the mirror of $\mathbb{X} = \mathsf{T}^*(\mathbb{S}^1)^n = \mathbb{T}$ correspond to one-dimensional local systems on the base $\mathbb{B} = \{x \in \mathbb{T} : |x| = 1\} = (\mathbb{S}^1)^n$ or, equivalently, one-dimensional representations of its fundamental group. This fundamental group can be identified with N because we can turn every λ_v into a loop $\ell_v : [0, 1] \to \mathbb{B} : s \mapsto \lambda_v(e^{2\pi i t})$. Therefore, the mirror of $\mathbb{T}$ can naturally be seen as $\mathrm{Hom}(N, \mathbb{C}^*) = \mathbb{T}^\vee$. This observation will be the starting point in our search for other mirror pairs.

8.2 Toric Varieties

One way to extend mirror symmetry beyond the complex torus is to look at toric varieties. These form a certain class of varieties that contain a complex torus as an open dense subset and can be defined combinatorially. The theory of toric varieties is well developed and contains many illustrative examples of important concepts in algebraic geometry. For more information we refer to [89, 57].

Let $\mathbb{T}, \mathbb{T}^\vee$ be the dual tori from the previous section. Denote the corresponding dual lattices by $M = \mathrm{Hom}(\mathbb{T}, \mathbb{C}^*) = \mathrm{Hom}(\mathbb{C}^*, \mathbb{T}^\vee)$ and $N = \mathrm{Hom}(\mathbb{T}^\vee, \mathbb{C}^*) =$

$\mathrm{Hom}(\mathbb{C}^*, \mathbb{T})$, and their tensor products with $\mathbb{R}$ by $M_{\mathbb{R}} = M \otimes_{\mathbb{Z}} R$ and $N_{\mathbb{R}} = N \otimes_{\mathbb{Z}} R$. The natural pairing between these two lattices and vector spaces will be denoted by $\langle\ ,\ \rangle$. We identify M, N with $\mathbb{Z}^n$ and $M_{\mathbb{R}}$, $N_{\mathbb{R}}$ with $\mathbb{R}^n$ such that this pairing is equal to the standard inner product.

The coordinate ring $\mathscr{O}(\mathbb{T})$ is isomorphic to the group ring $\mathbb{C}[M] = \mathbb{C}[X_1^{\pm 1}, \ldots, X_n^{\pm 1}]$ and we will write X^u for $X_1^{u_1} \cdots X_n^{u_n}$ when $u \in M = \mathbb{Z}^n$. Similarly, $\mathscr{O}(\mathbb{T}^\vee) = \mathbb{C}[N] \cong \mathbb{C}[Y_1^{\pm 1}, \ldots, Y_n^{\pm 1}]$ and we write Y^v for the monomial corresponding to $v \in N = \mathbb{Z}^n$.

Definition 8.2 An *integral cone* in $N_{\mathbb{R}}$ is a subset of the form

$$\sigma = \{\lambda_1 v_1 + \cdots + \lambda_k v_k \mid \lambda_i \in \mathbb{R}_{\geq 0}\} \subset M_{\mathbb{R}},$$

where $v_i \in N$. Every cone has a *dual cone*

$$\sigma^\vee := \{u \in M_{\mathbb{R}} \mid \forall\, v \in \sigma,\ \langle u, v\rangle \geq 0\},$$

and for each $u \in \sigma^\vee$ we can define the subset

$$\tau_u = \{v \in \sigma \mid \langle u, v\rangle = 0\}.$$

These subsets are also integral cones and they are called the *faces* of σ.

Definition 8.3 A *fan* Δ is a collection of integral cones in $N_{\mathbb{R}}$ such that

- if $\sigma \in \mathcal{F}$ and τ is a face of σ then $\tau \in \Delta$,
- if $\sigma_1, \sigma_2 \in \Delta$ then $\sigma_1 \cap \sigma_2$ is a face of both σ_1 and σ_2.

We will denote the subset of all k-dimensional cones by Δ_k.

A fan is a combinatorial gadget that can be used to define a variety. Each cone defines an affine subring of $\mathbb{C}[M]$,

$$R_\sigma = \mathbb{C}[\sigma^\vee \cap M] = \mathbb{C}[X_1^{u_1} \ldots X_n^{u_n} \mid u \in \sigma^\vee \cap M],$$

and an affine variety $\mathbb{U}_\sigma = \mathbb{V}(R_\sigma)$. If τ_u is a face of σ then $\tau^\vee = \sigma^\vee + \mathbb{R}u$ and therefore $R_{\tau_u} = R_\sigma[X^{-u}]$ is a localization of R_σ. From a geometric point of view, $\mathbb{U}_\tau$ is an open subset of $\mathbb{U}_\sigma$. If we have two cones σ_1, σ_2 that share a common face τ, we can glue them together using the diagram $\mathbb{U}_{\sigma_1} \leftarrow \mathbb{U}_\tau \rightarrow \mathbb{U}_{\sigma_2}$. If we do this for all cones in the fan we get a variety.

Definition 8.4 The variety $\mathbb{X}_\Delta = \cup_{\sigma \in \Delta} \mathbb{U}_\sigma$ is called the *toric variety* associated to Δ.

Example 8.5 Let $M = \mathbb{Z}^2$, set $v_1 = (1, 0)$, $v_2 = (0, 1)$ and $v_3 = (-1, -1)$ and define $\sigma_i = \mathbb{R}_{\geq 0} v_j + \mathbb{R}_{\geq 0} v_k$, where $\{i, j, k\} = \{1, 2, 3\}$. Furthermore, we set $\tau_i = \mathbb{R}_{\geq 0} v_i = \sigma_j \cap \sigma_k$ and note that $\sigma_1 \cap \sigma_2 \cap \sigma_3 = 0$.

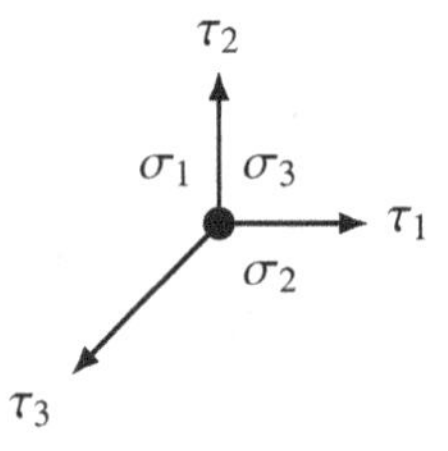

The set $\Delta = \{\sigma_1, \sigma_2, \sigma_3, \tau_1, \tau_2, \tau_3, 0\}$ is a fan and we have

$$R_{\sigma_1} = \mathbb{C}[YX^{-1}, X^{-1}], \quad R_{\sigma_2} = \mathbb{C}[XY^{-1}, Y^{-1}], \quad R_{\sigma_3} = \mathbb{C}[X, Y].$$

These are isomorphic to polynomial rings, so the $\mathbb{U}_{\sigma_i}$ are all isomorphic to $\mathbb{A}^2 = \mathbb{C}^2$. They are glued together using the transition functions that coincide with the transitions between the three standard affine planes in the projective plane,

$$\begin{array}{cccc} & \mathbb{U}_{\sigma_3} & \mathbb{U}_{\sigma_2} & \mathbb{U}_{\sigma_1} \\ \mathbb{X}_\Delta & (x, y) & (xy^{-1}, y^{-1}) & (yx^{-1}, x^{-1}) \\ \mathbb{P}^2 & (x : y : 1) & (xy^{-1} : 1 : y^{-1}) & (1 : yx^{-1} : x^{-1}) \end{array}$$

so we can conclude that $\mathbb{X}_\Delta \cong \mathbb{P}^2$.

Because all rings R_σ are contained in $\mathbb{C}[M]$, all the affine open $\mathbb{U}_\sigma$ contain $\mathbb{V}(\mathbb{C}[M]) = \mathbb{T}$. This open subset is dense and it is an orbit for the action of the group $G = \mathbb{C}^{*n} \cong \mathbb{T}$ induced from the action of G on all of the R_σ by $g \cdot X^u := g_1^{u_1} \cdots g_n^{u_n} X^u$. This geometric characterization also goes the other way.

Theorem 8.6 *Every normal n-dimensional variety $\mathbb{X}$ that has a faithful action of $\mathbb{T} = \mathbb{C}^{*n}$ with a dense orbit comes from a fan Δ. There is a bijection between cones $\sigma \in \Delta$ and the orbits $\mathbb{O}_\sigma \in \mathbb{X}/\mathbb{T}$ such that* $\dim \sigma = n - \dim O_\sigma$.

Sketch of the proof Fix a point p in the dense torus orbit of $\mathbb{X}$ and let N_+ be the set of one-parameter subgroups $\lambda_v \in \mathtt{Grp}(\mathbb{C}^*, \mathbb{T})$ for which the limit $\lim_{t\to 0} \rho_u(t)p$ exists in $\mathbb{X}$. In N_+ we define two elements as equivalent if they give rise to a limit in the same $\mathbb{T}$-orbit. This equivalence relation divides N_+ into equivalence classes E, one for each orbit. For each class E we look at the closure of all positive linear combinations

$$\sigma_E = \overline{\{\textstyle\sum \lambda_i v_i \mid \lambda_i \in \mathbb{R}_{\geq 0}, v_i \in E\}}.$$

This is a cone and all the σ_E form the required fan Δ. To show that $\mathbb{X}_\Delta \cong \mathbb{X}$ we can construct a torus-equivariant morphism that maps a point q in the dense torus of $\mathbb{X}_\Delta$ to p. This morphism will be surjective because the orbit of p is dense in $\mathbb{X}$ and injectivity can be deduced from the normality of $\mathbb{X}$.

For more details on the correspondence between orbits and cones see [89, Section 3.1]. □

Example 8.7 (The affine plane) Look at $\mathbb{A}^2$ and let $g = (g_1, g_2) \in \mathbb{C}^* \times \mathbb{C}^*$ act by $g \cdot (x, y) = (g_1 x, g_2 y)$. Fix the point $(1, 1)$ inside the dense torus $\mathbb{C}^* \times \mathbb{C}^* \subset \mathbb{A}^2$. The limit $\lim_{t \to 0}(t^{k_1}, t^{k_2})$ exists if $k_1, k_2 \geq 0$, so $N_+ = \mathbb{Z}_{\geq 0} \times \mathbb{Z}_{\geq 0}$. It splits into four equivalence classes $\{(0,0)\}, \{0\} \times \mathbb{Z}_{>0}, \mathbb{Z}_{>0} \times \{0\}, \mathbb{Z}_{>0} \times \mathbb{Z}_{>0}$ corresponding to the limit points $(1,1), (1,0), (0,1), (0,0)$.

$(1, 0)$

$(0, 0)$

$(0, 1)$

$(1, 1)$

The corresponding fan is

$$\Delta = \{\{(0,0)\},\ \{0\} \times \mathbb{R}_{\geq 0},\ \mathbb{R}_{\geq 0} \times \{0\},\ \mathbb{R}_{\geq 0} \times \mathbb{R}_{\geq 0}\}.$$

Suppose that $\mathbb{T} = \mathbb{C}^{*n}$ and that $\pi\colon \tilde{\mathbb{X}} \to \mathbb{X}$ is a surjective $\mathbb{T}$-equivariant map of toric varieties such that $\pi|_{\mathbb{T}} = \mathbb{1}_{\mathbb{T}}$. If two one-parameter subgroups are equivalent for $\tilde{\mathbb{X}}$ then they are also equivalent for $\mathbb{X}$, so the cones for $\tilde{\mathbb{X}}$ are subcones of the cones of $\mathbb{X}$, and therefore the fan of $\tilde{\mathbb{X}}$ is a subdivision of the cones of $\mathbb{X}$. The reverse is also true: if we subdivide a fan into smaller cones, we create a new toric variety that maps surjectively onto the old one.

Example 8.8 (The blow-up of the affine plane) The simplest example occurs when we subdivide the cone of the affine plane into two.

In $\mathbb{A}^2$, the two-dimensional cone corresponds to the zero orbit. In $\mathrm{Bl}\mathbb{A}^2$, this cone is subdivided into three: two two-dimensional cones, each corresponding to a zero-dimensional orbit, and one one-dimensional cone corresponding to a one-dimensional orbit, which is isomorphic to $\mathbb{C}^*$. Together these orbits form a $\mathbb{P}^1$ that maps onto the zero point. In other words, we have blown up the zero point $\mathbb{A}^2$ to a projective line.

Example 8.9 (The A_{k-1}-singularity) Consider the fan Δ in $\mathbb{R}^2$ whose only two-dimensional cone is $\sigma = \mathbb{R}_{\geq 0}(1,0) + \mathbb{R}_{\geq 0}(1,k)$. The dual cone σ is given by $\sigma^\vee = \mathbb{R}_{\geq 0}(0,1) + \mathbb{R}_{\geq 0}(k,-1)$ and therefore

$$\begin{aligned}\mathbb{C}[\mathbb{U}_\sigma] &= \mathbb{C}[X^iY^j \mid i \geq 0,\ i - kj \geq 0]\\ &= \mathbb{C}[X^kY^{-1}, Y, X] \cong \mathbb{C}[U, V, W]/(UV - W^k).\end{aligned}$$

This ring is also the ring of invariants for the group action of $G = \langle g \mid g^k \rangle \cong \mathbb{Z}/k\mathbb{Z}$ on $\mathbb{C}^2$ via $g(x, y) = (\zeta x, \zeta^{-1}y)$, with ζ the primitive kth root of unity. Therefore, $\mathbb{X}_\Delta$ is isomorphic to the affine quotient $\mathbb{C}^2/\!\!/G$ and the zero-dimensional torus orbit corresponding to σ is the singularity.

We can subdivide the cone σ into k smaller cones by adding rays in the directions $v_l = (1, l)$ with $0 < l < k$. This gives us a new fan with two-dimensional cones $\sigma_l = \mathbb{R}_{\geq 0}(1, l) + \mathbb{R}_{\geq 0}(1, l + 1)$ with $0 \leq l < k$. One can easily verify that $\mathbb{C}[\mathbb{U}_{\sigma_l}] \cong \mathbb{C}[X, Y]$ for all σ_ℓ, so the new toric variety $\tilde{\mathbb{X}}$ is smooth.

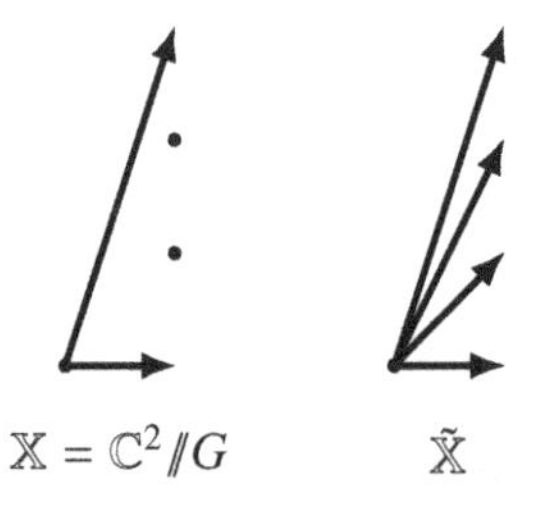

$\mathbb{X} = \mathbb{C}^2/\!\!/G$ $\qquad\qquad$ $\tilde{\mathbb{X}}$

Each extra ray gives a torus orbit isomorphic to $\mathbb{C}^*$, while each σ_l gives a zero-dimensional torus orbit. Together they form a string of $\mathbb{P}^1$'s that map onto the singular point via the projection map $\pi\colon \tilde{\mathbb{X}} \to \mathbb{X}$. The map $\pi\colon \tilde{\mathbb{X}} \to \mathbb{X}$ is called a *resolution of* $\mathbb{X}$: it is surjective, proper, almost everywhere one-to-one and its domain $\tilde{\mathbb{X}}$ is smooth.

The previous theorem and examples show us that the torus orbits in a toric variety are in one-to-one correspondence with the cones in the fan. The dense torus orbit itself corresponds to the zero cone, while the one-dimensional cones correspond to the $(n - 1)$-dimensional orbits. For each one-dimensional cone σ we can define two objects:

- The *ray vector* v_σ is the generator (or smallest vector) of the semigroup $\sigma \cap N$.
- The *toric divisor* $\mathbb{D}_\sigma$ is the closure of the orbit $\mathbb{O}_\sigma$.

The latter is the $(n - 1)$-dimensional subvariety consisting of all orbits whose cone contains v_σ. As every nonzero cone contains at least one ray vector, the complement of the dense torus is the union of all toric divisors

$$\mathbb{X}_\Delta \setminus \mathbb{T} = \bigcup_{\sigma \in \Delta_1} \mathbb{D}_\sigma.$$

We can use this observation to get a hint of what the mirror of a toric variety can be. Let us look at the relative Fukaya category ${}^{\mathrm{cu}}\mathtt{Fuk}^{\pm}(\mathbb{X}_\Delta, \mathbb{D})$ where $\mathbb{D} = \sum_\sigma \mathbb{D}_\sigma$ is the sum of the toric divisors. This category is a curved deformation of the exact Fukaya category of the torus ${}^{\mathrm{ex}}\mathtt{Fuk}^{\pm}(\mathbb{T})$ defined over the ring $\mathbb{C}[[\hbar]]$. The objects are pairs $(\mathbb{L}, \rho)$ in ${}^{\mathrm{ex}}\mathtt{Fuk}^{\pm}(\mathbb{T})$, where $\mathbb{L}$ is the Lagrangian base $(\mathbb{S}^1)^n \subset \mathbb{C}^{*n} = \mathsf{T}^*(\mathbb{S}^1)^n$ and ρ is a representation of the fundamental group $\pi_1(\mathbb{L}) = \pi_1(\mathbb{T}) = N$, and if ρ is one-dimensional, $(\mathbb{L}, \rho)$ can identified with a point in $\mathrm{Hom}(N, \mathbb{C}^*) = \mathbb{T}^\vee$. Let us denote this point by $(y_1, \ldots, y_n) = (\rho(\ell_1), \ldots, \rho(\ell_n)) \in \mathbb{T}^\vee$, where the ℓ_i are the loops corresponding to the standard basis elements in $N = \pi_1(\mathbb{L}) = \mathbb{Z}^n$.

What happens if we turn on the curvature to go to the relative Fukaya category? If $v = v_\sigma$ is the ray vector of the divisor $\mathbb{D}_\sigma$ then we can consider its one-parameter subgroup $\lambda_v \colon \mathbb{C}^* \to \mathbb{T}$. Because the limit $t \to 0$ exists in $\mathbb{X}_\Delta$, the extension and restriction of λ_w to the disk $\mathbb{D} \subset \mathbb{C}$ will define a holomorphic disk with boundary on $\mathbb{L}$, which intersects $\mathbb{D}_\sigma$. For the object $(\mathbb{L}, \rho)$ this holomorphic disk contributes to μ_0 with a factor $\rho(\lambda_v|_{\mathbb{S}^1})\hbar$. The factor $\rho(\lambda_v|_{\mathbb{S}^1})$ comes from the parallel transport around the boundary of the disk and $\hbar$ measures the intersection with the divisor. From the mirror point of view this means that every point $y = (y_1, \ldots, y_n) \in \mathbb{T}^\vee$ picks up a curvature equal to $W_\hbar(y)$ where

$$W_\hbar = \sum_\sigma \hbar y^{v_\sigma}.$$

Note that there may be other contributions to μ_0 and there also may be contributions to $\mu_{>0}$, but if we are lucky the contributions we identified might be enough to determine the isomorphism class of the deformation. This seems to suggest that the B-mirror of the A-model for $(\mathbb{X}_\Delta, \mathbb{D})$ is a Landau–Ginzburg model of the form $(\mathbb{T}^\vee, W_\hbar)$.

Observation 8.10 *Adding a toric divisor $\mathbb{D}_v$ in the A-model introduces a dual monomial term y^v to the potential in the B-model (possibly with other deformations).*

This correspondence between toric divisors and monomial terms also seems to work in the other direction. If we look back at Examples 6.60 and 6.74, we see that by introducing the monomial terms x, x^{-1} in the A-model we get a directed Fukaya category that is equivalent to $\mathtt{Dmod}^\bullet \circ \!\Longrightarrow\! \circ$. In Section 4.3.2 we saw that this is equivalent to the derived category of $\mathbb{P}^1$. This is a one-dimensional toric variety corresponding to the fan

$$\Delta = \{0, \mathbb{R}_{\geq 0}, \mathbb{R}_{\leq 0}, \mathbb{R}\} \qquad \longleftarrow\!\bullet\!\longrightarrow$$

which has two toric divisors $\mathbb{D}_{+1} = \{0\}$ and $\mathbb{D}_{-1} = \{\infty\}$ corresponding to the ray vectors $\pm 1 \in N$.

Observation 8.11 *Adding a monomial term y^ν to the potential in the A-model corresponds to introducing a corresponding toric divisor $\mathbb{D}_\nu$ in the B-model.*

Take care: at this moment these observations are just hunches that suggest how A-models and B-models are connected. To get more precise statements we will need to introduce yet another kind of geometry.

8.3 Tropical Geometry

A Landau–Ginzburg model consists of a variety with a holomorphic function $(\mathbb{Y}, W)$. As we saw in the previous section, we are especially interested in the case where the variety is the complex torus $\mathbb{T}^\vee$ and $W_\hbar \colon \mathbb{T}^\vee \to \mathbb{C}$ depends on a deformation parameter $\hbar$. Generically, the A-model $(\mathbb{Y}, W)$ will not depend on $\hbar$, so we will investigate what happens when $\hbar$ goes to 0, or when the parameter $t = \hbar^{-1}$ tends to infinity. This leads to a new kind of geometry called tropical geometry. For a survey of this field we refer to an overview article by Mikhalkin [175] and the introductory book by Maclagan and Sturmfels [181].

Consider a family of Laurent polynomials $W_t \in \mathbb{C}[Y_1^{\pm 1}, \dots, Y_n^{\pm 1}, t^{\pm 1}]$ and look at the hypersurface in $\mathbb{T}^\vee$ defined for a specific value $t \in \mathbb{R}_{>0}$. The topology of such a hypersurface is not always easy to visualize but one way to get a grip on it is by looking at the amoeba.

Definition 8.12 (Gelfand–Zelevinskiĭ–Kapranov [106]) Let $\log_t \colon \mathbb{T}^\vee \to \mathbb{R}^n \colon (y_1, \dots, y_n) \mapsto \frac{1}{\ln |t|}(\ln |y_1|, \dots, \ln |y_n|)$. The (rescaled) *amoeba* of W_t is the image of $W_t^{-1}(0)$ under this log map:

$$\mathcal{A}_t := \log_t\{(y_1, \dots, y_n) \in \mathbb{C}^{*n} \mid W_t(y_1, \dots, y_n) = 0\}.$$

Example 8.13 The amoeba of $W_t = X^2Y^3 + X^2Y^2 + tX^2Y - 5XY^2 + 3XY - Y^2 + t^2Y + 1$ for $t = 2$ is the blob bounded by the curved lines shown below.

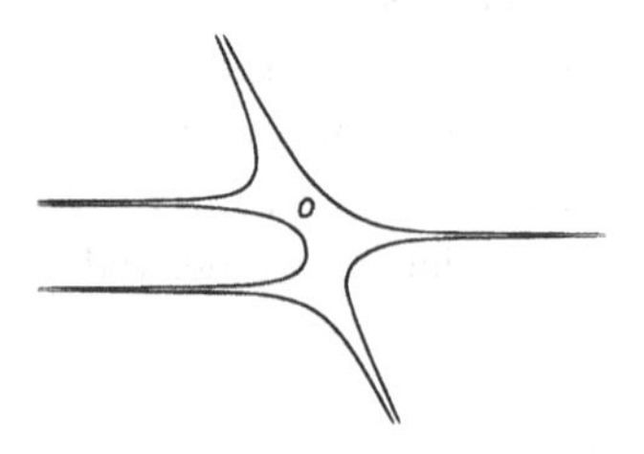

In general, an amoeba of a hypersurface in $\mathbb{C}^{*2}$ looks like a thickening of some piecewise linear graph with infinite legs and if we let t go to infinity the thickening will shrink and the limit will be that graph.

To make this more precise write $W_t = \sum_{m\in\mathcal{M}} c_m t^{k_m} Y^m$ as a sum of monomials with an exponent $m \in \mathcal{M} \subset \mathbb{Z}^n$ and a coefficient that consists of a factor $c_m \in \mathbb{C}^*$ and an integral power $k_m \in \mathbb{Z}$ of t. If we let t approach infinity, the amoeba $\mathcal{A}_t$ approaches some polyhedral skeleton. The reason for this is that if t is large, at most points one of the monomials in $W_t(y)$ will dominate the sum and therefore $W_t(y)$ will not be zero. Only at locations where two monomials are equally large can they cancel out to make the sum zero. Each monomial $c_m t^{k_m X^m}$ defines a linear function on $\lambda_m\colon \mathbb{R}^n \to \mathbb{R}\colon u \mapsto \langle u, m\rangle + k_m$ and the points of $\mathcal{A}_t$ concentrate near the locations where the maximum of $\{\lambda_m\}_{m\in\mathcal{M}}$ is attained by two of the functions.

Definition 8.14 The *tropical function* associated to $W_t = \sum_{m\in\mathcal{M}} c_m t^{k_m} Y^m$ is the map

$$W_{\text{trop}}\colon \mathbb{R}^n \to \mathbb{R}\colon u \mapsto \max_{m\in\mathcal{M}} \lambda_m(u).$$

It is a piecewise linear function on $\mathbb{R}^n$ and it divides $\mathbb{R}^n$ into cells $\bar{B}_m$, where the function W_{trop} is equal to the linear function λ_m for a given monomial $m \in \mathcal{M}$.

The *tropical hypersurface* H_{trop} is the locus in $\mathbb{R}^n$ where W_{trop} is not smooth or, equivalently, where the maximum is attained by more than one linear function in W_{trop}. In other words, it is the complement of the interiors of all the $\bar{B}_m$.

Theorem 8.15 (Mikhalkin [175], Forsberg–Passare–Tsikh [86]) *Consider the Laurent polynomial $W_t \in \mathbb{C}[Y_1^{\pm1}, \ldots, Y_n^{\pm1}, t^{\pm1}]$. When $t \in \mathbb{C}^*$ tends to infinity, the amoeba $\mathcal{A}_t$ approaches H_{trop}. In particular, for $|t| \gg 0$ the complement of the amoeba in $\mathbb{R}^n$ will consist of blobs $(B_m)_t$, each corresponding to the locations where one of the monomials $c_m t^{k_m} x^m$ dominates the rest in W_t. In the limit, $(B_m)_t$ will become the convex polyhedral piece $\bar{B}_m$ where one of the linear functions λ_m dominates W_{trop}.*

The cells $\bar{B}_m$ are convex and bounded by hyperplanes of the form $\lambda_{m_1}(u) - \lambda_{m_2}(u) = 0$. The intersections between these hyperplanes can become quite complicated and therefore it is often interesting to impose some genericity condition.

Definition 8.16 The *Newton polytope* $\square$ of W_{trop} is the convex hull of $\mathcal{M} \subset \mathbb{Z}^n$ inside $\mathbb{R}^n$. We say that W_{trop} is *generic* if, for every monomial $m \in \square \cap \mathbb{Z}^n$, the cell $\bar{B}^m$ has a nonempty interior and the hyperplanes $\lambda_{m_1}(u) - \lambda_{m_2}(u) = 0$ all intersect transversally.

We will now concentrate on generic tropical curves, so $n = 2$. In that case, the Newton polytope is a polygon and the tropical curve is a graph embedded in $\mathbb{R}^2$ with some finite edges and some infinite ones, which we call legs. We can split the monomials into two types: those that are in the interior of the Newton polygon and those that are on the boundary. Let us denote the two sets by $\mathcal{I}$ and $\mathcal{B}$. For an interior monomial, $\bar{B}_m$ will be a bounded convex polygon, while for a boundary monomial, $\bar{B}_m$ will be an unbounded region sandwiched between two legs of the tropical curve that go to infinity. Therefore, the total number of infinite legs of the tropical curve will be equal to $\#\mathcal{B}$, while the number of holes in the compact part of the graph will be equal to $\#\mathcal{I}$.

In the *generic* situation, the vertices of the graph will be trivalent because a higher valency would imply that more than three linear functions $\lambda_i \colon \mathbb{R}^2 \to \mathbb{R}$ have the same value at a given point, contradicting the transversal intersection property. This implies that for each vertex $v \in H_{\text{trop}}$ we can draw a small triangle in $\square$ that connects the monomials whose regions bound v. This will give a triangulation of $\square$ that is dual to H_{trop}. Each of these triangles will contain exactly three lattice points, situated at the corners. Such triangles are called *elementary triangles*.

Because the amoeba $\mathcal{A}_t$ approaches H_{trop} for $t \gg 0$, the curve $\dot{\mathbb{S}} := W_t^{-1}(0) \subset \mathbb{T}^\vee$ can be thought of as a Riemann surface with punctures that wraps around H_{trop}. This tells us that the genus of the surface must be $\#\mathcal{I}$, while the number of punctures will be $\#\mathcal{B}$. If we cut the surface at the finite edges of the tropical curve, we get a decomposition into pairs of pants.

Example 8.17 Take the function $W_t = t^8 + tX^{-1}Y^{-1} + t^2Y^{-1} + tXY^{-1} + t^4Y$; the tropical function is

$$W_{\text{trop}} = \max\{8, 1 - X - Y, 2 - Y, 1 + X - Y, 4 + Y\}.$$

The tropical curve and the corresponding triangulation of the Newton polygon are shown below.

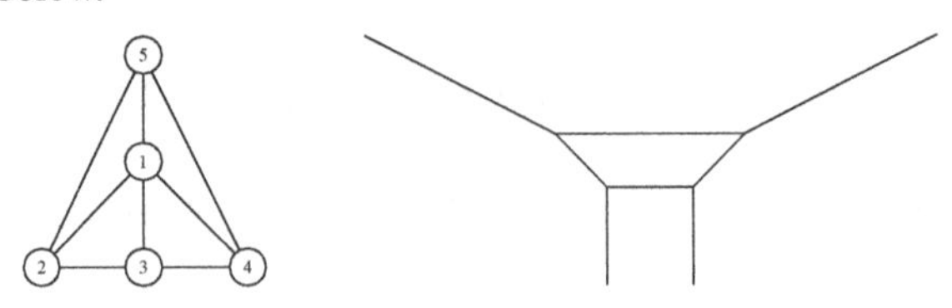

So, for generic t, the surface $\dot{\mathbb{S}} = W_t^{-1}(0) \subset \mathbb{C}^{*2}$ will have genus 1 and four punctures.

8.4 One, Two, Three, Mirror Symmetry

We will now study mirror symmetry in relation to toric and tropical geometry. Our starting point is a one-parameter family of functions $W_t\colon \mathbb{T}^\vee \to \mathbb{C}$ on an n-dimensional complex torus, which we can use to explore instances of mirror symmetry in different dimensions: $n-1, n, n+1$. For simplicity we will assume $\dim \mathbb{T}^\vee = 2$ so we get correspondences in dimensions 1, 2 and 3.

8.4.1 The Models

First we will define three A-models. In each case the symplectic geometry will not depend on the parameter t, so we can assume that t goes to infinity and go to the tropical limit. Therefore, the A-models will only depend on the Newton polytope, which for $n = 2$ is a polygon. We will denote it by $\square$ and use it as a subscript in the naming of the three A-models we will consider.

(1D) $\mathbb{Y}^1_\square := W_t^{-1}(0)$ is a punctured Riemann surface inside $\mathbb{T}^\vee = \mathbb{C}^{*2}$, which we can view as a Liouville submanifold with cylindrical ends near the punctures.

(2D) $\mathbb{Y}^2_\square := (\mathbb{T}^\vee, W_t)$ is a Landau–Ginzburg model on the complex torus $\mathbb{T}^\vee \cong \mathsf{T}^*(\mathbb{S}^1)^2$ with the standard exact symplectic structure and volume form.

(3D) $\mathbb{Y}^3_\square := \{(x, y, u, v) \in \mathbb{T}^\vee \times \mathbb{C}^2 \mid W_t(X, Y) + UV = 1\}$ is a hypersurface, which we can view as an exact symplectic submanifold of $\mathbb{T}^\vee \times \mathbb{C}^2 = \mathsf{T}^*(\mathbb{S}^1)^2 \times \mathbb{R}^2$. It can also be seen as a bundle of affine conic sections over $\mathbb{T}^\vee$ that become degenerate at $W_t = 1$.

As W_t is a function on the dual torus $\mathbb{T}^\vee$ its monomials sit in $N = \mathrm{Hom}(\mathbb{T}^\vee, \mathbb{C}^*)$, so the Newton polygon is a subset of $N_\mathbb{R}$. If we are in the generic case and go to the limit, the tropical function W_{trop} endows the Newton polygon with a triangulation, which we denote by $\boxtimes$. This triangulation will be the starting point for the construction of three B-models.

(3D) Put the Newton polygon at height 1 in a three-dimensional space $\square \times \{1\} \subset \mathbb{R}^3$, and take the cone $\sigma_\square = \mathbb{R}_{\geq 0}(\square \times \{1\})$. This determines an affine three-dimensional toric variety $\mathbb{X}^3_\square := \mathbb{V}(\mathbb{C}[\sigma^\vee_\square \cap M])$, which is singular unless the Newton polygon is an elementary triangle (in which case $\mathbb{X}^3_\triangle = \mathbb{A}^3$). Similarly to Example 8.9, we can resolve this singularity by subdividing the cone into smaller cones over the elementary triangles of the triangulation $\boxtimes$. This subdivision gives a smooth three-dimensional toric variety $\mathbb{X}^3_{\boxtimes}$ and a resolution $\pi\colon \mathbb{X}^3_{\boxtimes} \to \mathbb{X}^3_{\square}$. Such a resolution is called a *crepant resolution*.[1]

(2D) The ray vectors of $\mathbb{X}^3_{\boxtimes}$ are of the form $v = (a, b, 1)$, where (a, b) is a lattice point of the Newton polygon. If $m = (0, 0, 1)$ then $\langle m, v\rangle = 1$ for all ray vectors v. This means that the function $\ell = X^m$ is contained in all $\mathscr{O}(\mathbb{U}_\sigma)$ and hence it is a global function. The zero locus of this global function is equal to the complement of the dense torus $\mathbb{D} = \mathbb{X}^3_{\boxtimes} \setminus \mathbb{T}$, which is the union of the toric divisors $\mathbb{D}_v$.

Each toric divisor $\mathbb{D}_v$ is itself a smooth toric variety. Its fan can be constructed as follows. For each pair of ray vectors w_1, w_2 such that v, w_1, w_2 form an elementary triangle in the polygon, the set $\mathbb{R}_{\geq 0}(w_1 - v) + \mathbb{R}_{\geq 0}(w_2 - v)$ will be a cone in the plane $\mathbb{R}^2 \times \{0\}$. If v is an internal lattice point of $\square$ then the fan will cover the whole of $\mathbb{R}^2$ and $\mathbb{D}_v$ will sit in the zero fiber of $\pi\colon \mathbb{X}^3_{\boxtimes} \to \mathbb{X}^3_{\square}$. We will denote the divisor $\mathbb{D}_v$ as $\mathbb{X}^2_v$.

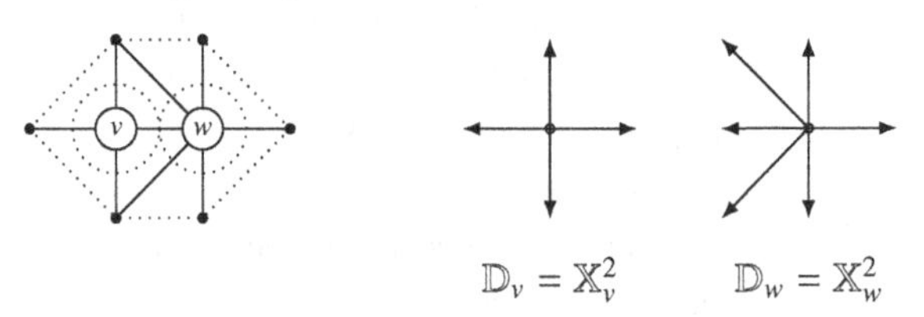

(1D) Finally, let us look at the Landau–Ginzburg model $\mathbb{X}^1_{\boxtimes} = (\mathbb{X}^3_{\boxtimes}, \ell)$. From a B-model perspective, this describes the singular locus of the zero locus of ℓ, which is one-dimensional. As we have already seen, $\ell^{-1}(0) = \mathbb{X}^2_{\boxtimes}$ is the union of smooth toric divisors and the singular locus will be the intersection between these divisors. Two divisors $\mathbb{D}_v$ and $\mathbb{D}_w$ share a common one-dimensional torus orbit (i.e. a $\mathbb{C}^*$) if their ray vectors span a two-dimensional cone in the fan of $\mathbb{X}^3_{\boxtimes}$. This means that in the triangulation of the Newton polygon, the two lattice points are joined by an edge. Three divisors will share a point if their ray vectors span a three-dimensional cone or the lattice points in the polygon form a triangle of the triangulation.

[1] A resolution $\tilde{\mathbb{X}} \to \mathbb{X}$ is crepant if the pullback of the canonical bundle on $\mathbb{X}$ is the canonical bundle on $\tilde{\mathbb{X}}$.

We can relate the geometry of the tropical curve with the same triangulation to the singular locus of $\ell^{-1}(0)$. The singular locus consists of a $\mathbb{C}^*$ for each edge of the tropical curve and a point for each vertex of the tropical curve. Each internal edge ends at two vertices, and the two points and the $\mathbb{C}^*$ together form a $\mathbb{P}^1$, while each leg has only one vertex so the resulting point and the $\mathbb{C}^*$ form an $\mathbb{A}^1$. We can conclude that the singular locus is a union of projective and affine lines, joined at triple points.

H_{trop} $\quad$ $\mathrm{Sing}\,\mathbb{X}^1_{\boxtimes}$

We are now ready to state mirror correspondences in three dimensions.

(1D) The category of matrix factorizations of the Landau–Ginzburg model $\mathbb{X}^1_{\boxtimes}$ and the wrapped Fukaya category of the punctured Riemann surface $\mathbb{Y}^1_{\square}$ are split-derived equivalent:

$$\mathrm{D}^{\pi}\mathrm{MF}^{\pm}\mathbb{X}^1_{\boxtimes} \cong \mathrm{D}^{\pi\mathrm{wr}}\mathrm{Fuk}^{\pm}\mathbb{Y}^1_{\square}.$$

(2D) For every internal lattice point v in the Newton polygon, the derived category of coherent sheaves of the divisor $\mathbb{X}^2_v$ embeds into the split-derived directed Fukaya category of the Landau–Ginzburg model $\mathbb{Y}^2_{\square}$,

$$\mathrm{D}\,\mathrm{Coh}^{\bullet}\,\mathbb{X}^2_v \subset \mathrm{D}^{\pi\ \mathrm{dr}}\mathrm{Fuk}^{\bullet}\,\mathbb{Y}^2_{\square}.$$

If $\mathbb{X}^2_v = \ell^{-1}(0)$ then the embedding is an equivalence.

(3D) The derived category of coherent sheaves on $\mathbb{X}^3_{\boxtimes}$ supported on $\mathbb{X}^2_v$ embeds into the split-derived exact Fukaya category of $\mathbb{Y}^3_{\square}$,

$$\mathrm{D}\,\mathrm{Coh}^{\bullet}{}_{\mathbb{X}^2_v}\,\mathbb{X}^3_{\boxtimes} \subset \mathrm{D}^{\pi\ \mathrm{ex}}\mathrm{Fuk}^{\bullet}\,\mathbb{Y}^3_{\square}.$$

There are several approaches to construct such equivalences. We will briefly sketch in each dimension how to tackle the problem.

8.4.2 Dimension 1

The first approach is to identify a set of split generators on both sides and show that their A_∞-endomorphism algebras are A_∞-isomorphic. An example of this for the one-dimensional case can be found in [1]. There the authors first look at the case where the Newton polygon is an elementary triangle. In this situation the A-model is the three-punctured sphere

$$\mathbb{Y}^1_\Delta = \{(x, y) \in \mathbb{C}^{*2} \mid (x + y + 1)t = 0\} = \mathbb{C}^* \setminus \{-1\},$$

which if we turn the punctures into cylindrical ends becomes a pair of pants. The B-model is the Landau–Ginzburg model

$$\mathbb{X}^1_\Delta = (\mathbb{A}^3, XYZ),$$

whose singular locus consists of the three coordinate axes.

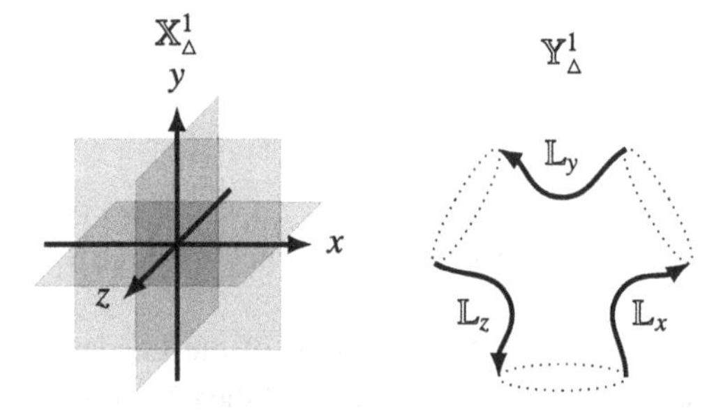

The wrapped Fukaya category of $\mathbb{Y}^1_\Delta$ has an easy set of conical Lagrangian submanifolds $\mathbb{L}_x, \mathbb{L}_y, \mathbb{L}_z$ given by three curves that each connect two cylindrical ends. These split-generate the Fukaya category because they cut open the sphere into two triangles. The category of matrix factorizations also has three easy split-generators which are the matrix factorizations corresponding to the three coordinate axes in $\mathbb{A}^3$:

$$M_X = R \underset{YZ}{\overset{X}{\rightleftarrows}} R\,, \quad M_Y = R \underset{XZ}{\overset{Y}{\rightleftarrows}} R\,, \quad M_Z = R \underset{XY}{\overset{Z}{\rightleftarrows}} R\,.$$

One can calculate that both sets of generators have A_∞-isomorphic endomorphism algebras, so they split-generate equivalent categories. More details can be found in Section 12.3 (see Theorem 12.21).

Remark 8.18 Abouzaid et al. [1] extended his technique to Riemann surfaces of higher genus and with more punctures by looking at covers of the pairs of pants. On the B-side this results in a stacky Landau–Ginzburg model, which is also equivalent to a quasi-projective one. In [36] the same technique is used to construct noncommutative mirrors, which can also be transformed into quasi-projective ones using moduli spaces of quivers. Another approach, worked out by Pascaleff and Sibilla in [204] and Lee in [155], is to use the correspondence for the three-punctured sphere as a building block for the other cases. Instead of looking for generators, we glue several copies of the A_∞-category for the three-punctured sphere together. On the A-side we can interpret this gluing as sewing cylindrical tubes together, while on the B-side we can see this gluing as turning the ambient space into a toric variety. This will be discussed in more detail in Chapter 9. Finally, a gluing approach by Nadler and Gammage [190, 94] relates the Fukaya category of the pair of pants and other punctured surfaces to the theory of microlocal sheaves.

Dimension 2

In the two-dimensional case the A-model is a Landau–Ginzburg model. This means that its objects are Lagrangian disks inside $\mathbb{T}^\vee$ with a boundary on $W_t^{-1}(0) = \mathbb{Y}^1_\square$. The image of $\mathbb{Y}^1_\square$ under the map $\log_t \colon \mathbb{T}^\vee \to \mathbb{R}^2$ is the amoeba of $W_t = \sum_{w \in \square} t^{a_w} Y^w$ and it consists of a blob with holes $(B_v)_t$. In the generic case each hole corresponds to a lattice point in the Newton polygon $v \in \square$. If we let t go to infinity, the hole corresponding to v becomes the polygonal region

$$\bar{B}_v = \{u \in \mathbb{R}^2 \mid \forall\, w - v \in \boxtimes,\ \langle w - v, u\rangle + a_w - a_v \geq 0\},$$

where the $w - v$ represent the edges in the triangulation $\boxtimes$ that are incident to v. Note that the $w - v$ are precisely the ray vectors of the $\mathbb{X}^2_v$. To get the two-dimensional correspondence (and its higher-dimensional analogs) Abouzaid [5] compared line bundles over $\mathbb{X}^2_v$ with Lagrangian disks that (in the limit $t \to \infty$) map onto $(B_v)_t$ under the map $\log_t$.

If $\mathbb{X}_\Delta$ is a toric variety and $d \colon S_\Delta \to \mathbb{Z}$ is a map from the set of ray vectors to $\mathbb{Z}$ then we can define a coherent sheaf $\mathcal{O}(d)$ over $\mathbb{X}_\Delta$,

$$\mathcal{O}(d)(\mathbb{U}_\sigma) = \mathbb{C}\{X^m \mid \forall\, v \in S_\Delta \cap \sigma,\ \langle v, m\rangle + d_v \geq 0\}.$$

From this notation it is clear that $\mathcal{O}(0) = \mathcal{O}$ is the structure sheaf of $\mathbb{X}_\Delta$ and, for every d, the $\mathcal{O}(\mathbb{U}_\sigma)$-module $\mathcal{O}(d)(\mathbb{U}_\sigma)$ is finitely generated. If $\mathbb{X}$ is smooth one can show that $\mathcal{O}(d)(\mathbb{U}_\sigma)$ is a free rank 1 module for every cone σ, so $\mathcal{O}(d)$ is a line bundle. The global sections are spanned by monomials coming from integral lattice points in the lattice polygon

$$P_d = \{m \in M_\mathbb{R} \mid \forall\, v \in S_\Delta,\ \langle v, m\rangle + d_v \geq 0\}.$$

If all the d_v are positive then P_d has the "same shape" as $\bar{B}_v$ (more precisely the normals to the edges are opposite). This means that there is a smooth map $\phi\colon \bar{B}_v \to P_d$ that identifies corners.

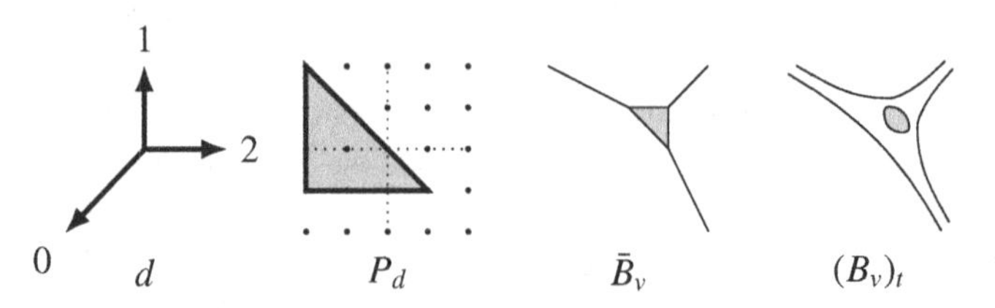

Split $\mathbb{T}^\vee$ as a product $(\mathbb{S}^1)^2 \times \mathbb{R}^2 = M_\mathbb{R}/M \times \mathbb{R}^2$ and construct the map

$$\tilde{\phi}\colon \bar{B}_v \to \mathbb{T}^\vee : u \mapsto (\phi(u) + M, u),$$

which wraps $\bar{B}_v$ around the real torus and is a section of $\log_t|_{\bar{B}_v}$. For t big enough, the component $(B_v)_t$ is close enough to $\bar{B}_v$ that this map can be tweaked to a Lagrangian section $\mathbb{L}_d$ of $\log_t|_{(B_v)_t}$, which sits on the boundary of $W_t^{-1}(0)$. This is an object in the directed Fukaya category of $\mathbb{Y}^2_\boxtimes$.

For the special case where $d = 0$, the polygon P_v is the zero point and ϕ becomes the constant map. This lifts to a constant section $\mathbb{L}_0$. The intersection points between $\mathbb{L}_0$ and $\mathbb{L}_d$ correspond to the integral points in P_d, which is indeed what we expect if $\mathbb{L}_0$ and $\mathbb{L}_d$ correspond to the line bundles $\mathscr{O}(0)$ and $\mathscr{O}(d)$. In [5] Abouzaid works out this idea to obtain an embedding of the A_∞-precategory of the line bundles over $\mathbb{X}^2_{\boxtimes,v}$ inside the directed Fukaya precategory of $\mathbb{Y}^2_\boxtimes$.

For toric varieties these line bundles split-generate the whole derived category of coherent sheaves [5, Prop. 1.3]. On the Fukaya side it is possible to show that the Lagrangian sections over all compact $(B_v)_t$ generate the directed Fukaya category, so in the case that there is only one internal lattice point, the embedding induces a derived equivalence. In [5] Abouzaid works out the details of this construction in all dimensions. The equivalence between the precategory of line bundles and the Fukaya category of Lagrangian sections is established via a sequence of intermediates that uses Morse theory on the $\bar{B}_v$.

Remark 8.19 The case of del Pezzo surfaces (two-dimensional Fano varieties) can also be found in [247]. A different approach by Auroux, Katzarkov and Orlov [11] relates the derived category of a projective plane blown up at $k \leq 8$ points to the directed Fukaya category of a Landau–Ginzburg model $W\colon \mathbb{M}_k \to \mathbb{C}$. While this approach is only worked out in dimension 2, it is more general in the sense that it also covers Fano varieties that are not toric (if $k \geq 4$).

Remark 8.20 We can also describe the A-model using partially wrapped Fukaya categories. Consider the cotangent bundle of the real torus $\mathsf{T}^*(\mathbb{S}^1)^2$ with its projection map $\pi\colon \mathsf{T}^*(\mathbb{S}^1)^2 \to (\mathbb{S}^1)^2$ and let N_ϵ be a small tubular neighborhood around $(\mathbb{S}^1)^2$.

Each edge e of the polygon P_d becomes a hyperplane in the real torus $(\mathbb{S}^1)^2 = M_\mathbb{R}/M$. The cotangent fibers above that line $F = \pi^{-1}(e)$ split naturally into two parts if we remove N_ϵ. One part consists of the covectors that have positive values on tangent vectors that point outside P_d and the other part of those with negative values. We denote these parts by E^+ and E^- and use the union of all E^+ as our set of stops σ.

The partially wrapped Fukaya category ${}^{\mathtt{wr}}\mathtt{Fuk}^\bullet{}_\sigma(\mathsf{T}^*(\mathbb{S}^1)^n)$ is derived equivalent with $\mathbb{X}^2_v$ because for each $\mathcal{O}(d)$ we can rescale the corresponding $\tilde{\phi}$ such that it lies inside N_ϵ and continue the embedding alongside the walls $E^+ \subset \sigma$. Below we illustrate the idea when P_d is one-dimensional instead of two-dimensional, so the cotangent bundle is a cylinder.

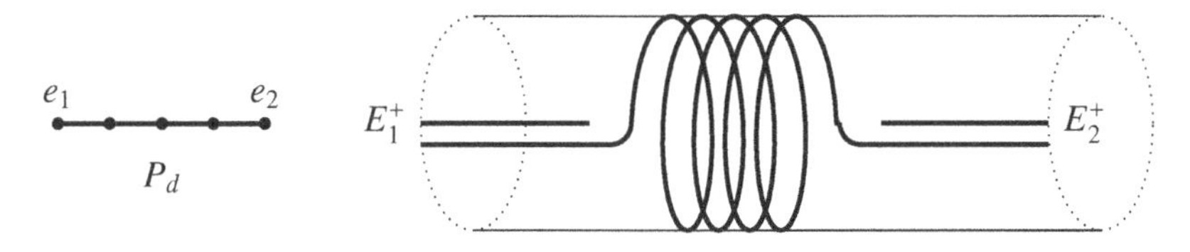

This approach is closely related to the constructible and microlocal sheaf approach to mirror symmetry by Nadler and Zaslow [195, 187]. For toric varieties, the constructible approach was first described by Bondal [38] and worked out by Fang et al. [80, 81]. For more information on the relation between Morse theory and microlocal sheaves we refer to Ganatra, Pardon and Shende [95].

Dimension 3

To study the three-dimensional case we have to look at the embedding of the toric divisor $\iota\colon \mathbb{D}_v = \mathbb{X}^2_v \to \mathbb{X}^3_\boxtimes$. Each line bundle $\mathcal{O}(d)$ on $\mathbb{X}^2_v$ can be considered as a sheaf on $\mathbb{X}^3_\boxtimes$ supported on $\mathbb{X}^2_v$ and therefore as an object $\iota^*\mathcal{O}(d) \in \mathtt{D\,Coh}^\bullet{}_{\mathbb{X}^2_v}\,\mathbb{X}^3_\boxtimes$. To establish a mirror symmetry correspondence we will have to find an object in $\mathbb{Y}^3_\square$ that behaves in the same way as $\iota^*\mathcal{O}(d)$. If one calculates the endomorphism ring in $\mathtt{D\,Coh}^\bullet{}_{\mathbb{X}^2_v}\,\mathbb{X}^3_\boxtimes$ then Serre duality tells us that it will look like $\mathbb{C} \oplus \mathbb{C}[3]$, which is isomorphic to the cohomology of a 3-sphere. Therefore, the idea is to lift the Lagrangian section $\mathbb{L}_d \in \mathbb{Y}^2_v$ to a Lagrangian sphere in $\mathbb{Y}^3_\square$.

As we saw above, each line bundle $\mathcal{O}(d)$ corresponds to a Lagrangian section $\mathbb{L}_d$ above $(B_v)_t$. This $\mathbb{L}_d$ has the topology of a disk in $\mathbb{C}^{*2}$ with a boundary on $W_t^{-1}(0)$. We can picture the three-dimensional space $\mathbb{Y}^3_\square$ as a bundle of conics of the form $\{(u,v) \in \mathbb{C}^2 \mid uv = W_t(x,y)\}$ above $\mathbb{C}^{*2}$ that degenerate to cones on $W_t^{-1}(0)$. This degeneration can be seen as a circle that shrinks to a point, so above the disk $\mathbb{L}_d$ we have a bundle of circles that shrink to points on the boundary, and hence the total space of this circle bundle is a 3-sphere. So far we have defined this sphere just topologically, but it is possible to see this sphere as an exact Lagrangian sphere $\mathbb{S}_d \subset \mathbb{Y}^3_\square$. The assignment $\iota^*\mathcal{O}(d) \to \mathbb{S}_d$ extends to a fully faithful embedding $\mathrm{D}\,\mathbf{Coh}^\bullet{}_{\mathbb{X}^2_v}\,\mathbb{X}^3_\boxtimes \subset \mathrm{D}^\pi\mathbf{Fuk}^\pm\mathbb{Y}^3_\square$.

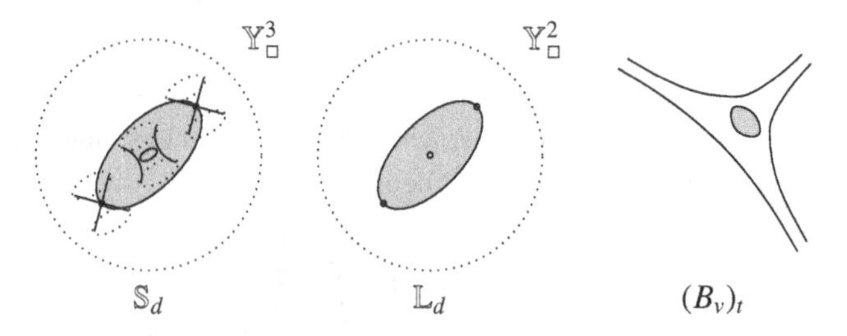

These ideas are worked out in detail in the Fano case by Seidel in [219] and in more generality by Abouzaid, Auroux and Katzarkov in [2].

8.5 Away from the Large Limit

The examples of mirror symmetry we have seen above are all instances of large volume limits. By this we mean that the areas as measured by the symplectic form do not come into play and hence we can rescale ω without changing the category. This also means that we can work over $\mathbb{C}$ instead of working over the Novikov field. On the B-side we have that the category only depends on the combinatorics of the toric variety.

The downside of the large volume limit is that the symplectic manifolds under consideration are necessarily noncompact. If we want to work with compact symplectic manifolds we have to complete our noncompact versions by adding in extra stuff. The easiest way to do this is to add in divisors and work with relative Fukaya categories.

Let us have a look at how this works for curves. In the large volume limit the symplectic variety is a punctured Riemann surface $\mathbb{Y}^1_\square$ determined by the zero locus of $W_t \colon \mathbb{C}^{*2} \to \mathbb{C}$. The Riemann surface is an exact Liouville manifold with cylindrical ends, one for each puncture or equivalently one for each infinite leg of the tropical curve. Each of these legs is determined by an elementary edge on the boundary of the Newton polygon of W_t. The easiest example to think of is the three-punctured sphere, which we can compactify to a projective line by embedding $X + Y + 1 = 0 \subset \mathbb{C}^{*2}$ in the projective plane.

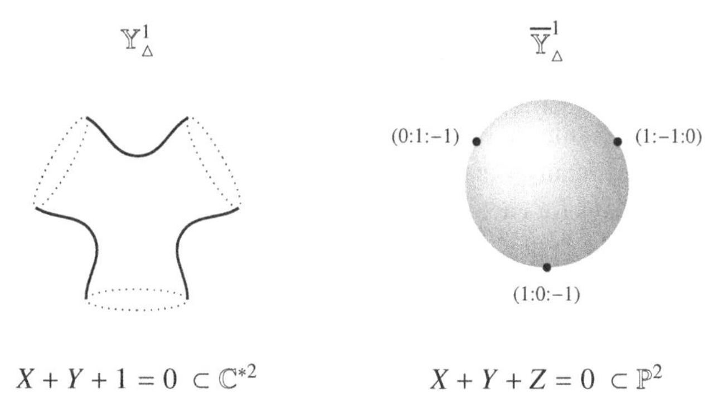

How can we do this for a general $\mathbb{Y}^1_\square$? The idea is to compactify $\mathbb{C}^{*2}$ to a toric variety $\mathbb{V}$ and look at the closure of $\mathbb{Y}^1_\square \subset \mathbb{C}^{*2} \subset \mathbb{V}$. The function W_t is a linear combination of monomials that correspond to the lattice points of the Newton polygon $\square$. We can interpret these monomials as sections of a line bundle $\mathscr{O}(d)$ over a toric variety whose ray vectors are inward-pointing normals to the sides of the Newton polygon. This variety, or an appropriate resolution, is the compactification we are looking for.

Example 8.21 Let us consider two examples:

(i) $W_t = t + X + Y + X^{-1}Y^{-1}$. Its Newton polygon is a triangle spanned by $m_1 = (1,0)$, $m_2 = (0,1)$, $m_3 = (-1,-1)$, and the inward-pointing normals to the edges are $n_1 = (2,-1)$, $n_2 = (-1,2)$, $n_3 = (1,1)$. Construct the toric variety $\mathbb{V}$ whose two-dimensional cones are $\mathbb{R}_{\geq 0}n_i + \mathbb{R}_{\geq 0}n_j$. This variety is unfortunately not smooth but it allows a resolution if we subdivide the cones. This results in a toric surface $\tilde{\mathbb{V}}$ such that $\tilde{\mathbb{V}} \setminus \mathbb{C}^{*2}$ consists of nine $\mathbb{P}^1$'s (one for each ray vector) that intersect each other in a cyclic way.

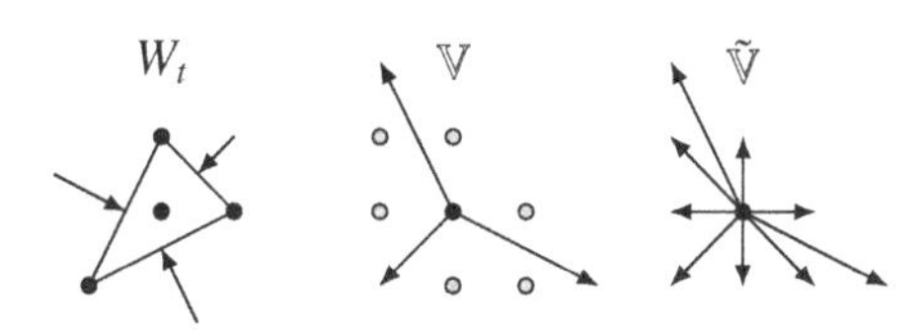

The function W_t can be seen as a section of $\mathscr{O}_{\mathbb{V}}(\mathbb{1})$ where $\mathbb{1}$ maps every ray vector n_i to 1, but also as a section of $\mathscr{O}_{\tilde{\mathbb{V}}}(\mathbb{1})$. For generic W_t, the zero locus $\bar{\mathbb{Y}}^1_\square$ intersects the three $\mathbb{P}_1$'s corresponding to the n_i at one point p_i, but it does not intersect the six extra $\mathbb{P}^1$'s.

(ii) For $W_t = t+X^2Y^{-1}+X^1Y^{-1}+Y^{-1}+X^{-1}Y^{-1}+X^{-1}+X^{-1}Y+X^{-1}Y^2+Y+X$, now the Newton polygon has three normal directions $m_1 = (1,0)$, $m_2 = (0,1)$, $m_3 = (-1,-1)$ and the corresponding toric variety is just $\mathbb{P}^2$, which is already smooth. There are three $\mathbb{P}^1$'s in this surface, and the compactification of $W_t^{-1}(0)$ intersects each $\mathbb{P}^1$ at three points because the corresponding equation is cubic or, equivalently, the tropical curve has three legs for each edge.

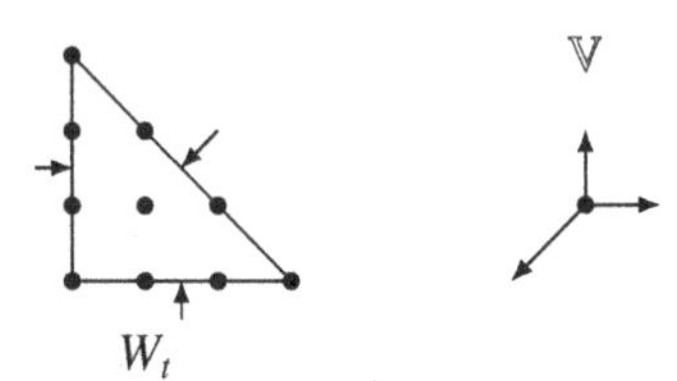

We can deform the wrapped Fukaya category by adding a divisor $\mathbb{D}$ that consists of one point for each cylindrical end, but as we saw in (ii), some cylindrical ends end on the same divisors in the compactification of $\mathbb{C}^{*2}$. This suggests that we should use the same deformation coefficient for these ends and this gives us a relative Fukaya category ${}^{\mathrm{cu}}\mathbf{Fuk}^{\pm}(\overline{\mathbb{Y}}^1_\square, \mathbb{D})$ over the ring $\mathbb{C}[\![q_1,\dots,q_r]\!]$, where r is the number of sides of the Newton polygon.

To get a mirror correspondence we also need to deform the B-model. Recall that the mirror of the punctured Riemann surface is given by a Landau–Ginzburg model $(\mathbb{X}^3_\square, \ell)$, where ℓ is the global function $X^m \in \mathscr{O}(\mathbb{X}^3_\square)$ with $m = (0,0,1) \in M$. The easiest way to deform this B-model is to add extra monomials to ℓ. These have to be taken from the global sections of the structure sheaf:

$$\mathscr{O}(\mathbb{X}^3_\square) = \mathbb{C}\{X^m \mid \forall\, u \in \square,\ \langle m, (u,1)\rangle \geq 0\}.$$

The monomials in $\mathscr{O}(\mathbb{X}^3_\square)$ are the integral points of the dual cone $\sigma^\vee_\square$. The ray vectors of this cone are dual to the faces of $\sigma_\square$, which are in one-to-one correspondence with the edges of $\square$. Therefore, the dual cone is spanned by vectors of the form $e_i = (n_i, c_i)$, where n_i is an inward-pointing normal vector to an edge of $\square$ and $c_i \in \mathbb{Z}$, which depend on the location of $\square$ in $\mathbb{R}^2$. These ray vectors are the "extremal monomials" that we can add. As a first attempt we can look at the Landau–Ginzburg model

$${}^{\mathrm{cu}}\mathtt{Tw}(\mathbb{X}^3_\square, \ell_q) \quad \text{with } \ell_q = \ell + \sum q_i X^{e_i}.$$

This gives a curved category over the ring $\mathbb{C}[[q_1, \ldots, q_r]]$, where r is the number of sides of the polygon $\square$. Note that this is precisely the number of parameters we used for the deformation of the A-model. Therefore, we expect a correspondence of these deformations, which leads to derived equivalences between the uncurved sectors. Of course we cannot expect a direct correspondence between the parameter spaces of both sides, so we expect that there is an automorphism α of $\mathbb{C}[[q_1, \ldots, q_r]]$ that allows us to translate between them.

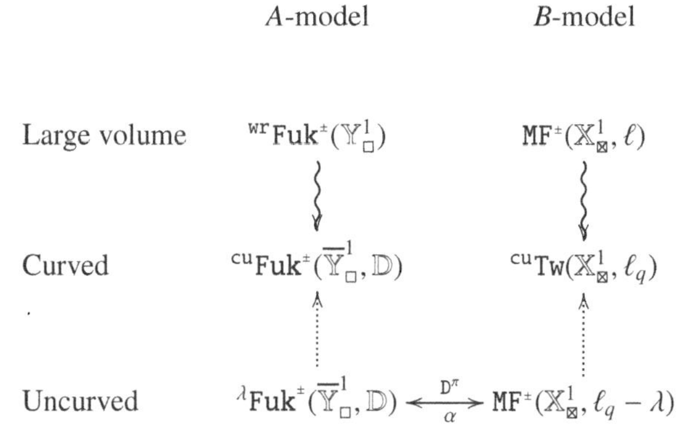

If we choose a specific symplectic structure ω on $\bar{\mathbb{Y}}^1_{\square}$, then we can work with the full Fukaya category over the Novikov ring on the A-side. In that case, we expect that there is a morphism $\beta\colon \mathbb{C}[[q_1, \ldots, q_r]] \to \mathbb{A}$ that allows us to make the translation. These heuristics have been partially worked out in several specific cases.

The Genus 2 Curve

Take the ring $\mathbb{C}[X, Y, Z]$ with the group action by $G = \langle g \mid g^5 \rangle$ such that $g(x, y, z) = (\zeta x, \zeta y, \zeta^3 z)$, with ζ a fifth root of unity. To determine the invariants, let us first look at $\mathbb{C}[X^{\pm 1}, Y^{\pm 1}, Z^{\pm 1}]^G$. This ring is freely generated by $U = XY^{-1}$, $V = X^2Y^3$, $W = XYZ$ and $U^iV^jW^k \in \mathbb{C}[X, Y, Z]$ if

$$i + 2j + k \geq 0, \quad -i + 3j + k \geq 0, \quad k \geq 0.$$

This implies that the ring of invariants is isomorphic to the affine toric variety spanned by the cone $\sigma = \mathbb{R}_{\geq 0}(1, 2, 1) + \mathbb{R}_{\geq 0}(-1, 3, 1) + \mathbb{R}_{\geq 0}(0, 0, 1)$. Therefore, $\mathbb{A}^3 /\!\!/ G \cong \mathbb{X}^3_{\square}$ with Newton polygon shown below.

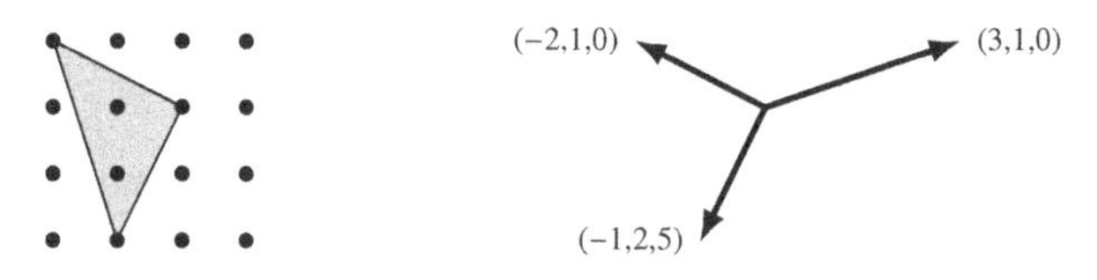

This is the same as $\mathbb{X}^3_\square$ for the Newton polygon with corners $(1,2)$, $(-1,3)$, $(0,0)$. If we fix any triangulation of this polygon we can reinterpret the $\mathbb{X}^3_\boxtimes$ as a resolution of the singular space $\mathbb{A}^3/\!\!/G$. On the other hand, the A-model $\mathbb{Y}^1_\square$ is a genus 2 curve with three punctures because the polygon has two internal lattice points and three on the boundary.

Filling in the punctures on the A-side corresponds to adding monomials according to the rays of the dual cone $\sigma^\vee$. These are $U^3V = X^5$, $U^{-2}V = Y^5$ and $U^{-1}V^{-2}W^5 = Z^5$, and hence we can see the potential as the invariant function $\ell_q = XYZ + q_1X^5 + q_2Y^5 + q_3Z^5$. On both the A-side and the B-side the parameters do not occur in infinite power series. This means that the mirror correspondence can be formulated in terms of ordinary $\mathbb{C}$-linear categories.

Theorem 8.22 (Seidel) *Let Σ_2 be a genus 2 curve viewed as a symplectic manifold and let $\tilde{\mathbb{X}}$ be a crepant resolution of $\mathbb{A}^3/\!\!/\mathbb{Z}_5$. Then we have the following equivalence:*

$$\mathtt{D}^\pi\mathtt{Fuk}^\pm(\Sigma_2) \stackrel{\infty}{=} \mathtt{D}^\pi\mathtt{MF}^\pm(\tilde{\mathbb{X}}, XYZ - X^5 - Y^5 - Z^5).$$

Sketch of the proof The theorem is proved by finding a generator whose endomorphism ring is a deformation of the exterior algebra Λ_3. By classifying the deformations of the exterior algebra one can show that these deformations are equivalent. Instead of $\tilde{\mathbb{X}}$ one can also work over the stack $[\mathbb{A}^3/\mathbb{Z}_5]$. For more information see [220]. □

Remark 8.23 This theorem has been generalized by Efimov [71] to curves of genus $g \geq 3$ by looking at an action of $\mathbb{Z}_{2g+1}$ on $\mathbb{A}^3$.

Elliptic Curves

For elliptic curves we can try a similar strategy and look at the $\mathbb{Z}_3$-action on $\mathbb{A}^3$. In this case, we get some extra features. First of all, the elliptic curve is a Calabi–Yau manifold so we can upgrade the Fukaya category to a $\mathbb{Z}$-graded version. On the other hand, the polynomial ℓ_q is homogeneous of degree 3 so we can consider graded matrix factorizations instead of ordinary matrix factorizations. Orlov's result allows us to interpret the category of graded matrix factorizations $\mathtt{grMF}^\bullet(\mathbb{C}[X,Y,Z], \ell_q)$ as the derived category of coherent sheaves of the zero locus of ℓ_q inside $\mathbb{P}^2$. This is again an elliptic curve $\mathbb{E}^\vee$. Therefore we should get a correspondence

$$\mathtt{D}^\pi\,\mathtt{Fuk}^\bullet(\mathbb{E}, \mathbb{D}) \stackrel{\infty}{=} \mathtt{grMF}^\bullet(\mathbb{C}[X,Y,Z], \ell_q) \stackrel{\infty}{=} \mathtt{D}\,\mathtt{Coh}^\bullet\,\mathbb{E}^\vee.$$

Unlike in the higher genus curve setting, there are infinitely many disks on the A-side, so we get power series in the deformation parameters.

The fact that the mirror of an elliptic curve is again an elliptic curve has been worked out in detail in many settings. Zaslow and Polishchuk were the first to give a description of the correspondence between the objects in the derived category of coherent sheaves and the Fukaya category [205]. Lekili and Perutz [162, 161] worked out the case of a relative Fukaya category with a single divisor and Lekili and Polishchuk for a divisor consisting of multiple points [164]. Their calculations start from a different representation of the elliptic curve from the one described above. The correspondence for the $\mathbb{E}^\vee \subset \mathbb{P}^2$ and noncommutative generalizations can be found in the work of Cho, Hong and Lau [55, 154].

Polar Duality

There is also an intriguing kind of symmetry between the A- and B-side models for the elliptic curve above. If we look at Example 8.21, we see that the elliptic curve $\mathbb{E} = \overline{\mathbb{Y}^1_\square}$ is given by the zero locus of W_t inside the compactification $\tilde{\mathbb{V}}$ from example (i), while the $\mathbb{E}^\vee$ is the zero locus of ℓ_q inside $\mathbb{P}^2$, which is in fact an example of the compactification in example (ii), so both W_t and ℓ_q cut out an elliptic curve in a toric variety coming from a fan associated to polygon with one internal lattice point.

Definition 8.24 A Newton polygon $\square$ is called *reflexive* if it has one internal lattice point $0 \in \mathbb{Z}^2 = M$. This implies that the edges of $\square$ are given by $\langle m, u\rangle + 1 = 0$, where u is a primitive vector in $\mathrm{Hom}(M, \mathbb{Z}) = M^\vee = N$. The *polar dual polygon* $\diamond \subset N_\mathbb{R}$ is the convex hull of these vectors. Note that because of the symmetry between u and m, the dual polygon $\diamond$ is again reflexive.

Theorem 8.25 *Up to a transformation of* $\mathsf{GL}_2(\mathbb{Z})$*, there are 16 reflexive polygons, which are listed below. The first six columns contain dual pairs, while the last two columns are self-dual (up to a linear transformation).*

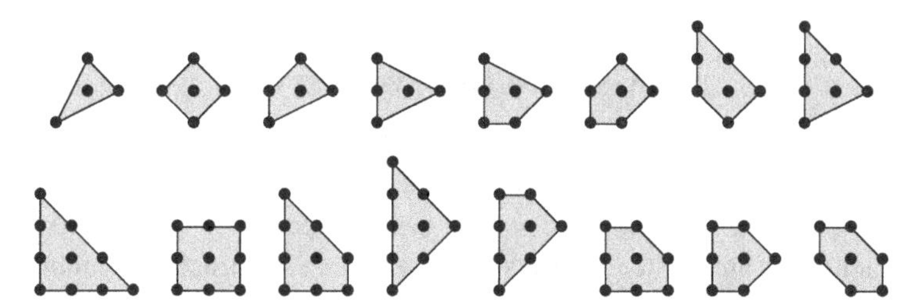

We start with a reflexive polygon $\square$. If we connect $v = (0,0)$ with the corners of this polygon we get the fan of a singular Fano variety, which we can resolve by connecting v to the other lattice points on the boundary. The resolution $\mathbb{V} = \mathbb{X}_v^2$ is a weak Fano variety, which is a slight generalization of a Fano variety. The global sections of the line bundle $\mathcal{O}_{\mathbb{V}}(\mathbb{1})$ are linear combinations of monomials that sit inside the dual polygon $\diamond$. If we pick a generic global section $W \in \mathcal{O}_{\mathbb{V}}(\mathbb{1})(\mathbb{V})$ then its zero locus defines an elliptic curve $\mathbb{E} \subset \mathbb{V}$, which is a one-dimensional Calabi–Yau variety. If we do the same trick for the dual polygon we get another weak Fano $\mathbb{V}^\vee$ with a section $W^\vee$, whose zero locus is a second elliptic curve $\mathbb{E}^\vee \subset \mathbb{V}^\vee$. For suitable symplectic structures and choices of global sections we expect to get a derived equivalence between the Fukaya category of $\mathbb{E}$ and the derived category of coherent sheaves of $\mathbb{E}^\vee$ and vice versa. If $\square$ is the first reflexive polygon from the list we get the correspondence we described in the previous subsection. For the second polygon, we get mirror symmetry for an elliptic curve embedded in $\mathbb{P}^1 \times \mathbb{P}^1$, which is worked out in [154].

The general idea was first proposed by Batyrev in [22]. Any reflexive polytope of dimension n defines an n-dimensional weak Fano variety with a line bundle. A generic section of this line bundle will define a hypersurface which is a Calabi–Yau manifold. The mirror of this Calabi–Yau manifold comes from the same construction with the dual polytope.

This proposal has been studied from many angles and generalized in many directions [24, 25]. In the classical approach to mirror symmetry, it has been used to confirm the conjecture about the Hodge numbers for important classes of mirror pairs [22, 23, 56]. From a homological mirror symmetry point of view, Seidel [221] worked out the case for a quartic hypersurface in $\mathbb{P}^3$ (the reflexive polygon is spanned by the basic coordinate vectors in $\mathbb{R}^3$ and $(-1,-1,-1)$. One dimension higher, the quintic hypersurface in $\mathbb{P}^4$, is the example that started off mirror symmetry in the first place [51]. The homological version in this case was proven by Nohara and Ueda in [194] and in all dimensions by Sheridan in [223].

LG–LG Mirror Symmetry

Instead of getting a correspondence between two Calabi–Yau varieties, we can also generalize this setting to obtain mirror symmetry between two Landau–Ginzburg models. Let $a = (a_{ij})$ be an invertible $(n \times n)$-matrix with coefficients that are positive integers. From a we can construct a polynomial in $R = \mathbb{C}[X_1, \dots, X_n]$,

$$W_a = \sum_i X_1^{a_{i1}} \dots X_n^{a_{in}}.$$

Each monomial in W_a is a character on the torus $\mathbb{T} \subset \mathbb{A}^n = \mathbb{V}(R)$. Inside $\mathbb{T}$ we have a finite subgroup G_a that fixes W_a and we take a group $\Gamma \subset G_a \subset \mathbb{T}$.

On the dual side we consider the polynomial ring $R^\vee = \mathbb{C}[Y_1, \dots, Y_n]$ and the dual polynomial

$$W_a^\vee = \sum_i Y_1^{a_{1i}} \dots X_n^{a_{ni}}.$$

The finite subgroup $G_a^\vee$ that fixes $W_a^\vee$ can naturally be seen as $\mathtt{Grp}(G_a, \mathbb{C}^*)$ and we denote the annihilator of Γ by $\Gamma^\vee$. This procedure gives us two stacky Landau–Ginzburg models

$$([\mathbb{A}_X^n/\Gamma], W_a) \quad \text{and} \quad ([\mathbb{A}_Y^n/\Gamma^\vee], W_a^\vee),$$

which should be mirror to each other. Note that just like before, each monomial in the potential corresponds to a toric divisor in $\mathbb{A}^n$ on the other side, but now an extra orbifold action is included to account for the fact that these monomials can be multiples of ray vectors.

The simplest case happens when $n = 1$ and Γ is trivial; then $W_a = X^a$, $W_a^\vee = Y^a$ and we get

$${}^{\mathrm{dr}}\mathtt{Fuk}^{\pm}(\mathbb{C}, X^a) \overset{\infty}{=} \mathtt{MF}^{\pm}(\mathbb{C}[Y] \star \mathbb{Z}_a, Y^a).$$

The left-hand side can be reinterpreted as the partially wrapped Fukaya category of $\mathbb{C}$ with a stops going to infinity or, equivalently, the disk with a points on the boundary. From this point of view we recover mirror symmetry for the linear quiver from Section 4.2.1.

This idea can also be used to study mirror symmetry for the cusp singularity $(\mathbb{A}^2, X^3 + Y^2)$ (see [161], and more generally [107]) and related instances of this duality have been studied by Ueda [247] and Lekili [169]. These stacky models can also be used in combination with arbitrary toric varieties [59].

8.6 Mirrors and Fibrations

We have seen how we can use the duality between tropical and toric geometry to find mirror pairs and prove instances of homological mirror symmetry. Historically this approach to the phenomenon is a late development. When mirror symmetry was first developed, it was a duality stated only in terms of compact Calabi–Yau manifolds, and it related numerical invariants that count pseudo-holomorphic maps to certain complex integrals on the mirror manifold. The relation between these invariants was originally formulated as a correspondence of generating functions and only later was it reinterpreted as an equivalence of categories.

To describe the relation between two mirror manifolds, Strominger, Yau and Zaslow [237] used T-duality (an idea from string theory) to formulate a conjecture on how mirror pairs are related, which we will discuss in below. For some mathematical takes on T-duality we refer to [19, 77].

8.6.1 SYZ-Fibrations

Recall that an n-dimensional Calabi–Yau Kähler manifold $\mathbb{M}$ comes with three structures: a complex structure $J\colon \mathsf{T}\mathbb{M} \to \mathsf{T}\mathbb{M}$, a symplectic structure $\omega \in \Omega^2\mathbb{M}$ and a complex volume form $\Omega \in \Omega^n_{\mathbb{C}}\mathbb{M}$.

Definition 8.26 A Lagrangian submanifold $\mathbb{L} \subset \mathbb{M}$ is called *special* with phase θ if the complex volume form $e^{-i\theta}\Omega$ restricts to a real volume form on $\mathbb{L}$. A *special Lagrangian torus fibration* is a map $\pi\colon \mathbb{M} \to \mathbb{B}$ such that each fiber is a special Lagrangian torus with the same phase.

Example 8.27 Consider the complex manifold $\mathbb{M} = \mathbb{C}/(\mathbb{Z} + i\mathbb{Z})$ with $\omega = dx \wedge dy$ where $z = x + iy$ is the standard coordinate on $\mathbb{C}$. For the complex volume form we can take $\Omega = dz$. This one-dimensional compact Calabi–Yau manifold has an obvious special Lagrangian torus fibration over the base $\mathbb{B} = \mathbb{R}/\mathbb{Z}$:

$$\pi\colon \mathbb{M} \to \mathbb{B}: x + iy \mapsto y.$$

The fibers of this map are horizontal circles, so the phase of the fibration is zero. If we project on the x-coordinate we get a fibration with phase $\pi/2$. More generally, for every $A = \left(\begin{smallmatrix} a & c \\ b & d \end{smallmatrix}\right) \in \mathsf{GL}(2, \mathbb{Z})$, the map $\pi \circ A^{-1}$ will be a special Lagrangian torus fibration with phase $\arg(a + bi)$.

We can also view this example in a different way. Take the circle $\mathbb{S}^1$ and look at the cotangent space $\mathsf{T}^*\mathbb{S}^1$. This space is a trivial vector bundle so it is possible to define a lattice $\Lambda^*_p \subset \mathsf{T}^*_p\mathbb{S}^1$ in each fiber. The cotangent bundle comes with a natural symplectic structure, which transfers naturally to the quotient bundle $\mathbb{M} = \mathsf{T}^*\mathbb{S}^1/\Lambda^*$. The special Lagrangians are precisely the fibers $\mathbb{L}_p = \mathsf{T}^*_p\mathbb{S}^1/\Lambda^*_p$.

On the other hand, the tangent bundle $\mathsf{T}\mathbb{S}^1$ comes with a natural complex structure and the lattice $\Lambda^*_p \subset \mathsf{T}^*_p\mathbb{S}^1$ defines a dual lattice $\Lambda_p \subset \mathsf{T}_p\mathbb{S}^1$. The dual of a lattice in a vector space V consists of the covectors $v \in V^*$ that map the lattice to $\mathbb{Z}$. If the lattice has a basis then the dual lattice is generated by the dual basis. If we quotient out the dual lattice Λ_p, we get a real torus $\mathbb{M}^\vee = \mathsf{T}_p\mathbb{S}^1/\Lambda_p$ and a pairing

$$\langle\, ,\, \rangle_p \colon \mathsf{T}_p^*\mathbb{S}^1/\Lambda_p^* \times \mathsf{T}_p\mathbb{S}^1/\Lambda_p \to \mathbb{R}/\mathbb{Z} \cong U_1.$$

The global result of this operation is a diagram

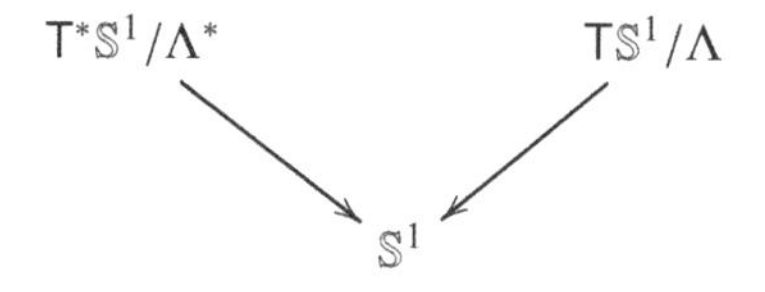

for which the fibers are pairwise dual.

As $\mathbb{M}$ and $\mathbb{M}^\vee$ form a mirror pair this suggests that mirror pairs should be related by dual Lagrangian fibrations. In general it is very hard to find special Lagrangian torus fibrations for compact Calabi–Yau manifolds [102, 101, 103], and in most cases they do not exist unless we relax the conditions on π. First of all, we should not demand that the base $\mathbb{B}$ is a smooth manifold and secondly we should not ask that every fiber is a special Lagrangian torus but only generically so [129]. This means that there is only a dense open subset of $\mathbb{B}$ where the fibers are dual special Lagrangian tori. For more background on special Lagrangian fibrations we refer to the lecture notes by Hitchin [113] and Joyce [128].

Conjecture 8.28 (Strominger–Zaslow–Yau [237]) *If two Calabi–Yau manifolds* $(\mathbb{M}, \mathbb{M}^\vee)$ *form a mirror pair then there exists a base* $\mathbb{B}$ *and maps*

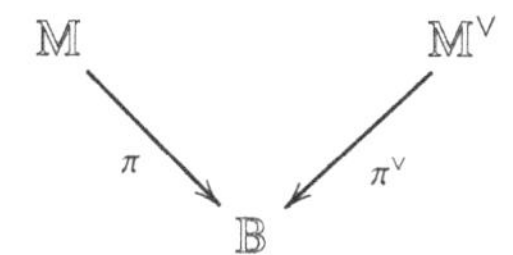

such that π, $\pi^\vee$ *are (possibly singular) special Lagrangian torus fibrations and for general points* $b \in \mathbb{B}$ *there are pairings* $\langle\, ,\, \rangle_b \colon \pi^{-1}(b) \times \pi^{\vee -1}(b) \to U_1$.

How does this conjecture relate to homological mirror symmetry? Remember that in the construction of the full Fukaya category, the objects of interest are Lagrangian submanifolds with a U_1-local system (see Remark 6.66). If $\mathbb{X}$ and $\mathbb{X}^\vee$ form a mirror pair then there is an equivalence $\mathtt{D}^\pi\mathtt{Fuk}(\mathbb{X}) \cong \mathtt{D}^\pi\mathtt{Coh}(\mathbb{X}^\vee)$. The skyscraper sheaves of $\mathbb{X}^\vee$ are objects in $\mathtt{D}^\pi\mathtt{Coh}(\mathbb{X}^\vee)$ with endomorphism ring equal to the exterior algebra. To first approximation the endomorphism ring of a Lagrangian submanifold $\mathbb{L}$ is equal to its cohomology ring, so we expect that the skyscraper sheaf will correspond to a Lagrangian *torus* with a local U_1-system.

This indicates that generically the moduli space of skyscraper sheaves of $\mathbb{X}^\vee$ should parametrize parallel Lagrangian tori on $\mathbb{X}$ with U_1-local systems. For each Lagrangian torus, the space of possible U_1-local systems is itself a real n-dimensional torus, so generically the space $\mathbb{X}^\vee$ is a union of n-dimensional real tori of skyscraper sheaves corresponding to the same Lagrangian torus. If we factor this out we get a map $\pi^\vee : \mathbb{X}^\vee \to \mathbb{B}$ for some n-dimensional space $\mathbb{B}$.

The hom-space between two skyscraper sheaves at different locations is zero, just like the hom-space between two nonintersecting Lagrangians. This suggests that the Lagrangian tori corresponding to two points in $\mathbb{B}$ do not intersect, and this n-dimensional family of n-tori will generically fill the whole of the $2n$-dimensional space $\mathbb{X}$. This gives rise to a projection map $\pi\colon \mathbb{X} \to \mathbb{B}$. The fiber $\mathbb{L}_b = \pi^{-1}(b)$ is a torus which we can write as $\mathbb{R}^n/\Lambda$ for a lattice $\Lambda \cong \pi_1(\mathbb{L}_b)$. The space of U_1 local systems on $\mathbb{L}_b$ is naturally equal to the dual torus:

$$\mathrm{Hom}(\Lambda, U_1) \cong \mathrm{Hom}(\Lambda, \mathbb{R}/\mathbb{Z}) \cong \frac{\mathrm{Hom}(\Lambda, \mathbb{R})}{\mathrm{Hom}(\Lambda, \mathbb{Z})} \cong \frac{\mathbb{R}^n}{\Lambda^*}.$$

This reasoning does not constitute a proof but gives an indication why, under certain conditions, the SYZ-conjecture might be true. For a different perspective on this see [245].

The SYZ-conjecture gives a nice geometrical rationale for the existence of mirror pairs but unfortunately it is not easy to prove nor is it clear whether it is true in full generality [130]. One of the reasons for this is that special Lagrangian submanifolds are in general very hard to construct, let alone families of them that cover almost the whole space. Furthermore, the base of the fibration is usually singular and it is not clear how we should construct the base and what kind of properties it must have.

In the Limit

To bypass these problems it is better to look at limiting situations, such as the large volume limit. In that case, we can use the duality between toric and tropical geometry. The complex torus has a natural Lagrangian torus fibration, $\log_t\colon \mathbb{T} \to \mathbb{R}^n$, and if $\mathbb{V}$ is a toric variety whose ray vectors are normal to the sides of a polytope $\square$, we complete this fibration to a moment map $\mu\colon \mathbb{V} \to \square$ such that $\mathbb{T}$ is mapped to the interior of $\square$ and the divisor of each ray vector to the corresponding face [89].

Now suppose that $W_t = 1 + tf$, where f is a generic function with reflexive Newton polytope $\square$. The fibers $W_t^{-1}(0) \subset \mathbb{C}^{*n}$ can be completed to a family $\bar{\mathbb{Y}}_t \subset \mathbb{V}$. The map $\mu|_{\bar{\mathbb{Y}}_t}$ is not a Lagrangian torus fibration for $t \neq 0$, but becomes one in the limit $t \to 0$, where the family degenerates to $\bar{\mathbb{Y}}_0 = \tilde{\mathbb{V}} \setminus \mathbb{C}^{*n}$, which is a union of toric varieties. It consists of a toric variety for every face of the polygon, which intersect if the faces meet. On the other hand, the polar dual gives rise to a second family $\bar{\mathbb{Y}}_t^\vee$, which can be viewed as mirror to the first. The limit is again a union of toric varieties, but now its intersection complex is dual to that of the first family because of the polar duality.

Example 8.29 If we start with the square with corners $(\pm 1, \pm 1)$, we get a family of elliptic curves in $\mathbb{P}^1 \times \mathbb{P}^1$ that degenerates to four $\mathbb{P}^1$'s. All curves have eight points in common: two on each $\mathbb{P}^1$ in $W_t^{-1}(0)$ coming from the zeros of f.

The polar dual is a square with corners $(\pm 1, 0)$ and $(0, \pm 1)$; this gives a family of elliptic curves inside a toric variety with four quotient singularities of order 2 (one for each corner of thc square). The family degenerates to four $\mathbb{P}^1$'s intersecting each other at the singular points of order 2. All curves have four points in common: one on each $\mathbb{P}^1$ coming from the zeros of f.

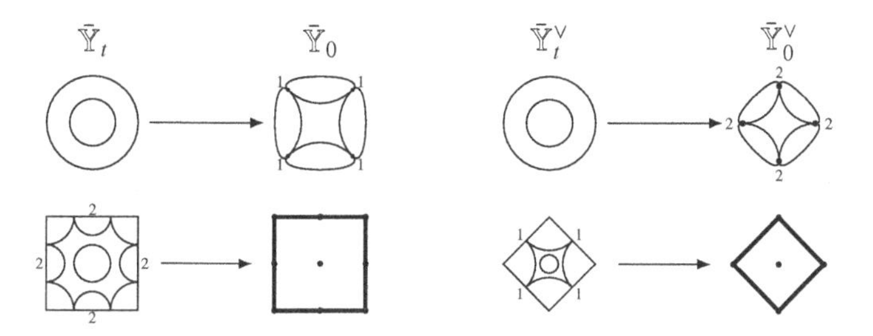

This idea is further explored and generalized in the Gross–Siebert research program. Basically, its aim is to study mirror symmetry through families of varieties that degenerate to unions of toric varieties that overlap in toric strata. The intersection complex of the degeneration, supplemented with some additional information, can then be used to reconstruct the mirror family via a discrete Legendre transform. This method uses many techniques from algebraic geometry, affine structures, log geometry and tropical geometry. For more information about this program we refer to [104, 105, 250].

8.7 Exercises

Exercise 8.1 Let $v_1, \ldots, v_4$ be primitive vectors in $N = \mathbb{Z}^2$ that divide the plane into four convex cones $\mathbb{R}_+ v_i + \mathbb{R}_+ v_{i+1}$ and let Δ be the corresponding fan. For each $d \in \mathbb{Z}^4$ we can define a line bundle $\mathcal{O}(d)$ on $\mathbb{X}_\Delta$. The Čech complex for $\mathcal{O}(d)$ can be graded by $M = N^\vee$. Give a geometrical way to calculate the sheaf cohomology for degree $m \in M$ in terms of the v_i and d_i.

Exercise 8.2 Let $\mathbb{X}$ be the complete toric variety corresponding to the toric diagram

Explain why you can see this as a projective plane blown up at two points or a $\mathbb{P}^1 \times \mathbb{P}^1$ blown up at one point. Find a tilting bundle for $\mathbb{X}$ which is a direct sum of five line bundles of the form $\mathcal{O}(d)$.

Exercise 8.3 Let $v_1, \dots, v_k \in \mathbb{Z}^2 \times \{1\}$ be the corners of a convex lattice polygon $\square$ placed at height 1 in $\mathbb{Z}^3$ and let $\mathbb{X}^3_\square$ be the toric variety corresponding to the cone $\sum \mathbb{R}_+ v_i$. Show that $\mathbb{X}^3_\square$ has an isolated singularity if and only if it has no extra lattice points on the edges and $\square$ is not an elementary triangle.

Exercise 8.4 Prove Pick's theorem, which states that the area of a lattice polygon is equal to $I + \frac{B}{2} - 1$, where I is the number of internal lattice points and B is the number of lattice points on the boundary.

Exercise 8.5 Fix two $(a, b) \in \mathbb{Z}^2$ and let $G = \langle g \mid g^n \rangle \cong \mathbb{Z}_n$ act on $\mathbb{C}^3$ by

$$g(x, y, z) = (e^{2a\pi i/n}x, e^{2b\pi i/n}y, e^{-2(a+b)\pi i/n}z).$$

Show that the quotient can be described as $\mathbb{X}^3_\square$ where $\square$ is a triangle with area $n/2$.

Exercise 8.6 Let $\mathbb{E}$ be an elliptic curve defined by the equation $Y^2 = X^3 + t^a X + t^b$. Describe how its tropicalization varies in terms of $a, b \in \mathbb{Z}$.

Exercise 8.7 In Section 8.4 we constructed mirrors in dimensions 1, 2, 3 for a lattice polygon. We can do something similar for an n-dimensional lattice polytope in dimensions $n-1, n, n+1$. Let us focus on $n = 1$. A one-dimensional lattice polytope is just a line segment L with length k, which we can put at height 1 in $\mathbb{Z}^2$.

(i) Show that $\mathbb{X}^{n+1}_\square$ is isomorphic to $\mathbb{V}(\mathbb{C}[X, Y, Z]/(XY - Z^k))$. Describe the exceptional fiber of the resolution $\mathbb{X}^{n+1}_\boxtimes \to \mathbb{X}^{n+1}_\square$.

(ii) The $(n-1)$-dimensional mirror correspondence predicts an equivalence

$$\mathtt{D}^\pi\mathtt{MF}^\pm(\tilde{\mathbb{X}}^{n+1}_\boxtimes, \ell) \cong \mathtt{D}^{\pi\mathrm{wr}}\mathtt{Fuk}^\pm\mathbb{Y}^{n-1}_\square.$$

Show that this works if you interpret the wrapped Fukaya category of a point as $\mathtt{vect}^\pm\mathbb{C}$.

(iii) For the n-dimensional version, we assume L has one internal lattice point. Show that this leads to a mirror correspondence that relates the derived category of $\mathbb{P}^1$ to a directed Fukaya category and give a strategy to prove it.

(iv) For the $(n + 1)$-dimensional version, describe the derived category of sheaves on $\mathbb{X}_{\boxtimes}^{n+1}$ supported on $\mathbb{P}^1$. Show that it is equivalent to the derived category of finite-dimensional modules of $\mathbb{C}[[X, Y]] \star \mathbb{Z}_2$ where the action on both X, Y is the sign action.

Exercise 8.8 Adapt the procedure for finding a mirror for the genus g curve to make it work for any lattice triangle with g internal lattice points and no lattice points on the boundary.

(i) Show that for $g = 2$ this always gives the same construction as Seidel.

(ii) Show that for $g = 3$ there is more than one possibility (i.e. different triangles up to an automorphism of the lattice). Which corresponds to Efimov's construction?

(iii) Work out the example for $g = 3$ for a triangle that does not correspond to Efimov's construction. Describe the group action and determine the different possible resolutions of the quotient singularity.

Exercise 8.9 The space $\mathbb{P}^1 \times \mathbb{P}^1$ can be described using homogeneous coordinates $((X : Y), (U : V))$. Show that the zero locus of $XYUV + q(X^2 + Y^2)(U^2 + V^2)$ defines an elliptic curve for generic q. Interpret this construction from the point of view of Batyrev duality.

Exercise 8.10 Work out LG–LG mirror symmetry for the simple singularity $\mathbb{X} := (\mathbb{A}^n, X_1^2 + \cdots + X_n^2)$. Determine the mirror $\mathbb{Y}$ and prove mirror symmetry in both directions (the A-model of $\mathbb{X}$ is equivalent to the B-model of $\mathbb{Y}$ and vice versa).

PART THREE

REFLECTIONS ON SURFACES

9
Gluing

In this chapter we will introduce an A_∞-category that acts as a model for the partially wrapped Fukaya category of a surface. The advantage of this *topological Fukaya category* is that it can be constructed combinatorially without the need to address the many technical issues that plague the definition of a genuine Fukaya category.

Instead, the category is simply defined as the derived category of a certain A_∞-algebra that comes from a subdivision of the surface into polygons. This A_∞-algebra is closely related to some special types of algebras that have been studied extensively by representation theorists: biserial algebras [88, 234] and gentle algebras [14, 208]. The Fukaya interpretation of this category offers a new perspective on the representation theory of these algebras in terms of curves on a surface [166].

It is also possible to construct these categories by gluing smaller pieces together in different ways [115, 65, 204, 189, 232]. We will study this phenomenon and illustrate it with several examples. These gluing methods can also be used to construct B-side mirrors for these categories in the form of Landau–Ginzburg models [204, 155]. This is an example of a broader program initiated by Kontsevich [144] to view the wrapped Fukaya category as a sheaf of categories over a Lagrangian skeleton.

Several versions of these categories have appeared in the literature [3, 156, 115, 65, 204, 36, 166, 155], using different conventions and restrictions on the types of surfaces that are allowed and whether one works in a $\mathbb{Z}_2$-graded or $\mathbb{Z}$-graded setting. In this chapter we will focus on the $\mathbb{Z}_2$-graded version, and the jump to the $\mathbb{Z}$-graded setting will be reserved for the next chapter.

9.1 Marked Surfaces

Definition 9.1 A *marked surface* $(\mathbb{S}, M)$ consists of a compact oriented surface $\mathbb{S}$, possibly with boundary components $\mathbb{B}_1, \ldots, \mathbb{B}_k$, and a finite collection of points $M \subset \mathbb{S}$ such that each connected component and each boundary component contains at least one point. These points are called *boundary or internal marked points*, depending on whether they lie on the boundary or not.

Definition 9.2 An *arc* is an embedded curve $a\colon [0,1] \to \mathbb{S}$ between marked points that does not meet marked points in between (i.e. $a^{-1}(M) = \{0,1\}$) and does not enclose a disk without internal marked points.[1] Two arcs are *isotopic* if there is a one-parameter family of arcs between them. Two arcs are *nonintersecting* if they do not intersect except at endpoints.

Definition 9.3 A collection of nonisotopic and nonintersecting arcs is called *generating* if every other arc can be isotoped to a sequence of connected arcs in the collection (disregarding orientation). We also say that these arcs *split the surface.*

Remark 9.4 If we cut the surface along a generating collection then the surface will split into polygons whose corners are marked points and whose edges are either arcs or pieces of the boundary. Note that for each polygon, all or all but one edge must be arcs. If this were not the case, a diagonal that separates two edges that are not arcs cannot be isotoped to a sequence of arcs. If every edge of the polygon is an arc then we call the polygon *closed*; if one edge is not an arc then we call the polygon *open.*

To a generating collection of arcs $\mathcal{A}$ we can associate a quiver $Q_{\mathcal{A}}$. The vertices of this quiver are the arcs, $(Q_{\mathcal{A}})_0 = \mathcal{A}$, and the arrows are the angles in the polygons between these arcs. By this we mean the following: draw a small circle around a marked point (or a half circle if the marked point is on the boundary). The arcs will cut this circle into small segments that we orient anticlockwise. For each segment that connects two arcs, we draw an arrow α in the quiver $Q_{\mathcal{A}}$.

Example 9.5 Below are three generating collections for the disk with six marked points on the boundary and two for an annulus with three marked points. We also draw the corresponding quivers.

[1] In other words, we do not want contractible arcs.

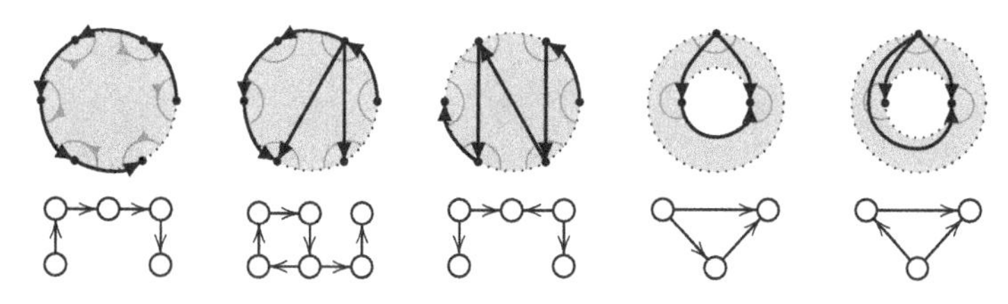

For each arc system, the quiver $Q_{\mathcal{A}}$ has a special property: there are at most two arrows leaving and two arrows arriving at each vertex. This means that for each arrow β there are at most two paths of length 2 starting with β. We call $\alpha\beta$ *angle-like* if they are consecutive angles in a circle around a marked point, and we call $\alpha\beta$ *face-like* if the two angles are consecutive angles in a polygon. We denote the set of angle-like paths by P_+ because they turn anticlockwise around a marked point and the set of face-like paths by P_- because they turn clockwise around a face.

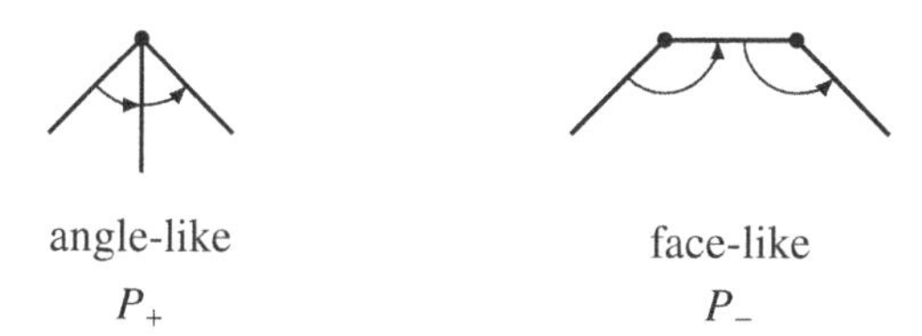

Definition 9.6 The *gentle A_∞-algebra* of an arc collection is the path algebra of $Q_{\mathcal{A}}$ divided by the ideal generated by the face-like paths:

$$\mathtt{Gtl}^{\pm}\mathcal{A} = \frac{\mathbb{C}Q_{\mathcal{A}}}{\langle \alpha\beta \mid \alpha\beta \in P_- \rangle}.$$

We give the algebra a $\mathbb{Z}_2$-grading using the following rule:

On $\mathtt{Gtl}^{\pm}\mathcal{A}$ we put an A_∞-structure defined by the rule that

$$\mu(\alpha\beta_1, \ldots, \beta_k) = \alpha \quad \text{and} \quad \mu(\beta_1, \ldots, \beta_k\gamma) = (-1)^{|\gamma|}\gamma$$

if $\beta_1, \ldots, \beta_l$ are the consecutive angles of an immersed polygon without internal marked points bounded by arcs.

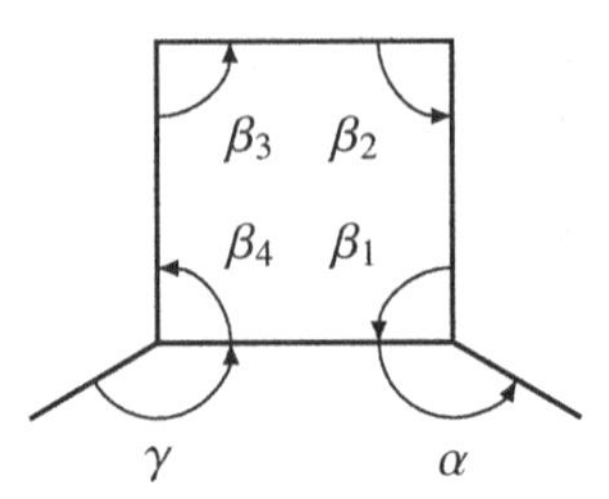

Lemma 9.7 ([115, 36]) *The rule above defines an A_∞-structure.*

Sketch of the proof First note that the product has the correct $\mathbb{Z}_2$-degree: if $\alpha_1, \ldots, \alpha_k$ sit in a polygon and all arcs have the same orientation around that polygon then all α_i have degree 1. If we delete these then the degree changes by k, which is the same as $2 - k$ in $\mathbb{Z}_2$. If we change the orientation of one of the arcs, the degrees of two α's change by 1 so the total degree is the same modulo 2.

Suppose that $\mu(\alpha_1, \ldots, \alpha_k, \mu(\alpha_{k+1}, \ldots, \alpha_l), \ldots, \alpha_n) \neq 0$; then we can distinguish four cases depending on whether the inner and outer μ are binary or not.

- If they are both binary this is just the associativity rule for the ordinary product.
- If the outer is binary we get $\mu(\alpha_1, \mu(\alpha_2\beta_1, \ldots, \beta_k)) - \mu(\mu(\alpha_1, \alpha_2\beta_1), \ldots, \beta_k) = 0$.

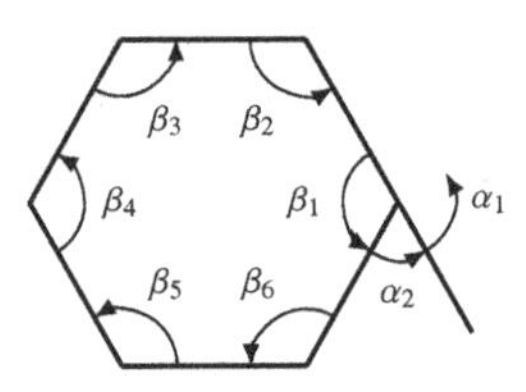

- If the inner is binary then the arc that sits in the middle of the inner splits the immersed polygon in two. This results in a nonzero term of two higher products that cancels it.
- If both are higher, these polygons glue together into a bigger immersed disk. This results in a nonzero term with an inner binary product that cancels it:

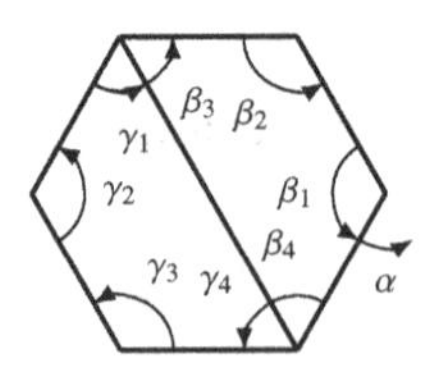

$$\mu(\alpha\beta_1, \beta_2, \mu(\beta_3\gamma_1, \gamma_2, \gamma_3, \gamma_4), \beta_4) - \mu(\alpha\beta_1, \beta_2, \beta_3\gamma_1, \gamma_2, \gamma_3, \mu(\gamma_4, \beta_4)) = 0.$$

In each case there are precisely two nonzero terms that cancel. □

Lemma 9.8 *Let $\mathcal{A}$ be an arc collection that splits the surface and* $(\mathtt{Gtl}^{\pm}\mathcal{A}, \mu)$ *its gentle A_∞-algebra.*

(i) *The higher products are trivial if and only if there are no closed polygons.*
(ii) *The gentle algebra is finite-dimensional if and only if there are no internal marked points.*

Sketch of the proof Every nonzero higher product comes from an immersed polygon that is bounded by arcs. This polygon must be made by gluing closed polygons together. The nonzero paths in the gentle algebra correspond to angles around the marked points. If a marked point is internal, these angles can turn around indefinitely. □

Example 9.9 Take a disk with k interior marked points and l marked points on the boundary. Connect one interior marked point to all other marked points. The quiver will be a cyclic quiver with loops on the vertices coming from arcs that connect two interior points. All arrows have degree 0 and the product between a loop and another arrow is zero. The gentle algebra is infinite but there are no higher products.

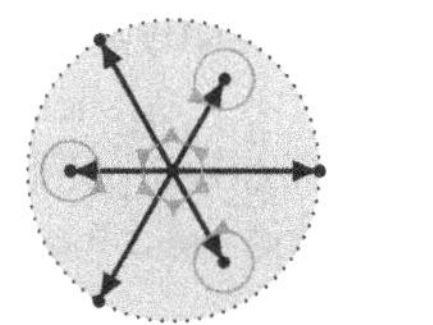
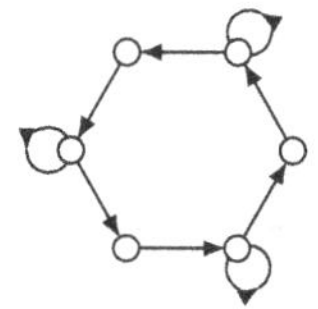

If we consider an annulus with k marked points on the inner circle and k marked points on the outer circle, we can connect them by arcs in a zigzag fashion. This gives a cyclic quiver of which the orientation of the arrows alternate. In this case, there are no relations nor higher products, so the gentle algebra is just the path algebra.

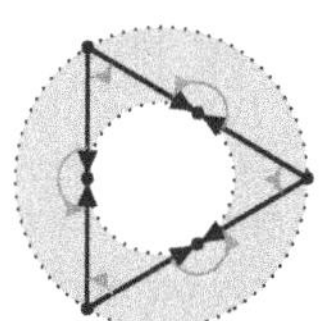
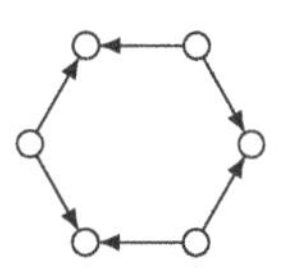

Remark 9.10 Originally, gentle algebras were introduced by Assem and Skowronski in [14]. It referred to a set of (ungraded) path algebras with relations $\mathbb{C}Q/I$ that satisfy the following conditions:

(G1) Q has no oriented cycles and every vertex has at most two incoming and at most two outgoing arrows.
(G2) $\mathcal{I}$ is generated by paths of length 2.
(G3) For each arrow $\alpha \in Q_1$ there is at most one path $\alpha\beta$, and at most one path $\gamma\alpha$ that belongs to $\mathcal{I}$.
(G4) For each arrow α there is at most one path $\alpha\beta$, and at most one path $\gamma\alpha$ that does not belong to $\mathcal{I}$.

These algebras were introduced because they have nice representation-theoretic properties that make the classification of their indecomposable representations possible. The connection with mirror symmetry was only discovered later [3, 36, 115, 166].

Theorem 9.11 (Lekili–Polishchuk[2]) *Every gentle algebra in the sense above comes from a surface.*

Sketch of the proof We construct the surface in steps.

(i) Consider all longest sequences $\alpha_1 \dots \alpha_k$ such that $\alpha_i\alpha_{i+1} \in \mathcal{I}$. For each such sequence we consider a $(k+2)$-gon and we label each edge except the last one by a vertex of Q as follows: $h(\alpha_1)$, $h(\alpha_2)$, …, $h(\alpha_k)$, $t(\alpha_k)$. Draw the arrows as small angles around the corners of the polygon. Note that by (G3), each arrow sits in exactly one polygon.
(ii) Consider all the longest sequences $\beta_1 \dots \beta_k$ such that $\beta_i\beta_{i+1} \notin \mathcal{I}$. For each such sequence we consider a marked point m_i and we label the corner of each polygon by the marked point whose sequence contains that angle.
(iii) Glue the polygons together along edges and corners with the same label. This gives an orientable surface because the angles all go clockwise around the polygons and anticlockwise around the marked points.
(iv) Orient the edges such that at each marked point the edges either all arrive or all leave.

This gives an arc system that cuts the surface into open polygons. The associated gentle A_∞-algebra is concentrated in degree 0, has no higher products and is isomorphic to $\mathbb{C}Q/\mathcal{I}$. □

Theorem 9.12 ([115, 36]) *The twisted completions of the gentle A_∞-algebras of two generating arc collections for the same marked surface are A_∞-quasi-equivalent.*

[2] This is a special case of the theorem by Lekili and Polishchuk [166] which deals with the $\mathbb{Z}$-graded version.

Sketch of the proof If $\mathcal{A}'$ is obtained from $\mathcal{A}$ by reversing the orientation of an arc a then we can embed $\mathtt{Gtl}^{\pm}\mathcal{A}'$ in $\mathtt{TwGtl}^{\pm}\mathcal{A}$ by identifying the reverse arc with the object $a[1]$ (remember that a is a vertex of $Q_{\mathcal{A}}$ and hence an object in the category $\mathtt{Gtl}^{\pm}\mathcal{A}$).

If $\mathcal{A}$ and $\mathcal{A}' = \mathcal{A} \cup \{a\}$ are both arc systems that split the surface then a will be contained in a polygon of $\mathcal{A}$ and we can find a sequence of arcs $a_1, \ldots, a_k$ that is isotopic to a.

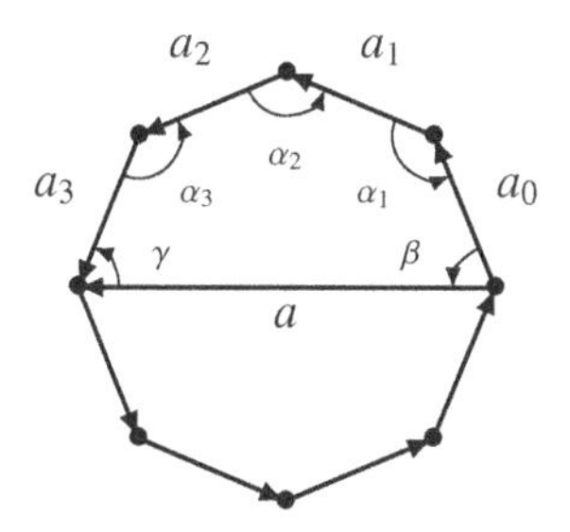

In the category $\mathtt{TwGtl}^{\pm}\mathcal{A}'$ the object a will be quasi-isomorphic to

$$M = (a_0 \oplus \cdots \oplus a_k, \alpha_1 + \cdots + \alpha_k).$$

Indeed, if β is the angle between a_0 and a, and γ between a and a_k, then $\beta \in \mathtt{TwGtl}^{\pm}\mathcal{A}'(M, a)$ and $\gamma \in \mathtt{TwGtl}^{\pm}\mathcal{A}'(a, M)$ will be inverses because

$$\hat{\mu}_2(\beta, \gamma) = \mu(\beta, \alpha_0, \ldots, \alpha_k, \gamma) = \mathbb{1}_a,$$
$$\hat{\mu}_2(\gamma, \beta) = \sum_i \mu(\alpha_i, \ldots, \alpha_k, \gamma, \beta, \alpha_0, \ldots, \alpha_{i-1}) = \mathbb{1}_{a_1} + \cdots + \mathbb{1}_{a_k} = \mathbb{1}_M.$$

Therefore, $\mathtt{Gtl}^{\pm}\mathcal{A}$ generates $\mathtt{TwGtl}^{\pm}\mathcal{A}'$.

So adding or deleting an extra arc does not change the twisted completion as long as both split the surface. By adding as many arcs as possible we can extend each arc collection to a triangulation of the surface. A theorem by Hatcher [110] tells us that all triangulations are related through moves that flip a diagonal in a quadrangle (i.e. remove one diagonal and replace it by the other diagonal).

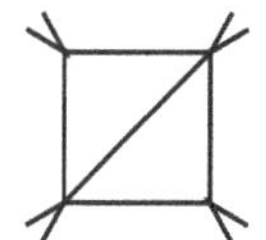 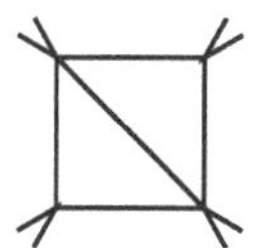

□

Definition 9.13 The *$\mathbb{Z}_2$-graded topological Fukaya category of a marked surface* $(\mathbb{S}, M)$ is the derived category of the gentle A_∞-category of any arc collection $\mathcal{A}$ that splits the surface:

$$^{\mathtt{top}}\mathtt{Fuk}^{\pm}(\mathbb{S}, M) := \mathtt{DGtl}^{\pm}\mathcal{A}.$$

Example 9.14 The three examples we studied in Chapter 4, the linear quiver, the Kronecker quiver and the cyclic quiver, are all examples that fit into this framework. Their derived categories are equivalent to the topological Fukaya categories of a disk with n marked points on the boundary, a cylinder with 1 marked point on each boundary and a disk with 1 internal marked point and n on the boundary.

Remark 9.15 There are different conventions in the literature to describe a marked surface depending on how the marked points and boundary components are treated:

(i) In [3], [156] and [36] the convention is the same as the one we use.
(ii) In [115] the marked points are cut out. An internal marked point becomes a small marked boundary circle and each boundary component with n marked points becomes a boundary component consisting of $2n$ segments that are marked and unmarked alternately. In this convention the endpoints of the arcs land on the marked segments of the boundary. Different arcs have different endpoints (but possibly on the same marked segments) and the angles are subsegments of these marked boundary segments. The angles are also known as Reeb chords.
(iii) In [166] the unmarked segments of the boundary are contracted to points and they are called stops (and sometimes confusingly also marked points) because they stop the flow of a Reeb chord.
(iv) In [204] the dual formalism of ribbon graphs is used (see Definition 9.32). The center of each polygon becomes a node of the ribbon graph and connects to each edge of the polygon to form a half edge of the ribbon graph.

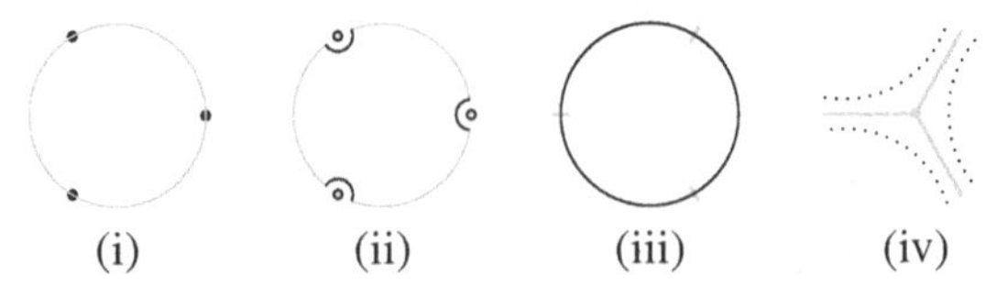

Remark 9.16 (Extras) There are some generalizations of the concept of a marked surface that sometimes come in handy.

- *Isotopic arcs.* If a, b are isotopic arcs then they bound a digon. If we extend the higher product rule to digons then this gives contributions $\mu_2(\alpha, \beta) = \pm\mathbb{1}_a$ and $\mu_2(\beta, \alpha) = \pm\mathbb{1}_b$. This means that a, b are isomorphic objects. The "gentle algebra" from the collection that contains both a and b is not a gentle algebra in the strict sense but it is an algebra that is Morita equivalent to the gentle algebra containing only a.

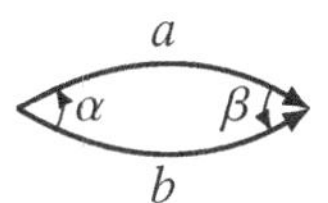

- *Punctures in the surface* can be modeled as an extra set of points P in the interior of $\mathbb{S} \setminus M$. Arcs are not allowed to contain, start, or end at these points and each polygon of a generating arc collection can contain at most one puncture. For the higher products we only use immersed polygons that do not cover the punctures.
- *Orbifold points* can be modeled as points o_i in the interior of $\mathbb{S} \setminus M$ with a degree $d_i \geq 2$. Arcs are not allowed to contain, start, or end at these points and each polygon of a generating arc collection can contain at most one orbifold point. For the higher product structure we demand that

$$\mu(\underbrace{\beta_1, \ldots, \beta_k, \ldots, \beta_1, \ldots, \beta_k}_{d_i}) = h(\beta_1)$$

  if $\beta_1, \ldots, \beta_k$ are the angles of a polygon with an orbifold point of order d_i. The other products can be calculated by lifting them to the orbifold universal cover.
- *Area.* If we assign to each polygon an area, we can weigh the higher μ with a factor e^{-A}, where A is the total area of the immersed polygon. In general, we can rescale these areas away with an automorphism of the gentle algebra.

9.2 Gluing Arcs to Strings and Bands

The reason why ${}^{\mathsf{top}}\mathbf{Fuk}^{\pm}(\mathbb{S}, M)$ is called the topological Fukaya category is because it contains objects that can be identified with homotopy classes of open and closed curves on the surface. In this section we will describe how to construct such objects as twisted complexes over the gentle A_∞-algebra. For more details we refer to [115, 36, 37].

Consider a sequence of arcs $a_1[i_1], \ldots, a_k[i_k]$, shifted in such a way that they form a path. Choose for each consecutive pair $a_u[i_u]$ and $a_{u+1}[i_{i+1}]$ an angle α_u

from a_u to a_{u+1} or from a_{u+1} to a_u that winds around their common marked point. Note that α_u can be seen as a degree 1 morphism from $a_u[i_u]$ to $a_{u+1}[i_{u+1}]$ or vice versa. From these data we can define two objects.

- An immersed curve $\mathbb{L}$ in $\mathbb{S}$ by stitching the arcs together using small chords around the marked points that follow the angles α_u.

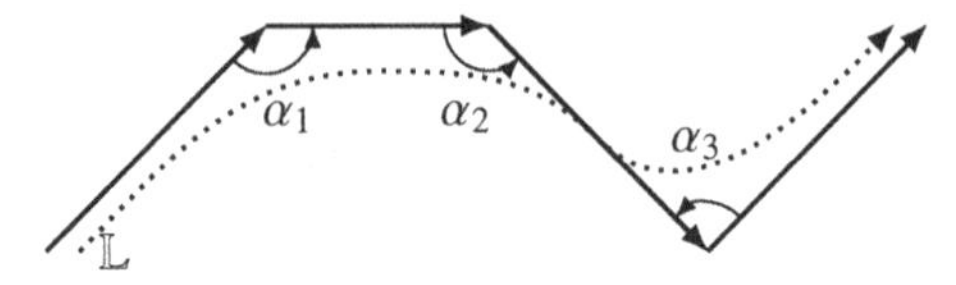

- A complex in $\mathtt{TwGtl}^{\pm}\mathcal{A}$,

$$(L, \delta) := \left(\bigoplus a_j[i_j], \sum_{j=1}^{k-1} \alpha_u \right).$$

The geometry of $\mathbb{L}$ and the algebraic properties of the twisted complex (L, δ) are closely intertwined.

Lemma 9.17 *Let $\mathbb{L}$ be a curve and (L, δ) its corresponding twisted complex.*

(i) *(L, δ) satisfies the Maurer–Cartan equation if and only if $\mathbb{L}$ does not contain a loop that is contractible in $\mathbb{S} \setminus M$.*
(ii) *Twisted complexes corresponding to isotopic curves are isomorphic in ${}^{\mathrm{top}}\mathbf{Fuk}^{\pm}(\mathbb{S}, M)$.*
(iii) *For every homotopy class we can find a sequence of arcs such that $\mathbb{L}$ is in that class.*

Sketch of the proof For the first statement, observe that if there is a contractible loop then $\mu(\alpha_u, \ldots, \alpha_v) \neq 0$, where the α's are the angles inside the contractible loop. If there is no contractible loop, all of the higher multiplications in the Maurer–Cartan equation are zero. Furthermore, all ordinary products are zero because there are just two situations for consecutive angles α_u, α_{u+1} in δ:

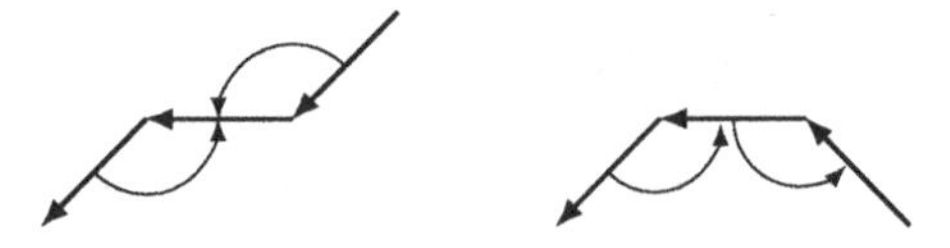

In the first situation the product is zero because they are not composable, in the second because of the relation in the gentle algebra.

For the second statement, consider two sequences that only differ on one polygon: the first one goes clockwise around the polygon and the second anticlockwise. These give two twisted objects (L_1, δ_1) and (L_2, δ_2).

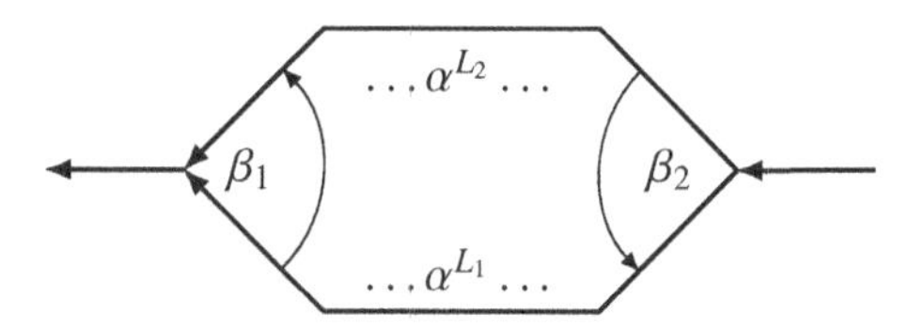

One can show that $u = \sum \mathbb{1}_a + \beta_1 \in \mathrm{Hom}^\bullet(L_1, L_2)$ and $v = \sum \mathbb{1}_a + \beta_2 \in \mathrm{Hom}^\bullet(L_2, L_1)$, where we sum over the common arrows in L_1 and L_2, are inverses. If we work out the twisted product we get extra higher-product terms from inserting angles of the polygon:

$$\hat{\mu}(u, v) = \sum_{a \in L_1 \cap L_2} \mathbb{1}_a + \sum \mu\left(\cdots \alpha_i^{L_2} \cdots, \beta_1, \cdots \alpha_i^{L_1} \cdots, \beta_2, \cdots \alpha_i^{L_2} \cdots\right)$$
$$= \sum_{a \in L_1 \cap L_2} \mathbb{1}_a + \sum_{a \in L_2 \setminus L_2} \mathbb{1}_a = \mathbb{1}_{L_2}.$$

Every two isotopic curves can be transformed into each other by a sequence of these moves.

For the third statement, isotope the curve such that it does not contain contractible loops and then substitute each segment of the curve that crosses a polygon by the sequence of arcs that goes around it anticlockwise or clockwise. Use the angles inside the polygon to stitch the arrows together. □

Definition 9.18 The *string object* $S(\mathbb{L})$ associated to a curve $\mathbb{L}$ between two marked points is the twisted object (L, δ) coming from a sequence of arcs that approximates the curve.

Lemma 9.19 *Let a be an arc between two different marked points p_1, p_2.*

(i) *If p_1, p_2 are both interior then* ${}^{\mathtt{top}}\mathbf{Fuk}^{\pm}(S(a), S(a)) \cong \frac{\mathbb{k}[\ell_1, \ell_2]}{(\ell_1 \ell_2)}$.
(ii) *If only p_1 is interior then* ${}^{\mathtt{top}}\mathbf{Fuk}^{\pm}(S(a), S(a)) \cong \mathbb{k}[\ell_1]$.
(iii) *If p_1, p_2 lie on the boundary then* ${}^{\mathtt{top}}\mathbf{Fuk}^{\pm}(S(a), S(a)) \cong \mathbb{k}$.

Sketch of the proof Choose an arc collection containing a and construct the gentle algebra $\mathtt{Gtl}^{\pm}\mathcal{A}$. In the first case we have

$${}^{\mathtt{top}}\mathbf{Fuk}^{\pm}(S(a), S(a)) = \mathbb{1}_a \cdot \mathtt{Gtl}^{\pm}\mathcal{A} \cdot \mathbb{1}_a.$$

This space is spanned by all nonzero cycles starting at e_a. There are two basic loops: the cycle ℓ_1 around the head of a and ℓ_2 around the tail of a. Their product is zero by the relations in $\mathtt{Gtl}^{\pm}\mathcal{A}$.

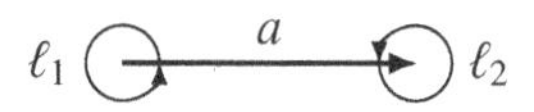

The other cases are similar. □

We can also make objects for closed curves. We start with a sequence of arcs $a_1[i_1], \dots, a_k[i_k]$ that form a cycle and stitch the arcs together by angles α_u to form a closed curve $\mathbb{L}$. But now we have an extra α_k to connect a_k and a_1. Choose a map $\lambda\colon \{1, \dots, k\} \to \mathsf{GL}_n(\mathbb{k})$ and define the pair

$$(L, \delta) := \left(\bigoplus a_j[i_j]^{\oplus n}, \sum_{j=1}^{k} \lambda_u \alpha_u\right).$$

We define the holonomy T_λ along the curve to be the product $\prod \lambda_i^{\pm 1}$ where the exponent is $+1$ if α_u turns in the direction of $\mathbb{L}$ and -1 if it turns in the opposite direction.

Lemma 9.20 *Let $\mathbb{L}$ be a curve, $\lambda\colon \{1, \dots, k\} \to \mathsf{GL}_n(\mathbb{k})$ and (L, δ) the corresponding twisted complex.*

(i) *The matrix δ is lower triangular if and only if not all the λ_u have the same exponent in the holonomy.*
(ii) *The matrix δ satisfies the Maurer–Cartan equation if and only if $\mathbb{L}$ does not contain a contractible loop.*
(iii) *Pairs corresponding to isotopic curves with conjugate holonomy are isomorphic in ${}^{\mathsf{top}}\mathsf{Fuk}^{\pm}(\mathbb{S}, M)$.*
(iv) *For every homotopy class we can find a sequence of arcs such that $\mathbb{L}$ is in that class.*

Sketch of the proof For the first statement note that all exponents are the same if and only if the α_i form an oriented cycle in $Q_{\mathcal{A}}$. The proofs of the second and fourth statements are the same as for string objects. For the third, note that if $G_i \in \mathsf{GL}_n(\mathbb{k})$ we can consider $G = \sum_j G_j \otimes \mathbb{1}_{a_j}$ as an element in $\mathrm{Hom}^\bullet(\bigoplus a_j[i_j]^{\oplus n}, \bigoplus a_j[i_j]^{\oplus n})$ with inverse $G^{-1} = \sum_j G_j^{-1} \otimes \mathbb{1}_{a_j}$. If $\lambda'_i = G_{i-1}\lambda_i G_i^{-1}$ then $(dG)_j = \lambda'_j G_j - G_{j-1}\lambda_j = 0$, so

$$G \in \mathsf{H}\,\mathrm{Hom}^\bullet\left(\bigoplus a_j[i_j]^{\oplus n}, \bigoplus a_j[i_j]^{\oplus n}\right) = {}^{\mathsf{top}}\mathsf{Fuk}^{\pm}((L, \delta_\lambda), (L, \delta_{\lambda'}))$$

is an isomorphism and $T_{\lambda'} = G_1 T_\lambda G_1^{-1}$. □

Definition 9.21 The twisted object (L, δ) is called the *band object* associated to $(\mathbb{L}, T_\lambda)$. We denote it by $B(\mathbb{L}, T_\lambda)$ to emphasize the fact that it only depends on the homotopy class of the curve and the conjugacy class of the holonomy.

Remark 9.22 Geometrically $(\mathbb{L}, T_\lambda)$ can be seen as a curve with a local system. We can also introduce local systems for string objects, but in that case we can use base change to turn all the matrices of λ into the identity, so the twisted object is just the direct sum of n string objects. As an open curve has a trivial fundamental group, the only local systems on such a curve are the trivial ones.

If we take a band object of a curve that runs l times through the same trajectory with holonomy T_λ, this can be seen as a band object that runs through the trajectory once but with holonomy $T_\lambda \otimes P$, where P is a cyclic permutation matrix.

Lemma 9.23 *If $\mathbb{L}$ is a simple closed curve and $\lambda, \lambda' \in \mathbb{k}^*$ then*

$$^{\mathtt{top}}\mathsf{Fuk}^{\pm}(B(\mathbb{L},\lambda), B(\mathbb{L},\lambda')) \cong \begin{cases} \mathbb{k}\iota \oplus \mathbb{k}\iota^*, & \lambda = \lambda', \\ 0, & \lambda \neq \lambda', \end{cases}$$

where $\iota = \mathbb{1}_B$ is the identity morphism and ι^* has degree 1.

Sketch of the proof Choose an arc collection such that $\mathbb{L}$ is isotopic to a path for which the connecting angles alternate in direction (i.e. consecutive angles do not compose):

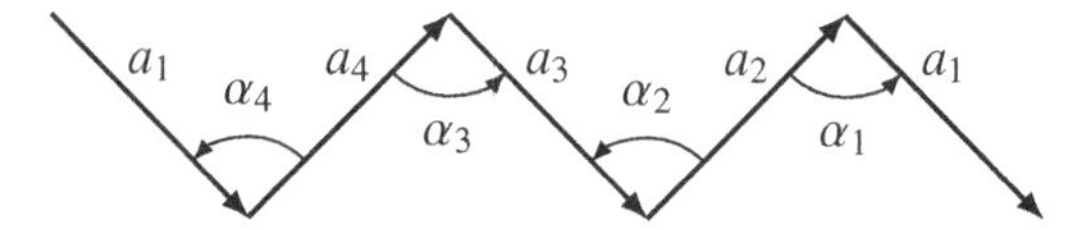

Construct a path homotopic to $\mathbb{L}$ and make

$$B = \left(\bigoplus_{j=1}^{k} a_j, \delta = \sum_{j=1}^{k} \lambda_j \alpha_j\right),$$

where α_j is the angle between a_j and a_{j+1} and the indices are taken modulo k. For simplicity we will assume that all marked points lie on the boundary. The cases with internal marked points are similar but a bit more computationally complex. Now

$$\mathrm{Hom}^\bullet(B,B) = \bigoplus_j \mathbb{k}\mathbb{1}_{a_j} + \mathbb{k}\alpha_j,$$

with the commutator as differential because there are no higher-order multiplications possible:

$$dx = \left[\sum_{j=1}^{k} \lambda_u \alpha_u, x\right].$$

We can easily calculate the homology. It is spanned by two elements

$$\iota = \mathbb{1}_B = \sum_a \mathbb{1}_a \quad \text{and} \quad \iota^* = \lambda_j \alpha_j.$$

Note that for the second homology class we can pick any $\lambda_j \alpha_j$ because in homology they are all the same up to a multiple: $d\mathbb{1}_{a_j} = \pm(\lambda_j \alpha_j - \lambda_{j-1}\alpha_{j-1})$. The second part of the proof is similar but now because $\lambda \neq \lambda'$ we have $d\mathbb{1}_B \neq 0$ and we can combine the $d\mathbb{1}_a$ to make a multiple of α_u. □

We will now investigate morphisms between two different band objects. To do this let $\mathbb{L}_1$ and $\mathbb{L}_2$ be two non-self-intersecting closed curves drawn on the surface $\mathbb{S}$ that have one intersection point p, and let $B_1 = B(\mathbb{L}_1, \lambda_1)$ and $B_2 = B(\mathbb{L}_2, \lambda_2)$ be two band objects associated to these curves.

Choose an arc collection such that the intersection point lies on one arc and the arc is oriented in the same way as the curves. To calculate the hom-space we can assume that locally the picture looks like

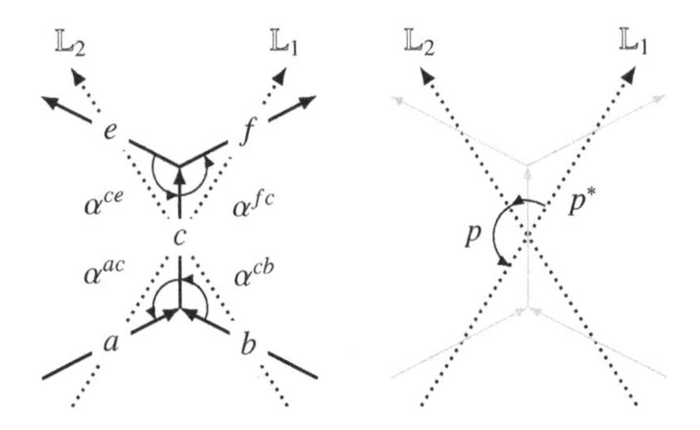

and there are no further overlaps. In this way,

$$B_1 := B(\mathbb{L}_1, \lambda_1) = (a \oplus c \oplus f \oplus \cdots, \lambda_1^{ac}\alpha_{ac} + \lambda_1^{fc}\alpha_{fc} + \cdots),$$
$$B_2 := B(\mathbb{L}_2, \lambda_2) = (b \oplus c \oplus e \oplus \cdots, \lambda_2^{cb}\alpha_{cb} + \lambda_2^{ce}\alpha_{ce} + \cdots),$$

where we used α_{ac} for the anticlockwise angle from c to a around their common marked point.

Lemma 9.24 *If two curves intersect at precisely one point p, and B_1, B_2 are the corresponding band objects with a one-dimensional local system then*

$$^{\mathtt{top}}\mathsf{Fuk}^{\pm}(B_1, B_2) = \Bbbk p^* \oplus 0 \quad \textit{and} \quad ^{\mathtt{top}}\mathsf{Fuk}^{\pm}(B_2, B_1) = 0 \oplus \Bbbk p$$

with $p^* := \mathbb{1}_c$ and $p := \lambda_1^{ac}\alpha_{ac} - \lambda_2^{cb}\alpha_{cb}$.

Sketch of the proof Again we will assume that all marked points lie on the boundary. The hom-complex has the form

$$\mathrm{Hom}(B_1, B_2) = \Bbbk\mathbb{1}_{cc} + \alpha_{ba} + \Bbbk\alpha_{ca} + \Bbbk\alpha_{bc} + \Bbbk\alpha_{ef} + \Bbbk\alpha_{cf} + \Bbbk\alpha_{ec}.$$

The differential is

$$x \mapsto (\lambda_2^{cb}\alpha_{cb} + \lambda_2^{ce}\alpha_{ce})x - x(\lambda_1^{ac}\alpha_{ac} + \lambda_2^{fc}\alpha_{fc})$$

and hence $\mathbb{1}_c$ is in the zeroth homology. We denote this morphism by p, where p stands for the intersection point. An easy calculation shows that this is the only element in the homology. There is also one element in $^{\mathtt{top}}\mathsf{Fuk}^{\pm}(B_2, B_1)$:

this is $\lambda_1^{ac}\alpha_{ac} - \lambda_2^{cb}\alpha_{cb}$, or equivalently $\lambda_1^{fc}\alpha_{fc} - \lambda_2^{ce}\alpha_{ce}$, because their sum is $d\mathbb{1}_c$. We will choose the former and denote it by p^*. This element has degree 1. □

Remark 9.25 This lemma can be extended to curves with several intersection points. Each intersection point will contribute an extra morphism in both directions and the degree of this morphism is determined by the orientations of the curves.

Now let us have a look at products. We start with two easy ones.

Lemma 9.26 *If p is an intersection point between two band objects B_1, B_2 then*

$$p \cdot p^* = \iota_{B_1}^* \quad \textit{and} \quad p^* \cdot p = -\iota_{B_2}^*.$$

Sketch of the proof This follows straight from the definitions of ι, ι^*, p and p^*. □

Suppose that we have a collection of closed curves $\mathbb{L}_1, \ldots, \mathbb{L}_k$ that bound a polygonal region of $\dot{\mathbb{S}}$ and suppose that there is an arc collection such that the local picture looks like this.

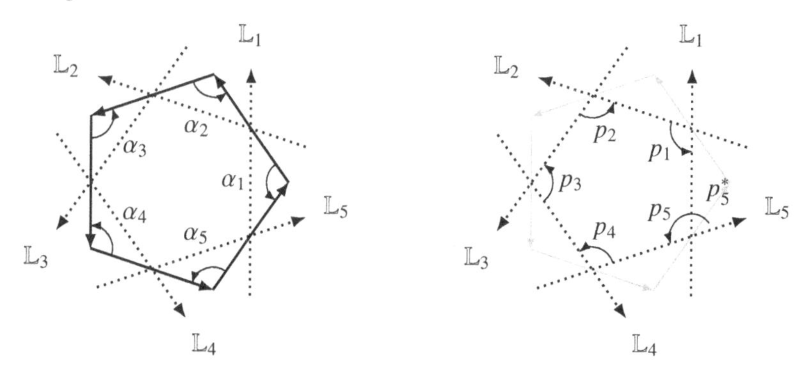

We denote the degree 1 morphism from $\mathbb{L}_{i+1}$ to $\mathbb{L}_i$ by p_i and the degree 0 morphism in the opposite direction by p_i^*.

Lemma 9.27 *In the situation above we have*

$$\mu(p_1, \ldots, p_{k-1}) = cp_k^*$$

with $c = \lambda_1 \cdots \lambda_k$ the holonomy around the cycle.

Sketch of the proof Each of the p_i has as expression $\lambda_i\alpha_i + \cdots$ where $\cdots$ is an angle outside the polygon (and hence plays no role in the calculation). The twist δ_i of $B(\mathbb{L}_i, \lambda_i)$ looks like $\lambda_i\alpha_i + \cdots$ (we write λ_i for the decoration of the

ith curve at angle α_i). To calculate $\mu(p_1, \dots, p_{k-1})$ we have to insert the δ_i in all possible ways between the morphisms and sum these products. The only nonzero product we can get is by inserting δ_k at the end:

$$\mu(\underbrace{\lambda_1\alpha_1 + \cdots}_{p_1}, \dots, \underbrace{\lambda_{k-1}\alpha_{k-1} + \cdots}_{p_{k-1}}, \underbrace{\lambda_k\alpha_k + \cdots}_{\delta_k}) = \lambda_1 \cdots \lambda_k \mathbb{1}_{a_k}.$$

This is the identity morphism of the kth arc and this corresponds to

$$cp_k^* \in {}^{\mathsf{top}}\mathbf{Fuk}^{\pm}(B_1, B_k).$$

Note that this calculation is not done in the minimal model, but after going to the minimal model we get the same result because every other tree that can contribute will contain a $\mu_2(\alpha_i, \alpha_{i+1})$, which is zero. □

There is however a second product associated to such a polygon that does depend on the minimal model we choose. To calculate products in the minimal model we will need a split of $\mathrm{Hom}^\bullet(B, B) = I^\bullet \oplus H^\bullet \oplus R^\bullet$ for each band object $B = (\oplus_i a_i, \sum \lambda_i \alpha_i)$. In order to do this we pick one arc a_m and one angle α_m, which we call the marked arc and angle. For the degree 0 part we put all $\mathbb{1}_a$ except $\mathbb{1}_{a_m}$ in R^0 and $\mathbb{1}_B$ in H^0. For the degree 1 part we put $\lambda_m\alpha_m$ in H^1 and all $\lambda_i\alpha_i - \lambda_{i+1}\alpha_{i+1}$ in I^0.

Lemma 9.28 *In the situation above we have*

$$\mu(p_1, \dots, p_{k-1}, p_k) := c\mathbb{1}_{\mathbb{L}_k} \;\; or \;\; 0.$$

The first case happens if the marked arc of $\mathbb{L}_k$ is part of the polygon; otherwise the product is zero.

Sketch of the proof In this case, the only nonzero product is the one that we get by inserting no δ_i's:

$$\mu(\underbrace{\lambda_1\alpha_1 + \cdots}_{p_1}, \dots, \underbrace{\lambda_k\alpha_k + \cdots}_{p_k}) = \lambda_1 \cdots \lambda_k \mathbb{1}_{a_k} = c\mathbb{1}_{a_k}.$$

Unfortunately, $\mathbb{1}_{a_k}$ is not a homology class in $\mathrm{Hom}(B(\mathbb{L}_k, \lambda), B(\mathbb{L}_k, \lambda))$, so if we go to the minimal model we have to perform a projection onto the homology. The degree 0 element in the homology $\mathrm{Hom}(B(\mathbb{L}_k, \lambda), B(\mathbb{L}_k, \lambda))$ is $\mathbb{1}_{\mathbb{L}_k} := \sum_{a \in \mathbb{L}_k} \mathbb{1}_a$, so $\lambda_1 \cdots \lambda_k \mathbb{1}_{a_k}$ will be projected onto a multiple of $\mathbb{1}_{\mathbb{L}_k}$ that depends on the split $\mathrm{Hom}(B(\mathbb{L}_k, \lambda), B(\mathbb{L}_k, \lambda)) = I \oplus R \oplus H$. As we put all $\mathbb{1}_a$ in R except $\mathbb{1}_{a_m}$, the corresponding projection will map all $\mathbb{1}_a$ to zero and $\mathbb{1}_{a_m}$ to $\mathbb{1}_{\mathbb{L}_k}$. □

Remark 9.29 This situation is similar to the geometric construction of the Fukaya category. Note that to define the hom-space between a Lagrangian and itself we have to choose a deformation. This deformation will intersect the original at two points, one of which corresponds to the identity. If this point lies on the boundary of the polygon there will be a nonzero product, but if it lies outside the polygon there will be no contribution.

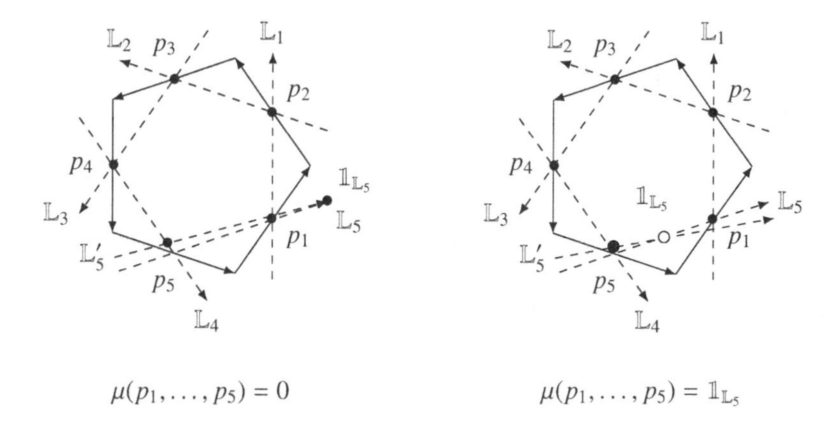

It is also important to note that the products above are not the only ones that are nonzero. A full calculation of all the products is however technically very complicated and we will not attempt this.

From the previous lemmas we can infer that the morphisms between band objects are given by intersection points and the products between these intersection points are counting immersed disks that are bounded by the curves. This is precisely the behavior we expect of a Fukaya-type category. Therefore we can interpret ${}^{\mathtt{top}}\mathsf{Fuk}^{\pm}(\mathbb{S}, M)$ as an algebraically defined substitute for the Fukaya category. If $(\mathbb{S}, M)$ is a marked surface then we can construct a Liouville manifold with cylindrical ends and a set of stops $(\mathbb{X}, \sigma)$.

(i) Cut a disk around each internal marked point. This gives a surface with boundary on which we can put the structure of a Liouville domain $\mathbb{X}_0$.
(ii) On the boundaries coming from an internal marked point we glue a cylinder without stops.
(iii) On a boundary component with k marked points we glue a cylinder with k stops. These stops are lines that end on the boundary between the marked points.

Theorem 9.30 *The Fukaya category of the marked surface* $(\mathbb{S}, M)$ *is* A_∞*-equivalent to the partially wrapped Fukaya category of* $(\mathbb{X}, \sigma)$*:*

$$ {}^{\mathtt{top}}\mathsf{Fuk}^{\pm}(\mathbb{S}, M) \stackrel{\infty}{=} {}^{\mathtt{wr}}\mathsf{Fuk}^{\pm}_{\sigma}\mathbb{X}. $$

Sketch of the proof We can turn each arc into a Lagrangian with conical ends by extending it to infinity near the marked points. The morphism spaces between these Lagrangians in the partially wrapped Fukaya category are the same as those between the arcs in ${}^{\mathtt{top}}\mathtt{Fuk}^{\pm}(\mathbb{S}, M)$ because every Reeb flow corresponds to an angle. The higher products also coincide because the immersed disks coincide with the polygons on $\mathbb{S}$. Finally, because the arc collection splits the surface into polygons, one can use the generation criterion Theorem 6.77 to show that the conical Lagrangians generate the Fukaya category. For more details see the appendix of [36]. □

In this identification the strings correspond to open Lagrangians and the bands to closed Lagrangians with a local system. Therefore, it makes sense to define the (partially) wrapped Fukaya category of a marked surface as the category generated by the string objects and the exact Fukaya category as the subcategory generated by the band objects:

- ${}^{\mathtt{wr}}\mathtt{Fuk}^{\pm}(\mathbb{S}, M) := \langle S(\mathbb{L}) \mid S(\mathbb{L}) \text{ is a string object}\rangle = {}^{\mathtt{top}}\mathtt{Fuk}^{\pm}(\mathbb{S}, M)$,
- ${}^{\mathtt{ex}}\mathtt{Fuk}^{\pm}(\mathbb{S}, M) := \langle B(\mathbb{L}, \lambda) \mid B(\mathbb{L}, \lambda) \text{ is a band object}\rangle \subset {}^{\mathtt{top}}\mathtt{Fuk}^{\pm}(\mathbb{S}, M)$.

Remark 9.31 Traditionally the objects in the exact Fukaya category are closed curves with a spin structure and a Morse function [218]. The spin structure is needed for the signs and the Morse function is to define endomorphisms of $\mathbb{L}$ as intersection points between the curve and its Hamiltonian deformation by the Morse function. These data can be drawn as three special points $\circ$, $*$, $\times$ on the curve $\mathbb{L}$.

- $\circ$ is the minimum of the Morse function and corresponds to $\iota = \mathbb{1}_{\mathbb{L}}$. In our setting this is equivalent to the choice of the marked arc if we put $\circ$ just before the marked arc.
- $*$ is the maximum of the Morse function and corresponds to $\iota^* = \lambda_m \alpha_m$. In our setting this is equivalent to the choice of the marked angle if we put $*$ in the middle of the segment corresponding to α_m.
- $\times$ is the position where the spin structure flips. In our terminology, a curve with a nontrivial spin structure is one with a local system whose holonomy is -1. This can be realized by setting all λ_i to 1 except for one that is -1. We place $\times$ on the segment of the angle with the -1.

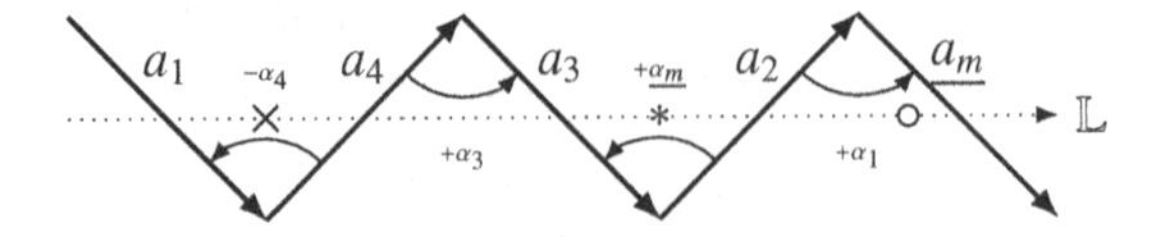

9.3 Gluing Fukaya Categories over Graphs

Marked surfaces can be constructed by gluing polygons together over common edges. This procedure can also be mimicked on the level of the Fukaya categories. In this section we will investigate different ways to glue A_∞-categories together. In general, gluing A_∞-categories or dg-categories together uses many techniques from higher algebra, such as model theory [239] and homotopy (co)limits [31, 68]. We will only illustrate these ideas in the special case of our topological Fukaya categories. For more details about the general framework we refer to the work of Toën [243, 244].

Our aim is to glue topological Fukaya categories together over a specific type of one-dimensional space that is obtained by taking a finite number of closed, half-open and open intervals together and identifying some of the endpoints. We will call these spaces *graphs with legs* Γ. The endpoints are called the nodes and the intervals the edges. These edges come in three varieties: closed, half open and open, according to the type of interval they come from. Half-open edges will sometimes be referred to as legs and open edges as lone edges, because they do not connect to any nodes.

Each open subset of Γ with a finite number of connected components can again be interpreted as a graph with legs. To exclude opens with an infinite number of components, we will redefine $\mathtt{Open}(\Gamma)$ as the category of all opens with a finite number of components, together with the inclusions as morphisms.

If we have two such opens Υ_1, Υ_2 intersecting in $\Upsilon_{12} := \Upsilon_1 \cap \Upsilon_2$, we have a gluing diagram in $\mathtt{Open}(\Gamma)$:

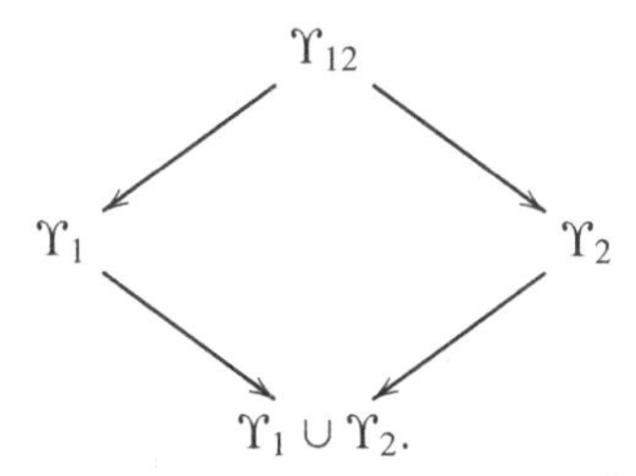

In categorical terms we say that $\Upsilon_1 \cup \Upsilon_2$ is the colimit of $\Upsilon_1 \leftarrow \Upsilon_{12} \rightarrow \Upsilon_2$ because every possible way to complete this diagram factors through it.

Gluing over Ribbon Graphs

Definition 9.32 Let $\mathcal{A}$ be a generating arc collection for the surface $(\mathbb{S}, M)$ and assume that all boundary segments are arcs in $\mathcal{A}$.

The *ribbon graph* $\Gamma_{\mathcal{A}}$ is the graph with legs that has a node at the center of each polygon, a closed edge for each internal arc and a half-open edge for each arc on the boundary. The edges connect to the nodes of the polygons that the arcs bound.[3]

To each open $\Upsilon \in \mathtt{Open}(\Gamma)$ we can associate a new marked surface $(\mathbb{S}_\Upsilon, M_\Upsilon)$ by gluing together the polygons corresponding to vertices in the subgraph along arcs corresponding to closed edges in Υ. If Υ contains lone edges then we add a disk with two marked points for each lone edge. The corners of the polygons are the marked points of $\mathbb{S}_\Upsilon$.

Theorem 9.33 *There is a covariant functor from the category* $\mathtt{Open}(\Gamma)$ *to the category of derived* A_∞*-categories with* A_∞*-functors as morphisms that assigns to each* Υ *the topological Fukaya category of* $(\mathbb{S}_\Upsilon, M_\Upsilon)$*:*

$$\mathscr{R}\colon \mathtt{Open}(\Gamma) \to \mathtt{DCAT}_\infty : \Upsilon \mapsto {}^{\mathtt{top}}\mathtt{Fuk}^{\pm}(\mathbb{S}_\Upsilon, M_\Upsilon).$$

Sketch of the proof Let $\phi\colon \Upsilon_1 \to \Upsilon_2$ be a morphism in $\mathtt{Open}(\Gamma)$. Pick the arc collection $\mathcal{A}_i$ for $(\mathbb{S}_{\Upsilon_i}, M_{\Upsilon_i})$ that consists of all edges of the polygons. The morphism $\phi\colon \Upsilon_1 \to \Upsilon_2$ induces a map $\mathcal{A}_1 \to \mathcal{A}_2$ between the arc collections because every edge of Υ_1 sits in a unique edge of Υ_2. This map is not necessarily an injection: two arcs in $\mathcal{A}_1$ can be mapped to the same arc in $\mathcal{A}_2$.

As every nontrivial angle is uniquely associated to a polygon and two incoming arcs (or dually a node and two edges), the map $\phi\colon \Upsilon_1 \to \Upsilon_2$ also induces a map between the angles, so we have morphism of gentle algebras

$$\mathcal{F}\colon \mathtt{Gtl}^{\pm}\mathcal{A}_1 \to \mathtt{Gtl}^{\pm}\mathcal{A}_2.$$

This morphism also preserves the A_∞-structure because if an angle is present in both surfaces this means that the polygon is present in both surfaces. Therefore, any immersion of polygons in $\mathbb{S}_{\Upsilon_2}$ will also exist in $\mathbb{S}_{\Upsilon_1}$. By going to the derived categories we get a strict A_∞-functor

$$\mathscr{R}(\Upsilon_i) := \mathtt{DGtl}^{\pm}\mathcal{A}_i,$$
$$\mathscr{R}(\phi) := \mathtt{D}\mathcal{F}\colon \mathscr{R}(\Upsilon_1) \to \mathscr{R}(\Upsilon_2).$$

The composition rule is clear from the construction. □

The functor $\mathscr{R}$ looks suspiciously like a presheaf but with a covariant functor instead of a contravariant one or, in other words, a precosheaf. It even makes sense to say that $\mathscr{R}$ is a cosheaf of A_∞-categories. This means that, just as global sections of a sheaf can be constructed from local sections agreeing on

[3] This graph is called a ribbon graph because its natural embedding in $(\mathbb{S}, M)$ allows us to thicken the edges to ribbons.

overlaps, the category $\mathscr{R}(\Upsilon_1 \cup \Upsilon_2)$ is a universal construction that fits in the diagram

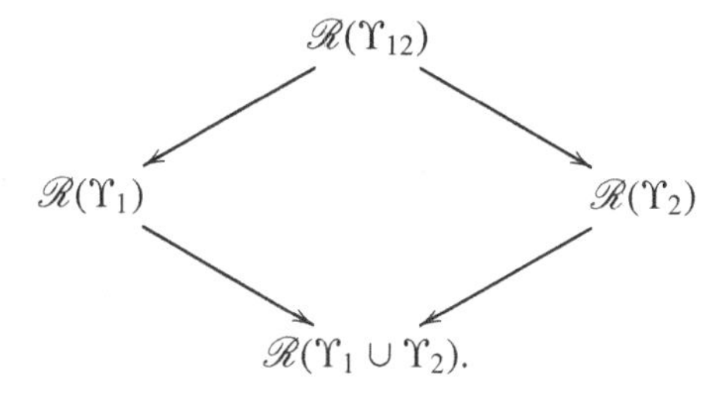

In mathematical terms, $\mathscr{R}(\Upsilon_1 \cup \Upsilon_2)$ is the *homotopy colimit* of $\mathscr{R}(\Upsilon_1) \leftarrow \mathscr{R}(\Upsilon_{12}) \rightarrow \mathscr{R}(\Upsilon_2)$. This is a more advanced version of the classical colimit in categories which is compatible with quasi-isomorphisms and twisted completions. How this works precisely is beyond the scope of these notes. For more information we refer to the work of Haiden, Katzarkov and Kontsevich [115], Dyckerhoff and Kapranov [65], Sibilla, Treumann and Zaslow [232], Nadler [189, 188, 191], Pascaleff and Sibilla [204].

Example 9.34 We can see the cylinder with one marked point on each boundary as two triangles glued together.

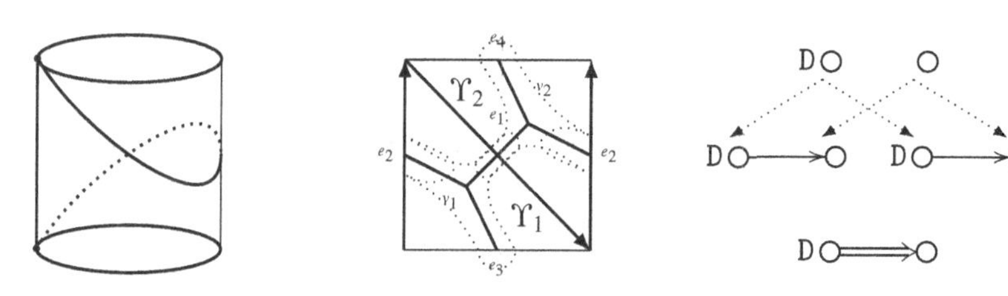

The ribbon graph has two trivalent nodes v_1, v_2 connected by two edges e_1, e_2 and there are two more open edges e_3 connected to v_1 and e_4 connected to v_2. If we take as open cover

$$\begin{aligned}
\Upsilon_1 &= \{v_1\} \cup \mathrm{Int}(e_1) \cup \mathrm{Int}(e_2) \cup \mathrm{Int}(e_3),\\
\Upsilon_2 &= \{v_1\} \cup \mathrm{Int}(e_1) \cup \mathrm{Int}(e_2) \cup \mathrm{Int}(e_4),\\
\Upsilon_{12} &= \mathrm{Int}(e_1) \cup \mathrm{Int}(e_2),
\end{aligned}$$

then the diagram of Fukaya categories consists of the derived categories of two linear quivers, both containing the derived category of a quiver with two vertices and no arrows. The homotopy colimit is the derived category of the two linear quivers glued together over the two vertices. This is the Kronecker quiver.

Remark 9.35 The idea of gluing Fukaya categories using ribbon graphs is a one-dimensional realization of the philosophy of Kontsevich that wrapped Fukaya categories of certain noncompact symplectic manifolds should be constructed by looking at a *Lagrangian skeleton.* This is a possibly singular Lagrangian submanifold that is a deformation retract of the symplectic manifold. This gives rise to a cosheaf of categories over this skeleton and the corresponding homotopy colimit is the wrapped Fukaya category of the symplectic manifold. By looking at the combinatorial structure of the skeleton, it is then possible to view a wrapped Fukaya category as a colimit of certain easy A_∞-categories. In our situation, the ribbon graph is the Lagrangian skeleton and the easy pieces correspond to neighborhoods of the nodes of the graph (which give derived categories of the linear quiver $\mathtt{Dmod}^{\pm} L_n$) and edges (which give derived categories of $\mathbb{C}$).

Gluing over Tubular Graphs

Definition 9.36 Let $(\mathbb{S}, M)$ be a marked surface. A simple closed curve is an embedding of the circle $\mathbb{S}^1$ that is not homotopic to a small circle around one of the internal marked points. A *collection of cuts* C is a set of simple closed curves that are pairwise nonintersecting and not homotopic.

Definition 9.37 To a collection of cuts C we associate a graph with legs called the *tubular graph* Γ_C. Its nodes are the connected components of the complement of the simple curves and its edges are the simple curves and the internal marked points.

The edges corresponding to the simple curves connect to the nodes of the components they bound. The other edges can be considered as half-open edges because they only connect to the node of the component in which the marked point is contained.

Just like in the case of ribbon graphs, we associate to each open $\Upsilon \in \mathtt{Open}(\Gamma_C)$ a new marked surface. It is the union of the connected components corresponding to the nodes in Υ, glued together by the closed curves of the closed edges in Υ. We add a cylinder for each open edge and shrink all the open ends that are not glued to internal marked points.[4] We denote this new marked surface by $(\mathbb{S}^\Upsilon, M^\Upsilon)$.

Theorem 9.38 *There is a contravariant functor from the category* $\mathtt{Open}(\Gamma)$ *to the category of derived* A_∞*-categories with* A_∞*-functors as morphisms that assigns to each* Υ *the topological Fukaya category of* $(\mathbb{S}^\Upsilon, M^\Upsilon)$*:*

[4] This last step is not needed in the formalism where the marked points are in fact marked boundary components. See Remark 9.15.

$$\mathscr{T}\colon \mathtt{Open}(\Gamma) \to \mathtt{DCAT}_\infty \colon \Upsilon \mapsto {}^{\mathtt{top}}\mathtt{Fuk}^{\pm}(\mathbb{S}^{\Upsilon}, M^{\Upsilon}).$$

Sketch of the proof Choose an arc collection $\mathcal{A}$ that generates $(\mathbb{S}, M)$. If $\Upsilon \in \mathtt{Open}(\Gamma_C)$ then every arc a will split as an alternating union of segments a_i that are contained in $(\mathbb{S}^{\Upsilon}, M^{\Upsilon})$ and segments b_i that are not contained in $(\mathbb{S}^{\Upsilon}, M^{\Upsilon})$. Some of the a_i arcs might be isotopic in $(\mathbb{S}^{\Upsilon}, M^{\Upsilon})$ but together they will split the surface $\mathbb{S}^{\Upsilon}$. Call this collection $\mathcal{A}^{\Upsilon}$ and set $\mathscr{T}(\Upsilon) := \mathtt{DGtl}^{\pm}\mathcal{A}^{\Upsilon}$.

If $\phi\colon \Upsilon_1 \to \Upsilon_2$ is a morphism in $\mathtt{Open}(\Gamma_C)$ then every arc $a \in \mathcal{A}^{\Upsilon_2}$ breaks into subarcs $a_i \in \mathcal{A}^{\Upsilon_2}$. The map

$$\mathcal{T}\colon \mathtt{Gtl}^{\pm}\mathcal{A}^{\Upsilon_2} \to \mathtt{DGtl}^{\pm}\mathcal{A}^{\Upsilon_1} \colon a \mapsto \bigoplus_{a_i \subset a} a_i$$

can be turned into an A_∞-morphism as follows: If α is an angle in $\mathbb{S}^{\Upsilon_2}$ that is also present in $\mathbb{S}^{\Upsilon_1}$ then we set $\mathcal{T}(\alpha) = \alpha$, where the latter is considered as a map between the correct summands of $\mathcal{T}(h(a))$ and $\mathcal{T}(t(a))$. If α is not present then $\mathcal{T}(\alpha) = 0$.

For the higher parts of $\mathcal{T}$, note that the angles between the arcs in $\mathcal{A}^{\Upsilon_1}$ come in two types: angles that are also present in $\mathcal{A}^{\Upsilon_2}$ and new angles around the closed curves that we have contracted. These can be seen as chords in $\mathbb{S}^{\Upsilon_2}$ between arcs. If $a_0 \xleftarrow{\alpha_1} \cdots \xleftarrow{\alpha_k} a_k$ is a sequence of angles in $\mathbb{S}^{\Upsilon_2} \setminus \mathbb{S}^{\Upsilon_1}$ and β is a chord such that they bound an immersed polygon, then we set $\mathcal{T}(\alpha_1, \ldots, \alpha_k) = \pm\beta$.

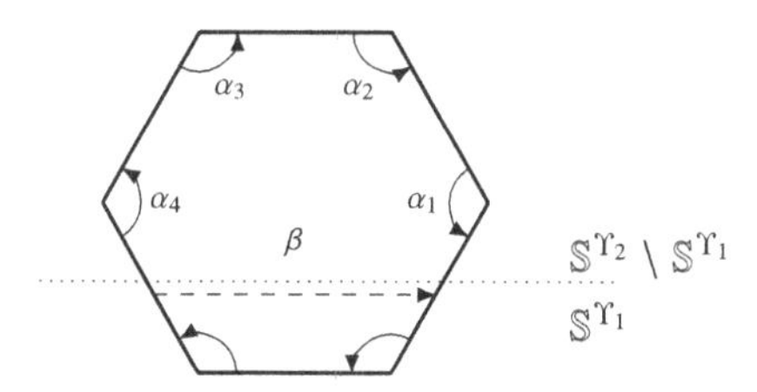

It is easy to check that this assignment turns $\mathscr{T}$ into an A_∞-structure. By going to the derived categories we get a functor

$$\mathscr{T}(\Upsilon_i) := \mathtt{DGtl}^{\pm}\mathcal{A}_i,$$
$$\mathscr{T}(\phi) := \mathtt{D}\mathcal{T}\colon \mathscr{T}(\Upsilon_2) \to \mathscr{T}(\Upsilon_1).$$

The composition rule is clear from the construction. □

Remark 9.39 Just like in the case of ribbon graphs, this A_∞-functor can be thought of as a sheaf and the category $\mathscr{T}(\Upsilon_1 \cup \Upsilon_2)$ can be constructed from the diagram

$$\mathscr{T}(\Upsilon_1) \longrightarrow \mathscr{T}(\Upsilon_1 \cap \Upsilon_2) \longleftarrow \mathscr{T}(\Upsilon_2),$$

but now as a *homotopy limit*. For this to work properly one has to work with ind-completed enlargements of the Fukaya category.[5] For more information we refer to the work of Pascaleff and Sibilla [204]. Note that owing to different conventions, this type of gluing is referred to there as a gluing over closed subgraphs.

Example 9.40 Consider again the cylinder with one marked point on each boundary component. Take the simple closed curve that cuts the cylinder in two.

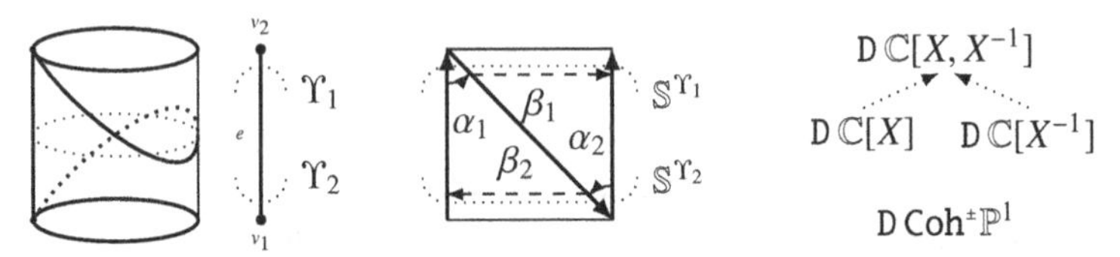

The tubular graph consists of two nodes v_1, v_2 of valency 1 joined by an edge e. Each node corresponds to a punctured disk with one marked point on the boundary, while the overlap is a sphere with two internal marked points. If we take as open cover

$$\Upsilon_1 = v_1 \cup \mathrm{Int}(e),$$
$$\Upsilon_2 = v_2 \cup \mathrm{Int}(e),$$
$$\Upsilon_{12} = \mathrm{Int}(e),$$

then the surfaces $\mathbb{S}^{\Upsilon_1}$, $\mathbb{S}^{\Upsilon_2}$ are two disks with one marked point on the boundary and one internal marked point. The surface $\mathbb{S}^{\Upsilon_1 \cap \Upsilon_2}$ is an open cylinder (or sphere with two marked points).

If we pick the same arc collection as in Example 9.34, the collection $\{a_1, a_2\}$ restricts to two arcs in all three surfaces. Apart from the original angles $\alpha_1, \alpha_2 \colon a_1 \to a_2$, there are two more angles in $\mathbb{S}^{\Upsilon_1 \cap \Upsilon_2}$, labeled $\beta_1, \beta_2 \colon a_2 \to a_1$. The angle β_1 exists in $\mathbb{S}^{\Upsilon_2}$ but not in $\mathbb{S}^{\Upsilon_1}$. For β_2 the opposite is true:

$$\mathtt{Gtl}^{\pm}\mathcal{A}_1 = \circ \overset{\alpha_1,\alpha_2}{\underset{\beta_2}{\rightleftarrows}} \circ \Big/ (\alpha_1\beta_2 - \mathbb{1}_{a_2}, \beta_2\alpha_1 - \mathbb{1}_{a_1}),$$

$$\mathtt{Gtl}^{\pm}\mathcal{A}_2 = \circ \overset{\alpha_1,\alpha_2}{\underset{\beta_1}{\rightleftarrows}} \circ \Big/ (\alpha_2\beta_1 - \mathbb{1}_{a_2}, \beta_1\alpha_1 - \mathbb{1}_{a_1}),$$

$$\mathtt{Gtl}^{\pm}\mathcal{A}_{12} = \circ \overset{\alpha_1,\alpha_2}{\underset{\beta_1,\beta_2}{\rightleftarrows}} \circ \Big/ (\alpha_1\beta_2 - \mathbb{1}_{a_2}, \beta_2\alpha_1 - \mathbb{1}_{a_1}, \alpha_2\beta_1 - \mathbb{1}_{a_2}, \beta_1\alpha_1 - \mathbb{1}_{a_1}).$$

[5] For instance, if ${}^{\mathrm{top}}\mathtt{Fuk}^{\pm}(\dot{\mathbb{S}}) \overset{\infty}{\cong} \mathtt{DMod}^{\pm} Q$ then the ind-completion is $\mathtt{DMOD}^{\pm} Q$.

Therefore, $\mathtt{Gtl}^{\pm}\mathcal{A}_1$ is Morita equivalent to $\mathbb{C}[\alpha_2\beta_2] = \mathbb{C}[X]$ and $\mathtt{Gtl}^{\pm}\mathcal{A}_2$ to $\mathbb{C}[\alpha_1\beta_1] = \mathbb{C}[Y]$. Finally, in $\mathtt{Gtl}^{\pm}\mathcal{A}_{12}$, $\alpha_1\beta_1$ and $\beta_2\alpha_2$ are inverses so this algebra is Morita equivalent to $\mathbb{C}[X,Y]/(XY-1)$.

This implies that we are gluing the derived categories of $\mathbb{C}[X]$ and $\mathbb{C}[X^{-1}]$ over the derived category of $\mathbb{C}[X,X^{-1}]$. The homotopy limit is the derived category of two affine lines glued together over $\mathbb{C}^*$, or in other words the $\mathbb{Z}_2$-graded derived category of $\mathbb{P}^1$.

9.4 Gluing and Mirror Symmetry

We will now use the last gluing technique to look at mirror symmetry for punctured Riemann surfaces. The idea is to construct one building block of mirror symmetry and then glue copies of this building block together. Our basic ingredient is the *pair of pants* or the *sphere with three marked points* $(\mathbb{S}^2, \{0, 1, \infty\})$. Its mirror symmetry was worked out by Abouzaid, Auroux, Efimov, Katzarkov and Orlov in [1].

Theorem 9.41 *The topological Fukaya category of the pair of pants is derived equivalent to the Landau–Ginzburg model* $(\mathbb{C}^3, XYZ)$*:*

$$^{\mathtt{top}}\mathtt{Fuk}^{\pm}(\mathbb{S}^2, \{0,1,\infty\}) \overset{\infty}{\cong} \mathtt{MF}^{\pm}(\mathbb{C}^3, XYZ).$$

Sketch of the proof We split the sphere into two triangles. This gives an arc collection with three arcs $\{x, y, z\}$. The quiver looks like the following:

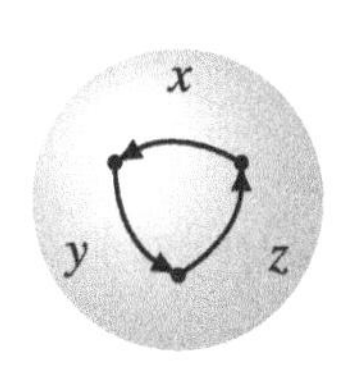

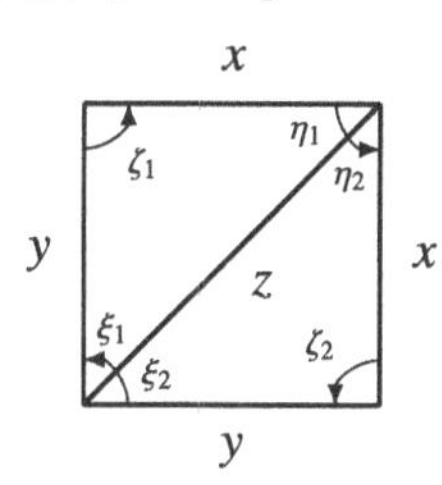

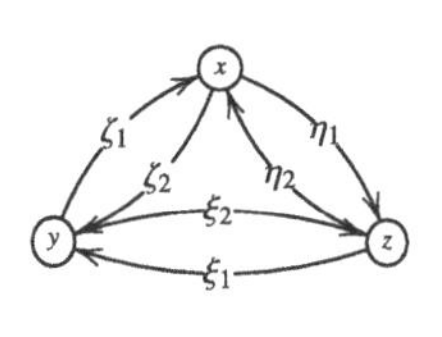

All the arrows have degree 1 and, for instance,

$$\mu(\zeta_1, \xi_1, \eta_1) = \mu(\eta_2, \xi_2, \zeta_2) = \mathbb{1}_x,$$
$$\mu(\xi_1, \eta_1, \zeta_1) = \mu(\zeta_2, \eta_2, \xi_2) = \mathbb{1}_y,$$
$$\mu(\eta_1, \zeta_1, \xi_1) = \mu(\xi_2, \zeta_2, \eta_2) = \mathbb{1}_z.$$

On the B-side we have three obvious matrix factorizations

$$M_X := \left(R^{\oplus 2}, \begin{pmatrix} 0 & X \\ YZ & 0 \end{pmatrix}\right), \quad M_Y := \left(R^{\oplus 2}, \begin{pmatrix} 0 & Y \\ XZ & 0 \end{pmatrix}\right), \quad M_Z := \left(R^{\oplus 2}, \begin{pmatrix} 0 & Z \\ ZX & 0 \end{pmatrix}\right),$$

where $R = \mathbb{C}[X, Y, Z]$. The Hom-space from M_Z to M_Y is $\mathrm{Mat}_{2\times 2}(R)$ with the differential

$$d\begin{pmatrix} a & b \\ c & d \end{pmatrix} = \begin{pmatrix} 0 & Y \\ XZ & 0 \end{pmatrix}\begin{pmatrix} a & b \\ c & d \end{pmatrix} - \begin{pmatrix} a & -b \\ -c & d \end{pmatrix}\begin{pmatrix} 0 & Z \\ XY & 0 \end{pmatrix} = \begin{pmatrix} Yc+bXY & Yd-aZ \\ XZa-dXY & XZb+cZ \end{pmatrix}.$$

The homology of this space is given by $\mathbb{C}[X]\xi_1$ with $\xi_1 = \begin{pmatrix} 0 & -1 \\ X & 0 \end{pmatrix}$. Similar calculations show the results in the following table.

	M_X	M_Y	M_Z
M_X	$\mathbb{C}[Y]\mathbb{1} + \mathbb{C}[Z]\mathbb{1}$	$\mathbb{C}[Z]\begin{pmatrix} 0 & -1 \\ Z & 0 \end{pmatrix}$	$\mathbb{C}[Y]\begin{pmatrix} 0 & -1 \\ Y & 0 \end{pmatrix}$
M_Y	$\mathbb{C}[Z]\begin{pmatrix} 0 & -1 \\ Z & 0 \end{pmatrix}$	$\mathbb{C}[Z]\mathbb{1} + \mathbb{C}[X]\mathbb{1}$	$\mathbb{C}[X]\begin{pmatrix} 0 & -1 \\ X & 0 \end{pmatrix}$
M_Z	$\mathbb{C}[Y]\begin{pmatrix} 0 & -1 \\ Y & 0 \end{pmatrix}$	$\mathbb{C}[X]\begin{pmatrix} 0 & -1 \\ X & 0 \end{pmatrix}$	$\mathbb{C}[X]\mathbb{1} + \mathbb{C}[Y]\mathbb{1}$

We denote the six degree 1 generators by $\xi_1, \xi_2, \dots, \zeta_1, \zeta_2$ depending on the variable occurring in the matrix. Only products with the same Greek letter are nonzero because, for instance,

$$\zeta_1\eta_1 = \begin{pmatrix} 0 & -1 \\ Z & 0 \end{pmatrix}\begin{pmatrix} 0 & -1 \\ Y & 0 \end{pmatrix} = \begin{pmatrix} -Y & 0 \\ 0 & -Z \end{pmatrix} = d\begin{pmatrix} 0 & 0 \\ 1 & 0 \end{pmatrix}.$$

This allows us to conclude that the six degree 1 morphisms behave exactly like the angles in the arc collection above, so if $M = M_x \oplus M_y \oplus M_z$ then

$$\mathtt{MF}^{\pm}(M, M) \cong \mathtt{Gtl}^{\pm}\mathcal{A}.$$

This isomorphism of algebras extends to an A_∞-isomorphism. For instance, to show that $\mu(\zeta_1, \xi_1, \eta_1) = \mathbb{1}_{M_x}$, pick a split such that $\begin{pmatrix} 0 & 0 \\ 0 & 1 \end{pmatrix} \in R$ and $\begin{pmatrix} -X & 0 \\ 0 & -Y \end{pmatrix} = d\begin{pmatrix} 0 & 0 \\ -1 & 0 \end{pmatrix} \in I$. Then

$$\begin{aligned} \mu(\zeta_1, \xi_1, \eta_1) &= \pi\begin{pmatrix} 0 & -1 \\ Z & 0 \end{pmatrix} h\left(\begin{pmatrix} 0 & -1 \\ Y & 0 \end{pmatrix}\begin{pmatrix} 0 & -1 \\ X & 0 \end{pmatrix}\right) + \pi h\left(\begin{pmatrix} 0 & -1 \\ Z & 0 \end{pmatrix}\begin{pmatrix} 0 & -1 \\ Y & 0 \end{pmatrix}\right)\begin{pmatrix} 0 & -1 \\ X & 0 \end{pmatrix} \\ &= \pi\begin{pmatrix} 0 & -1 \\ Z & 0 \end{pmatrix} h\begin{pmatrix} -X & 0 \\ 0 & -Y \end{pmatrix} + \pi h\begin{pmatrix} -Y & 0 \\ 0 & -Z \end{pmatrix}\begin{pmatrix} 0 & -1 \\ X & 0 \end{pmatrix} \\ &= \pi\begin{pmatrix} 0 & -1 \\ Z & 0 \end{pmatrix}\begin{pmatrix} 0 & 0 \\ -1 & 0 \end{pmatrix} + \pi\begin{pmatrix} 0 & 0 \\ -1 & 0 \end{pmatrix}\begin{pmatrix} 0 & -1 \\ X & 0 \end{pmatrix} \\ &= \pi\begin{pmatrix} 1 & 0 \\ 0 & 0 \end{pmatrix} + \pi\begin{pmatrix} 0 & 0 \\ 0 & 1 \end{pmatrix} = -\mathbb{1}. \end{aligned}$$

Instead of checking all higher nonzero products manually, it is better to show that the values of $\mu(\zeta_1, \xi_1, \eta_1)$ of $\mu(\eta_2, \xi_2, \zeta_2)$ uniquely characterize the A_∞-structure on the algebra $\mathtt{Gtl}^{\pm}\mathcal{A}$ up to A_∞-isomorphism. This can be done using deformation theory of A_∞-algebras and Hochschild cohomology and this will be explained in more detail in Chapter 12 (and, more precisely, in Theorem 12.21).

The final part of the proof is to show that these three matrix factorizations generate $\mathrm{MF}^{\pm}(\mathbb{C}^3, XYZ)$. By Theorem 7.66, $M_X[1]$ is also the category of singularities of $\mathbb{X} = W^{-1}(0)$, which consists of the three coordinate planes. Under this correspondence, $M_X[1]$ becomes $R/(X)$, which is the structure sheaf of the YZ-plane. Because the YZ-plane is smooth, the structure sheaf generates its derived category. Inside $\mathrm{D^bCoh}\,\mathbb{X}$, $R/(X)$ generates all sheaves supported on the YZ-plane. Therefore, $\mathrm{D^bCoh}\,\mathbb{X}$ is generated by $R/(X)$, $R/(Y)$ and $R/(Z)$ and hence the category of singularities is also generated by these objects. For more details of the proof see [1, 155]. □

Remark 9.42 The category of matrix factorizations describes the singular locus of $XYZ = 0$ inside $\mathbb{C}^3$. This locus consists of three affine lines joined together at the origin. For each point on the singular locus we can find a matrix factorization by constructing a resolution for its skyscraper sheaf and looking at its 2-periodic tail. For $p = (0, 0, z)$, we get

$$M_p = \begin{pmatrix} 0 & X & 0 & 0 \\ YZ & 0 & 0 & 0 \\ Z-z & 0 & 0 & X \\ 0 & z-Z & YZ & 0 \end{pmatrix} = \mathrm{Cone}(\zeta_1\zeta_2 - z).$$

This corresponds to a circle $\mathbb{L}_p$ around one of the three marked points for the pair of pants, equipped with a local system with holonomy z.

For the zero point we obtain a matrix factorization $R \otimes \Lambda_3$ with differential $d\phi := \iota_X\phi + \iota_Y\phi + \iota_Z\phi + (YZ\,dX + XZ\,dY + XY\,dZ) \wedge \phi$. This matrix factorization is quasi-isomorphic to

$$M_0 = \mathrm{Cone}\left(M_x \oplus M_y \oplus M_z \overset{\xi_1 \pm \cdots \pm \zeta_2}{\longrightarrow} M_x \oplus M_y \oplus M_z\right),$$

which corresponds to a curve $\mathbb{L}_0$ that wraps around the "crotch" of the pants and a local system with holonomy $\lambda = -1$.[6]

Finally, the three matrix factorizations M_X, M_Y, M_Z correspond to objects supported on two of the three affine lines. Note that M_p with $p = (0, 0, z)$ can be seen as a cone over M_X or M_Y but not over M_Z. Therefore, p is in the support of M_X, M_Y but not M_Z. On the A-side this is clear because $\mathbb{L}_p$ will intersect the arcs x, y but not z.

[6] In [71, 220] this local system is realized by a nontrivial spin structure.

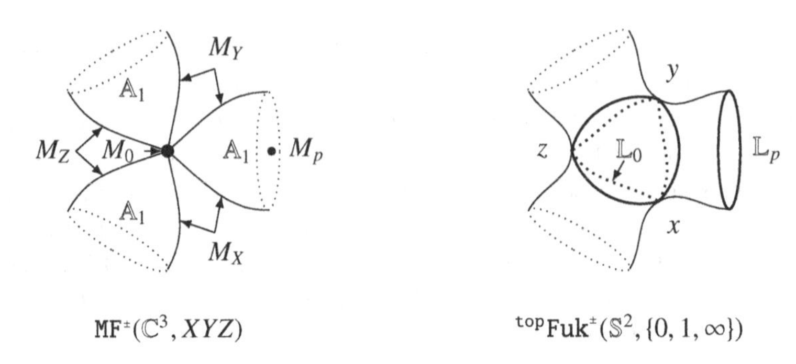

$\mathtt{MF}^{\pm}(\mathbb{C}^3, XYZ)$ $\qquad$ ${}^{\text{top}}\mathbf{Fuk}^{\pm}(\mathbb{S}^2, \{0, 1, \infty\})$

This mirror correspondence is also compatible with gluing. We will show how one can glue different pairs of pants together to obtain nonaffine Landau–Ginzburg models for curves with higher genus and more marked points.

Example 9.43 Let $\mathbb{U}_1, \mathbb{U}_2 = \mathbb{C}^3$ with $\mathbb{U}_i = \mathbb{V}(R_i)$ and $R_i = \mathbb{C}[X_i, Y_i, Z_i]$. To glue these affine spaces together we can use a transition function that maps the function $X_1Y_1Z_1$ to $X_2Y_2Z_2$ and inverts one of the coordinates. An example of such a transition is

$$X_2 = X_1Z_1, \quad Y_2 = Y_1Z_1, \quad Z_2 = Z_1^{-1}.$$

The overlap $\mathbb{U}_{12}$ is the variety associated to the ring $R_{12} = \mathbb{C}[X_1, Y_1, Z_1^{\pm 1}] = \mathbb{C}[X_2, Y_2, Z_2^{\pm 1}]$ and this ring contains the function $\ell = X_1Y_1Z_1 = X_2Y_2Z_2$.

The embedding $\mathbb{U}_{12} \to \mathbb{U}$ gives a restriction functor $\mathtt{MF}^{\pm}(R_1, \ell) \to \mathtt{MF}^{\pm}(R_{12}, \ell)$ that maps (M, d) to $(M \otimes_{R_1} R_{12}, d \otimes_{R_1} 1)$. What happens if we restrict the matrix factorizations M_X, M_Y, M_Z? If we invert Z, the morphisms ζ_1, ζ_2 become invertible and the matrix factorization M_Z becomes a zero object because

$$\mathbb{1}_{M_Z} = \begin{pmatrix} 0 & Z \\ XY & 0 \end{pmatrix}\begin{pmatrix} 0 & 0 \\ Z^{-1} & 0 \end{pmatrix} + \begin{pmatrix} 0 & 0 \\ Z^{-1} & 0 \end{pmatrix}\begin{pmatrix} 0 & Z \\ XY & 0 \end{pmatrix} = d\begin{pmatrix} 0 & 0 \\ Z^{-1} & 0 \end{pmatrix}.$$

So in $\mathtt{MF}^{\pm}(R_1, \ell)$ we have $M_X \cong M_Y[1]$ and $M_Z = 0$. Furthermore, the endomorphism ring of M_X is $\mathbb{C}[Z, Z^{-1}]\mathbb{1}_{M_X}$, so $\mathtt{MF}^{\pm}(\mathbb{C}[X, Y, Z, Z^{-1}], XYZ) \overset{\infty}{\cong} \mathtt{D}\mathbb{C}[Z, Z^{-1}]$, which is also the Fukaya category of the sphere with two marked points. If we glue $\mathbb{U}_1$ and $\mathbb{U}_2$ together over $\mathbb{U}_{12}$, we get a variety $\mathbb{X}$ with a global function $\ell = X_1Y_1Z_1 = X_2Y_2Z_2$, i.e. a nonaffine Landau–Ginzburg model.

The objects of $\mathtt{MF}^{\pm}(\mathbb{X}, \ell)$ are pairs $(\mathscr{P}, d)$ of a $\mathbb{Z}_2$-graded locally free sheaf $\mathscr{P}$ and a $d \in \operatorname{Hom}(\mathscr{P}, \mathscr{P})$ such that $d^2 = \ell$. The locally free sheaf is determined by a triple (R_1^n, R_2^n, U), where U is an invertible matrix $U \in \operatorname{Mat}_{n\times n}(R_{12})$. The differential is a pair of maps $(d_1, d_2) \in \operatorname{Mat}_n(R_1)\times\operatorname{Mat}_n(R_2)$ such that $(d_1^2, d_2^2) = (X_1Y_1Z_1, X_2Y_2Z_2)$ and $d_1 = Ud_2U^{-1}$ as matrices over R_{12}. With this in mind we can come up with four matrix factorizations:

	d_1	d_2	U
M_{Z_1}	$\begin{pmatrix} 0 & Z_1 \\ X_1Y_1 & 0 \end{pmatrix}$	$\begin{pmatrix} 0 & 1 \\ X_2Y_2Z_2 & 0 \end{pmatrix}$	$\begin{pmatrix} Z_1 & 0 \\ 0 & 1 \end{pmatrix}$
M_{Z_2}	$\begin{pmatrix} 0 & 1 \\ X_1Y_1Z_1 & 0 \end{pmatrix}$	$\begin{pmatrix} 0 & Z_2 \\ X_2Y_2 & 0 \end{pmatrix}$	$\begin{pmatrix} Z_1 & 0 \\ 0 & 1 \end{pmatrix}$
M_X	$\begin{pmatrix} 0 & X_1 \\ Y_1Z_1 & 0 \end{pmatrix}$	$\begin{pmatrix} 0 & X_2 \\ Y_2Z_2 & 0 \end{pmatrix}$	$\begin{pmatrix} Z_1^{-1} & 0 \\ 0 & 1 \end{pmatrix}$
M_Y	$\begin{pmatrix} 0 & Y_1 \\ X_1Z_1 & 0 \end{pmatrix}$	$\begin{pmatrix} 0 & Y_2 \\ X_2Z_2 & 0 \end{pmatrix}$	$\begin{pmatrix} Z_1^{-1} & 0 \\ 0 & 1 \end{pmatrix}$

The morphism space between two such objects $(\mathscr{P}, d_{\mathscr{P}})$ and $(\mathscr{Q}, d_{\mathscr{Q}})$ is the Čech complex of $\mathscr{H}om\,(\mathscr{P}, \mathscr{Q})$ with combined differential $d_C + d_{\mathrm{MF}}$:

$$\begin{aligned}
\check{\mathrm{C}}\mathrm{C} &:= (\,\mathrm{Mat}_{n\times m}(R_1) \oplus \mathrm{Mat}_{n\times m}(R_2) \oplus \mathrm{Mat}_{n\times m}(R_{12})), \\
d(\phi_1, \phi_2, \phi_{12}) &:= (d_{\mathscr{Q}1}\phi_1 - (-1)^{|\phi|}\phi_1 d_{\mathscr{P}1}, \\
&\qquad d_{\mathscr{Q}2}\phi_2 - (-1)^{|\phi|}\phi_2 d_{\mathscr{P}2}, \\
&\qquad d_{\mathscr{Q}1}\phi_{12} - (-1)^{|\phi|}\phi_{12} d_{\mathscr{P}1} + \phi_1 - U_{\mathscr{Q}}\phi_2 U_{\mathscr{P}}^{-1}).
\end{aligned}$$

Calculating the minimal model for $\mathtt{MF}^{\pm}(M, M)$ with $M = M_{z_1} \oplus M_{z_2} \oplus M_x \oplus M_y$, we end up with a quiver:

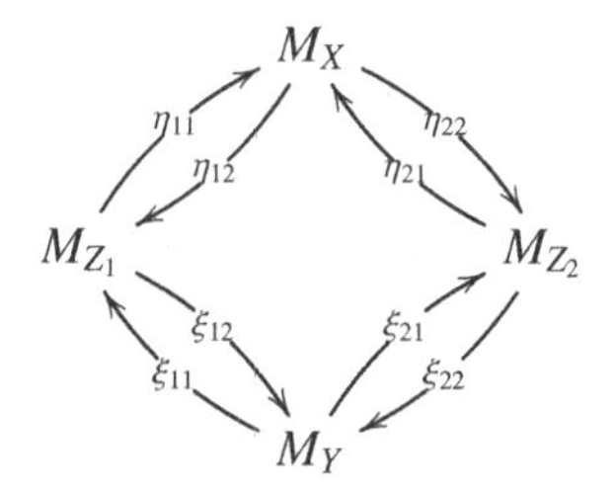

The ξ_{ij} and η_{ij} are all of the same form $(\phi_1, \phi_2, 0)$ with $\phi_1 \otimes_{R_1} R_{12} = \phi_2 \otimes_{R_{12}}$, e.g.

$$\xi_{11} = \left(\begin{pmatrix} 0 & -1 \\ X_1 & 0 \end{pmatrix}, \begin{pmatrix} 0 & -1 \\ X_2Z_2 & 0 \end{pmatrix}, 0\right).$$

A further determination of the A_∞-structure shows that $\mu(\xi_{12}, \eta_{12}, \xi_{22}, \eta_{22}) = \mathbb{1}_{M_x}$ and there is an A_∞-isomorphism between $\mathtt{MF}^{\pm}(M, M)$ and the endomorphism ring of the four sides of a square in the sphere with four marked points. Pictorially we get the following correspondence:

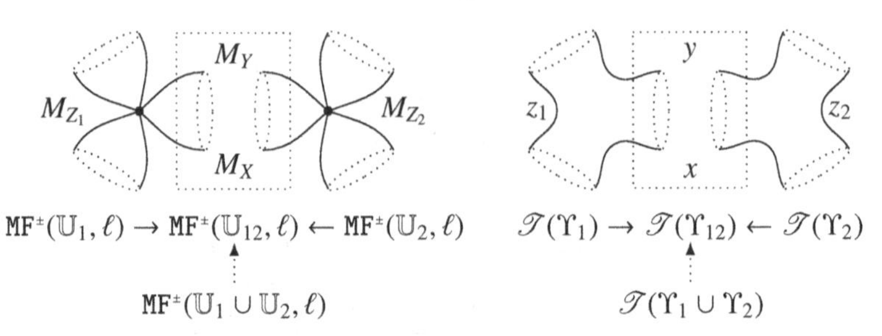

Both sides are homotopy limits of equivalent diagrams, so we get an equivalence

$$\mathtt{MF}^{\pm}(\mathbb{X}, \ell) \overset{\infty}{=} {}^{\mathtt{top}}\mathbf{Fuk}^{\pm}(\mathbb{S}^2, \{0, 1, \infty, p\}).$$

This idea extends to surfaces with higher genus and more marked points. Take a convex lattice polygon in $\mathbb{Z}^2$ with g internal lattice points and n segments on the boundary. We put the polygon a height $z = 1$ in $\mathbb{Z}^3$. Let $\mathcal{P}$ be the set of lattice points in the polygon, and for each $p \in \mathcal{P}$ let $(a_p, b_p, 1)$ be the coordinates of that lattice point.

Choose a triangulation $\mathcal{T}$ of the polygon into elementary triangles. These are triangles $v = \{r, s, t\} \subset \mathcal{P}$ that do not contain lattice points other than their corners or, stated differently, for which r, s, t form a basis of $\mathbb{Z}^3$. To each triangle we associate a subring

$$R_v = \mathbb{C}[X^i Y^j Z^k \mid a_p i + b_p j + k \geq 0,\ p \in v]$$

and an affine space $\mathbb{U}_v = \mathbb{V}(R_v)$. Because r, s, t form a basis of $\mathbb{Z}^3$, the ring R_v is isomorphic to a polynomial ring in three variables and $\mathbb{U}_v \cong \mathbb{C}^3$. The generators of R_v are the monomials $X^i Y^j Z^k$ such that the (i, j, k) form a dual basis for r, s, t. Denote these monomials by U_v, V_v, W_v. Because r, s, t all lie in the plane $z = 1$, we have the equation $Z = U_v V_v W_v$.

To each edge $e = \{r, s\}$ we assign the ring $R_e = \mathbb{C}[X^i Y^j Z^k \mid a_p i + b_p j + k \geq 0,\ p \in e]$, which is isomorphic to $\mathbb{C}[U_v, V_v, W_v, W_v^{-1}]$ because we dropped the third condition and as a consequence $\mathbb{U}_e = \mathbb{V}(R_e) \cong \mathbb{C}^2 \times \mathbb{C}^*$. If e is an edge of the triangle v then $R_v \subset R_e$ and dually $\mathbb{U}_e \subset \mathbb{U}_v$. If we glue all the $\mathbb{U}_v$ together by identifying the $\mathbb{U}_e$ over common edges, we get a three-dimensional smooth toric variety $\mathbb{Y}_{\mathcal{T}}$. All the rings are subrings of $\mathbb{C}[X^{\pm 1}, Y^{\pm 1}, Z^{\pm 1}]$ and they all contain the element $\ell = Z$. The latter can be seen as a global function on $\mathbb{Y}_{\mathcal{T}}$ and therefore we can consider the Landau–Ginzburg model $(\mathbb{Y}_{\mathcal{T}}, \ell)$.

Now look at the dual graph of the triangulation $\Gamma_{\mathcal{T}}$. It has a node for each triangle and a dual edge for each edge of the triangulation. Dual edges of edges that lie on the boundary of the polygon are half-open edges of the dual graph. Each node v comes with a Landau–Ginzburg model $(\mathbb{U}_v, \ell)$ and each edge with a model $(\mathbb{U}_e, \ell)$. If an edge e is connected to a node v we have an embedding $\mathbb{U}_e \to \mathbb{U}_v$.

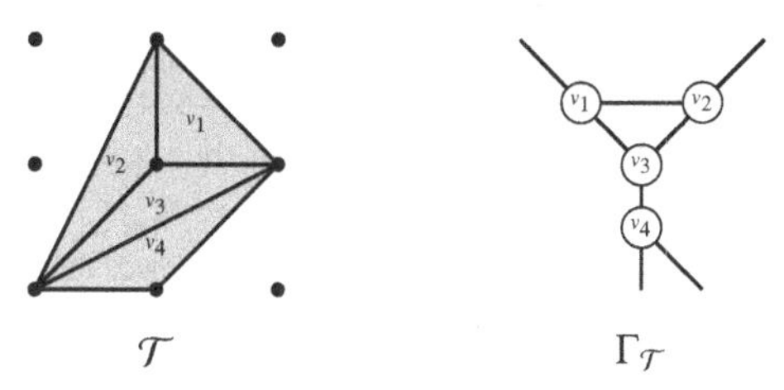

For each node v and edge e we have two categories of matrix factorizations and a restriction functor

$$\mathscr{M}(\mathbb{U}_v) := \mathtt{MF}^{\pm}(\mathbb{U}_v, \ell) \longrightarrow \mathscr{M}(\mathbb{U}_e) := \mathtt{MF}^{\pm}(\mathbb{U}_e, \ell).$$

This gives rise to diagram of A_∞-categories. The category of matrix factorizations for $(\mathbb{Y}_{\mathcal{T}}, \ell)$ can be interpreted as a homotopy limit of this diagram.

On the A-side of the story we construct a tubular thickening $\Sigma_{\mathcal{T}}$ of the dual graph $\Gamma_{\mathcal{T}}$. This gives a surface with punctures, for which the genus equals the number of holes in the graph and the number of punctures is equal to the number of legs of the tubular graph. In terms of the polygon, the genus equals the number of internal lattice points of the polygon and the number of legs is equal to the number of boundary lattice points. We can cover this graph with open stars around the nodes Υ_v and open edges Υ_e. If we apply the functor $\mathscr{T}$ we also get a diagram of Fukaya categories with ${}^{\mathtt{top}}\mathtt{Fuk}^{\pm}(\Sigma_{\mathcal{T}})$ as its homotopy limit:

$$\underbrace{\begin{array}{ccccc}
\mathscr{M}(\mathbb{U}_{e_1}) & & & & \mathscr{M}(\mathbb{U}_{e_2}) \\
\uparrow & & & & \uparrow \\
\mathscr{M}(\mathbb{U}_{v_1}) & \to & \mathscr{M}(\mathbb{U}_{e_3}) & \leftarrow & \mathscr{M}(\mathbb{U}_{v_2}) \\
\downarrow & & & & \downarrow \\
\mathscr{M}(\mathbb{U}_{e_4}) & \leftarrow & \mathscr{M}(\mathbb{U}_{v_3}) & \to & \mathscr{M}(\mathbb{U}_{e_5}) \\
 & & \downarrow & & \\
 & & \mathscr{M}(\mathbb{U}_{e_6}) & & \\
 & & \uparrow & & \\
\mathscr{M}(\mathbb{U}_{e_7}) & \leftarrow & \mathscr{M}(\mathbb{U}_{v_3}) & \to & \mathscr{M}(\mathbb{U}_{e_8})
\end{array}}_{\mathtt{MF}^{\pm}(\mathbb{Y}_{\mathcal{T}}, \ell)}
\;\stackrel{\infty}{=}\;
\underbrace{\begin{array}{ccccc}
\mathscr{T}(\Upsilon_{e_1}) & & & & \mathscr{T}(\Upsilon_{e_2}) \\
\uparrow & & & & \uparrow \\
\mathscr{T}(\Upsilon_{v_1}) & \to & \mathscr{T}(\Upsilon_{e_3}) & \leftarrow & \mathscr{T}(\Upsilon_{v_2}) \\
\downarrow & & & & \downarrow \\
\mathscr{T}(\Upsilon_{e_4}) & \leftarrow & \mathscr{T}(\Upsilon_{v_3}) & \to & \mathscr{T}(\Upsilon_{e_5}) \\
 & & \downarrow & & \\
 & & \mathscr{T}(\Upsilon_{e_6}) & & \\
 & & \uparrow & & \\
\mathscr{T}(\Upsilon_{e_7}) & \leftarrow & \mathscr{T}(\Upsilon_{v_3}) & \to & \mathscr{T}(\Upsilon_{e_8}).
\end{array}}_{{}^{\mathtt{top}}\mathtt{Fuk}^{\pm}(\Sigma_{\mathcal{T}})}$$

The idea is that the two diagrams are A_∞-equivalent and hence their limits are also equivalent.

Theorem 9.44 (Sibilla–Pascaleff, Lee) *If $\mathcal{T}$ is a triangulation of a convex lattice polygon with g internal points and n boundary points then*

$$\mathtt{MF}^{\pm}(\mathbb{Y}_{\mathcal{T}}, \ell) \overset{\infty}{=} {}^{\mathrm{top}}\mathbf{Fuk}^{\pm}(\Sigma_{\mathcal{T}}),$$

where $\Sigma_{\mathcal{T}}$ is a closed marked surface with genus g and n marked points.

Sketch of the proof To make the idea above work, we have to pass to suitable completions of these categories. After gluing in the completed setting we can establish the equivalence for the uncompleted versions by restricting to the subcategory of compact objects. For more details we refer to the work of Pascaleff and Sibilla [204]. Another approach by Lee identifies generators on both sides and can be found in [155]. □

9.5 Covers

Definition 9.45 Let $(\mathbb{S}_1, M_1)$ and $(\mathbb{S}_2, M_2)$ be two marked surfaces. A map $\pi\colon \mathbb{S}_1 \to \mathbb{S}_2$ is called a *covering* if it is a branched cover for which $\pi(M_1) = M_2$ and the branch points lie in M_2. The *degree of the cover* is $\pi^{-1}(p)$ for $p \in \mathbb{S}_2 \setminus M_2$.

If $\mathcal{A}_2$ is an arc collection that generates $(\mathbb{S}_2, M_2)$ then the pullback of $\mathcal{A}_2$ along π gives an arc collection $\mathcal{A}_1$ that generates $(\mathbb{S}_1, M_1)$. Every angle morphism between two arcs in $\mathcal{A}_1$ is mapped to an angle morphism in $\mathbb{S}_2$. Therefore, π can be turned into a strict A_∞-functor between the gentle categories,

$$\mathcal{F}_\pi\colon \mathtt{Gtl}^{\pm}\mathcal{A}_1 \to \mathtt{Gtl}^{\pm}\mathcal{A}_2\colon a \mapsto \pi(a),\ \alpha \mapsto \pi(\alpha),$$

which can be extended to a functor between the derived categories or the corresponding topological Fukaya categories. Similarly, the group of cover automorphisms of π,

$$G = \{\phi\colon \mathbb{S}_1 \to \mathbb{S}_1 \mid \pi\phi = \pi\},$$

has an action on the arc collection $\mathcal{A}_1$ which can also be turned into an action of strict A_∞-functors on $\mathtt{Gtl}^{\pm}\mathcal{A}_1$.

Now let us go in the opposite direction.

Definition 9.46 Let A be an A_∞-algebra and G be a finite group that acts as strict A_∞-isomorphisms on A. The *smash product* $A \star G$ is the algebra $A \star G :=$ $A \otimes \mathbb{C}G$ with A_∞-products

$$\mu(a_1 \otimes g_1, \ldots, a_k \otimes g_k) = \mu(a_1, g_1 \cdot a_2, \ldots, g_1 \cdots g_{k-1} \cdot a_k) \otimes g_1 \ldots g_k.$$

One can easily check that these μ satisfy the A_∞-axioms.

Suppose that we have a gentle A_∞-algebra $A = \mathbb{C}Q/I$ coming from an arc collection $\mathcal{A}$ on a marked surface $(\mathbb{S}, M)$ and an action of an *abelian* group G that fixes the vertices and scales each angle. This means that for every $\alpha \in Q_1$ there is a $\lambda_\alpha \in \widehat{G} = \mathtt{Grp}(G, \mathbb{C}^*)$ such that $g \cdot \alpha = \lambda_\alpha(g)\alpha$ and if $\alpha_1, \ldots, \alpha_k$ are the angles of a polygon then $\lambda_{\alpha_1} \cdots \lambda_{\alpha_k} = 1$.

Theorem 9.47 *In the setting above we have*

$$A \star G \cong \mathtt{Gtl}^{\pm}\tilde{\mathcal{A}},$$

where $\tilde{\mathcal{A}}$ is the pullback of $\mathcal{A}$ along a covering $\pi\colon \tilde{\mathbb{S}} \to \mathbb{S}$.

Sketch of the proof For each $\lambda \in \widehat{G}$ and each $a \in Q_0 = \mathcal{A}$ we define

$$e_\lambda := \frac{1}{|G|}\sum_{g\in\widehat{G}} \lambda(g)g \in \mathbb{C}G \quad \text{and} \quad a_\lambda := a \otimes e_\lambda \in A \star G.$$

The e_λ form a basis of orthogonal idempotents for $\mathbb{C}G$ and the a_λ form a complete set of orthogonal idempotents for $A \star G$. We use the a_λ as vertices of a new quiver $\tilde{Q}$. Furthermore, if $\alpha \in Q_1$ is an angle with $h(\alpha) = a$ and $t(\alpha) = b$ we can write

$$\begin{aligned}
\alpha \otimes 1 &= \Big(\sum_{\kappa\in\widehat{G}} a \otimes e_\kappa\Big)(\alpha \otimes 1)\Big(\sum_{\lambda\in\widehat{G}} b \otimes e_\lambda\Big)\\
&= \sum_{\kappa,\lambda\in\widehat{G}} (a \otimes e_\kappa)(\alpha \otimes e_{\kappa\lambda_\alpha})\Big(\sum_{\lambda\in\widehat{G}} b \otimes e_\lambda\Big)\\
&= \sum_{\kappa\in\widehat{G}} (a \otimes e_\kappa)(\alpha \otimes 1)(b \otimes e_{\kappa\lambda_\alpha})\\
&= \sum_{\kappa\in\widehat{G}} \underbrace{a_\kappa(\alpha \otimes 1)b_{\kappa\lambda_\alpha}}_{\alpha_\kappa}.
\end{aligned}$$

So $\alpha \otimes 1$ decomposes as $|G|$ new arrows α_κ in the quiver $\tilde{Q}$ with $h(\alpha_\kappa) = h(\alpha)_\kappa$ and $t(\alpha_\kappa) = t(\alpha)_{\kappa\lambda_\alpha}$. If $\alpha_1, \ldots, \alpha_k$ are the angles of a polygon in $\mathbb{S}$ then we can define for each $\kappa \in \widehat{G}$ a new polygon spanned by the angles

$$(\alpha_1)_\kappa, (\alpha_2)_{\kappa\lambda_{\alpha_1}}, \ldots, (\alpha_n)_{\kappa\lambda_{\alpha_1}\cdots\lambda_{\alpha_{k-1}}}.$$

If we glue all these polygons together we get a new surface $\tilde{\mathbb{S}}$ with an arc collection $\tilde{\mathcal{A}} = \tilde{Q}_0$ for which the gentle A_∞-algebra is isomorphic to $A \star G$. □

Example 9.48 Take the sphere with three marked points and construct a cover of degree 2 by a torus with three branch points. If we draw a triangle on the sphere we get an arc collection $\mathcal{A}_2 = \{x, y, z\}$, which pulls back to an arc collection $\mathcal{A}_1 = \{x_1, x_2, x_3, \ldots, z_1, z_2, z_3\}$. We can draw this arc collection on the torus as a periodic arc collection on the plane for which the fundamental domain consists of six triangles. The group of cover automorphisms is $\mathbb{Z}_3$ and it cyclically permutes the indices of the arcs.

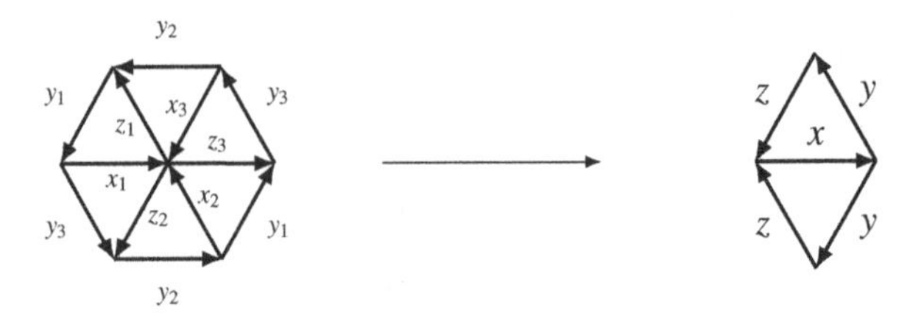

If we use the $G = \mathbb{Z}_3 = \langle g \mid g^3\rangle$-action on

$$A = \mathtt{Gtl}^{\pm}\mathcal{A} =$$ 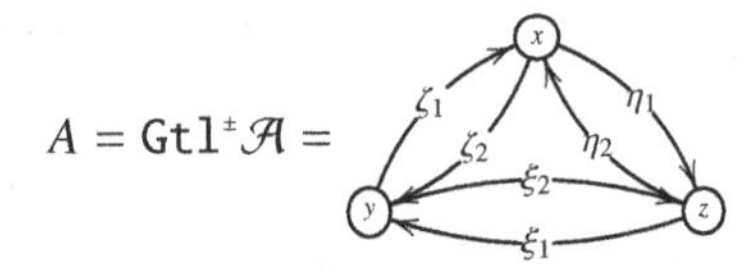

that assigns to each angle the same weight, $g \cdot \xi_i = e^{\frac{2\pi i}{3}}\xi_i, \ldots, \zeta_i = e^{\frac{2\pi i}{3}}\zeta_i$, then $A \star G$ is isomorphic to the gentle algebra of the cover above.

This covering technique allows us to produce stacky mirrors for punctured surfaces. Choose a finite abelian group G with a diagonal action on $\mathbb{C}^3$ such that the characters $\lambda_X, \lambda_Y, \lambda_Z \in \widehat{G}$ satisfy $\lambda_X\lambda_Y\lambda_Z = 1$. In this situation the function XYZ is invariant, so we can see it as a function on the stack $[\mathbb{C}^3/G]$ and look at its matrix factorizations. In algebraic terms, these are matrix factorizations of the noncommutative Landau–Ginzburg model $(\mathbb{C}[X, Y, Z] \star G, XYZ)$.

The ring $B = \mathbb{C}[X, Y, Z] \star G$ can be described by a quiver whose vertices v_ϕ are indexed by $\widehat{G}$ and $3|G|$ arrows X_ϕ, Y_ϕ, Z_ϕ that connect v_ϕ to $v_{\phi\lambda_X}$, $v_{\phi\lambda_Y}$ and $v_{\phi\lambda_Z}$. The relations between these arrows come from the commutation relations and look like

$$X_{\phi_{\lambda_Y}}Y_\phi - Y_{\phi_{\lambda_X}}X_\phi = 0, \quad Y_{\phi_{\lambda_Z}}Z_\phi - Z_{\phi_{\lambda_Y}}Y_\phi = 0, \quad Z_{\phi_{\lambda_X}}X_\phi - X_{\phi_{\lambda_Z}}Z_\phi = 0.$$

Just like every arrow has $|G|$ counterparts in the quiver, every standard matrix M_X, M_Y, M_Z factorization has $|G|$ versions: e.g.

$$M_{X_\phi} = \left(v_\phi B \oplus v_{\phi\lambda_X}B[1], \begin{pmatrix} 0 & Z_{\phi\lambda_Y\lambda_X}Y_{\phi\lambda_X} \\ X_\phi & 0 \end{pmatrix}\right).$$

The basic morphisms $\zeta_1 \colon M_X \to M_Y$ and $\zeta_2 \colon M_X \to M_Y$ turn into morphisms

$$\zeta_{1,\phi} \colon M_{X_\phi} \to M_{Y_{\phi\lambda_Z}} \quad \text{and} \quad \zeta_{2,\phi} \colon M_{Y_\phi} \to M_{X_\phi}.$$

These $2|G|$ morphisms turn into $|\widehat{G}/\langle\lambda_Z\rangle|$ cycles in the quiver.

Theorem 9.49 *If $G \subset \mathrm{SL}_3(\mathbb{C})$ acts diagonally on $\mathbb{C}^3$ then*

$$\mathtt{D}^{\pi}\mathtt{MF}^{\pm}([\mathbb{C}^3/G], XYZ) \overset{\infty}{\cong} {}^{\mathtt{top}}\mathtt{Fuk}^{\pm}(\mathbb{S}, M),$$

where $\mathbb{S}$ is a surface with $n = \frac{|G|}{|\langle\lambda_X\rangle|} + \frac{|G|}{|\langle\lambda_Y\rangle|} + \frac{|G|}{|\langle\lambda_Z\rangle|}$ marked points and Euler characteristic $n - |G|$.

Sketch of the proof Just like for the pair of pants, one can show that the M_{X_ϕ}, M_{Y_ϕ} and M_{Z_ϕ} split-generate $\mathtt{D}^{\pi}\mathtt{MF}^{\pm}([\mathbb{C}^3/G], XYZ)$. Furthermore, the endomorphism ring of these objects will be the smash product of the endomorphism ring of $M_X \oplus M_Y \oplus M_Z$ with G. From the mirror symmetry of the pair of pants, we know that this is the smash product of a gentle algebra with G. This is the gentle algebra of a covering. This surface will consists of $2|G|$ triangles and $3|G|$ arcs. The number of marked points is equal to the number of minimal cycles consisting products of ξ's, or η's or ζ's. As we have already noted above, the number of such cycles is $n = \frac{|G|}{|\langle\lambda_X\rangle|} + \frac{|G|}{|\langle\lambda_Y\rangle|} + \frac{|G|}{|\langle\lambda_Z\rangle|}$. □

Example 9.50 Let $\mathbb{Z}_{2g+1}$ act on $\mathbb{C}[X, Y, Z]$ with weights

$$(e^{2\pi i/2g+1}, e^{2\pi i/2g+1}, e^{-4\pi i/2g+1}).$$

This gives a stacky Landau–Ginzburg model that is mirror to a surface $(\mathbb{S}, M)$ with three marked points and Euler characteristic $3 - (2g + 1) = 2 - 2g$, which implies that the genus is g. We can draw this surface as a $2(2g + 1)$-gon glued together.

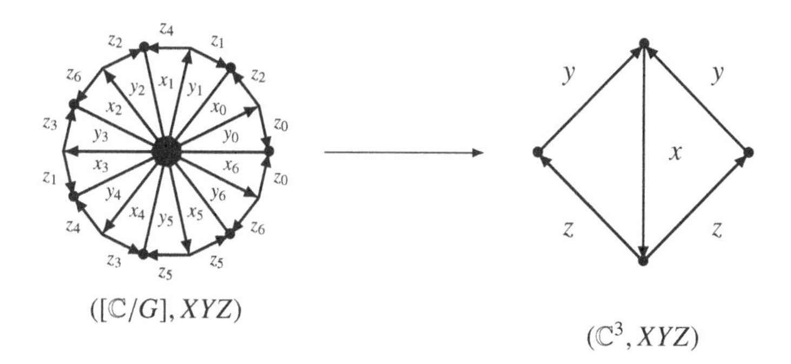

$([\mathbb{C}/G], XYZ)$ $(\mathbb{C}^3, XYZ)$

9.6 Dimer Models

We can expand this construction to include other noncommutative mirrors. In order to do this we have to introduce some technology.

Definition 9.51 A *dimer model* $Q = (\mathbb{S}, M, \mathcal{A})$ consists of a marked surface and an arc collection that splits the surface, such that the arcs around each polygon are oriented cyclically. We will consider Q as a quiver embedded in the surface $\mathbb{S}$ with $Q_0 = M$ and $Q_1 = \mathcal{A}$. Every polygon gives a cycle in Q, which we gather in a set Q_2. This set is the union of two subsets: Q_2^+ which contains the arcs that are oriented anticlockwise and Q_2^- which contains the clockwise cycles.

Remark 9.52 The notion of a dimer model originally comes from statistical physics and was defined as a bipartite graph embedded on a surface [137, 132]. This graph is the dual ribbon graph of our arc collection. In each clockwise polygon we put a white node and in each anticlockwise a black node. Nodes of adjacent faces are connected by an edge.

Definition 9.53 On the set of dimer models we have a duality called *specular duality* [121] or *dimer duality* [36, 35]. The procedure goes as follows:

(i) **cut** open the dimer Q along the arcs,

(ii) **flip** the orientation of the arcs on the clockwise polygons,

(iii) **glue** everything back together again with matching orientations.

This gives a new dimer model $Q^\vee$ and the procedure is an involution: $Q^{\vee\vee} = Q$. The dimer $Q^\vee$ is also called the *mirror dimer* of Q.

Example 9.54 Start with a dimer that splits a torus into two triangles. If we apply dimer duality we get a dimer on a sphere with three marked points.

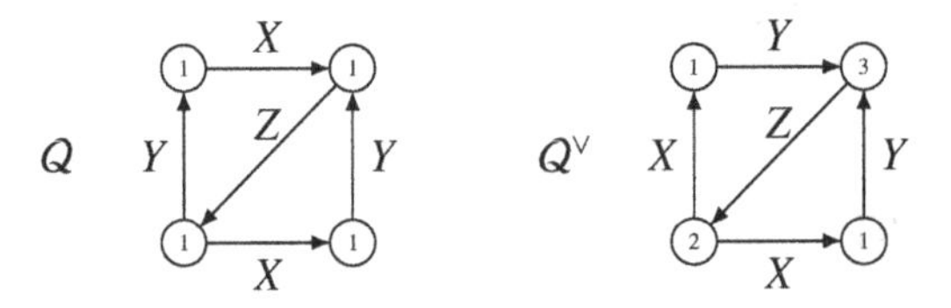

Example 9.55 Below are four more examples of this procedure from [36]. On the top row are four dimers. The first two are embedded in a torus, the third in a surface with genus 2 and the fourth in a sphere. The corresponding mirrors are on the bottom row and they are all embedded in a torus. Note that the first two dimers are isomorphic to their mirrors, but in a nontrivial way.

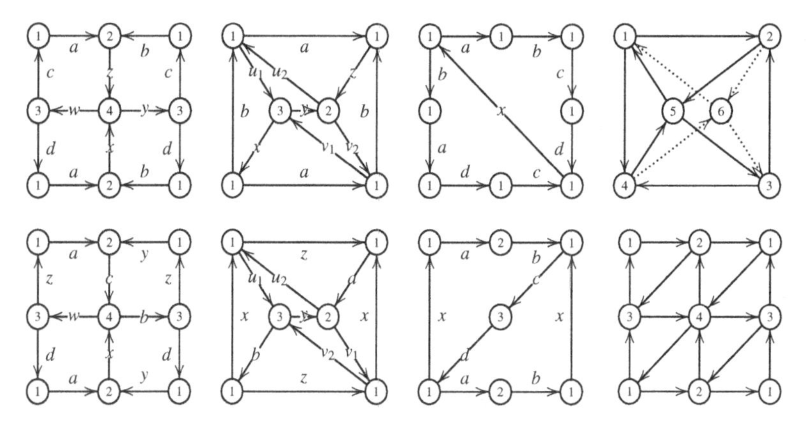

It is important to note that dimer duality messes up the genus and number of marked points of the dimer. To keep track of this we need the notion of a *zigzag path*.

Definition 9.56 A *zigzag path* is a path $a_1 \dots a_k$ in Q such that $a_i a_{i+1}$ follow each other in a positive cycle if i is even and in a negative cycle if i is odd. In other words, the path is constructed by alternately turning right and left on the surface. Starting from an arrow a we can construct two zigzag paths: the *zigzag ray* which starts by first turning left and then turning right and the *zagzig ray* which starts by turning right and then left.

Under the duality operation the arrows of a zigzag cycle will all be incident to the same marked point in the dual dimer, so the duality operation swaps marked points and zigzag cycles.

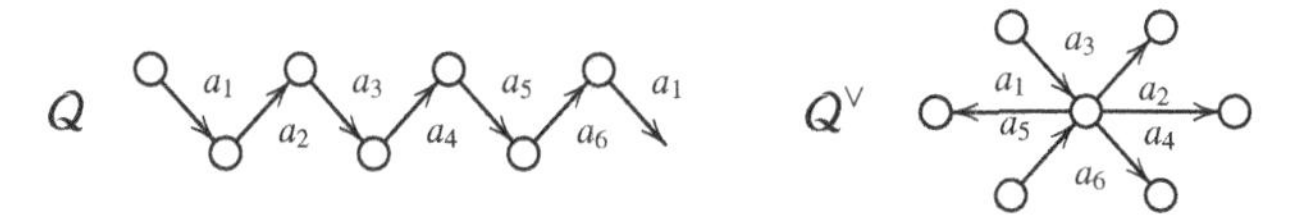

Lemma 9.57 *If Q is a dimer with n marked points, z zigzag cycles and Euler characteristic χ then $Q^\vee$ will be a dimer with z marked points, n zigzag cycles and Euler characteristic $\chi - n + z$.*

Definition 9.58 To a dimer Q we can associate a noncommutative algebra called its *Jacobi algebra*. It is the path algebra of the quiver Q modulo relations of the form $l_a - r_a$, where a is an arrow and $l_a a$ and $r_a a$ are the two cycles in Q_2 that contain a:

$$\mathtt{Jac}\, Q := \frac{\mathbb{C}Q}{\langle l_a - r_a \mid a \in Q_1,\ al_a \in Q_2^+,\ ar_a \in Q_2^- \rangle}.$$

l_a a r_a

These relations imply that for every arrow, going back along the cycle on the left is the same as going back along the cycle on the right.

Lemma 9.59 *The Jacobi algebra* $\mathtt{Jac}\, Q$ *has a central element* ℓ *such that for each* $v \in Q_0$, $v\ell$ *is a cycle in* Q_2.

Sketch of the proof Fix for each v a cyclic path ℓ_v that starts at v and represents a cycle from Q_2. Because of the relations, ℓ_v as an element in $\mathtt{Jac}\, Q$ does not depend on the specific choice of cyclic path. Furthermore, if p is any path, one can use the relations to show that $\ell_{h(p)}p = p\ell_{t(p)}$ and therefore $\ell := \sum_{v\in Q_0} \ell_v$ is central. □

Definition 9.60 The *Landau–Ginzburg model of* Q is the pair $(\mathtt{Jac}\, Q, \ell)$.

Example 9.61 Start with the dimer on a torus that splits it into two triangles (from Example 9.54). The Jacobi algebra of Q is the polynomial ring in three variables,

$$\mathtt{Jac}\, Q = \frac{\mathbb{C}\langle X, Y, Z\rangle}{\langle XY - YX, XZ - ZX, YZ - ZY\rangle},$$

and the central element is $\ell = XYZ$. Its Landau–Ginzburg model is mirror to the pair of pants and therefore $Q^\vee$ sits on a sphere with three marked points.

This construction is also compatible with the coverings. Start with the one-punctured torus and enlarge the fundamental domain of the torus by choosing two vectors $v_1, v_2 \in \mathbb{Z}^2$ and considering the surface $\mathbb{S} = \mathbb{R}^2/\mathbb{Z}v_1 + \mathbb{Z}v_2$. For the marked points we take the integral points $M = \mathbb{Z}^2/\mathbb{Z}v_1 + \mathbb{Z}v_2$ and for the arc collection we take arrows that start at a marked point and have direction vectors $(1, 0)$, $(0, 1)$ or $(-1, -1)$. Call these arrows x_m, y_m, z_m according to their direction and starting point $m \in M$. In total there are $k = |\det(v_1 v_2)|$ marked points and $3k$ arrows, which split the torus into $2k$ triangles. Let us denote this dimer model by Q_{v_1,v_2}.

Note that $M = \mathbb{Z}^2/\mathbb{Z}v_1 + \mathbb{Z}v_2$ is actually a group. If we set $G = \mathrm{Hom}(M, \mathbb{C}^*)$ then M is the character group of G and we can define an action of G on $\mathbb{C}[X, Y, Z]$ by $g \cdot X = g(1, 0) \cdot X$, $g \cdot Y = g(0, 1) \cdot Y$ and $g \cdot Z = g(-1, -1) \cdot Z$. Here $(1, 0)$, $(0, 1)$, $(-1, -1)$ are considered as elements in M, so we can apply $g \in G = \mathrm{Hom}(M, \mathbb{C}^*)$ to them.

Lemma 9.62 *The Jacobi algebra of the dimer model* Q_{v_1,v_2} *is isomorphic to* $\mathbb{C}[X, Y, Z] \star G$.

Proof The vertices of the quiver underlying $\mathbb{C}[X, Y, Z] \star G$ are the characters of G and hence they correspond to M. The arrows in the quiver are of the form X_ϕ, Y_ϕ, Z_ϕ, where $\phi \in \widehat{G}$, so they naturally correspond to the x_m, y_m, z_m. This gives a natural surjection from $\mathbb{C}Q$ to $\mathbb{C}[X, Y, Z] \star G$, which factors through $\mathtt{Jac}\, Q$ because the relations in $\mathtt{Jac}\, Q$ are all of the form $x_\phi y_\psi - y_\phi x_\psi$ (or similar versions with other pairs from $\{x, y, z\}$). These relations become $[X, Y]e_\psi = 0$ under the map.

The relations in the Jacobi algebra imply that each path is uniquely determined by its starting point and the number of x's, y's and z's. Therefore, no two paths in the Jacobi algebra can be mapped to the same element in $\mathbb{C}[X, Y, Z] \star G$, so the map is also injective. □

If we view these stacky Landau–Ginzburg models as noncommutative Landau–Ginzburg models, we can interpret mirror symmetry as an instance of dimer duality. Each arrow a in the quiver $Q = Q_{v_1,v_2}$ has a natural matrix factorization of $\ell \in \mathtt{Jac}\, Q$ associated to it:

$$M_a = t(a)\mathtt{Jac}\, Q \underset{r_a\cdot}{\overset{a\cdot}{\rightleftarrows}} h(a)\mathtt{Jac}\, Q\,.$$

As we saw in Theorem 9.49, the endomorphism ring of $\bigoplus_a M_a$ is A_∞-isomorphic to the gentle algebra of an arc collection on a surface. If one does the calculation, this arc collection is precisely that of the mirror dimer of $Q^\vee$:

$$\mathrm{End}_{\mathtt{MF}^\pm(\mathtt{Jac}\, Q,\ell)} \bigoplus_{a\in Q_1} M_a \overset{\infty}{=} \mathtt{Gtl}^\pm Q^\vee.$$

Example 9.63 Let $v_1 = (g+1, g)$ and $v_2 = (g, g+1)$. In that case, $M = \mathbb{Z}/v_1\mathbb{Z} + v_2\mathbb{Z} \cong \mathbb{Z}/(2g+1)\mathbb{Z} = \langle(1, 1)\rangle$ and denote a dual generator in $\mathrm{Hom}(M, \mathbb{C}^*)$ that maps $(1, 1)$ to $e^{\frac{2\pi i}{2g+1}}$ by u. We have $u\cdot X = e^{\frac{2\pi i}{2g+1}} X$, $u \cdot Y = e^{\frac{2\pi i}{2g+1}} Y$, $u \cdot Z = e^{-2\frac{2\pi i}{2g+1}} Z$.

Q

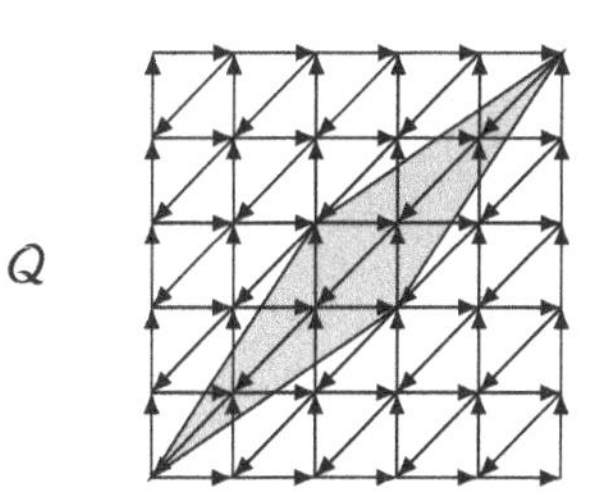

$Q^\vee$

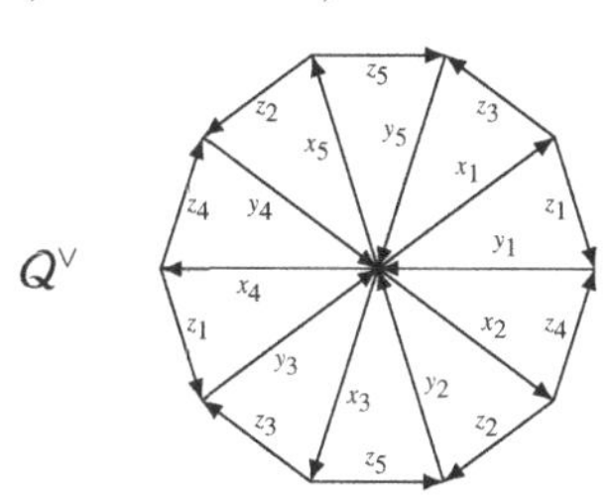

From the dimer point of view we get a dimer Q with genus 1 and $2g+1$ marked points that is dual to a dimer $Q^\vee$ with genus g and 3 marked points. This is the same as Example 9.50.

From this example one might expect that such dimer duality is always true for any dimer. Unfortunately, this is not the case. The crucial issue is that in some cases the Jacobi algebra is not nice enough (e.g. finite global dimension) and therefore we have to put extra conditions on the dimer to ensure that it is.

These are called *consistency conditions*. Many different consistency conditions have been studied in the literature [180, 63, 42, 126, 33] but we will focus on one condition in particular.

Definition 9.64 A dimer model Q is called *zigzag consistent* if for every arrow $a \in Q_1$ the zigzag ray and zagzig ray do not intersect in the universal cover (apart from at a itself).

Example 9.65 Let us make a table of consistency properties for the eight dimers in Example 9.55.

Z	¬Z	Z	¬Z
Z	¬Z	¬Z	Z

For the second one in the top row, one can easily check that the zig and zag rays that start at z also meet at x.

Theorem 9.66 *If Q is a zigzag consistent dimer then*

$$\mathrm{End}_{\mathtt{MF}^{\pm}(\mathtt{Jac}\,Q,\ell)} \bigoplus_{a \in Q_1} M_a \overset{\infty}{=} \mathtt{Gtl}^{\pm} Q^{\vee}.$$

Sketch of the proof If M_a and M_b are two matrix factorizations coming from arrows and $ax_1 \dots x_k b$ is a zigzag path then we can construct a morphism from M_a to M_b using the opposite paths of the zigzag path. These are the two paths z_+ and z_- that run back along the positive and negative cycles that bound the zigzag path.

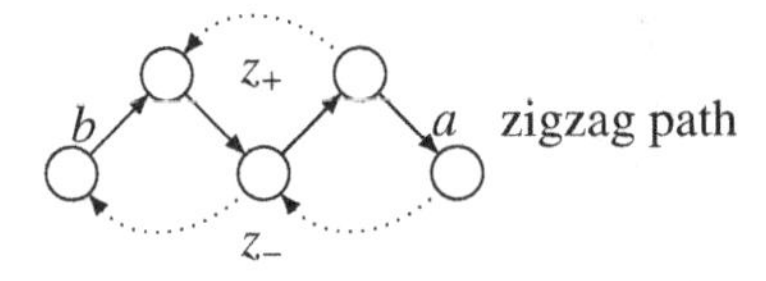

This morphism looks like

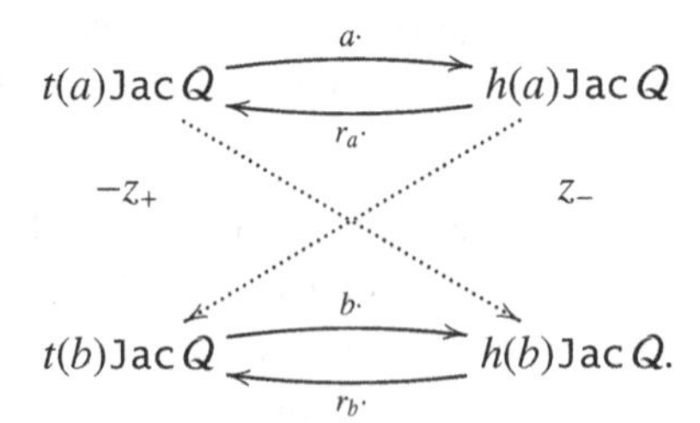

Under the dimer duality construction, zigzag cycles become marked points and zigzag paths can be seen as angles that turn around them or, in other words, paths in the gentle algebra of $Q^\vee$. If the dimer is zigzag consistent, one can show that this interpretation can be extended to an A_∞-isomorphism between $\mathrm{End}_{\mathtt{MF}^\pm(\mathtt{Jac}\,Q,\ell)} \bigoplus_{a\in Q_1} M_a$ and $\mathtt{Gtl}^\pm Q^\vee$. For more details of the proof we refer to [36]. □

Example 9.67 Take a torus with two marked points and tile it with two squares. The dual dimer is a sphere with four marked points. One can easily check that the first dimer is consistent: in the universal cover the zigzag rays are staircases in different directions, so they can only meet at one arrow. The second dimer is not consistent because it sits on a sphere, so zigzag rays will always intersect multiple times.

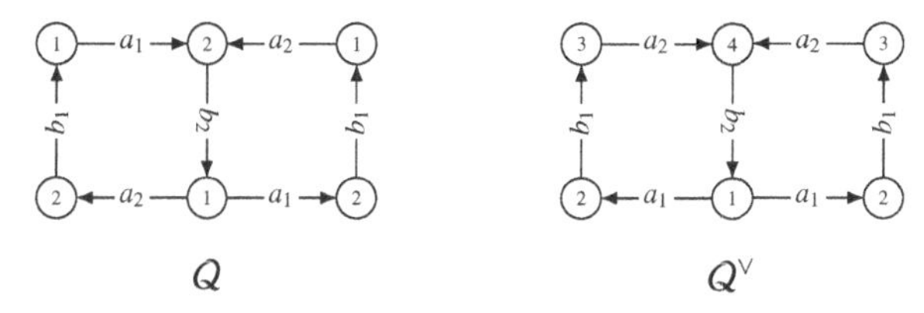

The Jacobi algebra is

$$\mathtt{Jac}\,Q := \circ \underset{b_1,b_2}{\overset{a_1,a_2}{\rightleftarrows}} \circ \,/\, I,$$

where I is generated by the four relations

$$a_1b_1a_2 - a_2b_1a_1, \quad a_1b_2a_2 - a_2b_2a_1, \quad b_1a_1b_2 - b_2a_1b_1, \quad b_1a_2b_2 - b_2a_2b_1.$$

The central element is $\ell = a_1b_1a_2b_2 + b_1a_2b_2a_1$. The noncommutative Landau–Ginzburg model $(\mathtt{Jac}\,Q, \ell)$ is mirror to the sphere with four marked points.

Just like the mirror of the pair of pants is constructed as a Landau–Ginzburg model sitting in a smooth three-dimensional space $\mathbb{A}^3$, these Jacobi algebras should be seen as noncommutative spaces. If the dimer model Q is consistent and its surface is a torus then $\mathtt{Jac}\,Q$ is in fact derived equivalent to a genuine commutative space, which is a three-dimensional toric Calabi–Yau variety. The normals to the edges of its Newton polygon are in one-to-one correspondences with the directions of the zigzag rays in the dimer (if we draw these in the universal cover of the dimer, where the fundamental domain is a unit square). For more details about the relation between three-dimensional toric Calabi–Yau varieties and dimer models we refer to [125, 127, 179].

Example 9.68 Take for instance the dimer Q from [35], which sits in a torus and has three marked points, while its mirror $Q^\vee$ sits in a sphere and has five marked points.

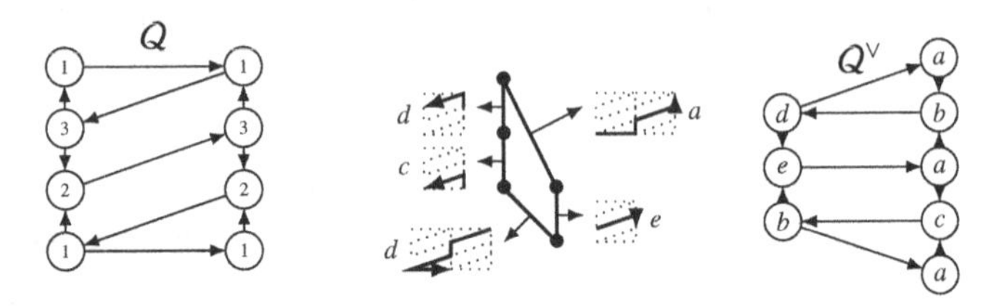

The dimer Q has five zigzag paths and in the center of the image we have drawn a Newton polygon $\square$ whose edges are normal to these zigzag paths. The Jacobi algebra $\mathtt{Jac}\, Q$ is derived equivalent to the three-dimensional toric variety $\mathbb{X}^3_{\boxtimes}$ from Section 8.4.1.

9.7 Mirrors Galore

Given a marked surface $(\mathbb{S}, M)$ with genus g and n internal marked points, it is possible to construct several different types of mirrors. Within each flavor there are also different possibilities that depend on some extra combinatorial data.

- A commutative Landau–Ginzburg model can be made from the triangulation of a Newton polygon. This gives rise to a $3d$-toric variety $\mathbb{X}_{\boxtimes}$ with a global function ℓ:

$$^{\mathtt{top}}\mathbf{Fuk}^{\pm}(\mathbb{S}, M) \cong \mathtt{D}^{\pi}\mathtt{MF}^{\pm}(\mathbb{X}_{\boxtimes}, \ell).$$

  The number of marked points is equal to the number of boundary lattice points and the genus of $\mathbb{S}$ is equal to the number of internal lattice points.
- A stacky Landau–Ginzburg model can be made from a finite abelian group $G \subset \mathrm{SL}_3(\mathbb{C})$. This gives rise to a $3d$-stack $[\mathbb{C}^3/G]$ with an invariant function $\ell = XYZ$:

$$^{\mathtt{top}}\mathbf{Fuk}^{\pm}(\mathbb{S}, M) \cong \mathtt{D}^{\pi}\mathtt{MF}^{\pm}([\mathbb{C}^3/G], \ell).$$

  The number of marked points is equal to the sum of the stabilizers of the functions X, Y and Z: $n = |G_X| + |G_Y| + |G_Z|$. The genus of $\mathbb{S}$ is equal to $\frac{1}{2}(|G| - |G_X| - |G_Y| - |G_Z|) + 1$.
- A noncommutative Landau–Ginzburg model can be made from a zigzag-consistent dimer model Q. This gives a Jacobi algebra $\mathtt{Jac}\, Q$ and a central element ℓ. Each arrow $a \in Q_1$ corresponds to a matrix factorization M_a of

$(\mathtt{Jac}\, Q, \ell)$ and to an arc $a \in Q^\vee$, which gives an object in ${}^{\mathrm{top}}\mathtt{Fuk}^{\pm}(\mathbb{S}, M)$. We have an A_∞-isomorphism of A_∞-categories

$$\mathrm{D}^\pi\mathtt{MF}^{\pm}(\mathtt{Jac}\, Q, \ell) \supset \langle M_a \rangle \overset{\infty}{\cong} \langle L_a \mid a \in Q_0 \rangle \subset {}^{\mathrm{top}}\mathtt{Fuk}^{\pm}(\mathbb{S}, M).$$

The number of marked points n is equal to the number of zigzag cycles in Q and the genus of $\mathbb{S}$ is equal to $\frac{1}{2}(\#Q_1 - \#Q_2 - n) + 1$.

9.8 Exercises

Exercise 9.1 Let $\mathbb{S}$ be a marked surface with genus g, m boundary components with $(b_i)_{1\le i\le m}$ marked points and k internal marked points. Give a formula for the smallest number of arcs needed to split the surface.

Exercise 9.2 Find a surface with an arc collection whose gentle algebra is the path algebra of the quiver

$$\circ \underset{b_1}{\overset{a_1}{\rightrightarrows}} \circ \underset{b_2}{\overset{a_2}{\rightrightarrows}} \cdots \underset{b_k}{\overset{a_k}{\rightrightarrows}} \circ,$$

with relations $a_{i+1}a_i = b_{i+1}b_i = 0$ and all arrows have degree 0. Express its genus and number of boundary components in terms of k.

Exercise 9.3 Let γ be a curve that starts at two different marked points and self-intersects once, forming an incontractible loop. Determine the endomorphism ring of the corresponding string object in $\mathtt{DGtl}^{\pm}\mathcal{A}$.

Exercise 9.4 Take a polygon and triangulate it. Use the triangulation to write the Fukaya category of the polygon as a homotopy colimit of categories of the form $\mathrm{D}\circ\!\longrightarrow\!\circ$.

Exercise 9.5 Work out Example 9.40 for a cylinder with two marked points on each side.

Exercise 9.6 Consider the polygon spanned by $(1,0,1)$, $(0,1,1)$ and $(-1,-1,1)$ and triangulate it by connecting all corners with $v_0 = (0,0,1)$. The corresponding smooth three-dimensional variety $\mathbb{X}^3_{\boxtimes}$ is the total space of the anticanonical bundle of $\mathbb{P}^2$ and let $\mathbb{D} = \mathbb{P}^2$ be the divisor corresponding to v_0.

(i) Identify the genus and number of punctures of the surface $\mathbb{S}_{g,n}$ that is mirror to $(\mathbb{X}^3_{\boxtimes}, \ell)$.

(ii) Find a matrix factorization M of ℓ that corresponds to the image of the structure sheaf $\mathcal{O}_{\mathbb{D}}$ in the identification $\mathtt{MF}^{\pm}(\mathbb{X}^3_{\boxtimes}, \ell) = \mathtt{DSing}\, \ell^{-1}(0) = \frac{\mathtt{DCoh}\, \ell^{-1}(0)}{\mathtt{Perf}\, \ell^{-1}(0)}$. (Hint: glue matrix factorizations of the form M_X in $\mathtt{MF}^{\pm}(\mathbb{C}^3, XYZ)$.)

(iii) Using the homotopy limit interpretation, identify the curve in $\mathbb{S}_{g,n}$ corresponding to M.

Exercise 9.7 Check that the construction of the smash product $A \star G$ of an A_∞-algebra with a finite group is again an A_∞-algebra.

Exercise 9.8 Let $G = \mathbb{Z}_2 \times \mathbb{Z}_2$ act on $\mathbb{C}^3$ with

$$(a,b) \cdot (X,Y,Z) = ((-1)^a X, (-1)^b Y, (-1)^{a+b} Z).$$

Find the mirror for the stacky Landau–Ginzburg model $([\mathbb{C}^3/G], XYZ)$ and describe it as a cover over the three-punctured sphere. Also give the dimer model Q for which $\mathtt{Jac}\, Q \cong \mathbb{C}[X,Y,Z] \star G$.

Exercise 9.9 Let Q be a zigzag-consistent dimer and let $v \in Q_0$ be a vertex with two arrows a_1, a_2 arriving and two arrows b_1, b_2 leaving. *Mutate* the vertex by reversing these arrows and adding new arrows for the paths $b_i a_j$. Finally, remove all 2-cycles.

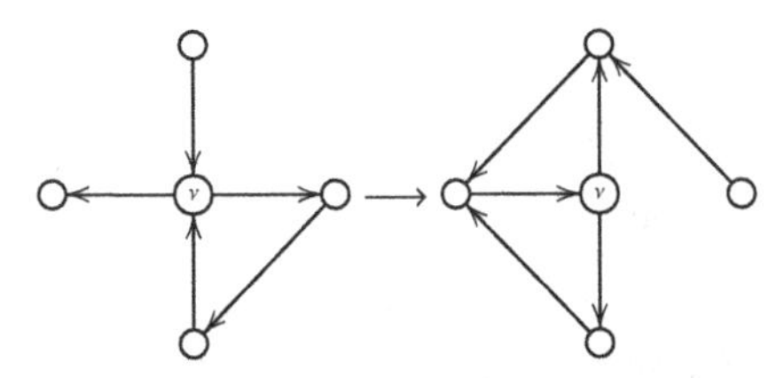

Show that the mutated dimer $\mu_v Q$ is also zigzag consistent.

Exercise 9.10 Let $G = \mathbb{Z}_3 = \langle g \mid g^3 \rangle$ act on $\mathbb{C}^3$ with

$$g \cdot (X,Y,Z) = (e^{2\pi i/3} X, e^{2\pi i/3} Y, e^{2\pi i/3} Z).$$

The algebra $\mathbb{C}[X,Y,Z] \star G$ can be seen as a path algebra with relations of the following quiver:

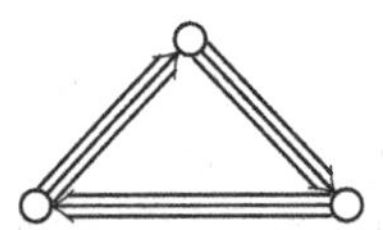

(i) Draw Q as a dimer Q on a torus such that $\mathtt{Jac}\, Q \cong \mathbb{C}[X,Y,Z] \star G$.
(ii) Show that the moduli space $\mathcal{M}_\theta(\mathtt{Jac}\, Q, \alpha)$ with dimension vector $\alpha = (1,1,1)$ and $\theta = (-2,1,1)$ is isomorphic to the variety from Exercise 9.6.
(iii) Find a matrix factorization of $(\mathtt{Jac}\, Q, \ell)$ that corresponds to the structure sheaf of the central $\mathbb{P}^2$.

10
Grading

In the previous chapter we used arc collections on marked surfaces to define $\mathbb{Z}_2$-graded Fukaya categories. In this chapter we will study how to upgrade this construction to the $\mathbb{Z}$-graded setting following the work of Haiden, Katzarkov and Kontsevich [115]. This will also enable us to get mirror correspondences with actual varieties instead of Landau–Ginzburg models. This was shown by Lekili and Polishchuk in [165].

10.1 Graded Surfaces

Definition 10.1 A *graded marked surface* $\vec{\mathbb{S}}$ consists of a triple $(\mathbb{S}, M, \eta)$ such that $(\mathbb{S}, M)$ is a marked surface and η is a smooth map that assigns to each point $p \in \dot{\mathbb{S}} := \mathbb{S} \setminus M$ a real one-dimensional subspace $\eta_p \subset \mathsf{T}_p\dot{\mathbb{S}}$. The map η is called a *line field* or *one-dimensional foliation* and it can be seen as a section of the bundle $\mathbb{P}\mathsf{T}\dot{\mathbb{S}}$ whose fiber at p is the projective line $\mathbb{P}(\mathsf{T}_p\mathbb{S})$. The *leaves* of η are the one-dimensional submanifolds $L \subset \dot{\mathbb{S}}$ such that $\mathsf{T}_pL = \eta_p$ for all $p \in L$.

A line field can be used to define graded arcs and curves. To do this we will temporarily fix a metric on $\dot{\mathbb{S}}$. This metric can be used to measure angles between tangent vectors on the surface and define a cover map

$$\dot{\mathbb{S}} \times \mathbb{R} \to \mathbb{P}\mathsf{T}\dot{\mathbb{S}} \colon (p, \theta) \mapsto V_p,$$

where $V_p \in \mathbb{P}(\mathsf{T}_p\mathbb{S})$ is the subspace that makes an angle θ with η_p. At each point $p \in \dot{\mathbb{S}}$ this map is the universal cover of $\mathbb{P}(\mathsf{T}_p\mathbb{S})$, and two points (p, θ_1), (p, θ_2) are mapped to the same point if they differ by a multiple of π. A grading of a curve will be a lift of the curve along this cover map. Concretely, this results in the following definition.

Definition 10.2 Let $\mathbb{L} \colon I \to \mathbb{S}$ be either an arc ($I = (0, 1)$) or a closed curve ($I = \mathbb{R}/\mathbb{Z}$). A grading of $\mathbb{L}$ is a map $\theta \colon I \to \mathbb{R}$ such that $\theta(t) \bmod \pi$ is equal to the angle between η_p at $p = \mathbb{L}(t)$ and the tangent line of $\mathbb{L}$. This angle is called the *phase*. The pair $\vec{\mathbb{L}} = (\mathbb{L}, \theta)$ is called a *graded arc* or curve . The nth *shift* of $\vec{\mathbb{L}}$ adds $n\pi$ to the phase: $\vec{\mathbb{L}}[n] := (\mathbb{L}, \theta + n\pi)$.

Remark 10.3 Although the definition we use seems to depend on our choice of metric, we can easily transport gradings between different metrics because different metrics result in homotopic cover maps $\dot{\mathbb{S}} \times \mathbb{R} \to \mathbb{P}\mathrm{T}\dot{\mathbb{S}}$.

Remark 10.4 If $\mathbb{L}\colon I \to \mathbb{S}$ is a curve and we fix $\theta(t)$ for one value of $t \in I$ then θ is fixed for all $u \in (0, 1)$ because we can transport the angle along the curve:

$$\theta(u) = \theta(t) + \int_t^u \frac{d\angle(\dot{\mathbb{L}}(t), \eta_{\mathbb{L}(p)})}{dt}\, dt.$$

Let us denote this contribution by $\Delta\mathbb{L}[t, u]$. For an arc $\mathbb{L}$ this allows us to construct a unique grading for every possible choice of $\theta(t) \in \angle(\dot{\mathbb{L}}(t), \eta_{\mathbb{L}(t)}) + \pi\mathbb{Z}$, so there is a bijection between the gradings of $\mathbb{L}$ and $\mathbb{Z}$.

For a closed curve the situation is more complicated. In general, if we transport the phase along a closed curve $\mathbb{L}$ the phase will have changed by a multiple of π. This number is called the *winding number* $w(\eta, \mathbb{L}) = \frac{\Delta\mathbb{L}[0,1]}{\pi}$. In order to define a grading, the winding number must be 0; otherwise $\theta\colon I \to \mathbb{R}$ cannot be well defined. Closed curves for which this is true are called *gradable*. Just like for arcs, there is a bijection between gradings on gradable closed curves and elements of $\mathbb{Z}$.

Example 10.5 If $V\colon \mathbb{S} \to \mathrm{T}\mathbb{S}$ is a vector field whose zeros are in M then we define $\eta_p = \mathbb{R}V_p$. The integral curves of the vector field can be graded by setting $\theta = 0$. To each marked point $m \in M$ we can assign an index: choose a small anticlockwise circle γ_m around m and measure its winding number around that circle. The index of m is

$$I_m = 1 - \frac{w(\eta, \gamma_m)}{2}.$$

Note that the small circle is gradable if the index of m is 1. More generally, if a curve on $\mathbb{S}$ bounds a disk $\mathbb{D}$, then it is gradable if the total sum of the indices of the marked points in $\mathbb{D}$ is equal to 1.

In general it is easy to determine the index of a marked point. If the integral curves of the vector field in a neighborhood of m circle around m then the index is 1. If not then we can split the neighborhood of the marked point into wedges and take the sum

$$1 - \frac{\#}{2} + \frac{\#}{2} + 0\frac{\#}{2}.$$

The famous *Poincaré index theorem* states that the sum of the indices is equal to the Euler characteristic of $\mathbb{S}$.

Example 10.6 If $\mathbb{S}$ is a Riemann surface with a complex 1-form ω that does not have zeros or poles outside M then we can define the line field as the one-dimensional subspace of $\eta_p \subset \mathsf{T}_p\mathbb{S}$ for which ω is real valued. Locally we can find a coordinate z such that $\omega = dz$ and then η_p is just the real axis of the tangent space.

Around a zero or a pole, ω looks like $z^k\,dz$ with $k \in \mathbb{Z}$ or $\lambda z^{-1}\,dz$. The corresponding line fields look like

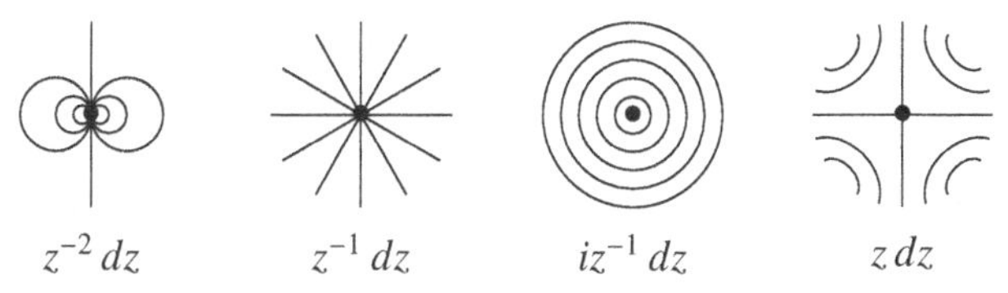

so the index is $-k$.

Example 10.7 We can also use a *quadratic differential* ϕ without zeros or poles outside M. This is a section of $(\mathsf{T}^*_{\mathbb{C}}\mathbb{S})^{\otimes 2}$. Locally we can represent this as an expression $\phi = f(z)\,dz^2$ and if $f(0) \neq 0$ we change coordinates such that $\phi = du^2$ and then we define η_p to be the real axis of the tangent space according to this new coordinate. Alternatively, we can define η_p as the line on which ϕ is real valued and positive, which for $\phi = f(z)\,dz^2$ means the direction $\mathbb{R}e^{-\theta i/2}$, where $\theta = \arg f(z)$. Around zeros and poles the line field will look like

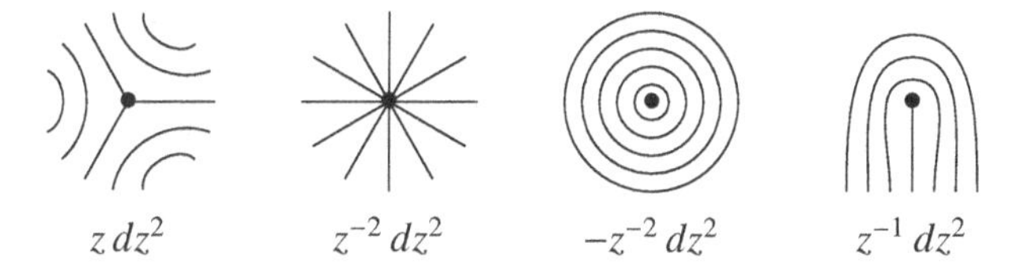

so the index of $z^k\,dz^2$ is $-k/2$. Note that, unlike for vector fields and 1-forms, the indices of these line fields can be half-integers.

Definition 10.8 We will say that the *parity* of a line field is *even* if all winding numbers of curves are even or, equivalently, if it comes from a vector field. If at least one curve has an odd winding number we say that the parity is odd. We denote the parity η of a line field by $\sigma(\eta) \in \mathbb{Z}/2\mathbb{Z}$.

Let $\mathcal{A}$ be an arc collection that splits a graded surface $\vec{\mathbb{S}}$. Choose a grading for each arc and denote the set of graded arcs by $\vec{\mathcal{A}}$. If α is an angle around the marked point p between the graded arcs $\vec{a} = (a, \theta_a)$ and $\vec{b} = (b, \theta_b)$, we set

$$\deg_{\mathcal{A}} \alpha := \frac{\Delta\alpha - (\theta_b(p) - \theta_a(p))}{\pi},$$

where $\Delta\alpha$ is the amount the line field changes as we move around a small circle segment corresponding to the angle α. This defines a grading of the gentle algebra that is compatible with the A_∞-structure.

Remark 10.9 Again note that this definition does not depend on the choice of our metric to measure the angles. Alternatively, it can be defined as follows. If $\alpha\colon \mathbb{L}_1 \to \mathbb{L}_2$ is an angle morphism, isotope $\mathbb{L}_1$ along the path α to a new curve $\mathbb{L}_1'$ such that the ends of $\mathbb{L}_1'$ and $\mathbb{L}_2$ match. During the isotopy we can also transport the grading of $\mathbb{L}_1$. The degree of α is the shift n such that the gradings of $\mathbb{L}_1'$ and $\mathbb{L}_2[n]$ coincide at the overlapping end.

Lemma 10.10 *The product μ_k has degree $2 - k$ for the $\deg_{\mathcal{A}}$-grading.*

Sketch of the proof For μ_2 it is clear that the degree is additive. If $\alpha_1, \dots, \alpha_k$ are the internal angles of a polygon then its product will be the identity morphism of an arc, so it has degree 0. To determine the total degree of the α_i, we can choose a metric for which the polygon is flat. The fact that the product has degree $2-k$ is now equivalent to the fact that the sum of the angles of a k-gon is $(k-2)\pi$. □

Example 10.11 In the previous chapter we saw that the gentle algebra came with a $\mathbb{Z}_2$-grading coming from the orientations of the arcs. This grading is not necessarily compatible with the grading coming from the line field.

Take for instance the Riemann sphere $\mathbb{S} = \widehat{\mathbb{C}}$ with the quadratic differential $\phi = (1 - z^2)^{-1}\, dz^2$. On the sphere we take three marked points $M = \{-1, 1, \infty\}$. The indices of these marked points are $\frac{1}{2}$, $\frac{1}{2}$ and 1 (for the last one use the substitution $z = u^{-1}$ and then $\phi = u^{-2}(u^2 - 1)^{-1}(du)^2$ so it has index $-(-2/2) = 1$).

For the arc collection we can take the completed real line $\widehat{\mathbb{R}} \subset \widehat{\mathbb{C}} = \mathbb{S}$, which is split into three arcs $a_1 = [\infty, -1]$, $a_2 = [-1, 1]$, $a_3 = [1, \infty]$. Because $z^{-1}(1 - z)^{-1}$ is positive on $(0, 1)$ the line field is parallel to a_2, while it is perpendicular to a_1, a_3. This means that we can choose $\theta_{a_1}, \theta_{a_3} = \pi/2$ and $\theta_{a_2} = 0$. If we take a half turn around $-1, 1$ the line field changes by $\frac{\pi}{2}$ so therefore we have

$$\deg_{\mathcal{A}} \alpha_{-1}^+ = 1, \quad \deg_{\mathcal{A}} \alpha_{-1}^- = 0,$$
$$\deg_{\mathcal{A}} \alpha_1^+ = 0, \quad \deg_{\mathcal{A}} \alpha_1^- = 1.$$

The line field does not change with respect to an angle that turns around ∞, and therefore $\deg_{\mathcal{A}} \alpha_\infty^+ = \deg_{\mathcal{A}} \alpha_\infty^- = 0$. The three arcs split the Riemann sphere into two triangles, and for each triangle the total degree of its inner angles is $1 + 0 + 0 = 3 - 2$, as required.

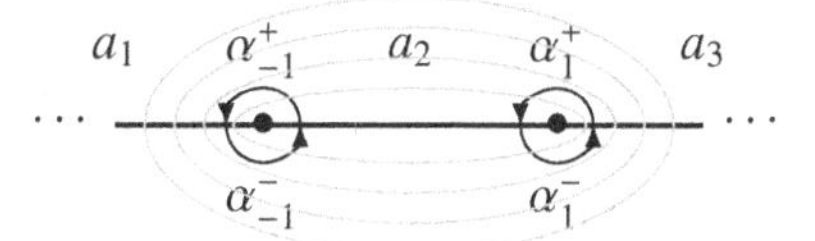

This grading is incompatible with the $\mathbb{Z}_2$-grading for the arc system because $\deg_{\mathbb{Z}_2} \alpha^+_{-1}\alpha^-_{-1} = 0$ as it is a morphism from a_1 to itself that turns a full turn around a marked point.

Example 10.12 (Making line fields for arbitrary gradings of the gentle algebra) When the arcs of an arc collection are also leaves of the line field then it is very easy to grade them: just take θ_a identical to zero. For these graded arcs it is also easy to determine the degrees of the angle morphisms. Around every marked point we can cut the surface into wedges and then the degree of the angle is a signed sum of the wedges that it crosses:

$$\deg \alpha = 1 \cdot \# \quad - 1 \cdot \# \quad + 0 \cdot \# \quad .$$

Vice versa, suppose that $\mathcal{A}$ is an arc collection with gentle algebra $\mathtt{Gtl}^\bullet \mathcal{A}$ and $\deg\colon Q_1 \to \mathbb{Z}$ is a degree function such that the sum of the degrees of the angles in a k-gon is $k - 2$. Then we can construct a line field that corresponds to this grading. To see why this is true, first note that by adding diagonals we can reduce to the case where all polygons are triangles . Now look at a triangle and let $\deg \alpha_1 + \deg \alpha_2 + \deg \alpha_3 = 3 - 2 = 1$. For the simplest case $(1, 0, 0)$, we can easily draw a line field on the triangle with that degree configuration.

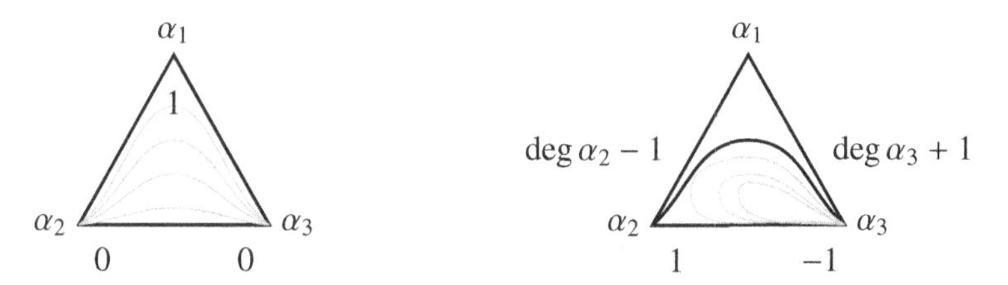

The other cases can be done by induction on $|\deg \alpha_1| + |\deg \alpha_2| + |\deg \alpha_3|$. Find two corners with degrees with opposite sign, e.g. $\deg \alpha_2 > 0$, $\deg \alpha_3 < 0$. Draw an extra line between them and fill the new digon with a teardrop line field as illustrated above. The remaining triangle should be filled with a line field with degrees at the corners equal to $\deg \alpha_1, \deg \alpha_2 - 1, \deg \alpha_3 + 1$.

Theorem 10.13 ([115]) *If $\vec{\mathcal{A}}$ and $\vec{\mathcal{B}}$ are two collections of graded arcs of $\vec{\mathbb{S}}$ that split the surface then* $\mathtt{DGtl}^\bullet \vec{\mathcal{A}}$ *and* $\mathtt{DGtl}^\bullet \vec{\mathcal{B}}$ *are equivalent A_∞-categories.*

Sketch of the proof The proof is completely similar to Theorem 9.12. First we observe that differently graded arcs correspond to shifts of objects. If two arcs are isotopic, we can transport the grading from one arc to the other along the isotopy and the second arc with the transported grading will be isomorphic to the first. Finally, we can identify a diagonal in a polygon with a cone over arcs on the boundary, appropriately graded such that the angle morphisms between these arcs all have degree 1, so adding, deleting or flipping a diagonal does not affect the derived category. The result now follows that two maximal arc collections are connected by a sequence of flips. □

10.2 Strings and Bands

Just as in the $\mathbb{Z}_2$-graded case, we can make string and band objects of the form

$$\left(\bigoplus a_j[i_j]^{\oplus n}, \delta = \sum_{j=1}^{k} \lambda_j \alpha_j\right),$$

but now the shifts i_j have to be chosen carefully: all the α_j must correspond to morphisms of degree 1. If we fix i_1 arbitrarily we have to set $i_{j+1} = i_j + \deg \alpha_j - 1$ for all $j \geq 1$. This is always possible for a string object, so every string object can be graded in precisely one way for each $i_1 \in \mathbb{Z}$.

Given a string object we can construct an arc that runs along the a_j outside small neighborhoods of the marked points. Inside these small neighborhoods, the arc first turns 90 degrees to the right, then it follows a circle segment of size α and then it turns 90 degrees again. If we transport the phase along this modified path, phases between consecutive a_j will match if the degree of α_j is 1 (because then it counterbalances the two right turns). The result is a graded arc that represents the string object. This is easiest to see when all the arcs have phase zero.

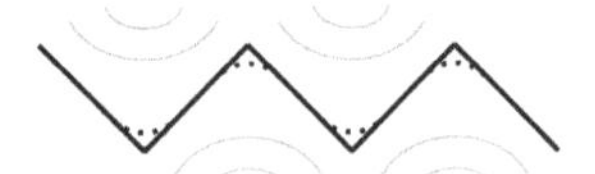

For band objects, the index parametrizing the arcs is cyclic, so i_{j+k} must be equal to i_j. This can only happen if $\sum_j (\deg \alpha_j - 1) = 0$. In that case, the curve obtained is gradable and the procedure described above results in a closed graded curve. Therefore, the band objects depend on a graded curve $\vec{\mathbb{L}}$ and a holonomy T_λ. Vice versa, if we have a graded arc or a closed curve (with local system) we can isotope it to a sequence of arcs and angles, and transport the grading along the isotopy. This allows us to identify the arc or closed curve with a string or band object. The identification also goes the other way.

Theorem 10.14 (Haiden–Katzarkov–Kontsevich) *Let $(\vec{\mathbb{S}}, M)$ be a graded surface with marked points and $\vec{\mathcal{A}}$ a graded arc collection that splits the surface. Every object in $\mathtt{D\,Gtl}^{\bullet}\vec{A}$ is isomorphic to a direct sum of string and band objects.*

Sketch of the proof The proof consists of two steps. In the first step, only surfaces without internal marked points are considered. In that case, it is possible to choose an arc collection for which the quiver has no oriented cycles, so there are no higher μ's and $\mathtt{Gtl}^{\bullet}\mathcal{A}$ is finite-dimensional. In this case, one can use standard techniques for ordinary (graded) gentle algebras (see [192, 197, 115]).

For the cases with internal marked points, one can cut out a small circle around that marked point and add a marked point on that boundary. Call this new surface $(\mathbb{S}', M')$ and grade it by restricting the line field. Every graded arc collection $\vec{\mathcal{A}}'$ for $(\mathbb{S}', M')$ can also be seen as an arc collection $\vec{\mathcal{A}}$ for $(\mathbb{S}, M)$. Just like in Theorem 9.38 there is an A_∞-functor $\phi\colon \mathtt{Gtl}^{\bullet}\mathcal{A}' \to \mathtt{Gtl}^{\bullet}\mathcal{A}$, which induces an essentially surjective functor on the derived categories (see [115]). Therefore, every object in $\mathtt{D\,Gtl}^{\bullet}\mathcal{A}$ comes from a direct sum of strings and bands in $\mathtt{D\,Gtl}^{\bullet}\mathcal{A}'$. Under $\mathtt{D}\phi$ a string or band in $(\mathbb{S}', M')$ is cut into pieces that are contained in $(\mathbb{S}, M)$ and these are string and band objects in $\mathtt{D\,Gtl}^{\bullet}\mathcal{A}$. □

Definition 10.15 The *topological Fukaya category of a graded surface* is

$$^{\mathtt{top}}\mathtt{Fuk}^{\bullet}(\vec{\mathbb{S}}, M) := \mathtt{D\,Gtl}^{\bullet}\vec{\mathcal{A}}.$$

10.3 Characterizing Graded Surfaces

Definition 10.16 Two line fields η_0 and η_1 are called *isotopic* if they can be connected by a path of line fields $\gamma\colon [0,1]\colon t \mapsto \eta_t$. Two graded surfaces $(\mathbb{S}_1, M_1, \eta_1)$ and $(\mathbb{S}_2, M_2, \eta_2)$ are called *equivalent* if there is a diffeomorphism $\phi\colon \mathbb{S}_1 \to \mathbb{S}_2$ that maps M_1 to M_2 and an isotopy between η_2 and $\phi_*\eta_1$.

Equivalent graded surfaces give rise to equivalent wrapped Fukaya categories because if $\mathcal{A}_1$ is an arc collection on $\mathbb{S}_1$ then $\mathcal{A}_2 = \phi(\mathcal{A}_1)$ will be an arc collection on $\mathbb{S}_2$ and we can use the isotopy between $\phi_*\eta_1$ and η_2 to transfer the gradings. Therefore, both $^{\mathtt{top}}\mathtt{Fuk}^{\bullet}(\vec{\mathbb{S}}_1, M_1)$ and $^{\mathtt{top}}\mathtt{Fuk}^{\bullet}(\vec{\mathbb{S}}_2, M_2)$ can be generated by the same graded gentle algebra.

It is not always easy to check whether two graded surfaces are equivalent. Therefore, it is important to find a set of invariants that characterize the equivalence class of the surface. This has been studied by Lekili and Polishchuk in [166].

Definition 10.17 Let $(\mathbb{S}, M)$ be a marked surface. We denote the set of line fields up to isotopy on $(\mathbb{S}, M)$ by

$$\mathrm{LF}(\mathbb{S}, M) := \pi_0(\Gamma(\mathbb{P}\mathrm{T}\dot{\mathbb{S}})).$$

On this set there is an action of the mapping class group

$$\mathrm{MCG}(\mathbb{S}, M) := \pi_0(\{\phi \in \mathrm{Diff}^+(\mathbb{S}, \mathbb{S}) \mid \forall\, m \in M,\ \phi(m) = m\}),$$

and we will denote the quotient of this group action by

$$\mathrm{MLF}(\mathbb{S}, M) := \frac{\mathrm{LF}(\mathbb{S}, M)}{\mathrm{MCG}(\mathbb{S}, M)}.$$

Let Γ be a set of small anticlockwise circles, one around each internal marked point and one around each boundary component of $\mathbb{S}$. The mapping class group acts trivially on marked points and hence also on the homotopy classes of Γ. Therefore, the winding numbers of a line field give a well-defined map $W\colon \mathrm{LF}(\mathbb{S}, M) \to \mathbb{Z}^\Gamma$ that factors through $\mathrm{MLF}(\mathbb{S}, M)$. By the Poincaré index theorem the image of this map lies inside the subspace

$$\{a \in \mathbb{Z}^\Gamma\colon\ \textstyle\sum_{\gamma\in\Gamma} 1 - \frac{a_\gamma}{2} = 2 - 2g\}.$$

Theorem 10.18 (Lekili–Polishchuk [166]) *Let $(\mathbb{S}, M)$ be a marked surface with genus g. Two line fields η_1 and η_2 correspond to the same point in* $\mathrm{MLF}(\mathbb{S}, M)$ *if and only if they have the (a) same parity, (b) $W(\eta_1) = W(\eta_2)$, and additionally,*

(c1) if $g = 1$ then $\mathrm{GCD}(\eta_1) = \mathrm{GCD}(\eta_2)$ with

$$\mathrm{GCD}(\eta) = \gcd\{w(\eta, \gamma) \mid \gamma \text{ is a closed curve such that } \mathbb{S} \setminus \operatorname{Im}\gamma \text{ is connected}\},$$

(c2) if $g \geq 2$ and $W(\eta_1) \subset (2 + 4\mathbb{Z})^\Gamma$ then $\mathrm{ARF}(\eta_1) = \mathrm{ARF}(\eta_2)$ with

$$\mathrm{ARF}(\eta) = \sum_{i=1}^{g} \frac{w(\eta, \alpha_i)}{2} \frac{w(\eta, \beta_i)}{2} \mod 2,$$

where α_i, β_i is a collection of closed curves on the surface that form a symplectic basis:

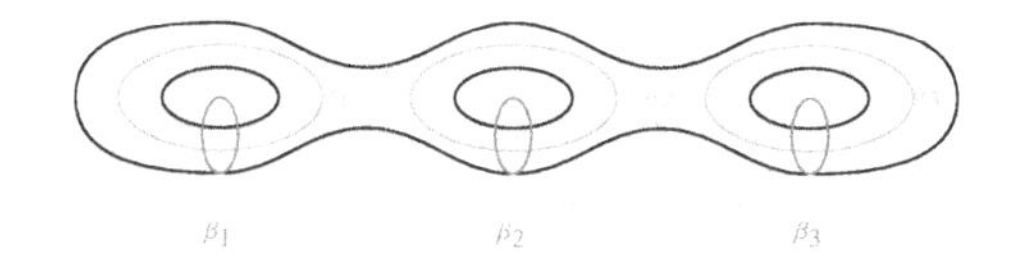

Remark 10.19 Because these invariants characterize the graded surface of an arc collection, they can be used to find nonobvious derived equivalences between gentle algebras. Gentle algebras with the same invariants will be derived equivalent but it is unclear whether the converse also holds. There might be derived equivalences between gentle algebras that do not come from equivalences of surfaces. For further details and the proof of the theorem we refer to [166].

10.4 Gradings and Matrix Factorizations

The $\mathbb{Z}$-gradings can be interpreted as line fields on the A-side. On the B-side we need to introduce $\mathbb{Z}$-gradings to categories of matrix factorizations. The μ_0 product in a curved algebra has degree 2, so this means that we have to find gradings for which the potential has degree 2. We will start with the pair of pants.

Let $(\mathbb{S}, M) = (\mathbb{P}^1, \{0, \infty, 1\})$. In the $\mathbb{Z}_2$-graded version we saw that it is mirror to the category of matrix factorizations of the Landau–Ginzburg model $(\mathbb{C}[X, Y, Z], XYZ)$. To turn this into a $\mathbb{Z}$-graded version we have to give the variables X, Y, Z a $\mathbb{Z}$-degree such that the potential has degree 2. So if we fix $\deg X$, $\deg Y$ then $\deg Z = 2 - \deg X - \deg Y$. The standard matrix factorizations can be graded

$$M_X = \left(R \oplus R[1 - \deg X], \left(\begin{smallmatrix} 0 & X \\ YZ & 0 \end{smallmatrix}\right)\right),$$
$$M_Y = \left(R \oplus R[1 - \deg Y], \left(\begin{smallmatrix} 0 & Y \\ XZ & 0 \end{smallmatrix}\right)\right),$$
$$M_Z = \left(R \oplus R[1 - \deg Z], \left(\begin{smallmatrix} 0 & Z \\ XZ & 0 \end{smallmatrix}\right)\right).$$

The degree of the morphism $\zeta_1 = \left(\begin{smallmatrix} 0 & -1 \\ Z & 0 \end{smallmatrix}\right) : M_X \to M_Y$ is $1 - \deg Y$ and the opposite morphism $\zeta_2 : M_Y \to M_X$ has degree $1 - \deg X$. Together they form the morphism $\zeta_1\zeta_2 = \left(\begin{smallmatrix} -Z & 0 \\ 0 & -Z \end{smallmatrix}\right) : M_Y \to M_Y$, which has degree $2 - \deg X - \deg Y = \deg Z$. On the A-side this corresponds to a full 2π-angle around the marked point $m_\zeta = 1 \in \{0, \infty, 1\}$. Using the method in Example 10.12, we can realize this by gluing two triangles together for which the inner angles have degrees $1 - \deg X$, $1 - \deg Y$, $1 - \deg Z$.

On the other hand, from Theorem 10.18 we know that every possible grading on the sphere with three marked points is characterized by the winding numbers around the marked points, and the degrees of the full angles around the marked points are $2 - 2I_m$, where I_m is the index of the marked point, so if we put $\deg X = 2 - 2I_{m_\xi}, \dots, \deg Z = 2 - 2I_{m_\zeta}$ we can realize each graded A-model by a graded B-model.

Theorem 10.20 (Graded pair of pants) *The graded marked surface*

$$(\mathbb{P}^1, \{0, \infty, 1\}, z^k(z-1)^l\, dz^2)$$

is mirror to the graded Landau–Ginzburg model $(\mathbb{C}[X, Y, Z], XYZ)$ *with degrees*

$$\deg X = 2 + k, \quad \deg Y = -k - l - 2, \quad \deg Z = 2 + l.$$

A-models for surfaces with more marked points and higher genera can be constructed by gluing pairs of pants together. This can be made compatible with the grading. The rule is that if we glue two pants overlapping in one leg, the degrees of the full angles must be the opposites of each other. If this is the case, we can isotope the line fields in such a way that they match up: for the pant leg with the negative winding number k we isotope it to a k loopy wedges. If we invert the coordinate the loopy wedges become k saddle wedges, which corresponds to a positive winding number, so the two match up on the overlap and we obtain a continuous line field.

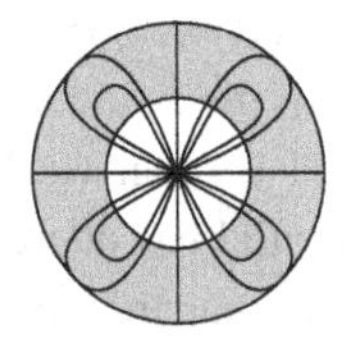

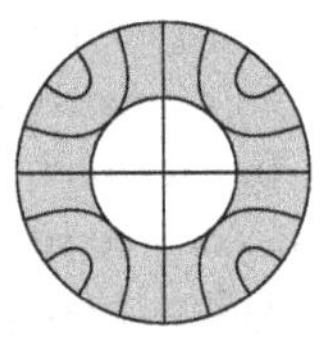

Mirrors for these surfaces could be constructed using a triangulated lattice polygon. In this way we get a three-dimensional toric variety $\mathbb{X}$ with a function ℓ. For each triangle Δ in the polygon we get a matrix factorization of the form $(\mathbb{C}[X_\Delta, Y_\Delta, Z_\Delta], X_\Delta Y_\Delta Z_\Delta)$, where X_Δ, Y_Δ, Z_Δ are monomials in $\mathbb{C}(U, V, W)$ and $X_\Delta Y_\Delta Z_\Delta = W$. If we choose $(\deg U, \deg V) \in \mathbb{Z}^2$ and set $\deg W = 2$ we can define a category of graded matrix factorizations over $(\mathbb{X}, \ell)$ that will be equivalent to the A-model of a graded surface made by gluing graded pairs of pants together.

Example 10.21 Consider the unit square and split it into two triangles. This gives a Landau–Ginzburg model that is mirror to the four-punctured sphere, or two pairs of pants glued together. We have written the monomials corresponding to the outer edges of each triangle: the monomial of the inner is UVW^{-1} or its inverse. Every outer edge corresponds to a marked point and to obtain the indices we have to look at the degrees of the monomials. The degrees of the inner edges determine the winding number of the corresponding curves on the surface.

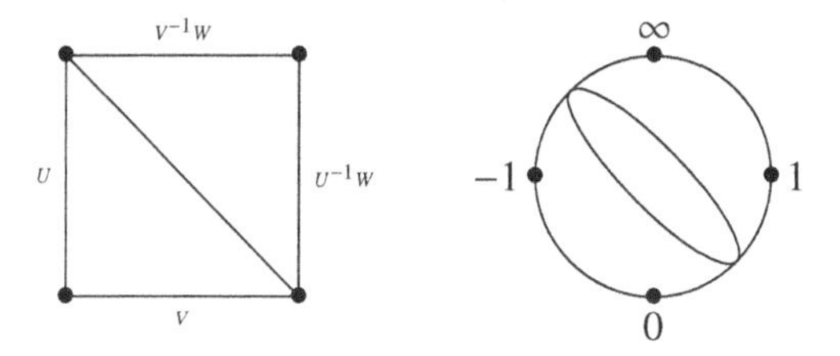

The line field in this example can be obtained from the quadratic differential

$$\phi = z^{\deg V-2}(z+1)^{\deg U-2}(z-1)^{-\deg U}\,dz^2.$$

In the previous chapter we also introduced noncommutative Landau–Ginzburg models. If Q is a consistent dimer model then we can grade its Jacobi algebra by assigning to each arrow in Q a degree R_a such that the sum of the degrees of the arrows in a face is 2. In this way $(\mathtt{Jac}, \ell)$ becomes a graded noncommutative Landau–Ginzburg model. The standard matrix factorizations coming from the arrows can easily be graded

$$M_a = t(a)\mathtt{Jac}(Q) \underset{r_a\cdot}{\overset{a\cdot}{\rightleftarrows}} h(a)\mathtt{Jac}(Q)[1-R_a]\,,$$

and the degree of the morphism $\alpha\colon \begin{pmatrix} 0 & -\mathbb{1}_{t(a)} \\ r_{ab} & 0 \end{pmatrix} : M_a \to M_b$ is $1-R_b$. For each k-gon $f \in Q_2$, the total degrees of the α inside this polygon is $\sum_{b\in f} 1-R_b = k-2$, so we can construct a polygon with a line field such that the edges are leaves. If we glue all these polygons together in the dual dimer way we get a graded marked surface with a graded arc collection and

$$\mathrm{End}^\bullet\left(\bigoplus_a M_a\right) \overset{\infty}{=} \mathtt{Gtl}^\bullet Q^\vee.$$

Example 10.22 Consider again the dimer Q from Example 9.67 and define $\deg a_1 = \deg a_2 = \deg b_1 = 0$ and $\deg b_2 = 2$. With this grading all angle morphisms get degree 1 except those arriving in b_2. In the dual quiver we can put loopy wedges in those angles and saddle wedges in the others.

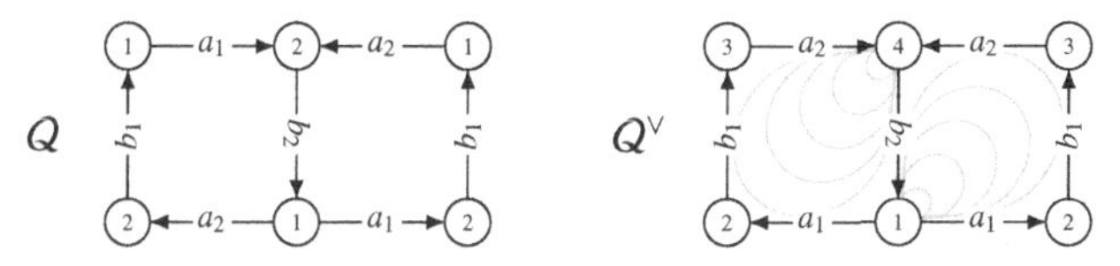

The indices of the four vertices in $Q^\vee$ are 1, 0, 0, 1, so this line field is isotopic to $z^{-2}\,dz^2$ if we identify these points with $0, +1, -1, \infty \in \widehat{\mathbb{C}}$.

10.5 Mirror Varieties

One advantage of using line fields is that we can go beyond the realm of matrix factorizations and construct mirrors that are actual varieties. This is only possible in special cases. In Section 4.3.2 we saw how we could interpret the derived category of $\mathbb{P}^1$ as the topological Fukaya category of a cylinder with one marked point on each boundary and a line field that is parallel to the tube. If we contract one or two of the boundary components to an internal marked point we get the derived categories of $\mathbb{A}^1$ or $\mathbb{C}^* = \mathbb{A}^1 \setminus \{0\}$.

If we allow singular varieties, these examples generalize to the setting where $(\vec{\mathbb{S}}, M)$ is a graded marked surface without boundary components such that the winding numbers around the marked points are either 0 (i.e. index 1) or 2 (i.e. index 0). Because of the Poincaré index theorem this can only happen if

- the genus is 0 and then there are two points with winding number 0,
- the genus is 1 and then there are no points with winding number 0.

Additionally, the points with winding number 0 can be changed to boundary components with one marked point. More details of these constructions can be found in [232, 165]; we will describe some examples.

Let us again start with the pair of pants. We can see this as $(\mathbb{P}^1, \{0, 1, \infty\})$ with the radial line field coming from the 1-form $z^{-1}\,dz$. This makes the winding numbers of 0 and ∞ equal to 0 and that of 1 equal to 2. We make the following claim.

Lemma 10.23 *The topological Fukaya category of the graded surface*

$$(\mathbb{P}^1, \{0, 1, \infty\}, z^{-1}\,dz)$$

is equal to the derived category of coherent sheaves of two intersecting affine lines:

$${}^{\mathtt{top}}\mathtt{Fuk}^\bullet(\mathbb{P}^1, \{0, 1, \infty\}, z^{-1}\,dz) \overset{\infty}{=} \mathtt{D}\,\mathtt{Coh}^\bullet\,\mathbb{A}^1 \cup_p \mathbb{A}^1 \overset{\infty}{=} \mathtt{D}\,\mathtt{Mod}^\bullet\,\mathbb{C}[X, Y]/(XY).$$

Sketch of the proof Let $R = \mathbb{C}[X, Y]/(XY)$ and consider the three modules R, $M_X = R/(X)$ and $M_Y = R/(Y)$. The latter two have resolutions of the form

$$R \xleftarrow{X} R \xleftarrow{Y} R \xleftarrow{X} R \longleftarrow \cdots,$$

which can be used to calculate the Ext-A_∞-algebra. It is generated by six morphisms, four of which are of degree 0:

$$\xi_1 : R \to R/(Y) \colon 1 \mapsto 1, \quad \xi_2 \colon R/(Y) \to R \colon 1 \mapsto X,$$
$$\eta_1 : R \to R/(X) \colon 1 \mapsto 1, \quad \eta_2 \colon R/(X) \to R \colon 1 \mapsto Y,$$

and two of which are of degree 1:

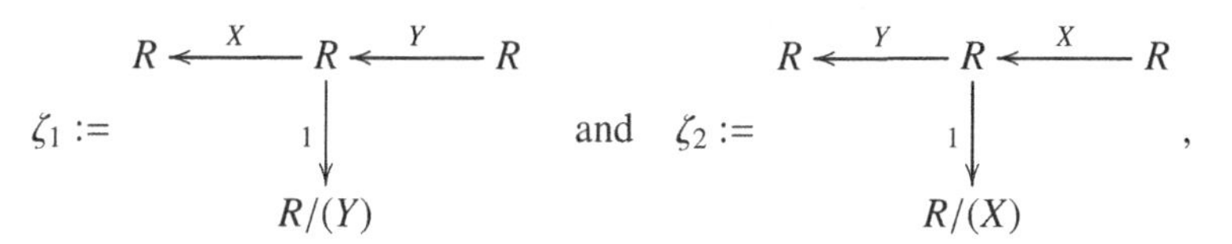

which also correspond to the extensions

$$R/(X) \xrightarrow{Y} R \xrightarrow{X} R/(Y) \quad \text{and} \quad R/(X) \xrightarrow{Y} R \xrightarrow{X} R/(Y)\,.$$

This Ext-algebra is A_∞-isomorphic to the standard gentle algebra that we associated to the pair of pants. The three modules generate $\mathtt{D\,Mod}^\bullet\,\mathbb{C}[X,Y]/(XY)$, so we get an equivalence of categories. □

We can picture this graded surface as $\mathbb{C}^* \setminus \{0\}$ with a line field that points radially away from the origin. In this picture, R corresponds to the real interval $(-\infty, 0)$, $R/(X)$ with $(0,1)$ and $R/(Y)$ with $(1,+\infty)$.

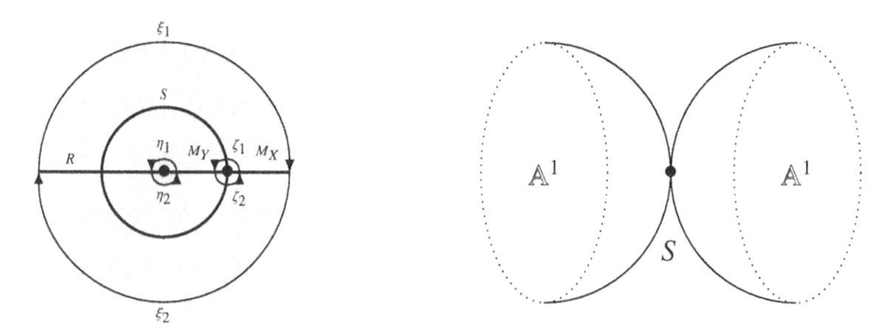

The skyscraper sheafs $R/(X, Y-\lambda)$ of points that lie in the first $\mathbb{A}_1 = \mathbb{V}(R/(X))$ correspond to band objects on a circle inside the unit disk. The skyscraper sheafs $R/(Y, X-\lambda)$ of points that lie in the second $\mathbb{A}_1 = \mathbb{V}(R/(Y))$ correspond to band objects on a circle inside the unit disk. The skyscraper sheaf of the singular point $S = R/(X,Y)$ corresponds to the string object on the unit circle.

Example 10.24 (Chains of lines) Let us consider $(\mathbb{P}^1, \{0, \infty, 1, \dots, k\})$ with the same line field.

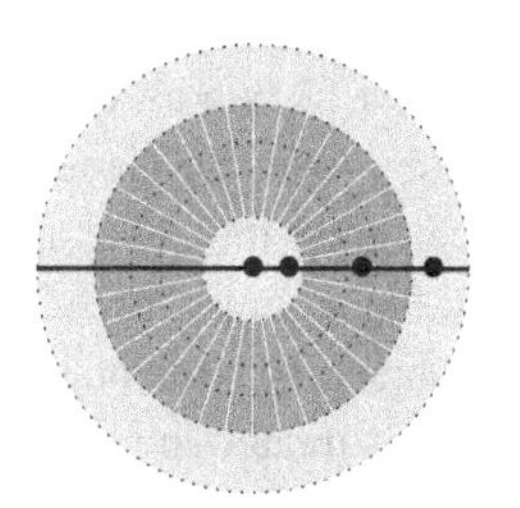

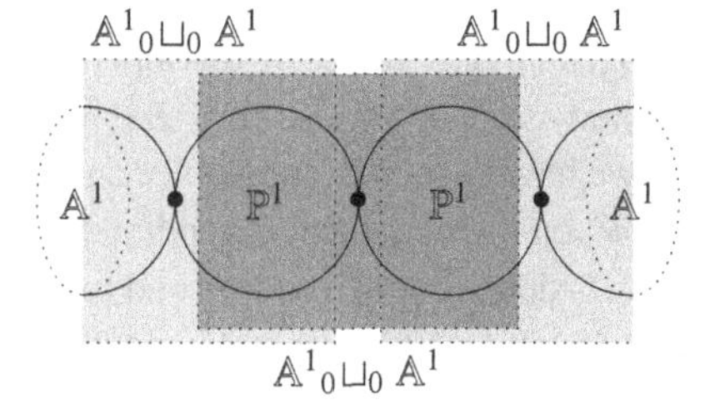

If we cut the surface into pieces along circles between the k marked points, we can see it as k pairs of pants glued together along band objects. On the B-side we glue $\mathbb{A}^1{}_0\sqcup_0 \mathbb{A}^1$ together over open $\mathbb{C}^*$'s. Two affine lines glued together make a projective line, so this results in a chain

$$\mathbb{A}^1{}_0\sqcup_\infty \mathbb{P}^1{}_0\sqcup_\infty \cdots {}_0\sqcup_\infty \mathbb{P}^1{}_0\sqcup_0 \mathbb{A}^1.$$

There is also a second way to glue: if we cut the surface along circles that go through the marked points, we can see it as cylinders with marked points on the boundaries glued together over string objects. The outer pieces are disks with one internal marked point and one on the boundary. This corresponds on the B-side to an affine line. The inner pieces are cylinders with one marked point on each boundary. This corresponds on the A-side to a projective line. The fact that we glue over string objects means that on the B-side we just identify one point, so we get the same chain. By omitting the end parts containing 0 and/or ∞, we can make chains with only one affine line or no affine line at all.

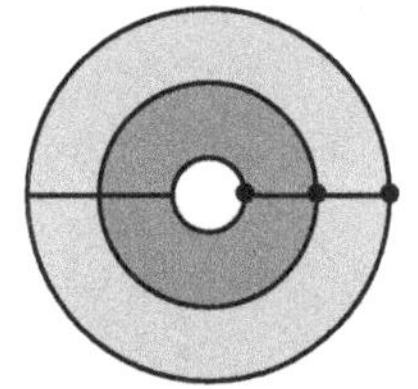

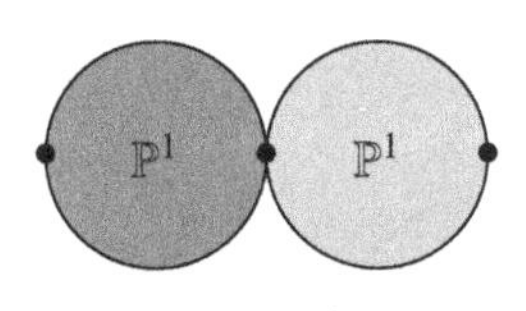

Example 10.25 (Necklaces of lines) Now consider the torus $\mathbb{C}/\mathbb{Z} + i\mathbb{Z}$ with $k > 1$ internal marked points and a horizontal line field. Again, we can cut the torus into cylinders with one internal marked point or cylinders with one marked point on each boundary. In both cases the result is a necklace of k projective lines.

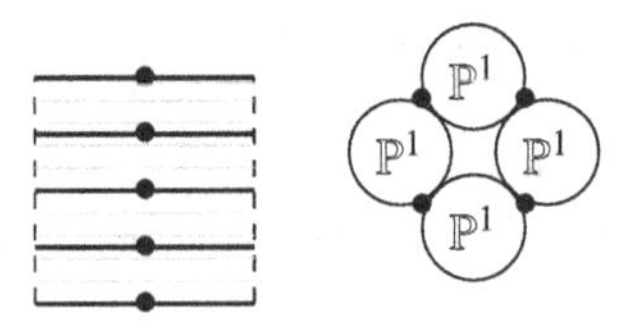

Example 10.26 If we consider the torus $\mathbb{C}/\mathbb{Z} + i\mathbb{Z}$ with one internal marked point and a horizontal line field, we have to glue two points of one projective line together. The result of this operation is a nodal curve $\mathbb{E}$, which can be described as the zero locus of $Y^2Z - XYZ - X^3 = 0$ in the projective plane $\mathbb{P}^2$. On $\mathbb{E}$ we have two interesting objects that generate the category: the skyscraper sheaf $\mathscr{S}$ on the nodal singular point $s = (0 : 0 : 1)$ and the structure sheaf $\mathscr{O}$. On the A-side these two correspond to a vertical string object from the marked point to itself and a horizontal band object. This example is treated in [162].

10.6 Mirror Orbifolds

If we only allow varieties on the B-side then we cannot find mirrors for surfaces with genus $g \geq 2$. With orbifolds or stacks we can get around this problem and also find new mirrors for the $g \geq 2$ cases.

On the B-side we will look at the singular orbifold $[\mathbb{A}^1{}_0\sqcup_0\mathbb{A}^1/\mathbb{Z}_k]_r$, where $\mathbb{Z}_k$ acts on the first line by multiplication with $e^{2\pi i/k}$ and on the second by $e^{2r\pi i/k}$, where $r \in \mathbb{Z}_k^*$. By definition, the derived category of this stack is the derived category $\mathtt{D\,Mod}^\bullet A$ with $A = \mathbb{C}[X,Y]/(XY) \star \mathbb{Z}_k$. This algebra is isomorphic to the path algebra of a quiver with relations. The quiver consists of a ring with k vertices v_i, each with two arrows leaving: $X_i\colon v_i \to v_{i+1}$ and $Y_i\colon v_i \to v_{i+r}$. The relations are $Y_{i+1}X_i = 0$ and $X_{i+r}Y_i = 0$.

To construct the A-model, we take two disks with an internal marked point and k marked points on the boundary. For the grading we use radial line fields. On the first disk we name the boundary arcs c_i, $i \in \{1,\dots,k\}$ in the anticlockwise direction and on the second we name the arcs c_{ri}, $i \in \{1,\dots,k\}$, where the indices are taken mod k. We glue the two disks together by identifying the arcs with the same name. The result is an oriented surface without boundary and a number of internal marked points. This number is $2+m$, where the first two are the centers of the disks and m is the number of orbits in $\mathbb{Z}_k$ under the action $u \mapsto u+1+r$. This number is also the number of cycles in the quiver consisting of an alternating sequence of X and Y arrows (zigzag paths). The result is a surface made up of two k-gons and m corners, so the genus of this surface is $\frac{k-m}{2}$. Furthermore, because the transport of the line field around an inner angle of the polygon is π, the total degree of a full turn around one of these is $2k/m$.

We can supplement the edges from the polygons with radial arcs from the centers of the disk to the marked points on the edges of the polygons to get an arc collection $\mathcal{A}$. Under the mirror correspondence the edges of the polygons correspond to the simple vertex representations $v_iA/(X,Y)$, while the radial arcs correspond to the modules $v_iA/(X)$ and $v_iA/(Y)$. These $3k$ modules generate $\mathtt{D\,Mod}^\bullet A$ and the Ext-ring of their direct sum is A_∞-isomorphic to the gentle algebra $\mathtt{Gtl}^\bullet\mathcal{A}$. By varying r we get different surfaces.

Example 10.27 Take $k = 5$ and $r = 3, 4$. In the first case we get $g = 2$ and in the second $g = 0$.

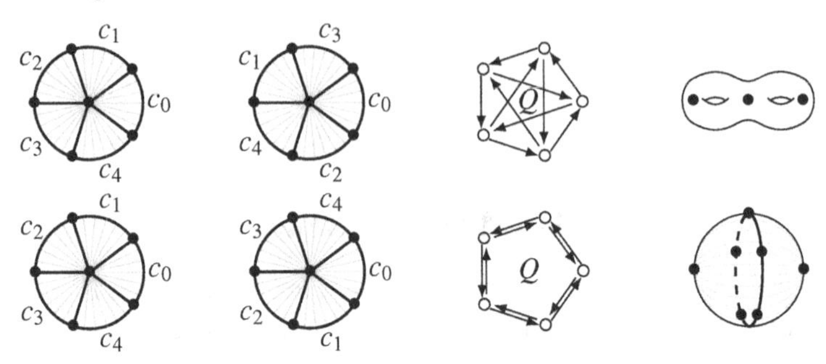

In general, if $r = k - 1$, the genus is always 0 and we get a mirror for $\mathbb{C}^* \setminus \{e^{2j\pi i/k} \mid j \in \mathbb{Z}\}$. This graded surface is equivalent to $\mathbb{C}^* \setminus \{1, \dots, k\}$ so the derived category of the orbifold singularity is equivalent to the derived category of the chain of two affine lines and $k - 1$ projective lines. We get a nontrivial correspondence between two B-models,

$$\mathsf{D}\,\mathsf{Coh}^\bullet[\mathbb{A}^1{}_0\sqcup_0 \mathbb{A}^1/\mathbb{Z}_k]_{k-1} \overset{\infty}{=} \mathsf{D}\,\mathsf{Coh}^\bullet\, \mathbb{A}^1{}_0\sqcup_0 \mathbb{P}^1{}_\infty\sqcup_0 \cdots {}_\infty\sqcup_0 \mathbb{P}^1{}_\infty\sqcup_0 \mathbb{A}^1.$$

One can construct many more similar mirror equivalences by gluing these orbifolds together to form chains or necklaces. This has been worked out in detail by Lekili and Polishchuk in [165].

10.7 Exercises

Exercise 10.1 Show that a closed surface $\mathbb{S}$ (without marked points) can only have a line field if its genus is 1.

Exercise 10.2 Let $\mathbb{S}$ be a closed surface with marked points M with a line field V on $\mathbb{S}\setminus M$. Assume that there is a set of leaves that forms an arc collection that splits the surface. Show that the Poincaré index theorem holds if we use the wedges formula to calculate the indices.

Exercise 10.3 Let $\mathbb{S}$ be a closed surface with marked points M and γ be a contractible loop in $\mathbb{S}$ surrounding a disk $\mathbb{U}$. What is the condition on the indices of the marked points inside $\mathbb{U}$ for γ to be gradable?

Exercise 10.4 Find a graded surface with arc collection whose gentle algebra is the path algebra of the quiver

$$\circ \overset{a_1}{\underset{b_1}{\rightrightarrows}} \circ \overset{a_2}{\underset{b_2}{\rightrightarrows}} \cdots \overset{a_k}{\underset{b_k}{\rightrightarrows}} \circ$$

with relations $a_{i+1}a_i = b_{i+1}b_i = 0$ such that $\deg a_i = 1$ and $\deg b_i = 0$.

Exercise 10.5 Consider the quiver

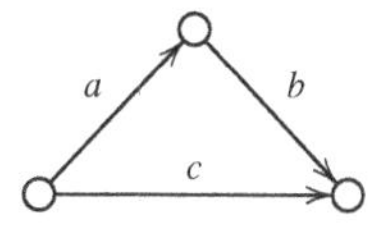

without relations. For each triple $(\deg a, \deg b, \deg c) \in \mathbb{Z}^3$, find a graded surface with an arc collection $\mathcal{A}$ for which the gentle algebra is the graded path algebra of this quiver. Describe the indecomposable objects in $\mathtt{D\,Gtl}^\bullet \mathcal{A}$ (make a distinction for the different triples).

Exercise 10.6 Construct a graded surface with genus 2 and a nonzero ARF invariant.

Exercise 10.7 Let $\square$ be a reflexive polygon with triangulation $\boxtimes$ and let v_0 be the internal lattice point in the polygon. An element $2v_0 \in N = \mathbb{Z}^3$ gives a grading $\deg_{2v_0}$ on $\mathbb{C}[M]$ for which $\deg_{2v_0} \ell = 2$. This means that we can use it to grade the category of matrix factorizations for $(\mathbb{X}^3_{\boxtimes}, \ell)$. Show that the corresponding graded marked surface is a torus with quadratic differential $(dz)^2$.

Exercise 10.8 Let Q be a consistent dimer on a torus. A perfect matching $\mathcal{P} \subset Q_1$ is a set of arrows such that every cycle in Q_2 contains exactly one arrow of $\mathcal{P}$. We can use this to define a grading $\deg_{2\mathcal{P}}$ on $\mathtt{Jac}\,(Q)$ that assigns degree 2 to each arrow in $\mathcal{P}$ and 0 to all the other arrows. This grading has the property that $\deg_{2\mathcal{P}} \ell = 2$, so we can use it grade the category of matrix factorizations of $(\mathtt{Jac}\,(Q), \ell)$. Construct a line field on $Q^\vee$ that is mirror to this grading.

Exercise 10.9 Let $\mathbb{X}$ be a necklace of k $\mathbb{P}^1$'s. Construct orbifolds stacks that are derived equivalent to $\mathbb{X}$. How many possibilities are there?

Exercise 10.10 Let $\mathbb{S}$ be a cylinder with k marked points on one boundary and k marked points on the other side. Equip $\mathbb{S}$ with a line field parallel to the axis of the cylinder. Its B-side mirror should be a projective line with two orbifold points. Construct its derived category as a derived category of sheaves of modules over a sheaf of algebras.

11
Stabilizing

Many problems in mathematics boil down to classifying certain objects with respect to an equivalence relation. Often it is not possible to classify all objects or to give the classifying space a nice geometrical structure. To address these problems it is interesting to impose some extra conditions on the space of objects that make the classification problem better behaved. Such conditions are known as stability conditions. We will look at the two main types of stability conditions that appear in representation theory: King stability conditions [138], which are used to construct moduli spaces of representations of algebras, and Bridgeland stability conditions [39], which can be used to investigate the structure of derived categories. In this chapter we will discuss these stability conditions for Fukaya categories of surfaces, gentle algebras and Jacobi algebras, investigate their relationships and interpret them geometrically.

11.1 The Grothendieck Group

One of the coarsest ways to classify objects is by means of the Grothendieck group [99]. In this section we will define this group for A_∞-categories and work out its structure in the case of topological Fukaya categories. For further information on the Grothendieck group and more general K-theory we refer to the book by Weibel [254]. The construction of the Grothendieck group for graded surfaces can be found in [115]. A similar calculation for Fukaya categories of closed surfaces was done by Abouzaid in [4].

Definition 11.1 Let $\mathtt{A}^\bullet$ be an A_∞-category closed under shifts and cones. The *Grothendieck group* $K_0\,\mathtt{A}^\bullet$ is the quotient of the free abelian group generated by symbols $[X]$, where X is an object in $\mathtt{A}^\bullet$, modulo the relations

- $[X] - [Y]$ if X and Y are quasi-isomorphic objects,
- $[X] + [X[1]]$,
- $[X] + [Y] - [\mathrm{Cone}(f)]$ if $f: X \to Y$ is a morphism of degree 1.

If $\mathtt{A}^\bullet$ is not closed under shifts and cones, we will write $K_0\,\mathtt{A}^\bullet$ for $K_0\mathtt{Tw}\,\mathtt{A}^\bullet$. Note that the last relation implies that if (M,δ) is a twisted complex then $[(M,\delta)] = [M]$.

Example 11.2 The Grothendieck group of $\mathtt{D\,vect}^\bullet$ is $\mathbb{Z}$ because every object is a direct sum of shifts of the one-dimensional vector space. Similarly, if G is a finite group then $\mathtt{D\,mod}^\bullet\,\mathbb{C}G$ is $\mathbb{Z}^k$, where k is the number of irreducible representations of G.

Example 11.3 Consider $R = \mathbb{C}[X]$. Every finitely generated module is a direct sum of copies of R and S_λ. The latter are cones of the maps $m_{X-\lambda}: R \to R: r \mapsto (X-\lambda)r$, so in the Grothendieck group they correspond to $[S_\lambda] = [R] - [R] = 0$. Therefore, $K_0\,\mathtt{Mod}^\bullet\,R \cong \mathbb{Z}$ measures the rank of the torsion-free part of the modules. On the other hand, if we look at the derived category of the finite-dimensional modules $\mathtt{D\,mod}^\bullet\,R$, we cannot perform the reduction above, so $K_0\,\mathtt{mod}^\bullet\,R$ is freely generated by an infinite number of generators $[S_\lambda]$.

Theorem 11.4 *If Q is a quiver without oriented cycles and $A = \mathbb{C}Q/I$ a path algebra with admissible relations, then $K_0\mathtt{D\,Mod}^\bullet\,Q \cong \mathbb{Z}^{Q_0}$.*

Sketch of the proof Because there are no oriented cycles, the vertex simples generate the finite-dimensional modules, so we have a surjective map $\phi: \mathbb{Z}^{Q_0} \to K_0\mathtt{D\,mod}^\bullet\,Q$. On the other hand, if $M^\bullet$ is a complex of modules then we can calculate

$$\alpha(M^\bullet) := \sum_i (-1)^i \,\mathrm{dimvect}\, H^i(M^\bullet).$$

Because $\alpha(M[1]^\bullet) = -\alpha(M^\bullet)$ and $\alpha(\mathrm{Cone}(f: M \to N)) = \alpha(M) + \alpha(N)$, we get a surjective morphism $\alpha: K_0\mathtt{D\,mod}^\bullet\,Q \to \mathbb{Z}^{Q_0}$. These two morphisms are inverses. □

Let $(\vec{\mathbb{S}}, M)$ be a graded marked surface and $\mathcal{A}$ an arc collection that splits it. Because these arcs generate the topological Fukaya category we get a surjective map

$$\mathbb{Z}^{\mathcal{A}} \to K_0\,{}^{\mathtt{top}}\mathtt{Fuk}^\bullet(\mathbb{S}, M).$$

If $a_1, \dots, a_k$ form a sequence of arcs that bound a polygon and $\alpha_i: a_i \to a_{i+1}$ are the inner angles, then we have the following isomorphism in the topological Fukaya category

$$(a_1[i_1] \oplus \cdots \oplus a_{k-1}[i_{k-1}], \alpha_1 + \cdots + \alpha_{k-2}) \cong a_k[i_k - 1],$$

where the i_j are chosen such that the $\alpha_j\colon a_j[i_j] \to a_{j+1}[i_{j+1}]$ all have degree 1. In the Grothendieck group this results in the relation

$$(-1)^{i_1}[a_1] + \cdots + (-1)^{i_k}[a_k] = 0.$$

Let $G_{\mathcal{A}}$ denote the abelian group generated by the a_i, subject to the relations above. If the grading comes from a vector field, it is compatible with the $\mathbb{Z}_2$-grading and then we can interpret the arcs as oriented line segments. The relation then just says that if the arcs form the boundary of a polygon then their sum is zero. An arc can be seen as a cycle in $\mathbb{S}$ relative to M, and therefore we can identify $G_{\mathcal{A}}$ with the relative first homology. If the grading does not come from a vector field one can still do something similar and interpret $G_{\mathcal{A}}$ as the homology of $\mathbb{S}$ with respect to M and coefficients in a certain sheaf over $\mathbb{S}$; for the details we refer to [115]. For us it suffices to define

$$H_1(\vec{\mathbb{S}}, M) := G_{\mathcal{A}}.$$

One can easily check that this definition does not depend on our choice of $\mathcal{A}$.

Theorem 11.5 (Haiden–Katzarkov–Kontsevich)

$$K_0\,{}^{\mathtt{top}}\mathtt{Fuk}^\bullet(\vec{\mathbb{S}}, M) \cong H_1(\vec{\mathbb{S}}, M).$$

Sketch of the proof If $(\mathbb{S}, M)$ has no internal marked points, we can find an arc collection without closed polygons. In that case, there are no relations in $G_{\mathcal{A}}$ and $\mathtt{Gtl}\mathcal{A}$ is the graded path algebra of a quiver with relations without oriented cycles or higher products, so we can use the previous theorem. For the general case we can see ${}^{\mathtt{top}}\mathtt{Fuk}^\bullet(\vec{\mathbb{S}}, M)$ as a localization of ${}^{\mathtt{top}}\mathtt{Fuk}^\bullet(\vec{\mathbb{S}}', M')$, where $(\vec{\mathbb{S}}', M')$ has no internal marked points. Using results from [99] one can relate the Grothendieck groups via exact sequences. For more details see [115]. □

If $\mathtt{A}^\bullet$ is a proper A_∞-category (i.e. the homology of all the hom-spaces is finite-dimensional) then we can equip its Grothendieck group with an extra structure: a bilinear form.

Definition 11.6 The *Euler form* of $\mathtt{A}^\bullet$ is the map

$$\langle\ ,\ \rangle\colon K_0\,\mathtt{A}^\bullet \times K_0\,\mathtt{A}^\bullet \to \mathbb{Z}\colon ([X],[Y]) \mapsto \sum_i (-1)^i \dim \mathtt{A}^i(X,Y).$$

Example 11.7 If Q is a quiver without oriented cycles and S_1, S_2 are two finite-dimensional representations then the bilinear form becomes

$$\langle S_1, S_2\rangle = \dim \mathrm{Hom}_{\mathbb{C}Q}(S_1,S_2) - \dim \mathrm{Ext}^1_{\mathbb{C}Q}(S_1,S_2)$$

because there are no higher Exts. The class of S_i in the Grothendieck group only depends on its dimension vector α_i:

$$\langle S_1, S_2 \rangle = \sum_{v \in Q_0} \alpha_1(v)\alpha_2(v) - \sum_{a \in Q_1} \alpha_1(t(a))\alpha_2(h(a)).$$

Example 11.8 Let $(\mathbb{S}, M)$ be a marked surface without internal marked points. If $B_1 = B(\mathbb{L}_1, \lambda_1)$ and $B_2 = B(\mathbb{L}_2, \lambda_2)$ are two band objects then the pairing for $K_0{}^{\mathtt{top}}\mathtt{Fuk}^{\pm}(\mathbb{S}, M)$ corresponds to the intersection pairing between the two curves:

$$\langle [B_1], [B_2] \rangle = \# \qquad - \#$$

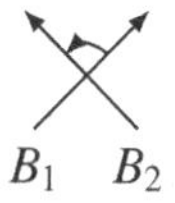

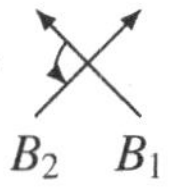

B_1 B_2 $\qquad$ B_2 B_1

Note that in this case the pair is antisymmetric with respect to the band objects (but not with respect to the string objects).

In many cases the Grothendieck group is not finitely generated but we can use the Euler form to project the group down to a smaller group.

Definition 11.9 We say that an $x \in K_0\,\mathtt{A}^\bullet$ is (left) numerically trivial if $\langle x, y \rangle$ is zero for every $y \in K_0\,\mathtt{A}^\bullet$. The (left) numerical Grothendieck group is the quotient of the Grothendieck group by the subgroup of numerically trivial elements.

Remark 11.10 It is also possible to define a right numerical Grothendieck group, which in general can be different. However, if the A_∞-category is smooth and proper then the left and right numerical Grothendieck groups coincide and are a finitely generated group. The proof of this result can be found in [240].

Example 11.11 Let $(\vec{\mathbb{S}}, M)$ be a graded marked surface and consider the subcategory $\mathtt{Bands}^\bullet \subset {}^{\mathtt{top}}\mathtt{Fuk}^\bullet(\vec{\mathbb{S}}, M)$ generated by the band objects. In general, the Grothendieck group $K_0\,\mathtt{Bands}^\bullet$ is not finitely generated. Two band objects B_1, B_2 with the same underlying simple curve will often give rise to different elements in $K_0\,\mathtt{Bands}^\bullet$ but they represent the same element in the numerical Grothendieck group because $\langle [B_1] - [B_2], - \rangle = 0$.

11.2 King Stability and Pair of Pants Decompositions

If Q is a quiver and $A = \mathbb{C}Q/I$ a path algebra with relations then one can try to classify representations by constructing a moduli space of semistable representations. Recall from Section 7.3.3 that given a dimension vector

$\alpha\colon Q_0 \to \mathbb{N}$ and a vector $\theta\colon Q_0 \to \mathbb{Z}$ we could construct a space $\mathbb{M}_\theta(A,\alpha) = \mathbb{P}(R_\theta)$ with

$$R_\theta := \bigoplus_{i=0}^{\infty} \{f \in \mathscr{O}(\mathsf{Rep}(A,\alpha)) \mid \forall g \in \mathsf{GL}_\alpha,\ g \cdot f = \textstyle\prod_{v \in Q_0} \det g^{\theta_v} f\}.$$

This space is empty unless $\theta \cdot \alpha = 0$ and its points correspond to isomorphism classes of direct sums of θ-stable representations, which are representations that have no proper subrepresentations with dimension vectors β with $\beta \cdot \theta \le 0$. This notion of stability is often referred to as King stability, as it was first studied in depth in this context by Alistair King in [138].

We will now use King stability to look at mirror symmetry for marked surfaces. In Chapter 9 we saw that one of the ways to construct a mirror was by using a dimer model Q and its Jacobi algebra. If the dimer model is zigzag consistent on a torus then the pair $(\mathtt{Jac}\,(Q), \ell)$ can be seen as a noncommutative Landau–Ginzburg model that is mirror to the dual dimer model.

Following [42, 125, 35] we will look at the space of representations of its Jacobi algebra $\mathtt{Jac} = \mathtt{Jac}\,(Q)$ for the fixed dimension vector $\alpha\colon Q_0 \to \mathbb{N}$ that maps every vertex to 1. The space of these representations $\mathsf{Rep}(\mathtt{Jac}\,(Q),\alpha)$ is an affine scheme with ring of coordinates

$$\mathbb{C}[\mathsf{Rep}(\mathtt{Jac}\,(Q),\alpha)] = \mathbb{C}[x_a \mid a \in Q_1]/\langle \textstyle\prod_{b \in l_a} x_b - \prod_{b \in r_a} x_b \mid a \in Q_a \rangle.$$

Each arrow is represented by a variable x_a and a representation ρ is called a *torus representation* if $\rho(a) = x_a$ is nonzero for every arrow. The torus representations form an open set of $\mathsf{Rep}(\mathtt{Jac}\,(Q),\alpha)$ and its closure is a component $\mathsf{Trep}(\mathtt{Jac}\,(Q),\alpha) \subset \mathsf{Rep}(\mathtt{Jac}\,(Q),\alpha)$ called the *scheme of all toric representations*. Given a stability condition θ, we can look at the moduli space of toric θ-stable representations $\mathbb{M}^t_\theta(J,\alpha) \subset \mathbb{M}_\theta(J,\alpha)$. If θ is *generic*, which means that $\theta \cdot \beta \neq 0$ for all $0 < \beta < \alpha$, this is a smooth toric variety. The structure of this toric variety for generic θ has been studied by Ishii and Ueda in [125]. To summarize their results we need some definitions.

Definition 11.12 Let Q be a consistent dimer on a torus and let $X, Y \in \mathtt{Jac}\,Q$ be two cyclic paths that span the homology of the torus.

- A *perfect matching* $\mathcal{P}$ is a set of arrows $a \in Q_1$ such that every face of the dimer $c \in Q_2$ contains precisely one arrow. To $\mathcal{P}$ we assign a grading such that $\deg_{\mathcal{P}}(a)$ is 1 if $a \in \mathcal{P}$ and 0 otherwise.
- The lattice point of $\mathcal{P}$ is the point $m_{\mathcal{P}} = (\deg_{\mathcal{P}} X, \deg_{\mathcal{P}} Y)$.
- The *matching polygon* is the convex hull of all $m_{\mathcal{P}}$, where $\mathcal{P}$ is a perfect matching.
- The *cosupport of a representation* ρ is the set of zero arrows $\{a \mid \rho(a) = 0\}$.

- A collection of perfect matchings $\{\mathcal{P}_1, \ldots, \mathcal{P}_k\}$ is called *θ-stable* if their union is the cosupport of a θ-stable representation. If $k = 2, 3$, we call these *stable pairs* or *stable triples*.

Theorem 11.13 (Ishii–Ueda) *Let Q be a consistent dimer on a torus and θ a generic stability condition.*

(i) $\mathbb{Y} = \mathbb{M}_0^t(\mathtt{Jac}\,(Q), \alpha)$ *is a three-dimensional affine toric variety corresponding to the cone over the matching polygon.*
(ii) *The stable pairs and stable triples define a triangulation of the matching polygon into elementary triangles.*
(iii) $\mathbb{Y}_\theta = \mathbb{M}_\theta^t(\mathtt{Jac}\,(Q), \alpha)$ *is the resolution of $\mathbb{Y}$ corresponding to this triangulation.*
(iv) $\mathbb{Y}_\theta$ *is derived equivalent to the Jacobi algebra:*

$$\mathtt{D\,Coh}^\bullet\, \mathbb{Y}_\theta \overset{\infty}{=} \mathtt{D\,Mod}^\bullet\, \mathtt{Jac}\,(Q).$$

(v) *The space $\mathbb{Y}_\theta$ comes with a natural function $\ell\colon \rho \mapsto \rho(\ell)$ and*

$$\mathtt{MF}^\pm(\mathbb{Y}_\theta, \ell) \overset{\infty}{=} \mathtt{MF}^\pm(\mathtt{Jac}\,(Q), \ell).$$

Sketch of the proof The first three parts are quite technical and rely heavily on the fact that Q is consistent and θ is generic. The fourth part follows from the fact that $\mathbb{Y}_\theta$ comes with a tilting bundle $\mathscr{V}_\theta$ whose endomorphism ring is $\mathtt{Jac}\,(Q)$. On each fiber, $\mathtt{Jac}\,(Q)$ acts like the corresponding representation $p \in \mathbb{Y}_\theta$. The bundle gives rise to a Fourier–Mukai functor $\mathrm{RHom}(-, \mathscr{V}_\theta)\colon \mathtt{DCoh}^\pm\mathbb{Y}_\theta \to \mathtt{D\,Mod}^\bullet\,\mathtt{Jac}$, which has an inverse $\otimes^L_{\mathtt{Jac}} \mathscr{V}_\theta$. This pair of functors can also be used to turn a matrix factorization (P, d) of $(\mathtt{Jac}\,, \ell)$ into a matrix factorization of $(\mathbb{Y}_\theta, \ell)$ and vice versa. For more details we refer to [125, 179, 29]. □

Example 11.14 Consider the quiver Q embedded on the torus as indicated in the illustration and fix cyclic paths $(x, y) = (a_9 a_8^{-1}, a_1^{-1} a_4^{-1} a_2)$.

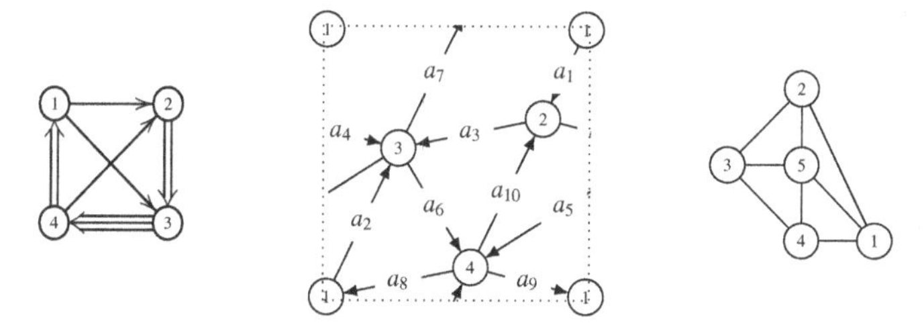

For $\theta = (-3, -1, 2, 2)$ the stable perfect matchings are given below.

Matching	Arrows	Lattice point
$\mathcal{P}_1$	$\{9,6,4\}$	$(1,-1)$
$\mathcal{P}_2$	$\{7,2,10\}$	$(0,1)$
$\mathcal{P}_3$	$\{3,8,5\}$	$(-1,0)$
$\mathcal{P}_4$	$\{1,6,5\}$	$(0,-1)$
$\mathcal{P}_5$	$\{3,2,4\}$	$(0,0)$

Their lattice points in the Newton polygon above right are indicated by the corresponding numbers. The singular space $\mathbb{Y}_\theta$ is the cone over a Hirzebruch surface Σ_1 and its resolution $\mathbb{Y}_\theta$ is the total space of the anticanonical bundle over that Hirzebruch surface.

The theorem also tells us that the toric variety is the union of torus orbits that parametrize representations with a given cosupport.

- The representations with empty cosupport form the dense three-dimensional torus orbit.
- Every stable perfect matching gives a two-dimensional torus orbit, whose closure is a toric divisor.
- Every stable pair gives a one-dimensional torus orbit, whose closure is a $\mathbb{P}^1$ or an $\mathbb{A}^1$.
- Every stable triple gives a fixed point.

Each triple T also defines an open subset $\mathbb{U}_T \subset \mathbb{Y}_\theta$, which is isomorphic to $\mathbb{C}^3$ and is the union of the torus orbits with cosupport contained in the stable triple.

Just like in Section 8.4.1, the zero locus of ℓ consists of all orbits except the dense orbit. The singular locus of $\ell^{-1}(0)$ is the union of all one- and zero-dimensional orbits. We can view the singular locus $\mathbb{S}_\theta = \mathsf{Sing}\, \ell^{-1}(0)$ as a union of $\mathbb{A}^1$'s and $\mathbb{P}^1$'s. For every internal edge we take a $\mathbb{P}^1$ and for every edge on the boundary of the matching polygon we take an $\mathbb{A}^1$. For each triangle we glue the $\mathbb{A}^1$'s and $\mathbb{P}^1$'s corresponding to its edges together at a point. For every open $\mathbb{U}_T$ the Landau–Ginzburg model $(\mathbb{Y}_\theta, \ell)$ restricts to a $(\mathbb{C}^3, XYZ)$, which is mirror to the pair of pants. Theorem 9.38 allows us to see the mirror of $(\mathbb{Y}_\theta, \ell)$ as a collection of pairs of pants glued together that forms a tubular surface around the dual graph of the triangulation.

Example 11.15 If we look at Example 11.14 we get the following pictures. The dual graph of the triangulation is a quadrangle with four legs.

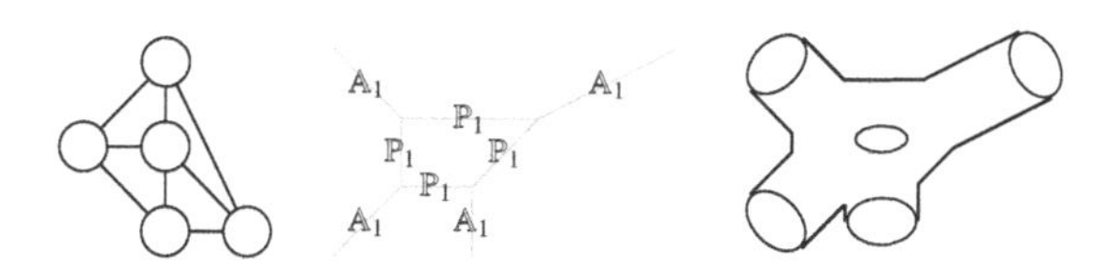

The space $\mathbb{S}_\theta = \mathsf{Sing}\,\ell^{-1}(0)$ consists of four $\mathbb{P}^1$'s and four $\mathbb{A}^1$'s, while the mirror surface is made by gluing four pairs of pants together.

On the other hand, by Theorem 9.66 the mirror can also be seen as the surface underlying the dual dimer $Q^\vee$, so we should be able to cut this surface into pairs of pants.

This can be done by looking at the perfect matchings on the dual dimer $Q^\vee$. Remember that this is a dimer on a marked surface $(\mathbb{S}, M)$ with genus g and m marked points. The genus is equal to the number of internal lattice points of the Newton polygon and the number of marked points is equal to the number of boundary lattice points. Therefore we can identify $\mathbb{S} \setminus M$ with a tubular thickening of the dual graph Γ_θ of the triangulation. But how do we draw the arc collection of the dual dimer on this tubular thickening?

Theorem 11.16 *Let Q be a consistent dimer model on a torus, θ a generic stability condition and $Q^\vee$ the dual dimer.*

(i) *If a is an arrow then we can draw a line on a thickening of the dual graph separating the stable matchings containing a from those not containing a.*

(ii) *If $c \in Q_2$ is a face then the lines coming from the arrows in c bound a polygonal region that sits on a subtree of the dual graph.*

(iii) *If we glue these regions together we get a tubular neighborhood of the dual graph for which the separation lines form the arc collection representing the dual dimer.*

Sketch of the proof The proof depends on the consistency of the dimer and is explained in detail in [37]. □

Example 11.17 If we return to the Hirzebruch example, Example 11.14, we can compile a list of all the arrows, the stable matchings in which they are contained and the separating lines.

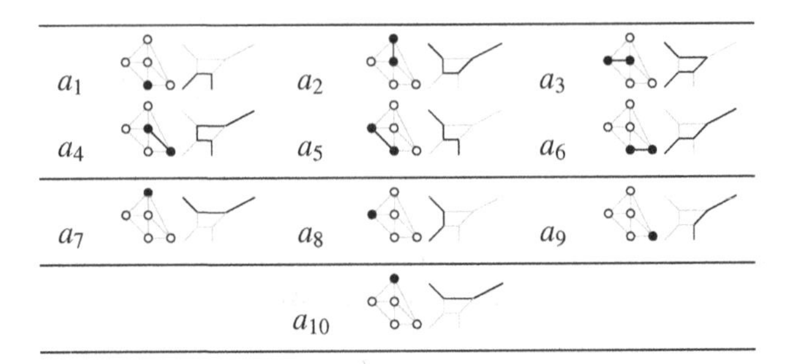

Now that we have put $Q^\vee$ on a tubular surface around the graph, take stable pair $\{\mathcal{P}_1, \mathcal{P}_2\}$ and can take the surface of $Q^\vee$ perpendicular to the corresponding edge in the dual graph. This gives a simple curve $\mathbb{L}_{\mathcal{P}_1,\mathcal{P}_2}$ that runs through all arrows in $(\mathcal{P}_1 \cup \mathcal{P}_2) \setminus (\mathcal{P}_1 \cap \mathcal{P}_2)$ and alternates between an arrow of $\mathcal{P}_1$ and an arrow of $\mathcal{P}_2$. We can easily identify these curves with objects on the B-side.

Lemma 11.18 *If ρ is a representation with cosupport $\mathcal{P}_1 \cup \mathcal{P}_2$, then under the identification*

$$\mathtt{DSing}^{\pm}\mathtt{Jac}\,/\langle\ell\rangle \overset{\infty}{=} \mathtt{MF}^{\pm}(\mathtt{Jac}, \ell) \overset{\infty}{=} {}^{\mathtt{top}}\mathbf{Fuk}^{\pm}(\mathbb{S}, M),$$

ρ corresponds to a direct sum of a band object with underlying curve $\mathbb{L}_{\mathcal{P}1,\mathcal{P}2}$ and its shift.

Sketch of the proof Because ρ is zero on $\mathcal{P}_1$ the corresponding matrix factorization M_ρ can be graded by $2\deg_{\mathcal{P}_1}$, so we know that it is a direct sum of string and band objects. It cannot contain any string objects because these have infinite-dimensional endomorphism rings, while

$$\mathrm{Hom}^{\pm}_{\mathtt{MF}}(M_\rho, M_\rho) \cong \mathrm{Mat}_{2\times 2}(\mathbb{C}[\epsilon]/(\epsilon^2))$$

is finite-dimensional. The latter also tells us that M_ρ is the direct sum of a band object of a simple curve and its shift.

Furthermore, we can calculate that for the standard matrix factorizations coming from the arrows M_a we have that $\mathrm{Hom}_{\mathtt{MF}^{\pm}}(M_a, M_\rho) = \mathbb{C} \oplus \mathbb{C}$ if $\rho(a) = \rho(\ell a^{-1}) = 0$ and is zero otherwise. Therefore, the bands of M_ρ intersect only the arrows in faces where $\mathcal{P}_1$ and $\mathcal{P}_2$ differ (because only then can a and ℓa^{-1} both be zero). Because the hom-space is $\mathbb{C}\oplus\mathbb{C}$ the band intersects those arrows in both directions. For more details see [37]. □

These results can be seen as some kind of SYZ-fibration: both $\mathbb{S}_\theta$ and the surface $(\mathbb{S}, M)$ can be projected onto the dual graph of the triangulation Γ_θ:

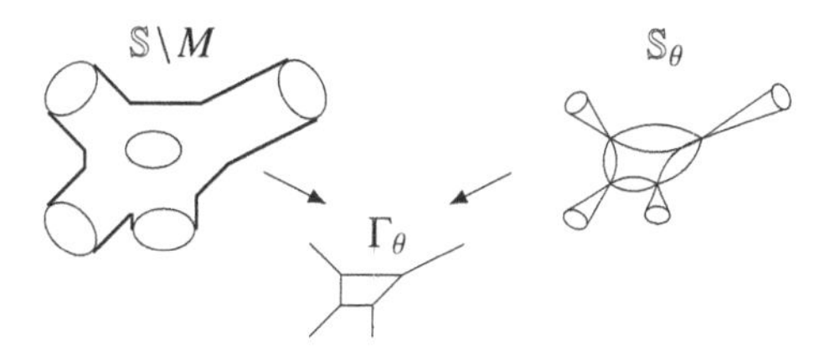

such that $\rho \in \mathbb{S}_\theta$ corresponds to a band object with underlying curve $\pi_A^{-1}\pi_B(\rho)$. In this diagram these projection maps are only defined up to homotopy because Γ_θ is only defined as a combinatorial object.

Finally, we can realize Γ_θ as an actual space using tropical curves. To a weight function $w\colon Q_1 \to \mathbb{Z}$ we can associate a stability condition

$$\theta_w\colon Q_0 \to \mathbb{Z}\colon v \mapsto \sum_{h(a)=v} w(a) - \sum_{t(a)=v} w(a).$$

This θ_w satisfies $\theta_w \cdot (1, \ldots, 1) = 0$ and every θ that satisfies this condition can be realized by some weight function w.

The weight function can also be used to define a polynomial

$$f_{w,t} = \sum_{\mathcal{P}} t^{\sum_{a\in\mathcal{P}} w(a)} X^{\deg_{\mathcal{P}} x} Y^{\deg_{\mathcal{P}} Y},$$

where the sum runs over all perfect matchings in Q.

Theorem 11.19 *If θ_w is generic then the dual graph of its triangulation is the same as the tropical curve of $f_{w,t}$.*

Sketch of the proof The tropical curve is the locus in $(\xi, \eta) \in \mathbb{R}^2$ where $F_{\mathsf{trop}}(\xi,\eta) = \max_{\mathcal{P}}\{L_{\mathcal{P}}(\xi,\eta)\}$, with

$$L_{\mathcal{P}}(\xi,\eta) = (\deg_{\mathcal{P}} x)\xi + (\deg_{\mathcal{P}} y)\eta + \sum_{a\in\mathcal{P}} w(a),$$

is reached by at least two functions $L_{\mathcal{P}}$. One can show that $\mathcal{P}$ is θ_w-stable if and only if there is a point (ξ,η) where $F_{\mathsf{trop}}(\xi,\eta) = L_{\mathcal{P}}(\xi,\eta)$ is the unique maximum and $\{\mathcal{P}_1,\mathcal{P}_2\}$ is a stable pair if there is a (ξ,η) such that $F_{\mathsf{trop}}(\xi,\eta) = L_{\mathcal{P}_1}(\xi,\eta) = L_{\mathcal{P}_2}(\xi,\eta)$. Therefore, the edges of the tropical curve are in bijection with the stable pairs. More details can be found in [35]. □

11.3 Bridgeland Stability

King stability is a notion that is suitable for representations of an algebra, but often we want to work with more general categories or A_∞-categories.

Therefore, we will have to expand our notion of stability. To do this we will first rephrase King stability in a way that is more suitable for generalization.

To every $\theta\colon Q_0 \to \mathbb{Z}$ we can associate a map called the *central charge*,

$$Z\colon \mathbb{Z}Q_0 \to \mathbb{C}\colon \alpha \mapsto \theta \cdot \alpha + i \sum_{v \in Q_0} \alpha_v.$$

This map allows us to assign to each representation ρ a complex number $Z(\rho)$ in the upper half-plane $\mathbb{H} = \{x \in \mathbb{C} \mid \operatorname{Re} x > 0\}$ that depends on its dimension vector. The argument $\phi(\rho) = \arg Z(\theta)$ is called the phase of the representation. Note that a representation is θ-stable in the original sense if (i) its phase is $\frac{\pi}{2}$ and (ii) all proper subrepresentations have a smaller phase than ρ. Therefore, it makes sense to drop the first condition.

Definition 11.20 A representation ρ is*(semi)stable with phase* $\phi(\rho)$ if every proper subrepresentation has a phase smaller than (or equal to) $\phi(\rho)$.

Theorem 11.21 *For every representation* $\rho\colon \mathbb{C}Q \to \operatorname{End}(W)$ *there is a filtration*

$$0 = V_0 \subset V_1 \subset \cdots \subset V_n = W$$

such that each factor $F_i = V_i/V_{i-1}$ *is a semistable* $\mathbb{C}$*-module with phase* ϕ_i *and the phases decrease:* $i < j \implies \phi_i \geq \phi_j$. *Such a filtration is called a Harder–Narasimhan filtration.*

Sketch of the proof We can use induction on the total dimension of W. For dimension 1 the statement is trivially true because such a representation is simple and we can take the filtration $0 \subset W$. Similarly, if W is semistable then we can take the same filtration.

If the representation is not semistable there is at least one subrepresentation V_1 with a bigger phase. Take the one with the biggest phase and look at the series $V_0 \subset V_1 \subset W$. Note that V_1 is semistable because if it contains a subrepresentation with a bigger phase then it was not the subrepresentation with the biggest phase. Similarly, W/V_1 cannot have a subrepresentation V' with phase bigger than $\phi(V_1)$, otherwise the pullback of V' under the map $\pi\colon W \to W/V_1$ would have a central charge $Z(V_1) + Z(V')$ and hence $\phi(\pi^{-1}(V')) > \phi(V_1)$. By the induction hypothesis, W/V_1 has a filtration

$$0 \subset V_2' \subset \cdots \subset V_k' = W/V_1$$

with decreasing phases. The pullbacks $V_i = \pi^{-1}(V_i')$ of these representations also have decreasing phases, so

$$0 \subset V_1 \subset V_2 \subset \cdots \subset V_k = \pi^{-1} V_k' = W$$

is the filtration we are looking for.

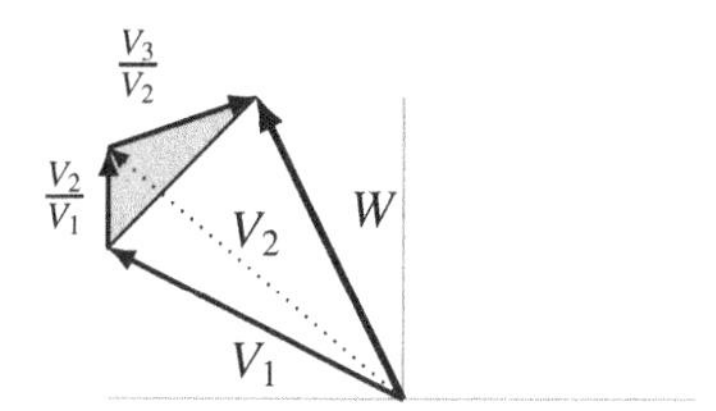

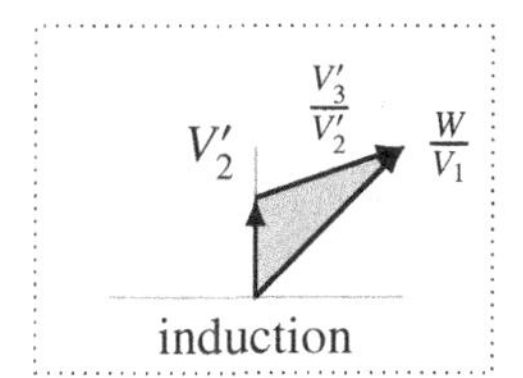

□

Remark 11.22 The existence of a Harder–Narasimhan filtration is not confined to the situation of representations of quivers. It is also important for the classification of vector bundles (the term actually originates there [193, 117]). If $\mathbb{E}$ is a curve and $\mathscr{V}$ is a vector bundle, we can define a central charge as

$$Z(\mathscr{V}) = -\deg \mathscr{V} + i \operatorname{rk} \mathscr{V}$$

and define its phase as the argument of $Z(\mathscr{V})$ chosen between $(0, \pi]$.

Remark 11.23 In the literature it is customary to work with the slope instead of the phase. The slope is the ratio $\mu = \deg \mathscr{V} / \operatorname{rk} \mathscr{V}$ (or $\mu = -\theta \cdot \alpha / \sum_v \alpha_v$ in the case of quivers [112, 207]), but this is equivalent as there is a monotone relation between them: $\phi = \arctan(\mu) + \frac{\pi}{2}$.

Unlike the slope, the phase has the advantage that we can extend it to the whole derived category of representations $\mathtt{Dmod}^\bullet Q$. We can interpret Z as a map from the (numerical) Grothendieck group to $\mathbb{C}$ and if ρ is a semistable representation with phase $\phi(\rho)$ then we say that $\rho[k]$ is a semistable object with phase $\phi(\rho) + nk\pi$. With this convention, all objects in $\mathtt{Dmod}^\pm Q$ can be seen as cones of semistable objects with increasing phase.

To transfer these ideas to the world of A_∞-categories, we will rephrase the notion of stability conditions for triangulated categories as developed by Bridgeland in [39] to the setting of twisted complete A_∞-categories. The key point in the translation is the switch from filtrations to twisted complexes. If $V_0 \subset \cdots \subset V_k$ is a filtration then V_k is a repeated cone of extensions, which can be written as a twisted complex

$$V_k = \left(\frac{V_k}{V_{k-1}} \oplus \cdots \oplus \frac{V_1}{V_0}, \alpha_k + \alpha_{k-1} + \cdots + \alpha_1 \right),$$

where $\alpha_i \colon \frac{V_i}{V_{i-1}} \to \frac{V_{i-1}}{V_{i-2}}$ corresponds to the extension

$$0 \to \frac{V_{i-1}}{V_{i-2}} \to \frac{V_i}{V_{i-2}} \to \frac{V_i}{V_{i-1}} \to 0.$$

So a filtration with decomposition factors with decreasing phases corresponds to a twisted complex with summands with increasing phases.

Definition 11.24 Let $\mathtt{A}^\bullet$ be a twisted complete minimal A_∞-category and fix a homomorphism $v\colon K_0\,\mathtt{A}^\bullet \to \Gamma$ where $\Gamma \cong \mathbb{Z}^N$ is a finitely generated free abelian group.

A *Bridgeland stability condition* consists of a pair $(Z, \mathcal{P})$, where $Z\colon \Gamma \to \mathbb{C}$ is a group morphism called the *central charge* and for each $\phi \in \mathbb{R}$ the *slice* $\mathcal{P}_\phi$ is a full additive subcategory of $\mathtt{A}^\bullet$, whose objects are called the *semistable objects with phase* ϕ. This pair satisfies the following axioms.

(SC1) The phase of the nonzero stable objects is the argument of the central charge:

$$\forall\, X \in \mathcal{P}_\phi \text{ we have } X \neq 0 \implies Z(X) \in \mathbb{R}_{>0}e^{i\phi}.$$

Here we write $Z(X)$ as shorthand for $Z(v[X])$.

(SC2) A shift adds π to the phase:

$$\mathcal{P}_{\phi+\pi} = \mathcal{P}_\phi[1].$$

(SC3) Nontrivial morphisms of degree 0 between stable objects increase the phase:

$$\forall\, X_1 \in \mathcal{P}_{\phi_1}, X_2 \in \mathcal{P}_{\phi_2} \text{ we have } \phi_1 > \phi_2 \implies \mathtt{A}^0(X_1, X_2) = 0.$$

(SC4) Every object is the iterated cone of stable objects with increasing phases:

$$\forall\, X \in \mathtt{A}^\bullet,\ \exists \phi_1 < \cdots < \phi_k, X_i \in \mathcal{P}_{\phi_i} \text{ such that } X \cong (X_1 \oplus \cdots \oplus X_k, \delta)\,.$$

One can show that for a given X the X_i are uniquely determined up to quasi-isomorphism. For each object X we set $\phi^-(X) = \phi_1$ and $\phi^+(X) = \phi_k$.

(SC5) The norms of the central charge of semistable objects cannot become arbitrarily small: there is a constant $C > 0$ such that for all nonzero semistable objects X,

$$\|v([X])\| \leq C|Z(X)|,$$

where $\|\cdot\|$ is the standard norm on $\mathbb{Z}^N$.

Remark 11.25 In the original definition, the group Γ was not present and the full Grothendieck group was used instead. A different property (local finiteness instead of SC5) was used to prove Theorem 11.28, the main result of [39]. Unfortunately, the Grothendieck group is not always finitely generated and this results in infinite-dimensional spaces of stability conditions. To fix this, a morphism ν can be used to bring down the size of the Grothendieck group and in [149] (SC5) was introduced as the *support property*. A further discussion on this property and its significance can be found in [30].

Remark 11.26 Note that in the module category $\mathtt{Mod}^\bullet R$, a map f is an injection if and only if the cone $\mathrm{Cone}(f\colon X[-1] \to Y) \in \mathtt{D\,Mod}^\bullet R$ is isomorphic to an object in $\mathtt{Mod}^\bullet R$ (or equivalently its cohomology is concentrated in degree 0). In that case, f embeds X in Y and the cone is the cokernel of f. However, in an A_∞-category there is no notion of a subobject or a quotient object. We can solve this problem with the help of the slices.

For a general A_∞-category $\mathtt{A}^\bullet$ with a stability condition $\sigma = (Z, \mathcal{P})$, we define its *heart* $\heartsuit_\sigma$ as the subcategory containing all objects X with $\phi^+(X), \phi^-(X) \in (0, \pi]$. If $X, Y \in \heartsuit_\sigma$ and $f\colon X \to Y$ is a map of degree 0 then we say that f is an *embedding* if

$$\mathrm{Cone}(f\colon X[-1] \to Y) \in \heartsuit_\sigma.$$

In that case, we say that X is a subobject of Y. The heart and the central charge are enough to reconstruct all the slices: for $\phi \in (0, \pi]$ the slice $\mathcal{P}_\phi$ will contain all $Y \in \heartsuit_\sigma$ with $Z(Y) \in \mathbb{R}_{>0}e^{i\phi}$ without subobjects X with $Z(X) \in \mathbb{R}_{>0}e^{i\phi'}$ and $\phi' > \phi$. This idea leads to a different axiomatization of a stability condition as a pair $(Z, \heartsuit)$. For more details see [40].

Example 11.27 Let Q be the linear quiver and look at $\mathtt{D\,mod}^\bullet Q$. In Chapter 4, we saw that we could interpret this as the category of edges and diagonals of an $(n+1)$-gon. If we embed this $(n+1)$-gon as a convex $(n+1)$-gon in $\mathbb{C}$ and use the horizontal line field on $\mathbb{C}$ then we could see every indecomposable object as a vector in $\mathbb{C}$ together with a phase angle. Mapping each line to its vector defines a central charge and we take for $\mathcal{P}_\phi$ the category of all direct sums of indecomposables with phase ϕ.

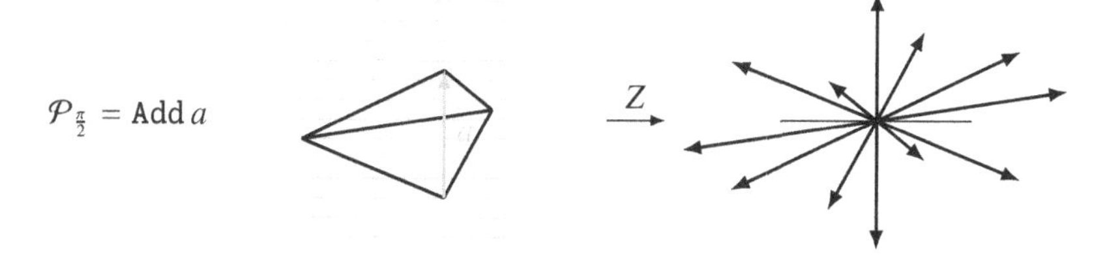

We can easily check that the five stability conditions hold: (SC1) and (SC2) follow from the definition of the grading. (SC3) is true because all morphisms can be seen as positive angles in $(0, \pi)$ and (SC4) holds because we can take the indecomposable summands ordered according to phase with trivial δ. Finally, (SC5) holds because there are only finitely many indecomposables.

Now move one of the points of the convex polygon inwards until the polygon becomes concave. What happens with the stability condition? The phases of the two edges meeting at that point become gradually the same and then they swap order. If that happens, the diagonal that moves outside the polygon has a smaller phase than the first edge, so the map between them runs between objects with decreasing phase. We can solve this problem by declaring that diagonal unstable.

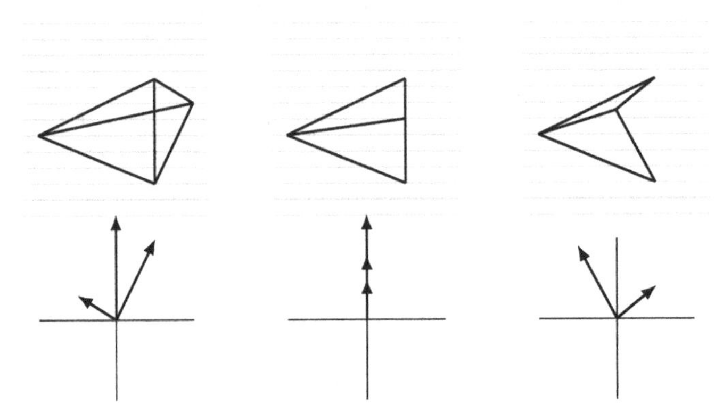

If we do this, the Harder–Narasimhan property still holds: the unstable diagonal is the cone over the two edges and because their phases are swapped it is a cone between objects with increasing phases.

This example suggests that given a stability condition $(Z, \mathcal{P})$ it is possible to change the central charge continuously in all directions and change the slices $\mathcal{P}_\phi$ along the way. This intuition can be neatly expressed in the following theorem.

Theorem 11.28 (Bridgeland) *Given an $\mathcal{A}_\infty$-category $\mathtt{A}^\bullet$ and a map $v\colon K_0\,\mathtt{A}^\bullet \to \Gamma$ then we can give the set of stability conditions $\mathsf{Stab}_\Gamma\,\mathtt{A}^\bullet$ the structure of a complex manifold such that the forgetful map*

$$\mathsf{Stab}_\Gamma\,\mathtt{A}^\bullet \to \mathtt{Grp}(\Gamma, \mathbb{C})\colon (Z, \mathcal{P}) \to Z$$

is a local diffeomorphism.

Sketch of the proof The proof of the original theorem by Bridgeland for triangulated categories, without the extra map v, and using locally finite stability conditions, can be found in [39]. For more information about the connection with the support property see [30, Appendix A]. Its main aim is to keep the phases of the stable objects under control when we change Z slightly. If $K_0\,\mathtt{A}^\bullet$ is finitely generated we can use $\Gamma = K_0\,\mathtt{A}^\bullet$ and drop the subscript. □

Remark 11.29 In general it is not easy to describe the whole stability manifold of a category, but given one particular stability condition it is sometimes possible to describe the connected component in which it is contained. In some rare cases it is also possible to show that the stability manifold is connected and then we get a full description.

Remark 11.30 If $\mathcal{F}\colon \mathtt{A}^\bullet \to \mathtt{A}^\bullet$ is an autoequivalence of the category, it will induce a homeomorphism on the stability manifold. Therefore, the stability manifold comes with an action of the automorphism group of the category and one can also look at the quotient space $\mathsf{Stab}(\mathtt{A}^\bullet)/\mathsf{Aut}\,\mathtt{A}^\bullet$. While the space of stability conditions itself is often contractible, this quotient space can have a rich topological structure.

11.4 Stability Conditions and Quadratic Differentials

In Example 11.27 we saw a way to construct stability conditions by identifying the polygon with a piece of the complex plane. We will now generalize this idea to surfaces with 1-forms or quadratic differentials. The idea of using quadratic differentials to construct stability conditions was first pioneered by Bridgeland and Smith [44] and Sutherland [233] for Fukaya categories of three-dimensional manifolds. The case of Fukaya categories of surfaces was worked out by Haiden, Katzarkov and Kontsevich in [115]. In this section we show some examples of their construction and in the next section we will discuss their main result.

Let us first look at 1-forms, because this is a little easier. Let $(\vec{\mathbb{S}}, M)$ be a graded surface without boundary, with a grading coming from a 1-form ω whose zeros and poles lie in M. To each object L in ${}^{\mathtt{top}}\mathsf{Fuk}^\bullet(\vec{\mathbb{S}}, M)$ we can associate a number by integrating ω along its underlying curve:

$$Z(L) = \int_0^1 \omega(\dot{\mathbb{L}}(t))\,dt,$$

where $\mathbb{L}(t)$ is a parametrization of the curve underlying the object L. This number only depends on the homology class of $\mathbb{L}$, so $Z(L)$ is well defined but it can be infinite. This happens when L is a string that starts or ends at a pole in ω. If Γ is the subgroup of $H_1(\vec{\mathbb{S}}, M)$ generated by the homology classes of strings and bands that avoid these poles, then we can see the map Z as a central charge $Z\colon \Gamma \to \mathbb{C}$. Furthermore, if we denote the subcategory generated by all string and band objects that avoid these poles by

$${}^{\mathtt{fin}}\mathsf{Fuk}^\bullet(\vec{\mathbb{S}}, M) \subset {}^{\mathtt{top}}\mathsf{Fuk}^\bullet(\vec{\mathbb{S}}, M),$$

then there will be a natural morphism $\nu\colon {}^{\text{fin}}\mathsf{Fuk}^\bullet(\vec{\mathbb{S}}, M) \to \Gamma$ that assigns to each string or band object its homology class.

To pick out the semistable objects, we first define the length of the curve as

$$\ell(\mathbb{L}) := \int_0^1 |\omega(\dot{\mathbb{L}}(t))|\, dt.$$

By the triangle inequality we have $|Z(L)| \leq \ell(\mathbb{L})$ and we say that $\mathbb{L}$ is a *special curve* if the equality holds. The metric on $\mathbb{S} \setminus M$ that comes with this notion of length is flat because at each point in $p \in \mathbb{S} \setminus M$ we can find a local complex coordinate z for which $\omega = dz$. If we write $z = x + iy$, the formula for the length is just the Euclidean length.

Therefore, we can think of these special curves as straight lines. The phase function θ of a graded Lagrangian is constant and equal to the argument of $Z(L)$ up to a multiple of π. Note that this is a property that depends on the specific curve, not its homotopy class. Some homotopy classes contain a special curve while others do not. If an object L can be represented by a special curve $\mathbb{L}$ then we will call it *semistable with phase* $\phi = \theta$. In this way we get a slicing $\mathcal{P}_\phi$.

Given any curve $\mathbb{L}$ we can try to minimize its length by varying the curve. If we do this the curve will try to straighten up, but sometimes it will get stuck behind a marked point. If that happens, the result will be a sequence of broken line segments and we can see the string or band object as a cone over the broken pieces. Because the angle morphisms have degree 1 but turn over an angle of more than π, the summands in this cone will have increasing phase, so this will correspond to a Harder–Narasimhan filtration.

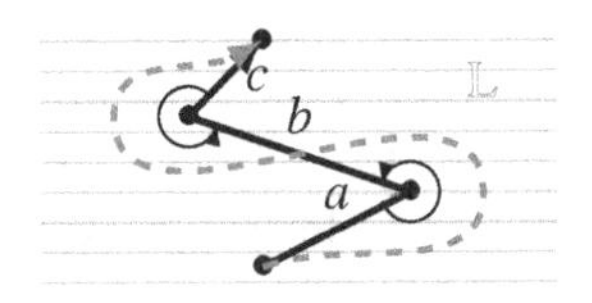

The final thing that needs to be checked is the support property. For this we refer to [115, Section 5.4].

Example 11.31 Let $\mathbb{S}$ be the Riemann sphere with four marked points $\{0, 1, 2, \infty\}$ and take the 1-form $\omega = dz/z = d\ln z$. The leaves of the line field corresponding to ω are radial lines. An arc has infinite length if it starts or ends at $0, \infty$. An arc that starts at 1, ends at 2 and turns once around the annulus $1 < |z| < 2$ has a central charge $\int_{\mathbb{L}} d\ln z = \ln 2 + 2\pi i = u$. It can be represented by the special Lagrangian curve $\gamma(t) = e^{ut}$ with phase $\theta = \arctan(2\pi/\ln 2)$; hence this arc is a stable object.

The arc that starts and ends at 1 and turns around 2 but not around the annulus has central charge 0, so it cannot be stable. It breaks down as $(a \oplus a[1], \alpha)$, where a is the simple straight arc from 1 to 2 and α is a full turn around 2.

If we use a quadratic differential ϕ instead of a 1-form, we can do the same thing, provided we take the square root of ϕ. Because we are working over the complex numbers, the square root is not a 1-form on $\mathbb{S}$ but on a double cover $\tilde{\mathbb{S}}$. Each curve $\mathbb{L}$ will lift to $\tilde{\mathbb{L}}$ and we can take the integral upstairs.

Example 11.32 Let $\mathbb{S} = \widehat{\mathbb{C}}$ be the Riemann sphere with three marked points $\{-1, 1, \infty\}$ and take the quadratic differential $\phi = dz^2/(1 - z^2)$. The horizontal line field consists of a straight line between -1 and 1, surrounded by ellipses (see Example 10.11). If we pull back ϕ along the map $z = \frac{1}{2}(u+u^{-1})$, it becomes $-u^{-2}\,du^2$, so we can take the square root: $\sqrt{\phi} = iu^{-1}\,du$. The map is a double cover that is branched at $z = \pm 1$ but not at $z = \infty$, so this cover is a sphere with four marked points and it can be identified with $(\mathbb{P}^1, i\,du/u)$ with marked points $\{0, 1, -1, \infty\}$.

The straight line between -1 and 1 lifts to two halves of the unit circle between 1 and -1. The total charge of this lift is 2π. Each ellipse lifts to two circles, one inside the unit circle and one outside, which gives a total charge of 4π. The corresponding string and band objects have phase 0 because the charge is real. Using the charges and the phases from the double cover, we get a stability condition for ${}^{\mathtt{fin}}\mathtt{Fuk}^{\bullet}(\vec{\mathbb{S}}, M)$.

Every quadratic differential ϕ on $(\mathbb{S}, M)$ defines two things: a horizontal foliation and a metric. The leaves of this foliation are the geodesics for this metric that have constant phase 0. There are different kinds of leaves depending on their endpoints:

- *Closed leaves* have no endpoints.
- *Finite leaves* run between two marked points that are of the form $z^k dz^2$ with $k \geq -1$ (i.e. they have positive winding number). These leaves have finite length.
- *Ray leaves* start at a marked point with $k \geq -1$ and end at one with $k < -1$. These leaves are finite at one end and infinite at the other.
- *Infinite leaves* run between two marked points that are of the form $z^k dz^2$ with $k < -1$.

As illustrated in Example 10.7, each marked point of the form $z^k dz^2$ with $k \geq -1$ has $2 + k$ finite ends of leaves arriving (the tangent directions $e^{r2\pi i/k+2}$). So there is an upper bound on the numbers of rays and finite leaves.

Having finite leaves is a special condition because it requires two finite ends to meet up. One can show that if at least one of the marked points has a positive winding number, it is possible to rotate the quadratic differential slightly to $e^{i\epsilon}\phi$ such that there are no closed or finite leaves. In that situation we can cut the surface along the ray leaves. The angle between two consecutive ray leaves emanating from a marked point is π, so together they form a straight line. Therefore, we end up with pieces that are homeomorphic with strips

$$S_r = (\{x + iy \in \mathbb{C} \mid 0 < y < r\}, (dz)^2) \quad \text{with } r \in (0, +\infty].$$

If $r < \infty$ we call the strip finite and then it is bounded on each side by two rays meeting at a marked point $m_1 = x_1, m_2 = x_2 + ri$. If $r = \infty$ we call the strip infinite and then only the lower line consists of two rays.

Example 11.33 Let us return to the previous example. The quadratic differential $\phi = dz^2/(1 - z^2)$ has one finite leaf but if we tweak it to $e^{i\epsilon} dz^2/(1 - z^2)$, this leaf breaks into two rays that go to ∞. If we cut open the rays we get one finite strip with on each side of a marked point. Each side is bounded twice by the same ray.

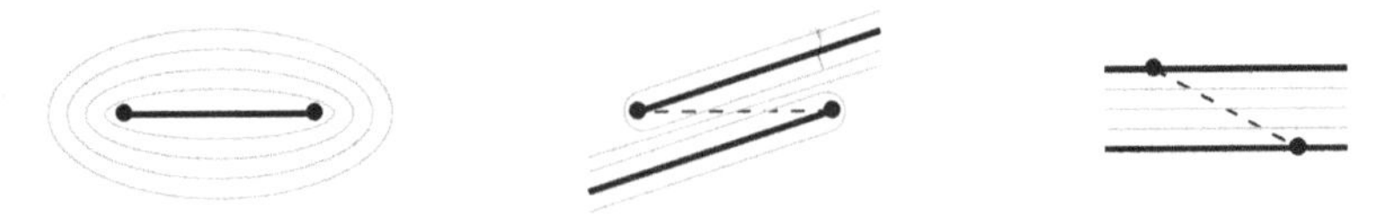

Example 11.34 Consider the Riemann sphere $\widehat{\mathbb{C}}$ with marked points $\{0, \infty\}$ and quadratic differential $\phi = z(dz)^2$. This has three ray leaves that start from 0 and go to ∞. If we cut it open we get three infinite strips. Now add one extra marked point at $1 \in \widehat{\mathbb{C}}$. This breaks the leaf in the real direction in two and we

get a finite leaf. By rotating ϕ slightly, this leaf misses 1 and becomes a ray. The wedge containing 1 splits into two pieces: a finite strip bounded on the other end by the leaf containing 1 (which is now two rays) and an infinite strip.

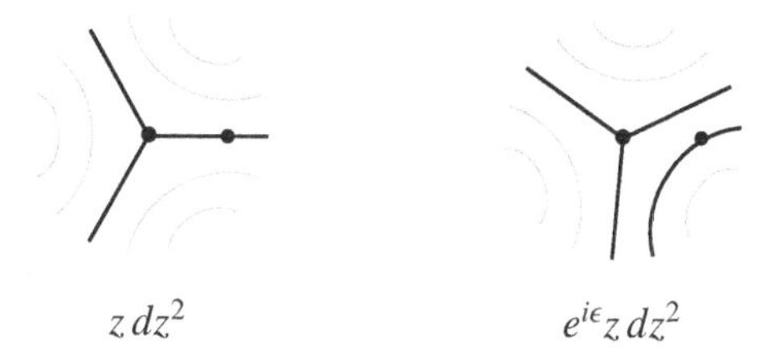

These two examples show us how to make stability conditions for graded surfaces with internal marked points but without boundary components. To include boundary components we need to look at quadratic differentials with exponential singularities.

Example 11.35 Consider $\mathbb{C}$ with the quadratic differential $\phi = e^{f(z)}dz^2$, where $f(z)$ is a polynomial of degree n. For $f(z) = z$, the line field for this quadratic differential consists of horizontal lines at heights $2k\pi$ and in between them curved lines that start from $+\infty + 2k\pi i$ turn in between and run back to $+\infty + 2(k+1)\pi i$.

If we compress $\mathbb{C}$ radially to fit inside the unit disk, each horizontal line will become a curve between -1 and 1 and in between these curves we have lobes starting and ending at 1. The lengths of these lobes are infinite. On the other hand, the length of any curve inside the circle that starts or ends at -1 will always be finite.

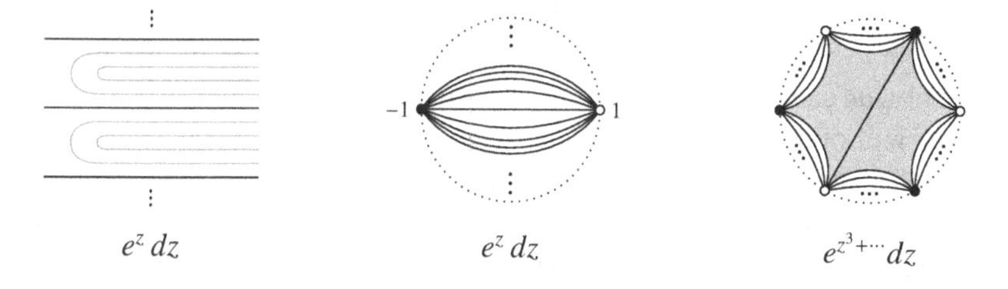

When $f(z) = z^n + \cdots$, this picture unfolds n times for $|z| \gg 0$. Generically, we have horizontal lines leaving the nth roots of -1 and arriving at the nth roots of 1. The lower-degree parts of $f(z)$ determine at which roots they arrive (in nongeneric cases there might also be lines between two roots of -1).

The length of any curve starting and ending at roots of -1 will be finite, so we can use the quadratic differential to calculate the central charge for any graded line segment on the disk with n marked points (being the roots of -1).

Furthermore, we can define stable curves as those for which the length equals the norm of the central charge. This gives a stability condition on the Fukaya category of the disk with n marked points.

Just like for the quadratic differentials on closed surfaces, we can do a strip decomposition of the interior of the disk together with the marked points. The ray leaves are the lines emanating from roots of -1 but as there are infinitely many of them, there will be an infinite number of strips. Only a finite number of them will be finite (in the picture above they are shaded gray). All others are infinite and can be ignored because all curves between marked points have representatives that lie outside these strips (so inside the gray area).

Remark 11.36 How do these stability conditions from exponential quadratic differentials relate to the basic stability conditions coming from embedding a polygon into the Euclidean plane? We can see the latter as examples of the former. Start with the polygon and draw horizontal lines through each corner. At each corner stick in a double series of infinite strips that form a bouquet of saddle wedges outside the polygon.

The result can be interpreted as a surface with an exponential quadratic differential. If we project it onto the disk, the horizontal rays become lines that start at a root of -1 and end at the root of -1 that lies between the two roots of -1 that are the corners of the edge of the polygon that the ray intersects.

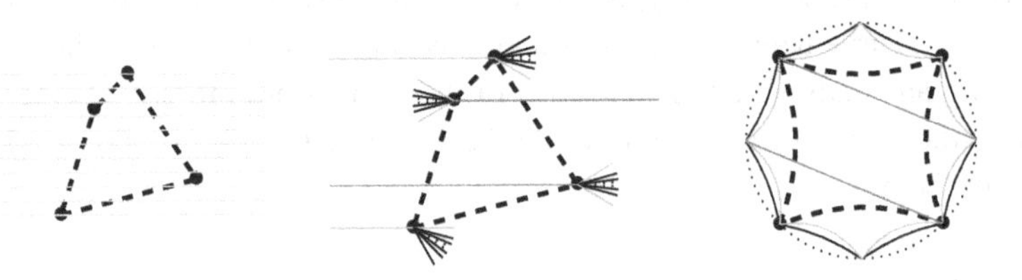

The polygon can be reconstructed from the quadratic differential by taking the convex hull of the marked points. This is the smallest subset containing these points that is closed under taking straight line segments (straight according to the metric of the quadratic differential). This set is also called the *core of the quadratic differential*. Some quadratic differentials will have a core that is a convex polygon, but not all.

Remark 11.37 Let γ be a curve in $\mathbb{C}$ whose image in the disk runs between two marked points on the circle. If we translate this curve by a constant c then the central charge and stability of this curve are not affected. Therefore, the quadratic differentials $\phi = e^{f(z)}dz^2$ and $\phi = e^{f(z-c)}dz^2$ define the same stability condition.

11.5 Stability Manifolds

We have seen that quadratic differentials can be used to construct stability conditions on Fukaya categories of graded surfaces. In most cases this idea is powerful enough to describe the whole stability manifold.

Suppose we have a marked compact surface without boundary $(\mathbb{S}, M)$. We turn $\mathbb{S}$ into a Riemann surface by giving it a complex structure $J\colon \mathsf{T}\mathbb{S} \to \mathsf{T}\mathbb{S}$ and add a quadratic differential $\varphi \in \mathsf{T}^{*\otimes 2}_{\mathbb{C}}\mathbb{S}$ that is well defined and nonzero on $\mathbb{S} \setminus M$. Furthermore, we assume that each of the marked points is at most an exponential singularity. This means that for each $m \in M$ we can find a local complex coordinate z such that $z = 0$ at m and

$$\varphi = e^{z^{-n}} z^k f(z)\, dz^2,$$

where $n \in \mathbb{Z}_{\geq 0}$, $k \in \mathbb{Z}$ and $f(z)$ is a holomorphic function with $f(0) \neq 0$.

- If $n > 0$ we call the marked point an *exponential pole of type* (n, k). In the neighborhood of such a point there are n special horizontal leaves that approach the point from a direction for which z^n is real and negative. These leaves have finite lengths near that point.
- If $n = 0$ we will call the marked point a *higher-order pole* if $k < -1$. Those are the marked points near which all horizontal leaves have infinite length.

The data $\mathbb{S}^\flat = (\mathbb{S}, M, J, \varphi)$ are sometimes referred to as a *flat surface with conical singularities*. Cut out a small disk around each exponential marked point of type (n, k) and put n new marked points at the intersection points of the special horizontal leaves. This gives a new marked surface that we can grade using the restriction of a line field coming from the quadratic differential. Denote this graded marked surface by $(\vec{\mathbb{S}}^\circ, M^\circ)$ and its relative homology group by $\Gamma = H_1(\vec{\mathbb{S}}^\circ, M^\circ)$.

Definition 11.38 The category $\mathsf{Fuk}^\bullet(\mathbb{S}^\flat)$ is the subcategory of ${}^{\mathrm{top}}\mathsf{Fuk}^\bullet(\vec{\mathbb{S}}^\circ, M^\circ)$ generated by the band objects and the string objects that do not end at higher-order poles. There is a natural morphism from $v\colon K_0\, \mathsf{Fuk}^\bullet(\mathbb{S}^\flat) \to \Gamma$ and we denote the space of stability conditions for this map by

$$\mathsf{Stab}(\mathbb{S}^\flat) := \mathsf{Stab}_\Gamma\, \mathsf{Fuk}^\bullet(\mathbb{S}^\flat).$$

Clearly, (J, φ) itself determines a stability condition $\sigma \in \mathsf{Stab}(\mathbb{S}^\flat)$. For the central charge we integrate the square root of the quadratic differential and take the limit where the small circles around the exponential poles go to zero. The slices $\mathcal{P}_\phi$ are the subcategories spanned by the objects that can be represented by a graded curve with constant phase ϕ.

Suppose that we have a family of flat surfaces $\mathbb{S}^{\flat}_t = (\mathbb{S}, M, J_t, \varphi_t)_{t\in[0,1]}$ that starts at $\mathbb{S}^{\flat}_0 = \mathbb{S}^{\flat}$ and keeps the types of the zeros and poles in M fixed. We also keep track of the finite incoming directions at the exponential poles and hence the marked points on the small circles. This means that $(\mathbb{S}^{\circ}, M^{\circ})$ stays the same and the gradings are isotopic. Moreover, we can transport the gradings of curves along the family and this gives an equivalence

$$\mathcal{I}\colon \operatorname{Fuk}^{\bullet}(\mathbb{S}^{\flat}_0) \to \operatorname{Fuk}^{\bullet}(\mathbb{S}^{\flat}_1).$$

For instance, for the family $(\mathbb{S}, M, J, e^{i2\pi t}\varphi)_{t\in[0,1]}$ this equivalence will correspond to the shift functor from $\operatorname{Fuk}^{\bullet}(\mathbb{S}^{\flat})$ to itself.

The functor $\mathcal{I}$ can be used to pull back the standard stability condition on $\operatorname{Fuk}^{\bullet}(\mathbb{S}^{\flat}_1)$ to one on $\operatorname{Fuk}^{\bullet}(\mathbb{S}^{\flat})$. This gives a point $\mathcal{I}^*\sigma_{\mathbb{S}^{\flat}_1} \in \operatorname{Stab}(\mathbb{S}^{\flat})$ that only depends on the homotopy class of the family. Furthermore, if $\tau_t\colon \mathbb{S} \to \mathbb{S}$ is a path of diffeomorphisms of $\mathbb{S}$ that fixes the marked points and $\tau_0 = \mathbb{1}_{\mathbb{S}}$, then one can check that the families $\tau_t^*\mathbb{S}^{\flat}_t$ and $\mathbb{S}^{\flat}_t$ will define the same stability condition.

Therefore, to describe all the stability conditions that can be obtained in this way we can look at the universal cover of the moduli space of all flat structures on $(\mathbb{S}, M)$ up to diffeomorphism. Let us denote the connected component containing $\mathbb{S}^{\flat}$ by $\operatorname{Flat}(\mathbb{S}^{\flat})$. Each point in this space corresponds to a family of flat structures starting at $\mathbb{S}^{\flat}$, so we get a map

$$\iota\colon \operatorname{Flat}(\mathbb{S}^{\flat}) \to \operatorname{Stab}(\mathbb{S}^{\flat})\colon [\mathbb{S}^{\flat}_t] \mapsto \mathcal{I}^*\sigma_{\mathbb{S}^{\flat}_1}.$$

Theorem 11.39 (Haiden–Katzarkov–Kontsevich) *The map ι is an open and closed embedding. In other words,* $\operatorname{Flat}(\mathbb{S}^{\flat})$ *is a connected component of* $\operatorname{Stab}(\mathbb{S}^{\flat})$.

Sketch of the proof For each flat surface $\mathbb{S}^{\flat}_0 = (\mathbb{S}, M, J, \varphi)$ we can find a factor $e^{i\epsilon} \in \mathbb{C}^*$ such that $(\mathbb{S}, M, J, e^{i\epsilon}\varphi)$ is made by gluing finite and infinite horizontal strips together with one marked point on each boundary. The vectors $a_j = h_j + iw_j$ connecting the marked points on each strip characterize the flat surface. Therefore, locally around $\mathbb{S}^{\flat}_0$ the space of flat structures is described by $e^{-i\epsilon}(h_j + iw_j)$.

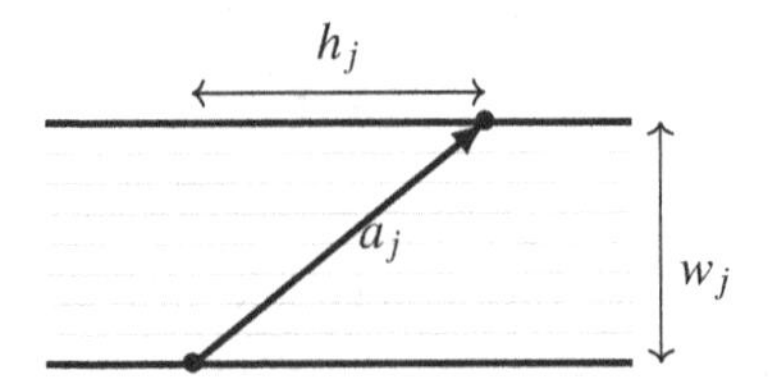

Each finite strip contains one stable arc that connects the two marked points on the opposite sides of the strip. Its central charge is the difference between the two marked points (viewed as complex numbers). Because this arc collection generates Γ these complex numbers $e^{-i\epsilon}(w_j + ih_j)$ determine the central charge of the stability conditions around $\mathbb{S}_0^\flat$. Therefore, ι is a local homeomorphism. To show that it is closed and injective we refer to [115]. □

Example 11.40 We return once more to the linear quiver with n vertices and its derived category, which is equivalent to the Fukaya category of the disk with $n+1$ marked points on its boundary. In our new terminology this category also corresponds to $\mathtt{Fuk}^\bullet(\mathbb{S}^\flat)$, where

$$\mathbb{S}^\flat = (\mathbb{S}^2, \{\infty\}, J_{\mathbb{P}^1}, e^{z^{n+1}} dz^2)$$

is the Riemann sphere with one exponential singularity at ∞ of order $n+1$.

To determine the space of all flat structures of this type, first note that all complex structures on $\mathbb{S}^2$ are homeomorphic to $\mathbb{P}^1$, so we can assume that the complex structure $J_{\mathbb{P}^1}$ remains fixed. Next we have to look at all quadratic differentials $\varphi = g(z)\,dz^2$, where $g(z)$ is an entire nonzero function with an exponential singularity of order $n+1$ at ∞. These quadratic differentials necessarily look like $e^{f(z)}\,dz^2$, where f is a polynomial of degree $n+1$.

On these quadratic differentials, we still have an action of the group of automorphisms of $\mathbb{P}^1$ that fix ∞. These are the affine maps $z \mapsto az + b$ and we can use these to make $f(z)$ monic and shift the coefficient of z^n to 0. This means that we can assume

$$f(z) = z^{n+1} + a_{n-1}z + \cdots + a_0.$$

Note that if we add $2\pi i$ to a_0 we get the same quadratic differential, so the space of quadratic differentials up to isomorphism is $\mathbb{C}^{n-1} \times \mathbb{C}/2\pi i\mathbb{Z}$ and its universal cover is $\mathbb{C}^n$. In [115] it is shown that the map $\iota\colon \mathsf{Flat}(\mathbb{S}^\flat) \to \mathsf{Stab}(\mathbb{S}^\flat)$ is an isomorphism:

$$\mathsf{Stab}(\mathtt{D\,Mod}^\bullet\, Q) = \mathbb{C}^n.$$

Example 11.41 The derived category of the projective line is equivalent to the Fukaya category of the cylinder with two marked points, one on each boundary. This category can be realized with a flat surface with two exponential singularities:

$$\mathbb{S}^\flat = (\mathbb{S}^2, \{0, \infty\}, J_{\mathbb{P}^1}, e^{z+z^{-1}} dz^2).$$

To describe $\mathsf{Flat}(\mathbb{S}^\flat)$ we can again assume that J is fixed and the poles remain at $\{0, \infty\}$. This implies that the quadratic differential looks like

$$e^{a_1 z + a_0 + a_{-1} z^{-1}} dz^2.$$

Up to an automorphism of $\mathbb{P}^1$ of the form $z \mapsto az$ we can assume $a_1 = 1$, so the space of quadratic differentials up to automorphisms is $\mathbb{C}/2\pi i\mathbb{Z} \times \mathbb{C}^*$ and its universal cover is $\mathbb{C} \times \mathbb{C}$. Again, [115] shows that the map $\iota\colon \mathsf{Flat}(\mathbb{S}^\flat) \to \mathsf{Stab}(\mathbb{S}^\flat)$ is an isomorphism:

$$\mathsf{Stab}(\mathrm{D}\,\mathrm{Coh}^\bullet\,\mathbb{P}^1) = \mathbb{C}^2.$$

Example 11.42 Consider the Riemann sphere with marked points $\{0, 1, \infty, p\}$ and the flat surface

$$\mathbb{S}^\flat = \left(\mathbb{S}^2, \{0, 1, p, \infty\}, J_{\mathbb{P}^1}, \frac{dz^2}{z^2}\right).$$

The metric coming from the quadratic differential turns $\mathbb{S}^2 \setminus \{0, \infty\}$ into a cylinder with a circumference of $\int_{\mathbb{S}^1} |\frac{dz}{z}| = 2\pi$. The points 0 and ∞ lie at the infinite ends.

The arcs in $(\mathbb{S}^2, \{0, 1, p, \infty\})$ with finite length are those that start and end at 1 and p and they generate $\mathbf{Fuk}^\bullet(\mathbb{S}^\flat)$ inside ${}^{\mathrm{top}}\mathbf{Fuk}^\bullet(\mathbb{S}^2, \{0, 1, p, \infty\})$. From Example 10.24 we know that the latter is equivalent to the derived category of $\mathbb{X} = \mathbb{A}^1{}_0 \sqcup_0 \mathbb{P}^1{}_\infty \sqcup_0 \mathbb{A}^1$ and therefore $\mathbf{Fuk}^\bullet(\mathbb{S}^\flat)$ corresponds to the subcategory of sheaves supported on the middle projective line

$$\mathbf{Fuk}^\bullet(\mathbb{S}^\flat) \overset{\infty}{=} \mathrm{D}\,\mathrm{Coh}^\bullet{}_{\mathbb{P}^1}\,\mathbb{X}.$$

To describe the stability conditions we have to determine $\mathsf{Flat}(\mathbb{S}^\flat)$. The quadratic differentials on $\mathbb{P}^1$ with poles of order 2 at 0 and ∞ is given by $a\frac{dz^2}{z^2}$ with $a \in \mathbb{C}^*$. Up to a diffeomorphism, all complex structures on $\mathbb{P}^1$ are the same and we can use a Möbius transformation to fix the first three marked points at $\{0, \infty, 1\}$. This uses up all freedom, so the complex structure is determined by the location of p. Therefore, the space of flat structures is the universal cover of $\mathbb{C}^* \times \mathbb{C} \setminus \{0, 1\}$, which is homeomorphic to the product of $\mathbb{C}$ with upper half-plane $\mathbb{C} \times \mathbb{H}^+$. By Theorem 11.39 this space forms a connected component of the stability manifold:

$$\mathbb{C} \times \mathbb{H}^+ \subset \mathsf{Stab}(\mathrm{D}\,\mathrm{Coh}^\bullet{}_{\mathbb{P}^1}\,\mathbb{X}).$$

We end with a theorem by Takeda [241] that gives a sufficient condition on the flat surface such that the flat structures give all stability conditions.

Theorem 11.43 (Takeda) *If $\mathbb{S}^\flat$ has no internal marked points and at least one boundary component with two marked points then* $\mathsf{Flat}(\mathbb{S}^\flat) = \mathsf{Stab}(\mathbb{S}^\flat)$.

11.6 Exercises

Exercise 11.1 Determine the Grothendieck group of the following categories:

(i) The topological Fukaya category of a cylinder with k marked points on one boundary and l on the other, with a vertical line field.
(ii) The topological Fukaya category of a closed surface with genus g and n marked points and an even line field.
(iii) $\mathtt{D\,Coh}^\bullet\,\mathbb{P}^1$.

Exercise 11.2 Let A, B be two finite-dimensional algebras over $\mathbb{C}$. Show that

$$K_0\,\mathtt{Mod}^\bullet\,A \oplus B \cong K_0\,\mathtt{Mod}^\bullet\,A \oplus K_0\,\mathtt{Mod}^\bullet\,B.$$

Exercise 11.3 Consider the following dimer model:

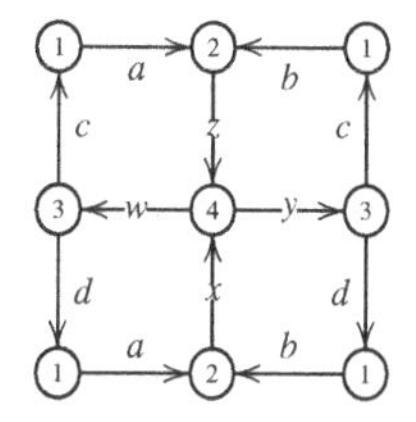

(i) Calculate its Newton polygon, determine all its perfect matchings and assign them to their corresponding lattice points.
(ii) Let $\alpha = (1,\dots,1)$. Determine the locus of all θ with $\theta\cdot\alpha$ such that $\mathbb{M}^t_\theta(\mathtt{Jac}\,Q,\alpha)$ is smooth.
(iii) Show that all smooth $\mathbb{M}^t_\theta(\mathtt{Jac}\,Q,\alpha)$ are isomorphic varieties.

Exercise 11.4 Consider the following dimer model on a torus:

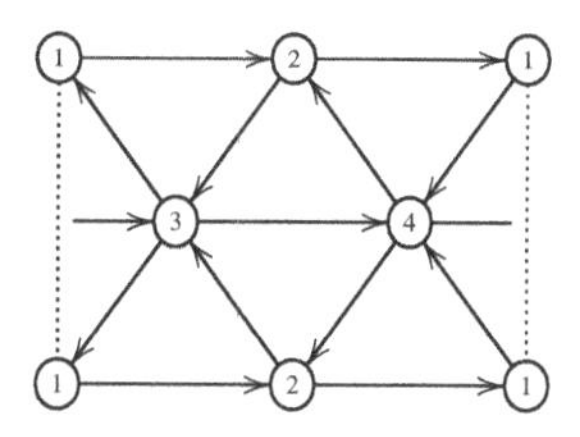

(i) Calculate its Newton polygon, determine all its perfect matchings and assign them to their corresponding lattice points.
(ii) Let $\alpha = (1,\dots,1)$. Determine the locus of all θ with $\theta\cdot\alpha$ such that $\mathbb{M}^t_\theta(\mathtt{Jac}\,Q,\alpha)$ is smooth.
(iii) Show that all smooth $\mathbb{M}^t_\theta(\mathtt{Jac}\,Q,\alpha)$ are isomorphic varieties.

Exercise 11.5 Use the dimer of Exercise 11.3.

(i) Find a set of weights $w\colon Q_1 \to \mathbb{Q}$ such that the corresponding character is $\theta = (-3,1,1,1)$.
(ii) Determine the stable perfect matchings.

(iii) Draw the associated tropical curve and draw each arrow of the dimer on the tropical curve.

Exercise 11.6 Let Q be the linear quiver with three vertices $① \leftarrow ② \rightarrow ③$. Consider the stability condition for which the charges of the simples are $Z(S_1) = 1 + i$, $Z(S_2) = i$ and $Z(S_3) = -1 + 1$, and their phases sit inside $(0, \pi)$. Look at all indecomposables of $\mathtt{mod}^\bullet\, Q$ and give Harder–Narasimhan filtrations for them.

Exercise 11.7 Show that if $A = \mathbb{C}Q/I$, where Q is a quiver without oriented cycles and $C\colon Q_0 \to \mathbb{R}{+}\mathbb{R}_{>0}i$ is any map, then there is precisely one Bridgeland stability condition on $\mathtt{D\,mod}^\bullet\, Q$ such that the simple S_v has charge $C(v)$ and phase in $(0, \pi)$.

Exercise 11.8 Let $\phi = (z^2{-}1)(dz)^2$ be a quadratic differential on $\mathbb{P}^1$. Construct an arc collection for the surface with marked points the poles of ϕ and grading coming from the quadratic differential. Determine its gentle A_∞-algebra.

Exercise 11.9 Let Q be the quiver $\circ \rightrightarrows \circ \rightrightarrows \circ$.

(i) Give a three-parameter family of quadratic differentials on $\mathbb{P}^1$ that give stability conditions on $\mathtt{D\,mod}^\bullet\, Q$.
(ii) Sketch a quadratic differential for which all simple representations of Q have the same central charge and phase.

Exercise 11.10 Let $\mathbb{X} = \mathbb{A}^1{}_0{\sqcup}_0\mathbb{P}^1$. Determine the space of stability conditions of $\mathtt{D\,Coh}^\bullet{}_{\mathbb{P}^1}\, \mathbb{X}$.

12
Deforming

In Chapter 9 we saw how we could construct mirrors for marked surfaces using gluing techniques. These categories correspond to partially wrapped Fukaya categories for exact manifolds. If we want to construct Fukaya categories of closed surfaces we have to look at deformations of these categories. In this chapter we will first have a closer look at the deformation theory of A_∞-algebras and its relation to Hochschild cohomology. For further details on this we refer to [147] and [153].

Afterwards, we will apply these concepts to the case of marked surfaces with internal marked points and we will discuss how to use this to construct mirror correspondences for closed surfaces. We will look at the case of surfaces of higher genus, which was worked out by Seidel [220] and Efimov [71]. Mirror symmetry for the genus 1 curve was originally treated by Polishchuck and Zaslow in [205] but only on the level of ordinary derived categories. Treatments including the higher products were done by Lekili and Perutz [162] and Lekili and Polishchuk [164]. Unlike for the higher-genus curves, in which we can work over the complex numbers, the theory for the elliptic curve needs formal power series to get the A_∞-structure working. For this part we will mainly follow the work of Cho, Hong and Lau [55, 154], because it is a versatile framework that offers many directions for generalization.

12.1 Deformation Theory for A_∞-Algebras

In many subjects in mathematics one is interested in deformations of objects. In general, if A is some kind of object defined over $\mathbb{C}$ (like an algebra, a module or a variety), a deformation is an object $A_\hbar$ defined over a ring $\mathbb{C}[\hbar]/(\hbar^{n+1})$ or $\mathbb{C}[[\hbar]]$ such that if we specialize $\hbar$ to zero we get A back. In the first case we call $A_\hbar$ an nth-order deformation, and in the second case a formal deformation. In what follows we will look at deformations of an A_∞-algebra or category and we will denote the ring over which we work by $\mathbb{C}[[\hbar]]$ and the ideal generated by the deformation parameter by $(\hbar)$.

Definition 12.1 A *deformation of an A_∞-algebra* (A, μ) is a set of products $\mu + \nu$ with $\nu_i \in \mathrm{Hom}^1(A[1]^{\otimes i}, A[1]) \otimes (\hbar)$, $i \geq 0$ that satisfy the curved A_∞-axioms.

Two deformations ν, ν' are *equivalent* if there is an A_∞-morphism $\mathbb{1} + \phi$ between $\mu + \nu$ and $\mu + \nu'$ with $\phi \in \mathrm{Hom}^0(A[1]^{\otimes i}, A[1]) \otimes (\hbar)$.

The problem of classifying deformations of an A_∞-algebra up to equivalence can be framed into a cohomology theory.

Definition 12.2 The *Hochschild cochain complex* of an A_∞-algebra A is the graded vector space

$$\mathsf{HC}^i(A) := \prod_{k \geq 0} \mathrm{Hom}^i_{\mathbb{k}}(A[1]^{\otimes k}, A[1])$$

with differential

$$(d\nu)_k(a_1, \ldots, a_k) = \sum_{i,j} (-1)^{\dagger \deg \nu} \mu(a_1, \ldots, a_{i-1}, \nu(a_i, \ldots, a_j), a_{i+1}, \ldots, a_k)$$
$$+ (-1)^{\dagger \deg \mu} \nu(a_1, \ldots, a_{i-1}, \mu(a_i, \ldots, a_j), a_{i+1}, \ldots, a_k).$$

The Hochschild complex also comes with a bracket, called the *Gerstenhaber bracket*, that turns it into a differentially graded Lie algebra or dgla:[1]

$$[\nu, \kappa](a_1, \ldots, a_k) = \sum_{i,j} (-1)^{\dagger \deg \kappa} \nu(a_1, \ldots, a_{i-1}, \kappa(a_i, \ldots, a_j), a_{i+1}, \ldots, a_k)$$
$$+ (-1)^{\dagger \deg \nu} \kappa(a_1, \ldots, a_{i-1}, \nu(a_i, \ldots, a_j), a_{i+1}, \ldots, a_k).$$

The homology of $\mathsf{HC}^\bullet(A)$ is called the *Hochschild cohomology* and is denoted by $\mathsf{HH}^\bullet(A)$.

Theorem 12.3 *Let A be an A_∞-algebra.*

(i) *The first-order deformations of A up to equivalence are described by* $\mathsf{HH}^1(A)$.
(ii) *The formal deformations of A are described by elements $x \in \mathsf{HC}^1(A) \otimes (\hbar)$ that satisfy the Maurer–Cartan equation*

$$\mathsf{MC}(x) := dx + \tfrac{1}{2}[x, x] = 0.$$

[1] A dgla is a cochain complex with a bracket that satisfies $[a, b] = (-1)^{|a||b|}[b, a]$, $d[a, b] = [da, b] + (-1)^{|a|}[a, db]$ and the graded Jacobian identity. Just like dg-algebras, dgla's have an ∞-generalization called L_∞-algebras.

On the space of Maurer–Cartan elements there is an action of the Lie algebra $\mathsf{HC}^0(A) \otimes (\hbar)$,

$$\phi \cdot x = \partial_x \mathsf{MC}(x)(\phi) = d\phi + \tfrac{1}{2}[\phi, x] + \tfrac{1}{2}[x, \phi],$$

which can be integrated to a Lie group action

$$\exp(\phi) \cdot x = x + \phi + [\phi, x] + \tfrac{1}{2}([\phi, [\phi, x]] + [\phi, dx]) + \cdots .$$

The orbits of this action are the equivalence classes.

Sketch of the proof The first part follows directly from the definition. The second part can be found in [147]. □

Remark 12.4 If $A = \Bbbk\mathbb{1} \oplus \bar{A}$ has a strict unit $\mathbb{1}$, we can define the *reduced Hochschild complex* $\mathsf{HC}^\bullet_{\mathrm{red}}(A)$ as the subcomplex of $\mathsf{HC}^i(A)$ of all the v that are zero as soon as one of the entries is $\mathbb{1}$. This complex describes the deformations that keep $\mathbb{1}$ as a strict unit. Because every A_∞-algebra with a weak unit is equivalent to one with a strict unit, one can show that $\mathsf{HC}^\bullet_{\mathrm{red}}(A)$ and $\mathsf{HC}^\bullet(A)$ are quasi-isomorphic.

Example 12.5 Let $E = \mathbb{C}[\xi]/(\xi^2)$ with $\deg \xi = 1$ and view it as a $\mathbb{Z}_2$-graded A_∞-algebra. The elements in the reduced Hochschild complex are determined by $v_k(\xi, \dots, \xi)$. If it is a multiple of ξ then the degree is 0; if it is a multiple of 1 the degree is 1. Any product with $v_k(\xi, \dots, \xi) = a_k\xi + b_k$ automatically satisfies the A_∞-axioms, so the differential is zero. If we set $a(X) = \sum a_k X^k$, $b(X) = \sum b_k X^k \in \mathbb{C}[[X]]$ then we can identify

$$\mathsf{HC}^\bullet_{\mathrm{red}}(E) = \mathbb{C}[[X]] \oplus \mathbb{C}[[X]].$$

The bracket on the complex can also easily be calculated. If v^1 corresponds to $(X^k, 0)$ and v^2 to $(0, X^l)$ we can fill $v^1_k(\xi, \dots, \xi) = \xi$ in l different locations in v^2_l; on the other hand, if we fill $v^2_l(\xi, \dots, \xi) = 0$ in v^1_k we get zero. Therefore, $[v^1, v^2]$ will correspond to $lX^{k+l-1} = X^k \partial_X X^l$. Therefore, it makes sense to identify $(a(X), 0)$ with the operator $a(X)\partial_X$ and $(0, b(X))$ with the function $b(X)$. In this way, $a(X)\partial_X \in \mathsf{HH}^0(E)$ acts on $b(X) \in \mathsf{HH}^1(E)$ as a normal derivation.

The bracket between two elements of type $b(X)$ is zero so all $b(X) \in \mathsf{HH}^1(E) \otimes (\hbar)$ are solutions to the Maurer Cartan equation. The derivation action of $\mathsf{HH}^0(E) \otimes (\hbar)$ integrates to an action

$$\exp(a(X)\partial_X) \cdot b(X) = b(e^{a(X)})$$

on $\mathsf{HH}^1(E) \otimes (\hbar)$. The orbits of this action correspond to powers $\hbar^i X^j$. Each such deformation corresponds to an A_∞-structure for which $\mu_j(\xi, \dots, \xi) = \hbar^i$ and all other products are zero.

Remark 12.6 Theorem 12.3 is a general feature of deformation theory. To each deformation problem one can associate a dgla $(L, d, [\ ,\ ])$. The first-order deformations are given by the first homology H^1L and the formal deformations by solutions of the Maurer–Cartan equation in $L^1 \otimes (\hbar)$. Quasi-isomorphic dgla's will give rise to isomorphic deformation problems: there will be a one-to-one correspondence between the equivalence classes of formal deformations and orbits of the exponentiated $L^0 \otimes (\hbar)$-action on the solutions of the Maurer–Cartan equation [203].

Analogous to the situation of dg-algebras, there is a notion of a minimal model of a dgla. Such an object is called an L_∞-algebra and it comes with higher brackets $[-, \dots, -]$ that satisfy suitable L_∞-axioms. The Maurer–Cartan equation expands to

$$\mathsf{MC}(x) = dx + \tfrac{1}{2!}[x, x] + \tfrac{1}{3!}[x, x, x] + \cdots = 0$$

and the Lie algebra action is defined accordingly: $\phi \cdot x = \partial_x \mathsf{MC}(\phi)$. For more information about this approach to deformation theory we refer to the online manuscript by Kontsevich and Soibelman [147]. A proof that for quasi-isomorphic L_∞-algebras there is a bijection between the equivalence classes of solutions to the Maurer–Cartan equation can be found in [71, Appendix A].

Remark 12.7 Originally, Hochschild cohomology was developed in the context of ordinary ungraded algebras only [119]. Because A is concentrated in degree 0, the space $\mathsf{HC}^k(A)$ only consists of $(k + 1)$-ary products, so we can also grade the complex according to the number of inputs. For this reason, $\mathsf{HH}^1(A)$ is better known as $\mathsf{HH}^2(A)_{\text{clas}}$ in the classical context and it describes the first-order algebra deformations of A.

The classical zeroth- and first-order classical Hochschild cohomologies also have nice interpretations: the former is the center, while the latter is the space of outer derivations:[2]

$$\begin{aligned}
\mathsf{HH}^0(A)_{\text{clas}} &= \{m_0 \in A \mid dm_0(a) = am_0 - m_0a = 0\} \\
&= Z(A), \\
\mathsf{HH}^1(A)_{\text{clas}} &= \frac{\{m_1 \colon A \to A \mid dm_1(a, b) = m_1(ab) - m_1(a)b - am_1(b) = 0\}}{\{dm_0 \colon A \to A \colon a \mapsto [a, m_0]\}} \\
&= \frac{\mathrm{Der}(A)}{\mathrm{Inn}(A)}.
\end{aligned}$$

[2] In the expressions for the classical Hochschild cohomology we used the sign convention from representation theory. In this way, m_1 is a derivation and m_2 an ordinary product.

Another interpretation of the classical Hochschild cohomology is the Ext-ring of A viewed as a bimodule itself, or as a right module over $A^e := A \otimes A^{\mathrm{op}}$:

$$\mathsf{HH}^\bullet(A)_{\mathrm{clas}} = \mathrm{Ext}^\bullet_{A^e}(A, A).$$

This implies that we can calculate the Hochschild cohomology of an algebra by using a bimodule resolution of the algebra.

Example 12.8 Let $A = \mathbb{C}[[X]]$; then A has the following bimodule resolution:

$$B^\bullet := A \otimes A \xrightarrow{X\otimes 1 - 1\otimes X} A \otimes A.$$

The complex $\mathrm{Hom}_{A^e}(B^\bullet, A)$ is $A \xrightarrow{0} A$, so $\mathsf{HH}^\bullet(A)_{\mathrm{clas}} = A \oplus A$. The classical zeroth cohomology is the center, which is A and the first is the derivations, which are of the form $A\partial_X$. If we revert to the A_∞-grading then $\mathsf{HH}^\bullet(A) = A\partial_X \oplus A$, so the Hochschild cohomology of $A = \mathbb{C}[[X]]$ is in fact the same as that of $E = \mathbb{C}[\xi]/(\xi^2)$ in Example 12.5. But now the deformation $\hbar^i X^j$ has a different interpretation: it deforms $\mathbb{C}[[X]]$ to the curved A_∞-algebra $(\mathbb{C}[[X]], \hbar^i X^j)$.

We can extend the example above to quivers. Let $A = \widehat{\mathbb{C}Q}/I$ be a completed path algebra with relations generated by paths of length 2. Let $Z_0 = Q_0$, $Z_1 = Q_1$ and Z_n be the set of paths $a_1 \dots a_n$ of length n for which all subpaths of length at least 2 sit in I, including the path itself. For each path p we define a bimodule $B[p] = Ah(p) \otimes [p] \otimes t(p)A$ and denote its generator by $[p]$. Now define

$$B_i = \bigoplus_{a_1 \dots a_n \in Z_i} B[a_1 \dots a_n] \quad \text{with } d[a_1 \dots a_n] = a_1[a_2 \dots a_n] - (-1)^k [a_1 \dots a_{n-1}]a_n.$$

Theorem 12.9 (Bardzell) *The complex $B_\bullet$ is a projective bimodule resolution of A, so*

$$\mathsf{HH}^\bullet(A)_{\mathrm{clas}} \cong \mathsf{H}\,\mathrm{Hom}^\bullet_{A^e}(B_\bullet, A).$$

Sketch of the proof It is easy to check $B[p]$ is projective and that d^2 is 0. Furthermore, in HB^0 an expression $p[h(a)]aq$ is the same as $pa[t(a)]q$, so we can bring everything to the left and therefore $HB^0 = \oplus_{v \in Q_0} Av = A$. To show that the rest of the homology is 0 see [21]. □

Example 12.10 Let us apply this to a cyclic quiver with vertices v_i and arrows $a_i \colon v_i \leftarrow v_{i+1}$, where $i \in \mathbb{Z}_n$ and $n \geq 2$. First consider the quiver without any relations. In that case, $Z_n = \{\,\}$ if $n \geq 2$ and the resolution only has two terms,

$$\bigoplus\nolimits_{v \in Q_0} Av[v]vA \xleftarrow{a[t(a)]-[h(a)]a \leftarrow [a]} \bigoplus\nolimits_{a \in Q_1} Ah(a)[a]t(a)A.$$

This implies that

$$\mathsf{HH}^0(A)_{\rm clas} = Z(A) = \mathbb{C}[\ell], \quad \mathsf{HH}^1(A)_{\rm clas} = \mathbb{C}[\ell]\delta \quad \text{and} \quad \mathsf{HH}^{>2}(A)_{\rm clas} = 0,$$

with $\ell = a_1 \dots a_n + \cdots + a_n \dots a_{n-1}$ and $\delta([a_i]) = a_i$.

Now let us look at the dual quiver algebra $E = \Lambda Q = \mathrm{Ext}^\bullet(S,S)$, with S the direct sum of the vertex simples. We can write E as the path algebra of a quiver with vertices v_i, arrows $\alpha_i\colon v_{i+1} \leftarrow v_i$ and relations $\alpha_{i+1}\alpha_i$. In this case, Z_k consists of all paths of length k. Because in ΛQ the only nonzero paths are vertices and arrows, there can only be morphisms $B_k \to E$ if $k = 0, 1 \bmod n$.

- If $k = rn$, a bimodule morphism $f\colon B_k \to E$ is determined by constants λ_i for which $f([\alpha_{i+rn-1}\dots\alpha_i]) = \lambda_i v_i$. The condition $df = 0$ implies that all λ_i are equal, while the image of d in $\mathrm{Hom}_{E^e}(B_{rk}, E)$ is zero, so

$$\mathsf{HH}^{rn}(E)_{\rm clas} = \mathbb{C}\left(\sum_i [\alpha_{i+rn-1}\dots\alpha_i \to v_i]\right) =: \mathbb{C}\Omega_{r,0}.$$

- If $k = rn+1$, a bimodule morphism $f\colon B_k \to E$ is determined by constants λ_i for which $f([\alpha_{i+rn}\dots\alpha_i]) = \lambda_i\alpha_i$. The condition $df = 0$ is automatically satisfied, while $d[\alpha_{i+rn-1}\dots\alpha_i \to v_i] = [\alpha_{i+rn}\dots\alpha_i \to \alpha_i] - [\alpha_{i+rn}\dots\alpha_i \to \alpha_i]$, so up to equivalence we can make all λ_i equal:

$$\mathsf{HH}^{rn+1}(E)_{\rm clas} = \mathbb{C}\left(\sum_i [\alpha_{i+rn}\dots\alpha_i \to \alpha_i]\right) =: \mathbb{C}\Omega_{r,1}.$$

If we identify $\ell^j\mathbb{1} \leftrightarrow \Omega_{r,0}$ and $\ell^i\delta \leftrightarrow \Omega_{r,1}$ we get a correspondence

$$\mathsf{HH}^0(A) = \mathbb{C}[\![\ell]\!]\delta = \mathsf{HH}^0(E),$$
$$\mathsf{HH}^{-1}(A) = \mathbb{C}[\![\ell]\!] = \mathsf{HH}^{-1}(E)$$

between the Hochschild cohomologies of A and E equipped with the A_∞-grading. For this grading, the $\Omega_{r,0}$ all have degree -1 and all the $\Omega_{r,1}$ degree 0. This correspondence is also compatible with the extra structure coming from the Gerstenhaber bracket: in both cases

$$[\ell^i\delta, \ell^j] = nj\ell^{i+j},$$

so $\ell^i\delta \in \mathsf{HH}^0(A)$ acts on $\mathsf{HH}^0(E)$ as the derivation $n\ell^{i+1}\partial_\ell$. Just like in Example 12.5 this implies that every deformation is equivalent to a deformation coming from one monomial and this leads to a correspondence between the Landau–Ginzburg model $(\widehat{\mathbb{C}Q}, \hbar\ell^j)$ and the A_∞-algebra $(E, \mu_{\hbar,j})$ whose only nonzero products are $\mu_{\hbar,j}(\alpha_i,\dots,\alpha_{i+jn-1}) = \hbar h(\alpha_i)$.

This is part of a general phenomenon called *Koszul duality*. Any completed path algebra $A = \widehat{\mathbb{C}Q}/I$ with relations and its Koszul dual A_∞-algebra

$\mathrm{Ext}^\bullet_A(S,S)$ have L_∞-quasi-isomorphic Hochschild complexes [134]. In Section 12.5 we will discuss this phenomenon in more detail.

The Hochschild cohomology is invariant under various other constructions, but before we look at this in more detail, we first need to extend the notion of Hochschild cohomology to A_∞-categories. This is not so difficult: if $\mathtt{A}^\bullet$ is an A_∞-category then we set

$$\mathsf{HC}^i(\mathtt{A}^\bullet) := \prod_{k\geq 0} \prod_{X_0,\dots,X_k\in\mathtt{A}^\bullet} \mathrm{Hom}^i_{\mathbb{C}}(\mathtt{A}^\bullet(X_{k-1},X_k)\otimes\cdots\otimes\mathtt{A}^\bullet(X_0,X_1),\mathtt{A}^\bullet(X_0,X_k))$$

and define d, $[\ ,\ ]$ in the same way as we did for algebras.

Theorem 12.11 *The following related A_∞-algebras and categories have quasi-isomorphic Hochschild complexes:*

(i) *quasi-A_∞-equivalent A_∞-categories,*
(ii) *an A_∞-category and its twisted completion,*
(iii) *(split-)derived equivalent A_∞-categories,*
(iv) *$A = \widehat{\mathbb{C}Q}/I$ and $E = \mathrm{Ext}^\bullet_A(S,S)$ with $S = A/\langle a \in Q_1\rangle$.*

Sketch of the proof These statements are well known in the context of dg-algebras, which is equivalent to the A_∞-setting. Proofs can be found in various sources, with different levels of generalities. For example, in [134] it is shown that these quasi-isomorphism are also compatible with extra structures such as the L_∞-structure from the Gerstenhaber bracket. □

Example 12.12 Let $\widehat{R} = \mathbb{C}[\![X_1,\dots,X_n]\!]$ be the completed polynomial ring in n variables concentrated in degree 0 and let $\Lambda = \mathrm{Ext}^\bullet_{\widehat{R}}(S,S)$ be the exterior algebra over a set of dual variables $\xi_1,\dots,\xi_n$ with degree 1. Every ξ_i can be seen as the derivation ∂_{X_i} on $\widehat{R}$ and we can interpret a product $g(X)\xi_{i_1}\dots\xi_{i_k}$ as a map $\nu_k\colon \widehat{R}^{\otimes k}\to\widehat{R}$:

$$\nu_k(f_1(X),\dots,f_k(X)) = g(x)\sum_{\sigma\in S_k}\frac{(-1)^\sigma}{k!}\partial_{X_{i_{\sigma(1)}}}f_1(X)\cdots\partial_{X_{i_{\sigma(k)}}}f_k(X).$$

This induces a map

$$F\colon \widehat{R}\otimes\Lambda\to\mathsf{HH}(\widehat{R})_{\mathrm{clas}}.$$

The famous *Hochschild–Kostant–Rosenberg theorem* [116] implies that F is an isomorphism of graded vector spaces.

The tensor product $\widehat{R}\otimes\Lambda$ can be turned into a dgla by setting $d=0$ and $[X_i,\xi_j]=\delta_{ij}$, $[X_i,X_j]=[\xi_i,\xi_j]=0$ and extending this to the whole of $\widehat{R}\otimes\Lambda$ via the graded Poisson rule $[a,bc]=[a,b]c+(-1)^{\deg b(\deg a-1)}[b,ac]$. For the original degree, the bracket has degree -1 but for $\widehat{R}\otimes\Lambda[1]$ it has degree 0.

Kontsevich's equally famous *formality theorem* [143, 242] states that F can be extended to an L_∞-isomorphism

$$F_\bullet : \widehat{R} \otimes \Lambda[1] \to \mathsf{HH}(\widehat{R}).$$

The Maurer–Cartan equation for $\widehat{R} \otimes \Lambda[1]$ becomes $[\psi, \psi] = 0$ with $\psi \in \widehat{R} \otimes \Lambda[1]^1 \otimes (\hbar) = \widehat{R} \otimes \Lambda^2 \otimes (\hbar)$. Solutions to this equation are called Poisson bivectors. The formality theorem implies that each Poisson bivector corresponds to a formal deformation of $\widehat{R}$.

If we work $\mathbb{Z}_2$-graded, we can also form deformations corresponding to elements in every $\widehat{R} \otimes \Lambda^k \otimes (\hbar)$ with k even. In particular, an element $f(X) \in \widehat{R} \otimes \Lambda^0 \otimes (\hbar)$ automatically satisfies $[f, f] = 0$ and it corresponds to a Landau–Ginzburg model $(\widehat{R}, f)$, which is a curved deformation of $\widehat{R}$.

To check when two of these curved deformations are equivalent, we have to look at the action of $\mathsf{HH}^0(\widehat{R}) = \mathsf{HH}^{\mathrm{odd}}(\widehat{R})_{\mathrm{clas}}$. Such an element is of the form $\sum_i g_i(X)\xi_i + K$, where K contains higher-degree powers of the ξ_i. If we bracket this with a function we get

$$\left[\sum_i g_i(X)\xi_i + K, f(X)\right] = \sum_i g_i(X)\partial_{X_i} f(X) + [K, f(X)].$$

The term $[K, f]$ still contains powers of the ξ_i, so to stay within the realm of functions we have to impose that $K = 0$. The element $\sum_i g_i(X)\xi_i$ acts as a derivation on $\widehat{R} = \mathsf{HH}^0(\widehat{R})_{\mathrm{clas}} \subset \mathsf{HH}^1(\widehat{R})$, and if we integrate this action we get the action of $\mathsf{Aut}\,\widehat{R}$. Therefore, two curved deformations f, f' are equivalent if they sit in the same $\mathsf{Aut}\,\widehat{R}$-orbit.

Example 12.13 The previous example can also be approached from the point of view of the exterior algebra. Because of Koszul duality there is an equivalence between $\mathsf{HH}^\bullet(\widehat{R})$ and $\mathsf{HH}^\bullet(\Lambda)$. This means that we can transport deformations of $\widehat{R}$ to deformations of Λ.

Consider the ring of differential forms $\Omega = \widehat{R}\langle dX_1, \ldots, dX_n\rangle$, viewed as a $\mathbb{Z}_2$-graded $\widehat{R}$-module, and set $\delta\omega = \iota_\eta\omega$, where $\eta = \sum_i X_i\partial_i$. This gives a twisted complex (Ω, δ) which can be seen as a resolution of S. Therefore, $E := (\mathrm{Hom}_{\widehat{R}}(\Omega, \Omega), [\delta, -])$ is a dg-model for Λ.

If we deform $\widehat{R}$ to the Landau–Ginzburg model $(\widehat{R}, \hbar f)$ with $f \in (X)$, the dg-algebra E deforms to a curved dg-algebra $E_f = (E, \hbar f \mathbb{1}_\Omega)$. On the other hand, like in Section 7.2.5, we can also deform (Ω, δ) into a matrix factorization of f. Write $f = X_1 f_1 + \cdots + X_n f_n$ and put $\gamma = f_1 dX_1 + \cdots + f_n dX_n$; then $\iota_\eta(\gamma) = f$ and for any $\omega \in \Omega$ we get

$$\begin{aligned}(\delta + \hbar\gamma\wedge)^2\omega &= \iota_\eta\iota_\eta\omega + \iota_\eta(\hbar\gamma \wedge \omega) + \hbar\gamma \wedge \iota_\eta\omega + \hbar^2\gamma \wedge \gamma \wedge \omega \\ &= 0 + \iota_\eta(\hbar\gamma) \wedge \omega + 0 = \hbar f\omega.\end{aligned}$$

The endomorphism ring of the matrix factorization $(\Omega, \delta + \hbar\gamma\wedge)$,

$$E_\gamma = (\mathrm{Hom}_{\widehat{R}}(\Omega, \Omega), [\delta + \hbar\gamma\wedge, -]),$$

is a deformation of E. The deformations E_γ and E_f are represented by the same element in the Hochschild cohomology $f = [\gamma\wedge, -] \in \mathsf{HC}(E)^1$ because, if we consider $\gamma\wedge$ as an element in $\mathrm{Hom}^1(\Omega, \Omega) = E[1]^0 = \mathrm{Hom}^0_{\mathbb{C}}(E[1]^{\otimes 0}, E[1]) \subset \mathsf{HC}(E)^0$, then

$$d_{\mathsf{HC}}(\gamma\wedge) = \delta\gamma - [\gamma\wedge, -] = f - [\gamma\wedge, -].$$

Remark 12.14 In general, the Hochschild–Kostant–Rosenberg theorem states that the Hochschild cohomology of a smooth affine variety $\mathbb{X}$ is equal to the ring of polyvector fields. Algebraically, if R is the coordinate ring of $\mathbb{X}$ then the Hochschild cohomology is $\mathsf{HH}_{\mathrm{clas}}(R) = \Lambda^\bullet_R \mathrm{Der} R$. For nonaffine varieties we can do a similar construction, but then the ring of polyvector fields is a sheaf of rings over $\mathcal{O}_{\mathbb{X}}$ and the Hochschild cohomology will be the cohomology of this sheaf.

12.2 A_∞-Extensions

The Hochschild cohomology is not only useful for studying infinitesimal deformations of an A_∞-algebra but it is also a handy tool for a related problem.

Definition 12.15 If $A, \cdot$ is a graded algebra then an *A_∞-structure on A* or an *A_∞-extension of A* is a set of products $\mu_i \colon A^{\otimes i} \to A$ that satisfy the A_∞-axioms for which $\mu_0 = \mu_1 = 0$ and $\mu_2(a, b) = (-1)^{\deg a} a \cdot b$. Two A_∞-structures μ, ν are called *equivalent* if there is an A_∞-functor $\mathcal{F} \colon (A, \mu) \to (A, \nu)$ with $\mathcal{F}_1 = \mathbb{1}_A$. An A_∞-structure on A is *trivial* if it is equivalent to a structure with $\mu_{\geq 3} = 0$.

The Hochschild complex of A comes with a filtration

$$F_r \mathsf{HC}^i(A) = \prod_{k \geq r+i+1} \mathrm{Hom}^i_{\mathbb{C}}(A[1]^{\otimes k}, A[1]),$$

which satisfies $dF_r\mathsf{HC}^i(A) \subset F_r\mathsf{HC}^{i+1}(A)$ and $[F_r\mathsf{HC}^i(A), F_s\mathsf{HC}^i(A)] \subset F_{r+s}\mathsf{HC}^i(A)$. An A_∞-structure on A can be seen as an element in $\mu \in F_1\mathsf{HC}^1(A)$ that satisfies the Maurer–Cartan equation, and two A_∞-structures are equivalent if they are connected via the action of an element $\exp(\phi)$ with $\phi \in F_1\mathsf{HC}^0(A)$. This action makes sense because the consecutive terms of $\exp(\phi) \cdot \mu$ sit further down in the filtration. Note that $\mathsf{HC}^0(A)$ and $\mathsf{HC}^1(A)$ also have parts with 0 or negative filtration degree but these have to be excluded because otherwise the Maurer–Cartan equation or the exponential action might not converge.

Remark 12.16 One can develop the notion of a filtered L_∞-algebra and look at Maurer–Cartan elements up to equivalence. As in Remark 12.6, a quasi-isomorphism of filtered L_∞-algebras will result in a bijection between the equivalence classes of Maurer–Cartan elements [71, Appendix A].

Example 12.17 Let $\widehat{R} = \mathbb{C}[[X_1, \dots, X_n]]$ and Λ be the exterior algebra over a set of dual variables $\xi_1, \dots, \xi_n$ with degree 1. Now take $f \in (X)^3$ and consider the matrix factorization $(\Omega, \delta + \gamma\wedge)$ for $(\widehat{R}, f)$. Because f does not contain linear terms, the minimal model of the endomorphism ring of this matrix factorization is isomorphic to Λ as a vector space, and because there are no quadratic terms, the product structure is also the same as Λ. The higher-order terms of f will induce an A_∞-structure on Λ, which we denote by Λ_γ. Note that according, to the filtration, we can think $f \in (X)^k$ as an element in $F_{k-2}\mathsf{HH}^1(\Lambda)$.

In [71, Section 8], Efimov showed that the equivalence class of Λ_γ allows us to reconstruct f up to a change of variables.

Theorem 12.18 (Efimov) *Let $f_1, f_2 \in (X)^3$ and choose γ_1, γ_2 to construct the endomorphism rings Λ_{γ_i} for $(\Omega, \delta + \gamma_i\wedge) \in \mathtt{MF}^{\pm}(\widehat{R}, f_i)$. If $\mathcal{F}_\bullet : \Lambda_{\gamma_1} \to \Lambda_{\gamma_2}$ is an A_∞-isomorphism with $\mathcal{F}_1 = \mathbb{1}_\Lambda$ then there is a change of variables $X_i \mapsto X_i + O_i$ with $O_i \in (X)^2$ that transforms f_1 into f_2.*

12.3 Deformation Theory for Gentle Algebras

Let $(\mathbb{S}, M)$ be a closed marked surface with genus g and n interior marked points. Choose an arc collection $\mathcal{A}$ that splits $(\mathbb{S}, M)$ and let Q be the quiver underlying the gentle A_∞-algebra $\mathtt{Gtl}^{\pm}\mathcal{A}$. We will use F for the set of polygons or faces in $\mathcal{A}$. Because of the Euler characteristic we have $|Q_0| = |\mathcal{A}| = 2g - 2 + |M| + |F| = 2g - 2 + n + |F|$, while $|Q_1| = 2|Q_0|$ because every vertex (arc) has two incoming arrows (angles). Every arrow $\alpha \in Q_1$ turns around exactly one marked point and one face. We will write n_i for the number of angles that turn around $i \in M \cup F$.

Let A be the $\mathbb{Z}_2$-graded algebra underlying the A_∞-algebra $\mathtt{Gtl}^{\pm}\mathcal{A}$ without the higher A_∞-structure. This is the path algebra of the quiver Q with quadratic relations and hence we can use the bimodule resolution of Bardzell to determine the Hochschild cohomology of A.

- For each $m \in M$, we can look at all the arrows $\alpha_1, \dots, \alpha_{n_m}$ that turn around m and construct a central element

$$\ell_m = \alpha_{n_m} \dots \alpha_1 + \dots + \alpha_1 \alpha_{n_m} \dots \alpha_2 \in Z(A)$$

and a derivation

$$\partial_m \colon A \mapsto A \colon \alpha \mapsto \begin{cases} \alpha & \text{if } \alpha \in \{\alpha_1, \dots, \alpha_{n_m}\}, \\ 0 & \text{otherwise.} \end{cases}$$

- For each $f \in F$ and $r \geq 1$, we can look at all the arrows $\alpha_1, \dots, \alpha_{n_f}$ that turn around f and in analogy to Example 12.10 construct maps

$$\ell_f^r := \sum_{i=1}^{n} [\alpha_{i+rn_f-1} \dots \alpha_i \to t(\alpha_i)] \in \mathsf{HH}^{rn_f}(A)_{\text{clas}},$$
$$\ell_f^r \partial_f := \sum_{i=1}^{n} [\alpha_{i+rn_f} \dots \alpha_i \to \alpha_i] \in \mathsf{HH}^{rn_f+1}(A)_{\text{clas}}.$$

Theorem 12.19 *The classical Hochschild cohomology of A has the following structure:*

- $\mathsf{HH}^0(A)_{\text{clas}} \cong \mathbb{C} \oplus \bigoplus_{m \in M, r \geq 1} \mathbb{C}\ell_m^r,$
- $\mathsf{HH}^1(A)_{\text{clas}} \cong \mathbb{C}^{|Q_0|+1} \oplus \bigoplus_{m \in M, r \geq 1} \mathbb{C}\ell_m^r \partial_{\ell_m},$
- $\bigoplus_{k \geq 2} \mathsf{HH}^k(A)_{\text{clas}} \cong \bigoplus_{f \in F, r \geq 1} \mathbb{C}\ell_f^r \oplus \mathbb{C}\ell_f^r \partial_f.$

Sketch of the proof The zeroth and first Hochschild cohomologies can easily be determined by hand. For the former, one can check that every central element is a linear combination of powers of the ℓ_m and that $\ell_m \ell_{m'} = 0$ if $m \neq m'$.

The first Hochschild cohomology is the space of outer derivations. A derivation is determined by its value for each arrow. We can map α_i to $f(\ell_m)\alpha_i$ if α_i turns around the marked point m. One can show that these are all possibilities for derivations on A.

If a is an arc (vertex) connecting to m then the inner derivation coming from commuting with $\ell_m^i \mathbb{1}_a$ will map the angle α_u with $h(\alpha_u) = a$ to $\ell_m^i \alpha_u$ and the arrow α_{u+1} with $t(\alpha_{u+1}) = a$ to $-\ell_m^i \alpha_{u+1}$. If $i > 0$, all angles around other marked points are mapped to zero. Therefore, up to an inner derivation we can assume that all α_i around are multiplied by the same nonconstant factor $\lambda_{m,i}\ell_m^i$.

For the constant factors, if we commute with $\mathbb{1}_a$ then the two angle arrows arriving at a are multiplied by 1 and those leaving a are multiplied by -1. So the dimension of the space of degree 0 outer derivations is

$$|Q_1| - (|Q_0| - 1) = |\mathcal{A}| + 1,$$

where the 1 comes from the fact that commuting with the identity $\mathbb{1} = \sum_a \mathbb{1}_a$ is trivial.

To calculate the higher Hochschild cohomology, one can use the bimodule resolution by Bardzell. Note that $B_{\geq 1}$ splits as a direct sum according to the polygon to which the path p for $B[p]$ belongs. Therefore, the calculation is essentially the same as in Example 12.10. For more details see [1, 36]. □

To convert this statement to the A_∞-setting we have to be careful. In the $\mathbb{Z}_2$-graded setting, the ℓ_m^r and the ℓ_f^r have A_∞-degree 1 so they make up $\mathsf{HH}^1(A)$, but because the latter are contained in different classical degrees and $\mathsf{HH}^\bullet(A) = \prod_i \mathsf{HH}^i(A)$, we are allowed to take formal sums in the ℓ_f but not in the ℓ_m. Therefore,

$$\mathsf{HH}^0(A) = \bigoplus_{m\in M} \mathbb{C}[\ell_m]\ell_m\partial_m \oplus \bigoplus_{f\in F} \mathbb{C}[[\ell_f]]\ell_f\partial_f \oplus \mathbb{C}^{|\mathcal{A}|+1},$$

$$\mathsf{HH}^1(A) = \frac{\mathbb{C}[\ell_m \mid m \in M][[\ell_f \mid f \in F]]}{(\ell_i\ell_j \mid i \neq j,\ i, j \in F \cup M)}.$$

This notation is chosen in such a way that the bracket action of $[\ell_i^j\partial_{\ell_i}, -]$ on $\mathsf{HH}^1(A)$ is precisely the derivation $\ell_i^j\partial_i$.

Remark 12.20 We can restore the asymmetry between the ℓ_m and ℓ_f by going to the formal completion $\widehat{A}$. In this way, all the ℓ_i become formal variables.

The gentle A_∞-algebra is an A_∞-extension of the algebra A. Using the Hochschild cohomology of A, we can obtain a nice criterion to check whether a given A_∞-extension is isomorphic to the gentle algebra.

Theorem 12.21 *Every A_∞-extension of A for which the higher product around every face is nonzero,*

$$\forall f \in F,\ \exists \lambda_f \in \mathbb{C} \setminus \{0\},\ \forall i \text{ we have } \mu_{n_f}(\alpha_{i+rn_f-1}, \ldots, \alpha_i) = \lambda_f t(\alpha_i),$$

is A_∞-isomorphic to $\mathtt{Gtl}^+\mathcal{A}$.

Sketch of the proof Such an extension corresponds to a Maurer–Cartan element

$$\sum_{f\in F} \lambda_f \ell_f + \text{ higher-order terms.}$$

The action $\ell_f^j\partial_f$ integrates to base change $\ell_f \mapsto \ell_f + O(\ell_f)$, which can be used to get rid of the higher-order terms. Rescaling the angle morphisms allows us to set all λ_f to 1. For more details see [36]. □

Theorem 12.22 *The Hochschild cohomology of the gentle A_∞-algebra* $\mathtt{Gtl}^+\mathcal{A}$ *is equal to*

$$\mathsf{HH}^0(\mathtt{Gtl}^{\pm}\mathcal{A}) = \bigoplus_{m\in M} \mathbb{C}[\ell_m]\ell_m\partial_m \oplus \mathbb{C}^{n+2g-1},$$

$$\mathsf{HH}^1(\mathtt{Gtl}^{\pm}\mathcal{A}) = \frac{\mathbb{C}[\ell_m \mid m \in M]}{(\ell_i\ell_j \mid i \neq j,\ i, j \in M)}.$$

Sketch of the proof As a graded vector space the Hochschild complex of A and $\mathtt{Gtl}^{\pm}\mathcal{A}$ are the same but the differentials are different:

$$d_{\mathtt{Gtl}^{\pm}\mathcal{A}} = d_A + [\mu_{\geq 3}, -].$$

Using spectral sequences [253], we can relate the cohomology of d_A to that of $d_{\mathtt{Gtl}^{\pm}\mathcal{A}}$. The result is that the latter becomes the cohomology of the complex $(\mathsf{HH}(A), [\sum_f \ell_f, -])$. For each face $f \in F$ this kills off all terms, including a power of ℓ_f and one constant term in $\mathsf{HH}_0(A)$. Therefore, the total number of constant terms is $|\mathcal{A}| + 1 - |F| = 2g - 1 + |M|$. □

Remark 12.23 Because $\mathtt{DGtl}^{\pm}\mathcal{A} = {}^{\mathtt{top}}\mathtt{Fuk}^{\pm}(\mathbb{S}, M)$ only depends on the marked surface, this calculation can be done for many different models of ${}^{\mathtt{top}}\mathtt{Fuk}^{\pm}(\mathbb{S}, M)$. For the commutative Landau–Ginzburg model $(\tilde{\mathbb{X}}, \ell)$, one can use the results from Lin and Pomerleano in [163]. For the noncommutative Landau–Ginzburg model $(\mathtt{Jac}\, Q, \ell)$, this calculation has been done by Wong in [257].

Remark 12.24 This result fits in a general geometrical interpretation of Hochschild cohomology on the A-side. Conjecturally, in nice cases the Hochschild cohomology of the wrapped Fukaya category of a Liouville manifold should match with its symplectic homology, which is generated by the Reeb loops. In the case of punctured surfaces, these Reeb loops are loops that wind around the punctures.

Our calculation of the Hochschild cohomology indicates that the deformations of the topological Fukaya category are parametrized by coefficients associated to powers of loops around the marked points: ℓ_m^r. Unfortunately, these deformations turn many objects in the topological Fukaya category into curved objects, not in the least all the arcs of our arc collection. Therefore, it is important to find objects that will not become curved under a deformation.

12.4 Filling the Pair of Pants

We are now ready to fill in the marked points to construct the Fukaya category of the closed surface. We will do this starting from the pair of pants or the sphere with three marked points. Recall from Chapter 9 that we can choose an arc collection that splits the sphere into two triangles and we can use this arc collection to construct its topological Fukaya category. We also saw that this category is equivalent to the category of matrix factorizations of the Landau–Ginzburg model $(\mathbb{C}^3, -XYZ)$ (we add a minus sign for later convenience):

$$\mathtt{D}^{\pi}{}^{\mathtt{top}}\mathtt{Fuk}^{\pm}(\mathbb{S}, \{0, 1, \infty\}) \overset{\infty}{=} \mathtt{D}^{\pi}\mathtt{MF}^{\pm}(\mathbb{C}^3, -XYZ).$$

Inside $\mathtt{MF}^{\pm}(\mathbb{C}^3, -XYZ)$ we have a special object corresponding to the simple representation $S = \mathbb{C}[X, Y, Z]/(X, Y, Z)$. As a matrix factorization, this object can be seen as

$$(\Omega, \delta + \gamma\wedge) \quad \text{with } \gamma = -XZ\, dY,$$

and its endomorphism ring is A_∞-equivalent to an A_∞-extension of the exterior algebra Λ_γ.

On the A-side, this object corresponds to the band object below equipped with a local system with holonomy -1 (a nontrivial spin structure). We will put the three special points $\circ$, $*$, $\times$ all on the same segment and conflate them to $\circ$ to make the picture less messy.

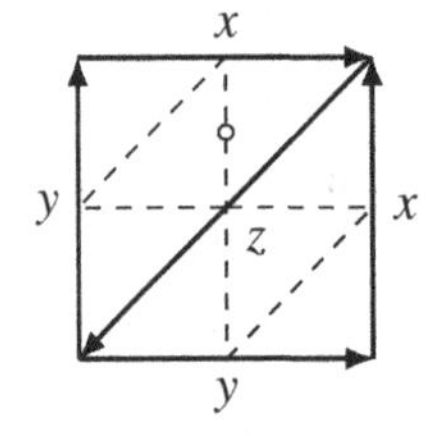

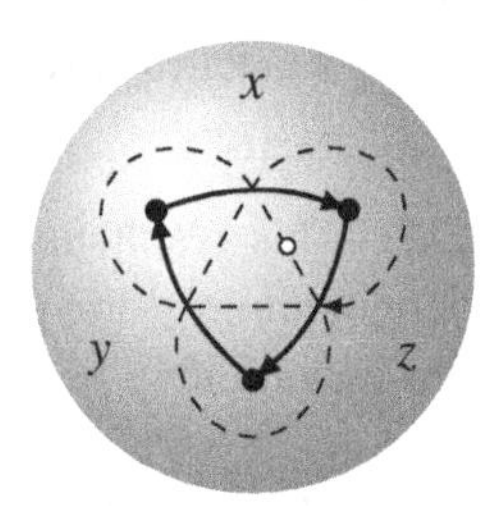

We will call this object the *zigzag Lagrangian*. As a twisted complex it looks like

$$\mathcal{Z} = (x \oplus y \oplus z \oplus x \oplus y \oplus z, \zeta_1 + \xi_2 - \eta_1 + \zeta_2 + \xi_1 + \eta_2).$$

Using Lemmas 9.23, 9.24, 9.26 and 9.27, one can check that $\mathtt{E}^{\pm}_{\mathcal{Z}} := {}^{\mathtt{top}}\mathtt{Fuk}^{\pm}(\mathcal{Z}, \mathcal{Z})$ is eight-dimensional and spanned by morphisms ι, ι^*, p_x, p_y, p_z, p_x^*, p_y^*, p_z^*, where the last six correspond to the three self-intersection points lying on the arcs x, y, z.

If we identify ι with 1 and p_x, p_y, p_z with $\xi, \eta, \zeta \in \Lambda_3$, then we can use products of the form $\mu(p_x, p_y) = -\mu(p_y, p_x) = p_z^*$ to identify p_x^*, p_y^*, p_z^* with $\eta\zeta, \zeta\xi, \xi\eta \in \Lambda_3$. Finally, the product $\mu(p_x, p_x^*) = \iota^*$ tells us that $\iota^* = \xi\eta\zeta$. This tells us that, as an algebra, $\Lambda_{\mathcal{Z}}$ is isomorphic to the exterior algebra Λ_3. Additionally, by Lemma 9.28 we know that ${}^{\mathtt{top}}\mathtt{Fuk}^{\pm}(\mathcal{Z}, \mathcal{Z})$ carries a nontrivial A_∞-structure. Because $\circ$ and $\times$ lie on the segment between x and z we get

$$\mu(p_x, p_y, p_z) = -\iota.$$

We can make the object corresponding to $\mathcal{Z}$ in $\mathtt{MF}^{\pm}(\mathbb{C}^3, -XYZ)$ using the cone construction and the fact that x, y, z correspond to M_X, M_Y, M_Z, and compare this to $(\Omega, \delta + \gamma\wedge)$:

$$\mathcal{Z} = \underbrace{R^{\oplus 6} \underset{\begin{pmatrix} -YZ & 0 & 0 & 0 & 0 & 0 \\ Z & -XZ & X & 0 & 0 & 0 \\ 0 & 0 & -XY & 0 & 0 & 0 \\ 0 & 0 & Y & -YZ & Z & 0 \\ 0 & 0 & 0 & 0 & -XZ & 0 \\ -Y & 0 & 0 & 0 & X & -XY \end{pmatrix}}{\overset{\begin{pmatrix} X & 0 & 0 & 0 & 0 & 0 \\ 1 & Y & 1 & 0 & 0 & 0 \\ 0 & 0 & Z & 0 & 0 & 0 \\ 0 & 0 & 1 & X & 1 & 0 \\ 0 & 0 & 0 & 0 & Y & 0 \\ -1 & 0 & 0 & 0 & 1 & Z \end{pmatrix}}{\rightleftarrows}} R^{\oplus 6}}_{\mathcal{Z}} \sim \underbrace{R^{\oplus 4} \underset{\begin{pmatrix} 0 & X & Y & Z \\ -YZ & 0 & XY & -XZ \\ XZ & -XY & 0 & YZ \\ -XY & -XZ & YZ & 0 \end{pmatrix}}{\overset{\begin{pmatrix} 0 & -YZ & XZ & -XY \\ X & 0 & Z & -Y \\ Y & -Z & 0 & X \\ Z & Y & -X & 0 \end{pmatrix}}{\rightleftarrows}} R^{\oplus 4}}_{(\Omega,\delta+\gamma\wedge)}.$$

It is a nice exercise to check these are quasi-isomorphic, and then we can conclude that

$$E_{\mathcal{Z}} \overset{\infty}{=} E_{\gamma} \overset{\infty}{=} \Lambda_{\gamma}.$$

Now we will fill in the marked points and look at what happens to $E_{\mathcal{Z}}$. Choose $a, b, c > 1$ in $\mathbb{N}$ and put orbifold points of orders a, b, c on the marked points. We denote the resulting surface by $\mathbb{S}_{abc}$. The orbifold characteristic of $\mathbb{S}_{abc}$ is

$$\chi = \frac{1}{a} + \frac{1}{b} + \frac{1}{c} - 3 + 2.$$

There are three possibilities:

- If $\chi > 0$ then we can see $\mathbb{S}_{abc}$ as the quotient of the sphere by a finite group of isometries.
- If $\chi = 0$ then we can see $\mathbb{S}_{abc}$ as the quotient of the Euclidean plane by a wallpaper group of isometries.
- If $\chi < 0$ then we can see $\mathbb{S}_{abc}$ as the quotient of the hyperbolic plane by a Fuchsian group of isometries.

In each case it is possible to define a relative Fukaya category for this surface. The objects of $^{\mathtt{rel}}\mathbf{Fuk}^{\pm}(\mathbb{S}_{abc})$ are band objects in $^{\mathtt{top}}\mathbf{Fuk}^{\pm}(\mathbb{S}, \{0, 1, \infty\})$ and the products count holomorphic maps $\psi\colon \mathbb{D} \to \mathbb{S}_{abc}$ that factor through the quotient construction. Each map is weighed by the holonomy of the local systems and a factor $\hbar^n$, where n is the number of points in $\mathbb{D}$ that are mapped onto the marked points.

With this in mind we can have a look at what happens to the zigzag Lagrangian. By going to the relative Fukaya category, its endomorphism algebra $\mathtt{E}^{\pm}_{\mathcal{Z}}$ deforms to a possibly curved A_∞-algebra $\mathtt{E}^{\hbar}_{\mathcal{Z}}$. Some of these deformed products can be read off from the picture. Note that on the sphere

p_x is the corner of two monogons that contain the orbifold point with order a. These monogons lift to a-gons in the universal cover. One of these a-gons has a $\circ$'s on it, so it acquires a sign $(-1)^a$; therefore, in analogy with Lemmas 9.27 and 9.28 we get

$$\mu_a(p_x, \dots, p_x) = (-1)^a \hbar \iota + \cdots,$$
$$\mu_{a-1}(p_x, \dots, p_x) = (\hbar + (-1)^a \hbar) p_x^* + \cdots,$$

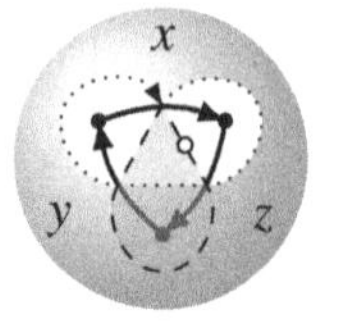

and similar products exist for p_y, p_z. Again we can distinguish three cases:

- If $\chi > 0$ then one of the a, b, c is 2, so the differential can become nonzero. Moreover, because the universal cover is a sphere there will also be a nontrivial μ_0.
- If $\chi = 0$ then there will be an orbifold point of order 2 or 3, so either the differential becomes nontrivial or the ordinary product deforms. Also, if $a = 3$ then $\mu_{a-1}(p_x, \dots, p_x)$ will get an infinite number of additional contributions from triangles with ever larger sizes.
- If $\chi < 0$ then there can be no polygons of arbitrary sizes bounded by the same angles because the curvature is negative. Therefore, all products are polynomials in $\hbar$ instead of power series and we can put $\hbar = 1$ and work over $\mathbb{C}$. If $a, b, c > 3$ then all extra products are higher products and if we put $\hbar = 1$ then $\mathtt{E}^{\hbar}_{\mathcal{Z}}$ becomes an A_∞-extension of $\mathtt{E}^{\pm}_{\mathcal{Z}} = \Lambda_3$.

Clearly, the easiest situation happens in the last case and we will look at a specific instance of this: $a = b = c = 2g + 1$ with $g \geq 2$.

Theorem 12.25 (Seidel ($g = 2$); Efimov ($g > 2$)) *The Fukaya category of $\mathbb{S}_{abc}$ with $a = b = c = 2g + 1$ is split-derived equivalent with a Landau–Ginzburg model:*

$$\mathtt{D}^{\pi}\,\mathtt{Fuk}^{\bullet}(\mathbb{S}_{abc}) \overset{\infty}{=} \mathtt{D}^{\pi}\mathtt{MF}^{\pm}(\mathbb{C}^3, -XYZ + X^{2g+1} + Y^{2g+1} + Z^{2g+1}).$$

Sketch of the proof Define $\gamma = -YZ\,dX + X^{2g}dX + Y^{2g}dY + Z^{2g}dZ$ such that

$$\delta\gamma = f = -XYZ + X^{2g+1} + Y^{2g+1} + Z^{2g+1}.$$

If we identify p_x, p_y, p_z with ξ, η, ζ and set $\hbar = 1$ then both $\mathtt{E}^{\hbar}_{\mathcal{Z}}$ and Λ_γ have the following nonzero products:

$$\mu_{2g+1}(\xi, \dots, \xi) = \mu_{2g+1}(\eta, \dots, \eta) = \mu_{2g+1}(\zeta, \dots, \zeta) = \iota = -\mu(\xi, \eta, \zeta).$$

Furthermore, there are no other nonzero products μ_k with $3 < k < 2g + 1$ and both A_∞-structures are G-invariant for the $G = \mathbb{Z}_{2g+1}$ action with weights 1, 1, -2 for ξ, η, ζ.

These properties are enough to show that both A_∞-extensions of Λ are A_∞-isomorphic. The idea is to look at the Hochschild cohomology classes and conclude they lie in the same orbit. The cohomology class of Λ_γ is $f = f \otimes 1 \in \widehat{R} \otimes \Lambda^{\text{even}} = \mathsf{HH}^1(\Lambda)$. Because of the products above, the class of $\mathrm{E}^\hbar_{\mathcal{Z}}$ is $f \otimes 1 + \cdots$, where the extra terms are G-invariant and restricted to certain degrees. By acting with appropriate elements in $\widehat{R} \otimes \Lambda^{\text{odd}} = \mathsf{HH}^0(\Lambda)$, these extra terms can be gauged away. For more details see [71].

To get the mirror correspondence we have to show that the two objects associated to these endomorphism algebras are split generators. For $(\Omega, \delta + \gamma\wedge)$ this follows from the generation criterion of Dyckerhoff [69] and the fact that f only has an isolated singularity at the origin. For $\mathcal{Z}$ one can use a generation criterion by Seidel [220]. Intuitively, this is clear from the fact that $\mathcal{Z}$ cuts the surface into polygons so that we can isotope each curve to a sequence of segments of $\mathcal{Z}$. □

This mirror correspondence can be lifted from the orbifold to a surface with genus g. Remember from Examples 9.50 and 9.63 that we have a branched cover between a marked surface $(\mathbb{S}_g, M)$ of genus g with three marked points and the pair of pants.

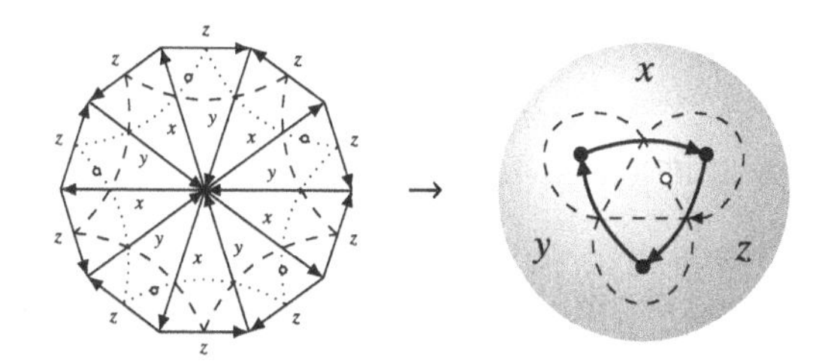

This cover has an automorphism group $G = \mathbb{Z}_{2g+1}$ and there is a mirror correspondence

$$\mathrm{D}^{\pi\,\mathrm{top}}\mathrm{Fuk}^{\pm}(\tilde{\mathbb{S}}, M) \overset{\infty}{\cong} \mathrm{D}^{\pi}\mathrm{MF}^{\pm}(\mathbb{C}[X, Y, Z] \star G, -XYZ).$$

The zigzag Lagrangian $\mathcal{Z}$ lifts to $2g + 1$ curves in $\mathbb{S}_g$ and the endomorphism ring of this collection $\tilde{\mathcal{Z}}$ is $\mathrm{E}^{\pm}_{\mathcal{Z}} \star G$, which deforms to $\mathrm{E}^{\hbar}_{\mathcal{Z}} \star G$. On the B-side, the function

$$f = -XYZ + X^{2g+1} + Y^{2g+1} + Z^{2g+1}$$

is invariant under the G-action, so we can use it to deform the Landau–Ginzburg model $(\mathbb{C}[X, Y, Z] \star G, -XYZ)$. The matrix factorization corresponding to $S \star G$ has an endomorphism ring $\Lambda_\gamma \star G$. Again we get a correspondence between two split-generators.

Theorem 12.26 (Seidel–Efimov) *The Fukaya category of the closed genus $g \geq 2$ surface is split-derived equivalent with*

- *a stacky Landau–Ginzburg model*

$$\mathtt{D}^\pi\mathtt{Fuk}^\pm(\mathbb{S}_g) \overset{\infty}{=} \mathtt{D}^\pi\mathtt{MF}^\pm([\mathbb{C}^3/G], -XYZ + X^{2g+1} + Y^{2g+1} + Z^{2g+1}),$$

  where $G = \mathbb{Z}_{2g+1}$ acts on $\mathbb{C}^3$ with weights $(1, 1, 2g-1)$,
- *a quasi-projective Landau–Ginzburg model*

$$\mathtt{D}^\pi\mathtt{Fuk}^\pm(\mathbb{S}_g) \overset{\infty}{=} \mathtt{D}^\pi\mathtt{MF}^\pm(\tilde{\mathbb{X}}, -XYZ + X^{2g+1} + Y^{2g+1} + Z^{2g+1}),$$

  where $\tilde{\mathbb{X}}$ is the crepant resolution of $\mathbb{C}^3 /\!\!/ G$ coming from the toric diagram

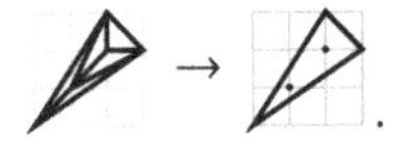

Sketch of the proof The proof of the stacky version follows the chain of equivalences

$$\begin{aligned}
\mathtt{D}^\pi\mathtt{Fuk}^\pm(\mathbb{S}_g) &\overset{\infty}{=} \mathtt{D}^\pi\Lambda_\gamma \star G \\
&\overset{\infty}{=} \mathtt{D}^\pi\mathtt{E}_\gamma^\hbar \star G \\
&\overset{\infty}{=} \mathtt{D}^\pi\mathtt{MF}^\pm(\mathbb{C}[X, Y, Z] \star G, -XYZ + X^{2g+1} + Y^{2g+1} + Z^{2g+1}) \\
&\overset{\infty}{=} \mathtt{D}^\pi\mathtt{MF}^\pm([\mathbb{C}^3/G], -XYZ + X^{2g+1} + Y^{2g+1} + Z^{2g+1}).
\end{aligned}$$

To get the quasi-projective version, we can use the derived equivalence between $\mathtt{D}^b\mathtt{Coh}\,\tilde{\mathbb{X}}$ and $\mathtt{D}^b\mathtt{Coh}\,[\mathbb{C}^3/G]$ to get an equivalence between the two categories of matrix factorizations. For more details we refer to [71]. □

Remark 12.27 Because of Lemma 9.62, we can also use the noncommutative Landau–Ginzburg model $(\mathtt{Jac}\,Q, f)$, where Q is the dimer model from Example 9.63 and f is the same function but now viewed as a central element in $\mathbb{C}[X, Y, Z] \star G \cong \mathtt{Jac}\,Q$.

Example 12.28 If the genus is 1, the same technique results in a threefold cover from a torus with three marked points to the sphere with three marked points. This means that the $2g + 1$-gons responsible for the products are triangles, so they also contribute to binary products. Furthermore, there are triangles of arbitrary sizes so these contributions become infinite sums.

To calculate the products we put a parameter $\hbar$ in each of the marked points. The product counts triangles with a given sequence of corners weighed by a factor $\hbar$ for each $\hbar$ in the triangle and a factor $\sigma = -1$ for each $\circ$ on the boundary (for the holonomy of the local system). This gives the following binary products:

$$\mu(\xi, \eta) = (1 + \sigma^3 \hbar^3 + \sigma^3 \hbar^6 + \cdots)\zeta^*,$$
$$\mu(\eta, \xi) = (\sigma + \sigma^2 \hbar^3 + \sigma^4 \hbar^6 + \cdots)\zeta^*,$$
$$\mu(\zeta, \zeta) = ((1 + \sigma^3)\hbar + (\sigma^3 + \sigma^6)\hbar^{10} + \cdots)\zeta^*$$

and cyclic variants. Note that for $\sigma = -1$ we get $\mu(\xi, \eta) = -\mu(\eta, \xi)$ and $\mu(\zeta, \zeta) = 0$. There are no products of the form $\mu(\xi^*, \eta^*)$ because these would come from a triangle with two obtuse angles. Therefore, after rescaling the starred variables, we can see that as an algebra $\mathrm{E}^{\hbar}_{\mathcal{Z}}$ is still isomorphic to the exterior algebra (but over $\mathbb{C}[[\hbar]]$).

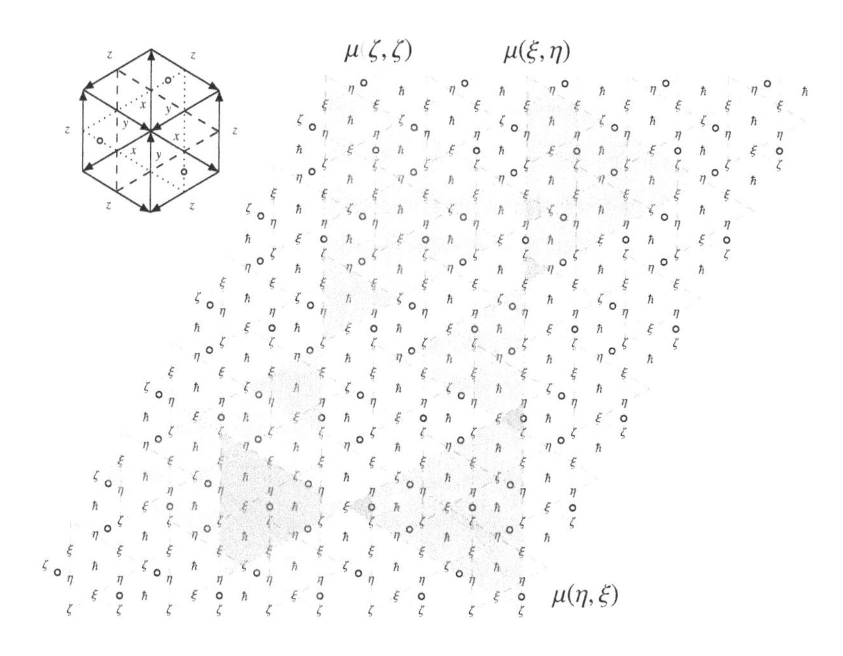

Now look at the triple products. They come from triangles that run through the points marked with a $\circ$:

$$\mu(\zeta, \zeta, \zeta) = (\sigma^3 \hbar + (\sigma^3 + 2\sigma^6)\hbar^{10} + \cdots)\iota,$$
$$\mu(\eta, \xi, \zeta) = (\sigma + 2\sigma^4 \hbar^6 + \cdots)\iota,$$
$$\mu(\xi, \eta, \zeta) = (\sigma^3 \hbar^3 + \sigma^6 \hbar^6 + \cdots)\iota,$$
$$\vdots$$
$$\mu(\zeta, \eta, \xi) = (\sigma^3 \hbar^3 + \sigma^3 \hbar^6 + \cdots)\iota.$$

So if $\mathtt{E}^{\hbar}_{\mathcal{Z}}$ is isomorphic to Λ_f then f is of the form

$$(\sigma + \cdots)ZYX + \cdots + (\sigma^3\hbar^3 + \cdots)XYZ \\ + (\sigma^3\hbar + (\sigma^3 + 2\sigma^6)\hbar^{10} + \cdots)(X^3 + Y^3 + Z^3).$$

After putting all permutations of XYZ together and rescaling it to 1 we can bring f into the form

$$f_\hbar = XYZ + \lambda_\hbar(X^3 + Y^3 + Z^3),$$

where $\lambda_\hbar \in (\hbar)$. If we pick a $\gamma_\hbar$ such that $\delta\gamma_\hbar = f_\hbar$ we get an equivalence of the following two deformations:

$$\mathtt{E}^{\pm}_{\mathcal{Z}} \rightsquigarrow \mathtt{E}^{\hbar}_{\mathcal{Z}} \quad \sim \quad \Lambda_{\gamma_0} \rightsquigarrow \Lambda_{\gamma_\hbar}.$$

Going to the twisted completion we get a mirror correspondence between ${}^{\mathtt{rel}}\mathtt{Fuk}^{\pm}(\mathbb{S}_{333})$ and $\mathtt{MF}^{\pm}(\mathbb{C}^3, f_\hbar)$, which we can smash with the group $G = \mathbb{Z}_3$ to get a mirror correspondence for the torus with three filled marked points. This is a $\mathbb{Z}_2$-graded correspondence, but the torus is Calabi–Yau, so the Fukaya category can be $\mathbb{Z}$-graded.

On the B-side, we see that $f_\hbar$ is homogeneous of degree 3, so we can construct the category of graded matrix factorizations. In this category there are shifts $S(i)$ corresponding to $(\Omega(i), \delta + \gamma_\hbar\wedge)$ and we have the identity $S(3) = S[2]$, where the shift on the left comes from the grading on $\mathbb{C}[X, Y, Z]$ and the shift on the right is the homological shift. Therefore, the category of matrix factorizations $\mathtt{grMF}^\bullet(\mathbb{C}^3, f_\hbar)$ is generated by S, $S(1)$, $S(2)$. As an algebra, the endomorphism ring of this object is $\Lambda \star G$, which can be seen as a path algebra of a quiver with relations.

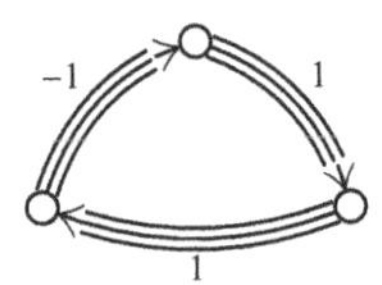

The triple arrows starting at each vertex represent versions of the ξ, η, ζ, which get different gradings depending on where they start: the first two copies still have degree 1 but the last, which ends at $S(3) = S[-2]$, has degree $1 - 2 = -1$. On the Fukaya side the zigzag Lagrangian lifts to three curves $\mathcal{Z}_1, \ldots, \mathcal{Z}_3$ on the torus, and we can choose their gradings in such a way that the degrees match. This gives an A_∞-isomorphism over $\mathbb{C}[[\hbar]]$

$$\mathrm{End}^\bullet_{\mathtt{Fuk}}(\mathcal{Z}_1 \oplus \mathcal{Z}_2 \oplus \mathcal{Z}_3)_\hbar \overset{\infty}{=} (\Lambda \star G)_\hbar.$$

If we go to the twisted completion (and work over the field $\mathbb{C}((\hbar))$), we get

$$\mathtt{D}^\pi\,\mathtt{Fuk}^\bullet(\mathbb{T}, M)_{\mathbb{C}((\hbar))} \overset{\infty}{=} \mathtt{D}^\pi\mathtt{grMF}(\mathbb{C}((\hbar))[X, Y, Z], f_\hbar).$$

By Orlov's theorem (Theorem 7.97) the latter is derived equivalent to $\mathbb{P}(\mathbb{C}(\!(\hbar)\!)[X,Y,Z]/(f_\hbar))$, which is the generic member of a family of elliptic curves. In the limit $\hbar = 0$ this family degenerates to $\mathbb{P}(\mathbb{C}[X,Y,Z]/(XYZ))$. This is a curve with three nodal singularities that forms a chain of three $\mathbb{P}^1$'s. This degeneration can be seen as three curves on the surface that contract to a point; these vanishing cycles correspond to the marked points on the A-side.

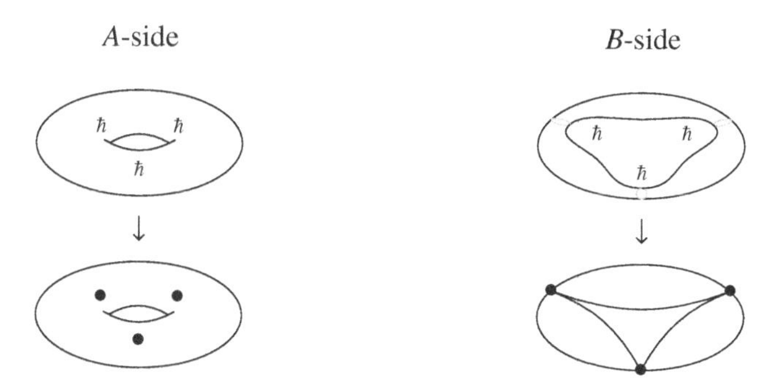

Remark 12.29 Mirror symmetry for the elliptic curve has been studied by many, some examples of which we give below.

- The original mirror correspondence of the torus (in terms of derived categories over $\mathbb{C}$ without focusing on the A_∞-structure) is due to Polishchuk and Zaslow [205].
- The construction we outlined above is worked out by Cho, Hong and Lau in [55], but they work with the full Fukaya category over the Novikov ring. This gives a different expression of $\lambda_\hbar$ in terms of an area parameter.
- The relative point of view for the torus has been worked out by Lekili and Perutz in the case of one marked point [162, 161] and by Lekili and Polishchuk in the case of several [164]. In the latter they work with a different system of Lagrangian branes that cut the torus into rectangles, each containing one marked point with its own deformation parameter t_i.

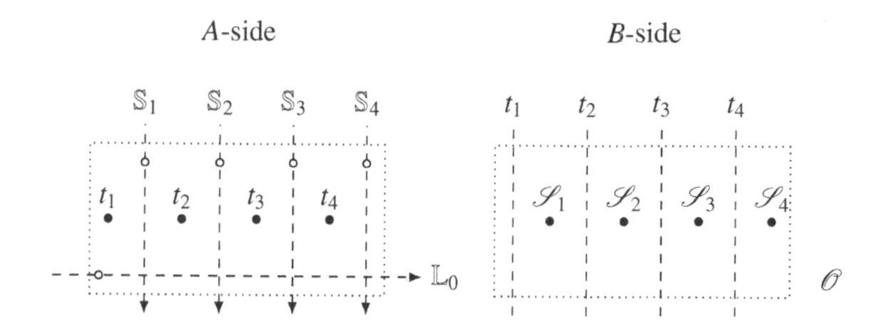

For $n = 1$ the mirror family can be seen as the Weierstrass family, defined by the equation

$$Y^2Z + XYZ = X^3 + a(\hbar)XZ^2 + b(\hbar)Z^3.$$

It degenerates to the nodal curve $Y^2Z + XYZ = X^3$ for $\hbar \to 0$. For the torus with n marked points the family is the Tate curve, which degenerates to a chain of n projective lines.

The object mirror to the Lagrangian $\mathbb{L}_0$ is the structure sheaf of the curve and the objects mirror to the $\mathbb{S}_i$ are skyscraper sheafs of points. Their endomorphism ring is an A_∞-extension of the following path algebra with relations

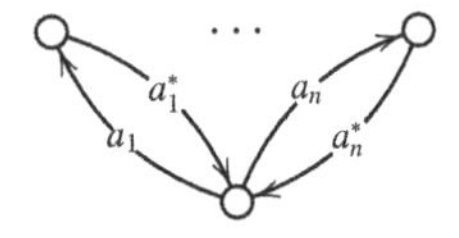

such that $a_i^* a_i = a_j^* a_j$ and $a_i a_j^* = 0$. Just like for the zigzag Lagrangian, this A_∞-extension is nontrivial even in the case $\hbar = 0$.

Note that there is a special symmetry in this construction. Each marked point on the A-side corresponds to a vanishing cycle on the B-side and each marked skyscraper sheaf on the B-side corresponds to a Lagrangian cycle on the A-side.

We end this section with a noncommutative mirror construction. In the calculation of the products for $\mathbb{S}_{333}$ we deliberately wrote the sign as a σ because now we want to do something slightly different. Instead of putting a nontrivial spin structure on $\mathcal{Z}$, we will equip it with a local system with holonomy $\rho \neq -1$. Denote this new Lagrangian brane by $\mathcal{Z}_\rho$, its lift to the torus by $\tilde{\mathcal{Z}}_\rho$ and its endomorphism ring by $\mathrm{E}^{\hbar}_{\mathcal{Z}_\rho}$.

To determine this endomorphism ring, we can use the same calculations as above but we need to substitute σ by ρ. The result is a new algebra structure on Λ because now $\mu(\zeta, \zeta)$ is not zero anymore and the coefficients of $\mu(\xi, \eta)$ are different. In fact there are three coefficients $\alpha, \beta, \gamma \in \mathbb{C}[[\hbar]]$ such that the binary products satisfy

$\cdot$	ξ	η	ζ
ξ	$\gamma\xi^*$	$\beta\zeta^*$	$\alpha\eta^*$
η	$\alpha\zeta^*$	$\gamma\eta^*$	$\beta\xi^*$
ζ	$\beta\eta^*$	$\alpha\xi^*$	$\gamma\zeta^*$

We denote this new algebra without the higher-order products by $\Lambda_{\alpha\beta\gamma}$, so $\mathtt{E}^{\hbar}_{\mathcal{Z}_\rho}$ is an A_∞-extension of $\Lambda_{\alpha\beta\gamma}$. The products in $\Lambda_{\alpha\beta\gamma}$ suggest that this algebra is the Koszul dual of another algebra: $\Lambda_{\alpha\beta\gamma} = \mathrm{Ext}_A(S,S)$ with

$$A = \mathtt{Skl}_{\alpha\beta\gamma} := \frac{\mathbb{C}\langle X,Y,Z\rangle}{\langle \alpha XY + \beta YX + \gamma Z^2, \alpha YZ + \beta ZY + \gamma X^2, \alpha ZX + \beta XZ + \gamma Y^2\rangle}.$$

Unfortunately, this is not always true, because this algebra might not have finite global dimension, but it holds for generic α, β, γ. These algebras are called three-dimensional *Sklyanin algebras* and they can be seen as nice noncommutative deformations of the polynomial ring in three variables.

These algebras have been studied by Artin, Tate and Van den Bergh [17] and they show that the center of such a ring is generated by one element ℓ of degree 3. The quotient $\mathtt{Skl}_{\alpha\beta\gamma}/\langle\ell\rangle$ has a nice geometrical description. It is the twisted coordinate ring of an elliptic curve. More precisely, there is an elliptic curve $\mathbb{E}$, a line bundle $\mathscr{L}$ and an automorphism $\tau\colon \mathbb{E} \to \mathbb{E}$ such that

$$B = \frac{\mathtt{Skl}_{\alpha\beta\gamma}}{\langle\ell\rangle} \cong \bigoplus_{r=0}^{\infty} \mathscr{L} \otimes \mathscr{L}^{\tau} \otimes \cdots \otimes \mathscr{L}^{\tau^{r-1}}(\mathbb{E})$$

and $u \cdot v = u \otimes \tau^{\deg u}(v)$ (τ acts by pull back on the line bundles). This can be interpreted in terms of noncommutative algebraic geometry. We can think of A as the homogeneous coordinate ring of a noncommutative projective plane $\mathbb{P}_{\alpha\beta\gamma}$ and inside this projective plane sits an elliptic curve. For both objects we can define the category of coherent sheaves as

$$\mathsf{Coh}\,\mathbb{P}_{\alpha\beta\gamma} := \frac{\mathtt{grMod}\,A}{\mathtt{grmod}\,A}, \quad \mathsf{Coh}\,\mathbb{E}_{\alpha\beta\gamma} := \frac{\mathtt{grMod}\,B}{\mathtt{grmod}\,B}.$$

It is important to stress that while $\mathsf{Coh}\,\mathbb{P}_{\alpha\beta\gamma}$ is not equivalent to the category of coherent sheaves of a commutative variety, the category $\mathsf{Coh}\,\mathbb{E}_{\alpha\beta\gamma}$ is equivalent to $\mathsf{Coh}\,\mathbb{E}$ where $\mathbb{E}$ is the elliptic curve above.

Surprisingly, this element ℓ is equal to the expression $f_\hbar$ we have determined using the triple products. If we calculate the endomorphism ring of the simple B-module $S = B/\langle X,Y,Z\rangle$ in $\mathtt{Sing}^\bullet B \overset{\infty}{=} \mathtt{D}^\pi\,\mathtt{MF}^\bullet(A, f_\hbar)$, we get an A_∞-extension $\mathtt{E}^{\hbar}_S$ of $\Lambda_{\alpha\beta\gamma}$ that is equivalent to $\mathtt{E}^{\hbar}_{\mathcal{Z}_\rho}$. Together with a noncommutative version of Orlov's theorem (Theorem 7.97), we obtain the following sequence of equivalences:

$$\begin{aligned}\mathtt{D}^\pi\,\mathsf{Fuk}^\bullet(\mathbb{T}, M)_{\mathbb{C}(\!(\hbar)\!)} &\overset{\infty}{=} \mathtt{D}^\pi\mathtt{E}^{\hbar}_{\tilde{\mathcal{Z}}_\rho} \overset{\infty}{=} \mathtt{D}^\pi\mathtt{E}^{\hbar}_{\mathcal{Z}_\rho} \star \mathbb{Z}_3 \overset{\infty}{=} \mathtt{D}^\pi\mathtt{E}^{\hbar}_S \star \mathbb{Z}_3\\ &\overset{\infty}{=} \mathtt{D}^\pi\,\mathtt{grMF}^\bullet(A, f_\hbar) \overset{\infty}{=} \frac{\mathtt{D}^\mathrm{b}\,\mathtt{grMod}^\bullet B}{\mathtt{D}^\mathrm{b}\,\mathtt{Perf}^\bullet B} \overset{\infty}{=} \frac{\mathtt{D}^\mathrm{b}\,\mathtt{grMod}^\bullet B}{\mathtt{D}^\mathrm{b}\,\mathtt{grmod}^\bullet B}\\ &\overset{\infty}{=} \mathtt{D}\,\mathsf{Coh}^\bullet\,\mathbb{E}.\end{aligned}$$

We can interpret this result in the following way. In both the commutative case and noncommutative case we can see the mirror of the torus as an elliptic curve sitting inside a projective plane as the zero locus of a homogeneous function

$$\mathbb{E} = \mathbb{P}(\mathbb{C}(\!(\hbar)\!)[X,Y,Z]/(f_\hbar)) \subset \mathbb{P}^2 = \mathbb{P}(\mathbb{C}(\!(\hbar)\!)[X,Y,Z]),$$
$$\mathbb{E} = \mathbb{P}(\mathtt{Skl}^{\hbar}_{\alpha\beta\gamma}/\langle f_q\rangle) \subset \mathbb{P}^2_{\alpha\beta\gamma} = \mathbb{P}(\mathtt{Skl}^{\hbar}_{\alpha\beta\gamma}).$$

These ideas are worked out by Cho, Hong and Lau in [154] using a much more general framework, which we will discuss in the next two sections.

12.5 Koszul Duality

We start with a short discussion on Koszul duality. For more information on this topic we refer to the work of Beilinson, Ginzburg and Soergel [27], and for its connection with A_∞-algebras to [167] and [249].

Let Q be a quiver, $\widehat{\mathbb{C}Q}$ its completed path algebra and ΛQ its dual quiver algebra. Recall that

$$\Lambda Q = \mathrm{Ext}^{\bullet}_{\mathbb{C}Q}(S,S) = \mathbb{C}Q^{\mathrm{op}}/\langle \alpha\beta \mid \alpha,\beta \in Q_1^{\mathrm{op}}\rangle,$$

with $S = \widehat{\mathbb{C}Q}/\langle a \in Q_1\rangle$. In words, for every arrow a in Q there is a dual arrow α in ΛQ in the opposite direction and all products $\alpha\beta$ are zero. This implies that a vector space ΛQ is just spanned by the vertices and the dual arrows.

The pairing between the arrows of Q and Q^{op} extends to a pairing between the paths: if $p = a_1 \dots a_k$ is a path in Q then we write

$$p^\vee = \alpha_k \otimes \cdots \otimes \alpha_1 \in (\mathbb{C}Q^{\mathrm{op}})_k = (\Lambda Q)^{\otimes k},$$

where the tensor products are taken over $\Bbbk = \mathbb{C}Q_0 = \mathbb{C}Q_0^{\mathrm{op}}$.

Choose a positive $\mathbb{Z}$-grading $\deg\colon Q_1^{\mathrm{op}} \to \mathbb{N}$ on the arrows of ΛQ and let μ be an A_∞-structure on ΛQ such that the vertices act as strict units if we consider $(\Lambda Q,\mu)$ as an A_∞-category.

We define a dual grading on $\widehat{\mathbb{C}Q}$ by the rule

$$\deg a = 1 - \deg \alpha$$

and a differential $d_\mu\colon \widehat{\mathbb{C}Q} \to \widehat{\mathbb{C}Q}$ such that for $a \in Q_1$ we have

$$d_\mu a = \sum_p \mu(p^\vee)(a)p,$$

where the sum runs over all paths in Q. In words, we can say that the coefficient of the path $a_1 \dots a_k$ in the $d_\mu a$ is the same as the coefficient of α in the products $\mu(\alpha_k,\dots,\alpha_1)$. Because there are no relations in $\widehat{\mathbb{C}Q}$, we can extend d_μ to the whole of $\widehat{\mathbb{C}Q}$ by the graded Leibniz rule. For this new grading, d_μ has degree 1 and $\widehat{\mathbb{C}Q}$ is zero in positive degree.

Theorem 12.30 (Koszul duality for A_∞-algebras) *The assignment $\mu \to d_\mu$ gives a one-to-one correspondence between*

- *A_∞-structures $(\Lambda Q, \mu)$ for which the vertices are strict units,*
- *dg-structures on $\widehat{\mathbb{C}Q}$.*

Sketch of the proof The key observation is that the A_∞-axioms for μ are equivalent to $d_\mu^2 = 0$. □

Definition 12.31 We call $A^! = (\widehat{\mathbb{C}Q}, d_\mu)$ the *Koszul dual* of $A = (\Lambda Q, \mu)$.

Example 12.32 Let $A = \Lambda_2 = \mathbb{C} \oplus \mathbb{C}\xi \oplus \mathbb{C}\eta \oplus \mathbb{C}\rho$ be the exterior algebra in two variables with $\rho = \xi\eta = -\eta\xi$. In that case, Q^{op} has one vertex and three loops ξ, η, ρ with $\mu(\xi, \eta) = -\mu(\eta, \xi) = \rho$. The Koszul dual is

$$\mathbb{C}\langle\langle X, Y, R\rangle\rangle \quad \text{with } dX = dY = 0,\ dR = XY - YX$$

and $\deg X = \deg Y = 0$, $\deg R = -1$. The homology of this dg-algebra is $\mathbb{C}[[X, Y]]$ concentrated in degree 0 and therefore $A^!$ and $\mathbb{C}[[X, Y]]$ are quasi-isomorphic algebras.

The products in Koszul dual A_∞-algebras are closely related and this is reflected in their representation theories.

Theorem 12.33 (Koszul duality for modules) *There is an equivalence between the category of finite-dimensional right A_∞-modules of A and twisted complexes of $A^!$ and vice versa:*

$$\mathtt{mod}^\bullet_\infty A \overset{\infty}{=} \mathtt{Tw}A^!, \quad \mathtt{mod}^\bullet_\infty A^! \overset{\infty}{=} \mathtt{Tw}A.$$

Sketch of the proof We briefly explain the case when A is just an algebra (i.e. the quiver has one vertex); the general case is similar but more cumbersome to write down. Every finite-dimensional A_∞-module of A can be seen as an A_∞-morphism $\mathcal{F}_\bullet\colon A^{\mathrm{op}} \to \mathrm{Mat}^\bullet_{n\times n}(\mathbb{C}) = \mathrm{End}_{\mathtt{grvect}^\bullet}(V)$, where V is a finite-dimensional graded vector space. Because A is positively graded we can assume these matrix maps are lower triangular.

Using the same idea as before, we can define an $A^!$-valued matrix

$$\delta_{\mathcal{F}} = \sum_p \mathcal{F}(p^\vee) \otimes p \in \mathrm{Mat}_{n\times n}(A^!) = \mathrm{End}_{\mathtt{grvect}^\bullet}(V) \otimes A^!.$$

The condition that $\mathcal{F}$ satisfies the A_∞-relations is equivalent to the condition that δ satisfies the Maurer–Cartan equation $d_\mu \delta_{\mathcal{F}} + \delta_{\mathcal{F}}^2 = 0$. Therefore we get a correspondence between V and the twisted complex $(A^! \otimes V, \delta_{\mathcal{F}})$.

In the opposite direction let S be the simple $A^!$-module on which the generators act as zero. Then one can show that $\mathrm{Ext}^\bullet_{A^!}(S,S) \overset{\infty}{=} A$. Because $A^!$ is completed, S generates all the finite-dimensional modules, so $\mathtt{mod}^\bullet_\infty A^! = \langle S \rangle \overset{\infty}{=} \mathtt{Tw}A$. □

The identification between $\mathtt{mod}^\bullet A$ and $\mathtt{Tw}A^!$ can also be used to map other categories to the category of twisted complexes of $A^!$. If $\mathtt{A}^\bullet$ is a subcategory of a larger category $\mathtt{C}^\bullet$ for which all the hom-spaces starting in A are finite-dimensional then we can see every object $X \in \mathtt{C}^\bullet$ as a right A_∞-module

$$\mathcal{F}_X\colon \mathtt{A}^\bullet \to \mathrm{End}^\bullet_{\mathbb{C}}(\mathtt{C}^\bullet(A,X))\colon \mathcal{F}_X(\alpha_1 \otimes \cdots \otimes \alpha_n)(\xi) := \mathcal{F}_X(\alpha_n, \ldots, \alpha_1, \xi).$$

Composing this with the identification above we get an A_∞-functor

$$\mathcal{K}_A\colon \mathtt{C}^\bullet \to \mathtt{TwA}^!.$$

Example 12.34 If we take the simple $\mathbb{C}[X,Y]$-module $S = \mathbb{C}[X,Y]/(X,Y)$ then we can consider its Ext-algebra $A = \mathrm{Ext}^\bullet_{\mathbb{C}[X,Y]}(S,S) = \Lambda_2$ as a subcategory of $\mathtt{mod}^\bullet \mathbb{C}[X,Y]$. This gives rise to a functor $\mathcal{K}_A\colon \mathtt{mod}^\bullet \mathbb{C}[X,Y] \to \mathtt{Tw}\,\Lambda_2^!$. Now we know that $\Lambda_2^!$ is quasi-isomorphic to the completed polynomial ring $\mathbb{C}[\![X,Y]\!]$, so we get a functor

$$\mathtt{mod}^\bullet \mathbb{C}[X,Y] \to \mathtt{Tw}\,\mathbb{C}[\![X,Y]\!].$$

This functor maps S to a twisted complex with $\dim \Lambda_2 = 4$ summands, which is the standard resolution of S as a $\mathbb{C}[\![X,Y]\!]$-module. Because the Ext-algebra between any other simple $\mathbb{C}[X,Y]$-module and S is zero, these other simples are mapped to zero complexes.

Example 12.35 If Q is a quiver without oriented cycles and $\mathtt{S}^\bullet$ is the subcategory of simples, then $A^! = \mathbb{C}Q$ and the functor $\mathcal{K}_{\mathtt{S}^\bullet}$ maps the module S_ρ of $\mathbb{C}Q$ to its resolution P_ρ, so $\mathcal{K}_{\mathtt{S}^\bullet}$ is an equivalence. If Q has oriented cycles then S_ρ is mapped to the resolution of $S_\rho \otimes_{\mathbb{C}Q} \widehat{\mathbb{C}Q}$.

Remark 12.36 The functor $\mathcal{K}_A\colon \mathtt{C}^\bullet \to \mathtt{TwA}^!$ should be thought of as some kind of localization. Every object X for which $\mathtt{C}^\bullet(A,X)$ is zero (in other words it is not close to A) is killed by $\mathcal{K}_A$.

If $\mathtt{A}^\pm$ is $\mathbb{Z}_2$-graded then the situation changes because now we can have products that contain vertices: e.g. $\mu(\alpha,\alpha,\alpha) = v$. Dually this would mean that $d_\mu v = a^3$, so in that case v is not a strict unit for the dg-algebra. Furthermore, the twisted complexes in the construction will not be lower triangular anymore because $\mathtt{A}^\pm$ is not positively graded. To address these problems we will construct a substitute for $A^!$ that is a Landau–Ginzburg model and the local functor will produce matrix factorizations instead of twisted complexes. We will work out the details in the case that is relevant for our situation.

12.6 The Mirror Functor

Let $\mathcal{B} = (\mathcal{B}_1, \rho_1) \oplus \cdots \oplus (\mathcal{B}_k, \rho_k)$ be a collection of band objects in a topological Fukaya category and choose markings $\circ$, $*$, $\times$ to fix the minimal model of its endomorphism A_∞-algebra $\mathrm{E}^{\pm}_{\mathcal{B}}$. As a vector space it is spanned by the following basis elements:

- the identity morphisms $\iota_1, \ldots, \iota_k$ corresponding to the $\circ$-markings on the bands,
- degree 1 morphisms $\alpha_1, \ldots, \alpha_l$, one for each intersection point,
- degree 0 morphisms $\alpha_1^*, \ldots, \alpha_l^*$, one for each intersection point that goes in the opposite direction of the α_i,
- degree 1 morphisms $\iota_1^*, \ldots, \iota_k^*$ corresponding to the $*$-markings on the bands.

We can consider $\iota_1, \ldots, \iota_k$ as the vertices of some quiver and $\alpha_1, \ldots, \alpha_l$ as the arrows. Let $v_1, \ldots, v_k$ and $a_1, \ldots, a_l$ be the vertices and arrows of the opposite quiver and denote the latter by Q. Note that, contrary to the construction of the Koszul dual, Q only contains duals to the unstarred half of $\mathrm{E}^{\pm}_{\mathcal{B}}$. From another perspective we can also say that we killed the starred half: as an algebra we can identify $\mathbb{C}Q = (\mathrm{E}^{\pm}_{\mathcal{B}})^! / I^*$, where $I^* \lhd (\mathrm{E}^{\pm}_{\mathcal{B}})^!$ is the ideal generated by the duals of the α_i^* and the ι_j^*.

In $\widehat{\mathbb{C}Q}$ we will define the following elements:

- for each vertex $v \in Q_0$ we set

$$\ell_v = \sum_p \mu(p^\vee)(v)p,$$

- for each vertex $a \in Q_1$ we set

$$r_a = \sum_p \mu(p^\vee)(a^*)p.$$

In both cases the sum runs over all paths in Q. These elements see the unstarred part of the Koszul differential

$$\ell_v = d_\mu(v) \bmod I^* \quad \text{and} \quad r_i = d_\mu(a_i^*) \bmod I^*.$$

Theorem 12.37 (Cho–Hong–Lau) *The pair*

$$(A, \ell) := \left(\frac{\widehat{\mathbb{C}Q}}{\langle r_1, \ldots, r_k \rangle}, \sum_v \ell_v \right)$$

is a noncommutative Landau–Ginzburg model.

Sketch of the proof A neat way to prove this is by extending the A_∞-structure on $\mathrm{E}^{\pm}_{\mathcal{B}}$ to an A_∞ structure on $A \otimes_{\Bbbk} \mathrm{E}^{\pm}_{\mathcal{B}}$ by demanding that

$$\mu(p_1 \otimes \gamma_1, \dots, p_k \otimes \gamma_k) = p_k \dots p_2 p_1 \otimes \mu(\gamma_1, \dots, \gamma_k).$$

One can easily check that this definition satisfies the A_∞-axioms. Now let $b = \sum_i a_i \otimes \alpha_i$; then

$$\begin{aligned}\sum_{k \geq 1} \mu_k(b, \dots, b) &= \sum_v \ell_v \otimes \iota_v + \sum_j r_j \otimes \alpha_i^* \\ &= \sum_v \ell_v \otimes \iota_v\end{aligned}$$

because the r_i are zero in A. The A_∞-axioms together with the fact that the ι_i are strict units tell us that

$$\begin{aligned}0 &= \sum \pm \mu_k(b, \dots, \mu_l(b, \dots, b), \dots, b) \\ &= \sum \pm \mu_k\left(b, \dots, b, \sum_v \ell_v \otimes \iota_v, b, \dots, b\right) \\ &= \sum \mu_2\left(b, \sum_v \ell_v \otimes \iota_v\right) - \mu_2\left(\sum_v \ell_v \otimes \iota_v, b\right) \\ &= \sum_j (\ell_{h(a)} a_j - a_j \ell_{t(a_j)}) \otimes \alpha_j,\end{aligned}$$

so $\sum_v \ell_v$ commutes with every a_j and hence it is a central element. For further information see [154]. □

If $\mathrm{E}^{\pm}_{\mathcal{B}}$ sits inside a bigger category $\mathsf{C}^{\pm}$, we can construct an A_∞-functor

$$\mathcal{F} \colon \mathsf{C}^{\pm} \to \mathtt{MF}^{\pm}(A, \ell) \colon X \mapsto (A \otimes \mathsf{C}^{\pm}(\mathcal{B}, X), d)$$

with $dx = \mu(x) + \mu(x, b) + \mu(x, b, b) + \cdots$. This differential squares to ℓ because the A_∞-relation tells us that

$$\begin{aligned}0 &= \sum \mu(\mu(x, b, \dots, b), b, \dots, b) + \sum (x, \mu(b, \dots, b), b, \dots, b) + \cdots \\ &= d^2 x + \mu(x, \ell \otimes 1) + \mu(x, b, \dots, b, \ell \otimes 1, b, \dots, b) + \cdots \\ &= d^2 - \ell x.\end{aligned}$$

The higher parts of the functor are given by

$$\mathcal{F}_k(c_1, \dots, c_k) = \sum \mu(c_1, \dots, c_k, b, \dots, b).$$

Theorem 12.38 (Cho–Hong–Lau) *If all hom-spaces $\mathsf{C}^{\pm}(\mathcal{B}, X)$ are finite-dimensional then $\mathcal{F}_\bullet \colon \mathsf{C}^{\pm} \to \mathtt{MF}^{\pm}(A, \ell)$ is an A_∞-functor.*

Sketch of the proof The proof is similar to Theorem 12.37 and can also be found in [154]. □

Example 12.39 Consider again the pair of pants with the zigzag Lagrangian $\mathcal{Z}$ described at the beginning of Section 12.4. If we look at $\mathtt{E}^{\pm}_{\mathcal{Z}}$ then we get $\ell = -XYZ$, $r_1 = XZ - ZX$, $r_2 = ZY - YZ$, $r_3 = YX - XY$. Therefore, the Landau–Ginzburg model is

$$(\mathbb{C}[\![X, Y, Z]\!], -XYZ).$$

Pick the arc x and look at its image under the functor $\mathcal{F} : {}^{\mathrm{top}}\mathbf{Fuk}^{\pm}(\mathbb{S}, \{0, 1, \infty\}) \to \mathtt{MF}^{\pm}(\mathbb{C}[\![X, Y, Z]\!], -XYZ)$. If we deform the arc a little, we see that there are two intersection points p, q that give rise to a morphism of degree 0 and another of degree 1.

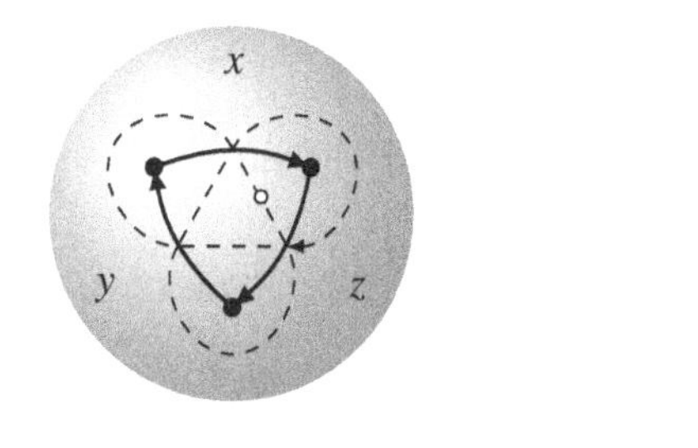

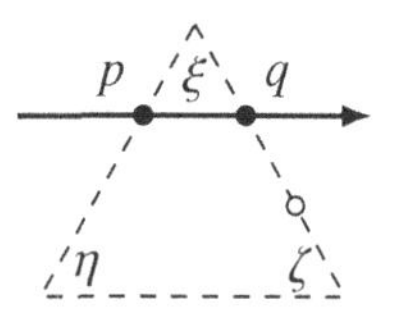

$$\mu(q, \xi) = p,$$
$$\mu(p, \eta, \zeta) = -q$$

From the picture we can read off two nontrivial products and therefore $dp = Xq$, $dq = -ZYp$, so $\mathcal{F}(x) = (\mathbb{C}[\![X, Y, Z]\!]^{\oplus 2}, \left(\begin{smallmatrix} 0 & X \\ -YZ & 0 \end{smallmatrix}\right))$. Band objects that circle around one of the marked points are mapped to zero because they will not intersect the zigzag Lagrangian.

Example 12.40 If we take the zigzag Lagrangian of the pair of pants but embed it in the sphere with three orbifold points, we get the Landau–Ginzburg model

$$(\mathbb{C}[\![X, Y, Z]\!], -XYZ + X^{2g+1} + Y^{2g+1} + Z^{2g+1}).$$

Under the mirror functor the zigzag Lagrangian itself is mapped to $(\mathbb{C}[\![X, Y, Z]\!] \otimes \Lambda_3, d)$ with

$$\begin{aligned}
d\iota &= X\xi + Y\eta + Z\zeta, \\
d\xi &= -YZ\iota + Z\eta - Y\zeta, \\
&\vdots \\
d\zeta^* &= Z\iota^* - Y\xi + X\eta, \\
d\iota^* &= (X^{2g} - YZ)\xi^* + (Y^{2g} - XZ)\eta^* + (Z^{2g} - XY)\zeta^*.
\end{aligned}$$

This is precisely $(\Omega, \delta + \gamma\wedge)$. Because the mirror functor is compatible with the cone construction and both $\mathcal{Z}$ and $(\Omega, \delta + \gamma\wedge)$ are generators, the mirror functor is an equivalence in this case.

Example 12.41 (Zigzag Lagrangians) Take a closed marked surface $(\mathbb{S}, M)$ with genus g and n internal marked points and an arc collection $\mathcal{A}$ that forms a dimer whose mirror we will denote by Q, so $\mathcal{A} = Q^\vee$. Following [154, Section 10] we will also need a sign map $\rho\colon (Q_{\mathcal{A}})_1 \to \{+1, -1\}$ such that every clockwise polygon with an odd number of corners contains precisely one angle arrow with a negative sign and all other angle arrows have a positive sign.

Start with an arc and continue in a zigzag fashion until the path repeats itself.

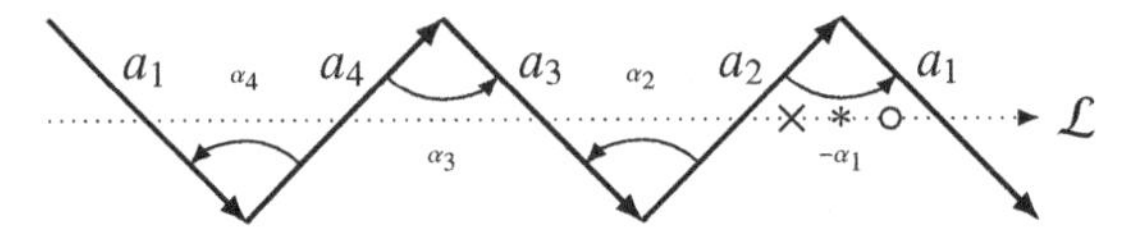

If we connect the midpoints of the arcs together, we end up with a closed curve. For each such curve we construct the complex

$$\mathcal{Z}_i := (a_1 \oplus \cdots \oplus a_k, \rho(\alpha_1)\alpha_1 + \cdots + \rho(\alpha_k)\alpha_k) \in {}^{\mathtt{top}}\mathtt{Fuk}^{\pm}(\mathbb{S}, M).$$

The *zigzag skeleton* $\mathcal{Z}$ is the direct sum of all the $\mathcal{Z}_i$. For each zigzag cycle we will also fix a marked arc a_1 and a marked angle α_1 for the construction of the minimal model. In accordance with Remark 9.31, we will put $\circ$, $*$ on the curve just behind a_1. We will also place a $\times$ for each α with $\rho(\alpha) = -1$.

Theorem 12.42 (Cho–Hong–Lau) *The Landau–Ginzburg model associated to $\mathcal{Z}$ consists of the completed Jacobi algebra of the dual dimer and the central element*

$$(\widehat{\mathtt{Jac}\,Q}, \ell).$$

Sketch of the proof Every every arc a is contained in precisely two polygons $a_1 \dots a_k$ and $b_1 \dots b_l$ with $b_1 = a_1 = a$; this gives rise to two nonzero μ's,

$$\mu(\alpha_2, \dots, \alpha_k) = -\mu(\beta_2, \dots, \beta_k) = \alpha^*,$$

with opposite sign because of the specific choice of ρ. Dually, $r_a = a_k \dots a_2 - b_l \dots b_2$, which are precisely the relations in the Jacobi algebra of the dual dimer in Definition 9.58. To determine ℓ it is sufficient to note that each $\circ$ sits in precisely one polygon, so ℓ is the sum of one polygonal cycle for each vertex, just like in Lemma 9.59. For further details see [154]. □

As we have seen in the example of the pair of pants, the mirror functor

$$\mathcal{F}\colon {}^{\mathtt{top}}\mathtt{Fuk}^{\pm}(\mathbb{S}, M) \to \mathtt{MF}^{\pm}(\widehat{\mathtt{Jac}\,Q}, \ell)$$

can never be an equivalence because the small circles around the marked points are disjoint from the zigzag Lagrangians, so they get killed. If we fill the marked points in the surface, these small circles will become obstructed so that they are also killed in the deformation process. This is what happens in the case of the pair of pants:

$$\mathcal{F}\colon {}^{\mathtt{top}}\mathtt{Fuk}^{\pm}(\mathbb{S}, M) \to \mathtt{MF}^{\pm}(\mathbb{C}[\![X,Y,Z]\!], -XYZ)$$

is not an equivalence, but if we fill the three marked points with orbifold points of order $r = 2g + 1$ then

$$\mathcal{F}\colon \mathtt{Fuk}^{\pm}(\mathbb{S}_{rrr}, M) \to \mathtt{MF}^{\pm}(\mathbb{C}[\![X,Y,Z]\!], -XYZ + X^{2g+1} + Y^{2g+1} + Z^{2g+1})$$

is an equivalence. In [154] several examples are worked out, but no general statement about when the mirror functor is an equivalence is known. There are however two main conditions to make the construction work as expected.

- The dimer Q has to be zigzag consistent because this is necessary for $\widehat{\mathtt{Jac}\,Q}$ to have nice homological properties. In that case, $\mathrm{Ext}^{\bullet}_{\widehat{\mathtt{Jac}\,Q}}(S,S)$ is spanned by the ι, ι^*, α, α^*. If Q is not consistent then the Ext-algebra will usually be infinite-dimensional.
- The dual dimer $Q^{\vee}$ also has to be zigzag consistent. In that case, the homology of the endomorphism ring of $\mathcal{Z}$ in the Fukaya category is spanned by the $\iota, \iota^*, \alpha, \alpha^*$. If $Q^{\vee}$ is not consistent then the monogons will cause obstructions and the digons a nontrivial differential.

In the doubly consistent case we conjecture that the mirror functor is an equivalence for closed surfaces. If $g > 1$ we can work over $\mathbb{C}$, but if $g = 1$ we will have to work over $\mathbb{C}(\!(\hbar)\!)$. The doubly consistent dimers for $g = 1$ have been classified [34, 121]. Their Jacobi algebras come from tilting bundles over weak toric del Pezzo surfaces and filling the marked points gives rise to nice noncommutative Sklyanin-type deformations of these surfaces [154, 178].

12.7 Exercises

Exercise 12.1 Let Q be a connected quiver without oriented cycles. Show that the classical Hochschild cohomology has the following form:

- $\mathsf{HH}^0_{\mathrm{clas}}(\mathbb{C}Q) = \mathbb{C}$,
- $\mathsf{HH}^1_{\mathrm{clas}}(\mathbb{C}Q) = \mathbb{C}^{1-\#Q_0+\sum_{a\in Q_1} p(a)}$, where $p(a)$ is the number of paths from $t(a)$ to $h(a)$,
- $\mathsf{HH}^{\geq 2}_{\mathrm{clas}}(\mathbb{C}Q) = 0$.

Exercise 12.2 Consider $\Pi = \mathbb{C}[X, Y] \star G$ with $G = \langle g \mid g^n \rangle$ and $gX = e^{2\pi i/n}$, $gY = e^{2\pi i/n}$.

(i) Show that we can write Π as the path algebra of the quiver

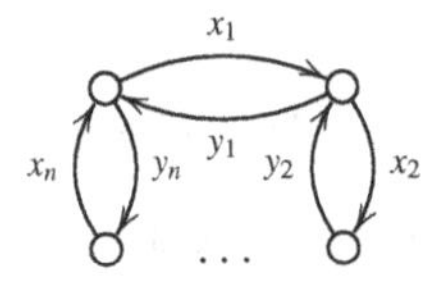

with relations $x_i y_i - y_{i+1} x_{i+1} = 0$.

(ii) Calculate the Hochschild cohomology of Π.

(iii) Choose $\lambda_i \in (\epsilon)$ and deform the relations to $x_{i+1} y_{i+1} - y_i x_i = \lambda_i$. Show that Π_λ can be seen as a deformation of Π and identify its cohomology class.

Exercise 12.3 Calculate the Hochschild cohomology of the topological Fukaya category of a disk with k points on the boundary and l internal marked points.

Exercise 12.4 Let $\mathbb{X}$ be a variety. There are three ways to determine the Hochschild cohomology of $\mathtt{D}\,\mathsf{Coh}^\bullet\,\mathbb{X}$.

(i) Take the generator $\mathscr{E}$ and calculate the Hochschild cohomology of the A_∞-algebra $\mathtt{E}^\bullet = \mathsf{Coh}^\bullet\,\mathbb{X}(\mathscr{E}, \mathscr{E})$.

(ii) Another characterization is that it is the Ext-ring of the structure sheaf of the diagonal in $\mathbb{X} \times \mathbb{X}$,

$$\mathsf{HH} = \mathsf{Coh}^\bullet\,\mathbb{X} \times \mathbb{X}(\mathscr{D}, \mathscr{D}),$$

with $\mathscr{D}(\mathbb{U} \times \mathbb{V}) = \mathscr{O}(\mathbb{U} \times \mathbb{V}) / \langle f \mid \forall\, x \in \mathbb{U} \cap \mathbb{V},\ f(x, x) = 0 \rangle$.

(iii) The Hochschild–Kostant–Rosenberg theorem says that it is also equal to the global sections of the sheaf of polyvector fields

$$\mathsf{HH}^i = \Gamma(\wedge^i \mathscr{T}).$$

Show that these three versions coincide for $\mathbb{P}^1$.

Exercise 12.5 Calculate the Hochschild cohomology of $\mathtt{MF}^\pm(\mathbb{C}^3, XYZ)$.

Exercise 12.6 Work out the mirror symmetry for the genus 3 curve by deforming the orbifold

$$\mathtt{MF}^\pm([\mathbb{C}^3/G], XYZ)$$

with $G = \langle g \mid g^7 \rangle$ and $g(X, Y, Z) = (e^{2\pi i/7} X, e^{4\pi i/7} Y, e^{8\pi i/7} Z)$.

Exercise 12.7 Find expressions for the power series that appear in the deformed binary product for the Fukaya category of the three-punctured torus in Example 12.28.

Exercise 12.8 Let $(\mathbb{T}, M)$ be the one-punctured torus and let $\mathcal{L}_0$, $\mathcal{L}_1$ be two curves that split the torus into a square with the puncture in the middle. Equip both with nontrivial spin structures. The endomorphism ring in the relative Fukaya category of $\mathcal{L}_0 \oplus \mathcal{L}_1$ is four-dimensional. It is spanned by the identity morphisms and two morphism $p\colon \mathcal{L}_0 \to \mathcal{L}_1$ and $p^*\colon \mathcal{L}_1 \to \mathcal{L}_1$. Calculate

$$\mu_3(p, p^*, p)$$

as a power series in the deformation parameter $\hbar$ for the puncture (see also [162]).

Exercise 12.9 Work out the mirror symmetry for the genus 1 curve by deforming the orbifold

$$\mathtt{MF}^{\pm}([\mathbb{C}^3/G], XYZ)$$

with $G = \langle g \mid g^4 \rangle$ and $g(X, Y, Z) = (iX, iY, -Z)$.

Exercise 12.10 Calculate the Ext-A_∞-algebra of the simple representation of the Sklyanin algebra

$$S = \mathtt{Skl}_{\alpha\beta\gamma}/\langle X, Y, Z\rangle.$$

You may assume that α, β, γ are generic such that $\mathrm{Ext}^{\geq 3}_{\mathtt{Skl}_{\alpha\beta\gamma}}(S, S) = 0$.

References

[1] M. Abouzaid, D. Auroux, A. I. Efimov, L. Katzarkov and D. Orlov, Homological mirror symmetry for punctured spheres. *J. Amer. Math. Soc.*, **26** (2013), 1051–1083.

[2] M. Abouzaid, D. Auroux and L. Katzarkov, Lagrangian fibrations on blowups of toric varieties and mirror symmetry for hypersurfaces. *Publ. Math. Inst. Hautes Études Sci.*, **123** (2016), 199–282.

[3] I. Assem, T. Brüstle, G. Charbonneau-Jodoin and P.-G. Plamondon, Gentle algebras arising from surface triangulations. *Algebra Number Theory*, **4** (2010), 201–229.

[4] M. Abouzaid, On the Fukaya categories of higher genus surfaces. *Adv. Math.*, **217** (2008), 1192–1235.

[5] M. Abouzaid, Morse homology, tropical geometry, and homological mirror symmetry for toric varieties. *Selecta Math. (N.S.)*, **15** (2009), 189–270.

[6] M. Abouzaid, A geometric criterion for generating the Fukaya category. *Publ. Math. Inst. Hautes Études Sci.*, **112** (2010), 191–240.

[7] M. Abouzaid, A cotangent fibre generates the Fukaya category. *Adv. Math.*, **228** (2011), 894–939.

[8] L. Abrams, Two-dimensional topological quantum field theories and Frobenius algebras. *J. Knot Theory Ramifications*, **5** (1996), 569–587.

[9] F. W. Anderson and K. R. Fuller, *Rings and Categories of Modules*, Graduate Texts in Mathematics, vol. 13. Springer, New York-Heidelberg, 1974.

[10] L. Angeleri Hügel, D. Happel and H. Krause, eds, *Handbook of Tilting Theory*, London Mathematical Society Lecture Note Series, vol. 332. Cambridge University Press, Cambridge, 2007.

[11] D. Auroux, L. Katzarkov and D. Orlov, Mirror symmetry for del Pezzo surfaces: vanishing cycles and coherent sheaves. *Invent. Math.*, **166** (2006), 537–582.

[12] V. I. Arnol'd, On a characteristic class entering into conditions of quantization. *Funkcional. Anal. i Priložen.*, **1** (1967), 1–14.

[13] M. Auslander, I. Reiten and S. O. Smalø, *Representation Theory of Artin Algebras*, Cambridge Studies in Advanced Mathematics, vol. 36. Cambridge University Press, Cambridge, 1997. Corrected reprint of the 1995 original.

[14] I. Assem and A. Skowroński, Iterated tilted algebras of type $\tilde{\mathbf{A}}_n$. *Math. Z.*, **195** (1987), 269–290.

[15] M. Abouzaid and P. Seidel, An open string analogue of Viterbo functoriality. *Geom. Topol.*, **14** (2010), 627–718.

[16] P. S. Aspinwall, D-branes on Calabi-Yau manifolds. In *Progress in String Theory*. World Sci. Publ., Hackensack, NJ, 2005, pp. 1–152.

[17] M. Artin, J. Tate and M. Van den Bergh, Some algebras associated to automorphisms of elliptic curves. In *The Grothendieck Festschrift, Vol. I*, Progress in Mathematics, vol. 86. Birkhäuser Boston, Boston, MA, 1990, pp. 33–85.

[18] M. Atiyah, Topological quantum field theories. *Publ. Math., Inst. Hautes Étud. Sci.*, **68** (1988), 175–186.

[19] D. Auroux, Mirror symmetry and T-duality in the complement of an anticanonical divisor. *J. Gökova Geom. Topol. GGT*, **1** (2007), 51–91.

[20] D. Auroux, A beginner's introduction to Fukaya categories. In *Contact and Symplectic Topology*, Bolyai Society Mathematical Studies, vol. 26. János Bolyai Math. Soc., Budapest, 2014, pp. 85–136.

[21] M. J. Bardzell, The alternating syzygy behavior of monomial algebras. *J. Algebra*, **188** (1997), 69–89.

[22] V. V. Batyrev, Dual polyhedra and mirror symmetry for Calabi-Yau hypersurfaces in toric varieties. *J. Algebraic Geom.*, **3** (1994), 493–535.

[23] V. V. Batyrev and L. A. Borisov, Mirror duality and string-theoretic Hodge numbers. *Invent. Math.*, **126** (1996), 183–203.

[24] V. V. Batyrev and L. A. Borisov, On Calabi-Yau complete intersections in toric varieties. In *Higher-Dimensional Complex Varieties (Trento, 1994)*. De Gruyter, Berlin, 1996, pp. 39–65.

[25] V. V. Batyrev, I. Ciocan-Fontanine, B. Kim and D. van Straten, Conifold transitions and mirror symmetry for Calabi-Yau complete intersections in Grassmannians. *Nuclear Phys. B*, **514** (1998), 640–666.

[26] A. A. Beilinson, The derived category of coherent sheaves on $\mathbf{P}^n$. *Selecta Math. Soviet.*, **3** (1983/84), 233–237. Selected translations.

[27] A. Beilinson, V. Ginzburg and W. Soergel, Koszul duality patterns in representation theory. *J. Amer. Math. Soc.*, **9** (1996), 473–527.

[28] A. Banyaga and D. Hurtubise, *Lectures on Morse Homology*, Kluwer Texts in the Mathematical Sciences, vol. 29. Kluwer Academic Publishers Group, Dordrecht, 2004.

[29] M. Bender and S. Mozgovoy, Crepant resolutions and brane tilings II: Tilting bundles. arXiv:0909.2013, 2009.

[30] A. Bayer, E. Macrìand P. Stellari, The space of stability conditions on abelian threefolds, and on some Calabi-Yau threefolds. *Invent. Math.*, **206** (2016), 869–933.

[31] M. Bökstedt and A. Neeman, Homotopy limits in triangulated categories. *Comp. Math.*, **86** (1993), 209–234.

[32] A. Bondal and D. Orlov, Reconstruction of a variety from the derived category and groups of autoequivalences. *Comp. Math.*, **125** (2001), 327–344.

[33] R. Bocklandt, Consistency conditions for dimer models. *Glasg. Math. J.*, **54** (2012), 429–447.

[34] R. Bocklandt, Generating toric noncommutative crepant resolutions. *J. Algebra*, **364** (2012), 119–147.

[35] R. Bocklandt, A dimer ABC. *Bull. Lond. Math. Soc.*, **48** (2016), 387–451.

[36] R. Bocklandt, Noncommutative mirror symmetry for punctured surfaces. *Trans. Amer. Math. Soc.*, **368** (2016), 429–469. With an appendix by Mohammed Abouzaid.

[37] R. Bocklandt, Strebel differentials and stable matrix factorizations. arXiv:1612.06800, 2016.

[38] A. Bondal, Derived categories of toric varieties. In *Convex and Algebraic Geometry*, Oberwolfach Conference Reports, vol. 3. EMS Publishing House, Zürich, 2006, pp. 284–286.

[39] T. Bridgeland, Stability conditions on triangulated categories. *Ann. of Math. (2)*, **166** (2007), 317–345.

[40] T. Bridgeland, Stability conditions on $K3$ surfaces. *Duke Math. J.*, **141** (2008), 241–291.

[41] M. Brion, Representations of quivers. In *Geometric Methods in Representation Theory. I*, Séminaires et Congrès, vol. 24. Soc. Math. France, Paris, 2012, pp. 103–144.

[42] N. Broomhead, Dimer models and Calabi-Yau algebras. *Mem. Amer. Math. Soc.*, **211** (2012).

[43] P. Balmer and M. Schlichting, Idempotent completion of triangulated categories. *J. Algebra*, **236** (2001), 819–834.

[44] T. Bridgeland and I. Smith, Quadratic differentials as stability conditions. *Publ. Math. Inst. Hautes Études Sci.*, **121** (2015), 155–278.

[45] R. Bott and L. W. Tu, *Differential Forms in Algebraic Topology*, Graduate Texts in Mathematics, vol. 82. Springer, New York-Berlin, 1982.

[46] R.-O. Buchweitz, Maximal Cohen-Macaulay modules and Tate-cohomology over Gorenstein rings. University of Hannover, 1987. `https://tspace.library.utoronto.ca/bitstream/1807/16682/1/maximal_cohen-macaulay_modules_1986.pdf`

[47] W. Crawley-Boevey, More lectures on representations of quivers. Preprint, University of Leeds, 1992. `http://ambio1.leeds.ac.uk/pure/staff/crawley_b/morequivlecs.pdf`

[48] W. Crawley-Boevey, DMV lectures on representations of quivers, preprojective algebras and deformations of quotient singularities. Preprint, University of Leeds, 1999. `www.math.uni-bielefeld.de/~wcrawley/dmvlecs.pdf`

[49] T. Coates, A. Corti, S. Galkin, V. Golyshev and A. Kasprzyk, Mirror symmetry and Fano manifolds. In *European Congress of Mathematics*. Eur. Math. Soc., Zürich, 2013, pp. 285–300.

[50] A. Caldararu, K. Costello and J. Tu, Categorical enumerative invariants, I: String vertices. arXiv:2009.06673, 2020.

[51] P. Candelas, X. C. de la Ossa, P. S. Green and L. Parkes, A pair of Calabi-Yau manifolds as an exactly soluble superconformal theory. *Nuclear Phys. B*, **359** (1991), 21–74.

[52] A. Cannas da Silva, *Lectures on Symplectic Geometry*, Lecture Notes in Mathematics, vol. 1764. Springer, Berlin, 2001.

[53] K. Cieliebak, A. Floer and H. Hofer, Symplectic homology. II. A general construction. *Math. Z.*, **218** (1995), 103–122.

[54] B. Chantraine, An introduction to Fukaya categories. Preprint, University of Nantes, 2014. `www.math.sciences.univ-nantes.fr/~chantraine-b/en/FukayaCategory.pdf`

[55] C.-H. Cho, H. Hong and S.-C. Lau, Localized mirror functor for Lagrangian immersions, and homological mirror symmetry for $\mathbb{P}^1_{a,b,c}$. *J. Differential Geom.*, **106** (2017), 45–126.

[56] D. A. Cox and S. Katz, *Mirror Symmetry and Algebraic Geometry*, Mathematical Surveys and Monographs, vol. 68. American Mathematical Society, Providence, RI, 1999.

[57] D. A. Cox, J. B. Little and H. K. Schenck, *Toric Varieties*, Graduate Studies in Mathematics, vol. 124. American Mathematical Society, Providence, RI, 2011.

[58] K. Costello, Topological conformal field theories and Calabi-Yau categories. *Adv. Math.*, **210** (2007), 165–214.

[59] D. A. Cox, Mirror symmetry and polar duality of polytopes. *Symmetry*, **7** (2015), 1633–1645.

[60] A. Căldăraru and J. Tu, Curved A_∞ algebras and Landau-Ginzburg models. *New York J. Math.*, **19** (2013), 305–342.

[61] A. Caldararu and J. Tu, Computing a categorical Gromov-Witten invariant. arXiv:1706.09912, 2019.

[62] A. Collinucci and T. Wyder, Introduction to topological string theory. Report No. KUL-TF-07/24, University of Leuven, 2007.

[63] B. Davison, Consistency conditions for brane tilings. *J. Algebra*, **338** (2011), 1–23.

[64] P. Deligne, P. Etingof, D. S. Freed, L. C. Jeffrey, D. Kazhdan, J. W. Morgan, D. R. Morrison and E. Witten, eds, *Quantum Fields and Strings: A Course for Mathematicians. Vols. 1, 2.* American Mathematical Society, Providence, RI; Institute for Advanced Study (IAS), Princeton, NJ, 1999. Material from the Special Year on Quantum Field Theory held at the Institute for Advanced Study, Princeton, NJ, 1996–1997.

[65] T. Dyckerhoff and M. Kapranov, Triangulated surfaces in triangulated categories. *J. Eur. Math. Soc. (JEMS)*, **20** (2018), 1473–1524.

[66] P. Dolbeault, Sur la cohomologie des variétés analytiques complexes. *C. R. Acad. Sci. Paris*, **236** (1953), 175–177.

[67] V. Drinfeld, DG quotients of DG categories. *J. Algebra*, **272** (2004), 643–91.

[68] D. Dugger, A primer on homotopy colimits. Preprint, University of Oregon, 2008. `https://pages.uoregon.edu/ddugger/hocolim.pdf`

[69] T. Dyckerhoff, Compact generators in categories of matrix factorizations. *Duke Math. J.*, **159** (2011), 223–274.

[70] A. I. Efimov, A proof of the Kontsevich-Soibelman conjecture. *Mat. Sb.*, **202** (2011), 65–84.

[71] A. I. Efimov, Homological mirror symmetry for curves of higher genus. *Adv. Math.*, **230** (2012), 493–530.

[72] D. Eisenbud and J. Harris, *The Geometry of Schemes*, Graduate Texts in Mathematics, vol. 197. Springer, New York, 2000.

[73] D. Eisenbud, Homological algebra on a complete intersection, with an application to group representations. *Trans. Amer. Math. Soc.*, **260** (1980), 35–64.

[74] D. Eisenbud, *Commutative Algebra*, Graduate Texts in Mathematics, vol. 150. Springer, New York, 1995. With a view toward algebraic geometry.

[75] G. Ellingsrud and S. A. Strømme, Bott's formula and enumerative geometry. *J. Amer. Math. Soc.*, **9** (1996), 175–193.

[76] B. Fantechi, Stacks for everybody. In *European Congress of Mathematics, Vol. I (Barcelona, 2000)*, Progress in Mathematics, vol. 201. Birkhäuser, Basel, 2001, pp. 349–359.

[77] B. Fang, Homological mirror symmetry is T-duality for $\mathbb{P}^n$. *Commun. Number Theory Phys.*, **2** (2008), 719–742.

[78] A. Floer, H. Hofer and K. Wysocki, Applications of symplectic homology. I. *Math. Z.*, **217** (1994), 577–606.

[79] A. Floer, Morse theory for Lagrangian intersections. *J. Differential Geom.*, **28** (1988), 513–547.

[80] B. Fang, C.-C. M. Liu, D. Treumann and E. Zaslow, T-duality and equivariant homological mirror symmetry for toric varieties. arXiv:0811.1228, 2008.

[81] B. Fang, C.-C. M. Liu, D. Treumann and E. Zaslow, The coherent-constructible correspondence and homological mirror symmetry for toric varieties. In *Geometry and Analysis. No. 2*, Advanced Lectures in Mathematics, vol. 18. Int. Press, Somerville, MA, 2011, pp. 3–37.

[82] K. Fukaya, Y.-G. Oh, H. Ohta and K. Ono, Canonical models of filtered A_∞-algebras and Morse complexes. In *New Perspectives and Challenges in Symplectic Field Theory*, CRM Proceedings and Lecture Notes, vol. 49. American Mathematical Society, Providence, RI, 2009, pp. 201–227.

[83] K. Fukaya, Y.-G. Oh, H. Ohta and K. Ono, *Lagrangian Intersection Floer Theory: Anomaly and Obstruction. Part I*, AMS/IP Studies in Advanced Mathematics, vol. 46. American Mathematical Society, Providence, RI; International Press, Somerville, MA, 2009.

[84] K. Fukaya, Y.-G. Oh, H. Ohta and K. Ono, *Lagrangian Intersection Floer Theory: Anomaly and Obstruction. Part II*, AMS/IP Studies in Advanced Mathematics, vol. 46. American Mathematical Society, Providence, RI; International Press, Somerville, MA, 2009.

[85] W. Fulton and R. Pandharipande, Notes on stable maps and quantum cohomology. In *Algebraic Geometry—Santa Cruz 1995*, Proceedings of Symposia in Pure Mathematics, vol. 62. American Mathematical Society, Providence, RI, 1997, pp. 45–96.

[86] M. Forsberg, M. Passare and A. Tsikh, Laurent determinants and arrangements of hyperplane amoebas. *Adv. Math.*, **151** (2000), 45–70.

[87] G. Friedman, Survey article: An elementary illustrated introduction to simplicial sets. *Rocky Mountain J. Math.*, **42** (2012), 353–423.

[88] K. R. Fuller, Biserial rings. In *Ring Theory (Proc. Conf., Univ. Waterloo, Waterloo, 1978)*, Lecture Notes in Mathematics, vol. 734. Springer, Berlin, 1979, pp. 64–90.

[89] W. Fulton, *Introduction to Toric Varieties*, Annals of Mathematics Studies, vol. 131. Princeton University Press, Princeton, NJ, 1993. The William H. Roever Lectures in Geometry.

[90] E. Getzler, Batalin-Vilkovisky algebras and two-dimensional topological field theories. *Comm. Math. Phys.*, **159** (1994), 265–285.

[91] P. Griffiths and J. Harris, *Principles of Algebraic Geometry*. Wiley Classics Library. John Wiley & Sons, New York, 1994. Reprint of the 1978 original.

[92] A. B. Givental, Equivariant Gromov-Witten invariants. *Int. Math. Res. Not. IMRN*, **2015** (1996), 613–663.

[93] E. Getzler and J. D. S. Jones, A_∞-algebras and the cyclic bar complex. *Illinois J. Math.*, **34** (1990), 256–283.

[94] B. Gammage and D. Nadler, Mirror symmetry for honeycombs. *Trans. Amer. Math. Soc.*, **373** (2020), 71–107.

[95] S. Ganatra, J. Pardon and V. Shende, Microlocal Morse theory of wrapped Fukaya categories. arXiv:1809.08807, 2018.

[96] S. Ganatra, J. Pardon and V. Shende, Structural results in wrapped Floer theory. arXiv:1809.03427, 2018.

[97] P. Gabriel and A. V. Roĭter, *Representations of Finite-Dimensional Algebras*, Encyclopaedia of Mathematical Sciences, vol. 73. Springer, Berlin, 1992. With a chapter by B. Keller.

[98] A. Grothendieck, Sur la classification des fibrés holomorphes sur la sphère de Riemann. *Amer. J. Math.*, **79** (1957), 121–138.

[99] A. Grothendieck, Groupes de classes des catégories abéliennes et triangulées. Complexes parfaits. In *Séminaire de Géométrie Algébrique du Bois-Marie 1965–66 SGA 5*. Springer, 1977, pp. 351–371.

[100] M. Gromov, Pseudo holomorphic curves in symplectic manifolds. *Invent. Math.*, **82** (1985), 307–347.

[101] M. Gross, Special Lagrangian fibrations. I. Topology. In *Integrable Systems and Algebraic Geometry (Kobe/Kyoto, 1997)*. World Sci. Publ., River Edge, NJ, 1998, pp. 156–193.

[102] M. Gross, Examples of special Lagrangian fibrations. In *Symplectic Geometry and Mirror Symmetry (Seoul, 2000)*. World Sci. Publ., River Edge, NJ, 2001, pp. 81–109.

[103] M. Gross, Special Lagrangian fibrations. II. Geometry. A survey of techniques in the study of special Lagrangian fibrations. In *Winter School on Mirror Symmetry, Vector Bundles and Lagrangian Submanifolds (Cambridge, MA, 1999)*, AMS/IP Studies in Advanced Mathematics, vol. 23. American Mathematical Society, Providence, RI, 2001, pp. 95–150.

[104] M. Gross and B. Siebert, Mirror symmetry via logarithmic degeneration data. I. *J. Differential Geom.*, **72** (2006), 169–338.

[105] M. Gross and B. Siebert, Mirror symmetry via logarithmic degeneration data, II. *J. Algebraic Geom.*, **19** (2010), 679–780.

[106] I. M. Gelfand, A. V. Zelevinskiĭ and M. M. Kapranov, A-discriminants and Cayley-Koszul complexes. *Dokl. Akad. Nauk SSSR*, 307 (1989), 1307–1311.

[107] M. Habermann, Homological mirror symmetry for invertible polynomials in two variables. arXiv:2003.01106, 2020.

[108] D. Happel, *Triangulated Categories in the Representation Theory of Finite-Dimensional Algebras*, London Mathematical Society Lecture Note Series, vol. 119. Cambridge University Press, Cambridge, 1988.

[109] R. Hartshorne, *Algebraic Geometry*, Graduate Texts in Mathematics, vol. 52. Springer, New York-Heidelberg, 1977.

[110] A. Hatcher, On triangulations of surfaces. *Topology Appl.*, **40** (1991), 189–194.

[111] A. Hatcher, *Algebraic Topology*. Cambridge University Press, Cambridge, 2002.

[112] L. Hille and J. A. de la Peña, Stable representations of quivers. *J. Pure Appl. Algebra*, **172** (2002), 205–224.

[113] N. Hitchin, Lectures on special Lagrangian submanifolds. In *Winter School on Mirror Symmetry, Vector Bundles and Lagrangian Submanifolds (Cambridge, MA, 1999)*, AMS/IP Studies in Advanced Mathematics, vol. 23. American Mathematical Society, Providence, RI, 2001, pp. 151–182.

[114] K. Hori, S. Katz, A. Klemm, R. Pandharipande, R. Thomas, C. Vafa, R. Vakil and E. Zaslow, *Mirror Symmetry*, Clay Mathematics Monographs, vol. 1. American Mathematical Society, Providence, RI; Clay Mathematics Institute, Cambridge, MA, 2003. With a preface by Vafa.

[115] F. Haiden, L. Katzarkov and M. Kontsevich, Flat surfaces and stability structures. *Publ. Math. Inst. Hautes Études Sci.*, **126** (2017), 247–318.

[116] G. Hochschild, B. Kostant and A. Rosenberg, Differential forms on regular affine algebras. *Trans. Amer. Math. Soc.*, **102** (1962), 383–408.

[117] D. Huybrechts and M. Lehn, *The Geometry of Moduli Spaces of Sheaves*, 2nd ed., Cambridge Mathematical Library. Cambridge University Press, Cambridge, 2010.

[118] M. Hazewinkel and C. F Martin, A short elementary proof of Grothendieck's theorem on algebraic vectorbundles over the projective line. *J. Pure Appl. Algebra*, **25** (1982), 207–211.

[119] G. Hochschild, On the cohomology groups of an associative algebra. *Ann. of Math. (2)*, **46** (1945), 58–67.

[120] L. Hille and M. Perling, Tilting bundles on rational surfaces and quasi-hereditary algebras. *Ann. Inst. Fourier (Grenoble)*, **64** (2014), 625–644.

[121] A. Hanany and R.-K. Seong, Brane tilings and specular duality. *J. High Energy Phys.*, **2012** (2012), 107.

[122] M. Henneaux and C. Teitelboim, BRST cohomology in classical mechanics. *Comm. Math. Phys.*, **115** (1988), 213–230.

[123] M. Hutchings, Lecture notes on Morse homology (with an eye towards Floer theory and pseudoholomorphic curves). Preprint, Berkeley, 2002. `https://math.berkeley.edu/~hutching/teach/276-2010/mfp.ps`

[124] D. Huybrechts, *Fourier-Mukai Transforms in Algebraic Geometry*. Oxford Mathematical Monographs. Clarendon Press, Oxford, 2006.

[125] A. Ishii and K. Ueda, On moduli spaces of quiver representations associated with dimer models. In *Higher Dimensional Algebraic Varieties and Vector Bundles*, RIMS Kôkyûroku Bessatsu, B9. Res. Inst. Math. Sci. (RIMS), Kyoto, 2008, pp. 127–141.

[126] A. Ishii and K. Ueda, A note on consistency conditions on dimer models. In *Higher Dimensional Algebraic Geometry*, RIMS Kôkyûroku Bessatsu, B24. Res. Inst. Math. Sci. (RIMS), Kyoto, 2011, pp. 143–164.

[127] A. Ishii and K. Ueda, Dimer models and the special McKay correspondence. *Geom. Topol.*, **19** (2015), 3405–3466.

[128] D. Joyce, Lectures on Calabi-Yau and special Lagrangian geometry. arXiv:math/0108088, 2001.

[129] D. Joyce, Singularities of special Lagrangian fibrations and the SYZ conjecture. *Comm. Anal. Geom.*, **11** (2003), 859–907.

[130] D. D. Joyce, *Riemannian Holonomy Groups and Calibrated Geometry*, Oxford Graduate Texts in Mathematics, vol. 12. Oxford University Press, Oxford, 2007.

[131] T. V. Kadeishvili, The algebraic structure in the homology of an $A(\infty)$-algebra. *Soobshch. Akad. Nauk Gruzin. SSR*, **108** (1982, 1983), 249–252.

[132] P. W. Kasteleyn, Dimer statistics and phase transitions. *J. Mathematical Phys.*, **4** (1963), 287–293.

[133] B. Keller, Introduction to A-infinity algebras and modules. *Homology Homotopy Appl.*, **3** (2001), 1–35.

[134] B. Keller, Derived invariance of higher structures on the Hochschild complex. Preprint, 2003. `https://webusers.imj-prg.fr/~bernhard.keller/publ/dih.pdf`

[135] B. Keller, A-infinity algebras, modules and functor categories. In *Trends in Representation Theory of Algebras and Related Topics*, Contemporary Mathematics, vol. 406. American Mathematical Society, Providence, RI, 2006, pp. 67–93.

[136] B. Keller, Derived categories and tilting. In *Handbook of Tilting Theory*, London Mathematical Society Lecture Note Series, vol. 332. Cambridge University Press, Cambridge, 2007, pp. 49–104.

[137] R. Kenyon, An introduction to the dimer model. In *School and Conference on Probability Theory*, ICTP Lecture Notes, XVII. Abdus Salam Int. Cent. Theoret. Phys., Trieste, 2004, pp. 267–304.

[138] A. D. King, Moduli of representations of finite-dimensional algebras. *Quart. J. Math. Oxford Ser. (2)*, **45** (1994), 515–530.

[139] A. Kapustin, L. Katzarkov, D. Orlov and M. Yotov, Homological mirror symmetry for manifolds of general type. *Cent. Eur. J. Math.*, **7** (2009), 571–605.

[140] H. Knörrer, Cohen-Macaulay modules on hypersurface singularities I. *Invent. Math.*, **88** (1987), 153–164.

[141] M. Kontsevich, Homological algebra of mirror symmetry. In *Proceedings of the International Congress of Mathematicians, Vol. 1, 2 (Zürich, 1994)*. Birkhäuser, Basel, 1995, pp. 120–139.

[142] M. Kontsevich, Triangulated categories and geometry. Course at the École Normale Supérieure, Paris, notes taken by J. Bellaıche, 1998.

[143] M. Kontsevich, Deformation quantization of Poisson manifolds. *Lett. Math. Phys.*, **66** (2003), 157–216.

[144] M. Kontsevich, Symplectic geometry of homological algebra. Preprint, IHES, 2009. `www.ihes.fr/~/maxim/TEXTS/Symplectic_AT2009.pdf`

[145] M. Kontsevich and A. L. Rosenberg, Noncommutative smooth spaces. In *The Gelfand Mathematical Seminars, 1996–1999*, Gelfand Math. Sem. Birkhäuser Boston, Boston, MA, 2000, pp. 85–108.

[146] M. Kontsevich and Y. Soibelman, Homological mirror symmetry and torus fibrations. In *Symplectic Geometry and Mirror Symmetry (Seoul, 2000)*. World Sci. Publ., River Edge, NJ, 2001, pp. 203–263.

[147] M. Kontsevich and Y. Soibelman, Deformation theory. *Livre en préparation*, 2002. https://people.maths.ox.ac.uk/beem/papers/kontsevich_soibelman_deformation_theory_1.pdf

[148] M. Kashiwara and P. Schapira, Categories and sheaves. Grundlehren der Mathematischen Wissenschaften [Fundamental Principles of Mathematical Sciences], vol. 332. Springer, Berlin, 2006.

[149] M. Kontsevich and Y. Soibelman, Stability structures, motivic Donaldson-Thomas invariants and cluster transformations. arXiv:0811.2435, 2008.

[150] M. Kontsevich and Y. Soibelman, Notes on A_∞-algebras, A_∞-categories and non-commutative geometry. In *Homological Mirror Symmetry*, Lecture Notes in Physics, vol. 757. Springer, Berlin, 2009, pp. 153–219.

[151] T. Y. Lam, *Lectures on Modules and Rings*, Graduate Texts in Mathematics, vol. 189. Springer, New York, 1999.

[152] L. Le Bruyn, *Noncommutative Geometry and Cayley-Smooth Orders*, Pure and Applied Mathematics (Boca Raton), vol. 290. Chapman & Hall/CRC, Boca Raton, FL, 2008.

[153] W. Lowen and M. Van den Bergh, The curvature problem for formal and infinitesimal deformations. arXiv:1505.03698, 2015.

[154] S. Lau, C.-H. Cho and H. Hong, Noncommutative homological mirror functor. *Mem. Amer. Math. Soc.*, 2019.

[155] H. M. Lee, *Homological Mirror Symmetry for Open Riemann Surfaces from Pair-of-Pants Decompositions*. ProQuest LLC, Ann Arbor, MI, 2015. Thesis (Ph.D.)–University of California, Berkeley.

[156] D. Labardini-Fragoso, Quivers with potentials associated to triangulated surfaces. *Proc. Lond. Math. Soc. (3)*, **98** (2009), 797–839.

[157] J. Lipman, Notes on derived functors and Grothendieck duality. In *Foundations of Grothendieck Duality for Diagrams of Schemes*, Lecture Notes in Mathematics, vol. 1960. Springer, Berlin, 2009, pp. 1–259.

[158] B. H. Lian, K. Liu and S.-T. Yau, Mirror principle. I. *Asian J. Math.*, **1** (1997), 729–763.

[159] V. A. Lunts and D. O. Orlov, Uniqueness of enhancement for triangulated categories. *J. Amer. Math. Soc.*, **23** (2010), 853–908.

[160] J.-L. Loday, *Cyclic Homology*, vol. 301. Springer, 2013.

[161] Y. Lekili and T. Perutz, Fukaya categories of the torus and Dehn surgery. *Proc. Natl. Acad. Sci. USA*, **108** (2011), 8106–8113.

[162] Y. Lekili and T. Perutz, Arithmetic mirror symmetry for the 2-torus. arXiv:1211.4632, 2012.

[163] K. H. Lin and D. Pomerleano, Global matrix factorizations. *Math. Res. Lett.*, **20** (2013), 91–106.

[164] Y. Lekili and A. Polishchuk, Arithmetic mirror symmetry for genus 1 curves with n marked points. *Selecta Math. (N.S.)*, **23** (2017), 1851–1907.

[165] Y. Lekili and A. Polishchuk, Auslander orders over nodal stacky curves and partially wrapped Fukaya categories. *J. Topol.*, **11** (2018), 615–644.

[166] Y. Lekili and A. Polishchuk, Derived equivalences of gentle algebras via Fukaya categories. *Math. Ann.*, **376** (2020), 187–225.

[167] D.-M. Lu, J. H. Palmieri, Q.-S. Wu and J. J. Zhang, Koszul equivalences in A_∞-algebras. *New York J. Math.*, **14** (2008), 325–378.

[168] D.-M. Lu, J. H. Palmieri, Q.-S. Wu and J. J. Zhang, A-infinity structure on Ext-algebras. *J. Pure Appl. Algebra*, **213** (2009), 2017–2037.

[169] Y. Lekili and K. Ueda, Homological mirror symmetry for K3 surfaces via moduli of A_∞-structures. arXiv:1806.04345, 2018.

[170] M. Markl, Transferring A_∞ (strongly homotopy associative) structures. *Rend. Circ. Mat. Palermo (2) Suppl.*, **79** (2006), 139–151.

[171] W. S. Massey, Some higher order cohomology operations. In *Symposium Internacional de Topología Algebraica International Symposium on Algebraic Topology*. Universidad Nacional Autónoma de México and UNESCO, Mexico City, 1958, pp. 145–154.

[172] W. S. Massey, *A Basic Course in Algebraic Topology*, Graduate Texts in Mathematics, vol. 127. Springer, New York, 1991.

[173] V. P Maslov, V. C. Bouslaev and V. I. Arnol'd, *Théorie des perturbations et méthodes asymptotiques*, Dunod, Paris, 1972.

[174] S. A. Merkulov, Strong homotopy algebras of a Kähler manifold. *Int. Math. Res. Not. IMRN*, **1999** (1999), 153–164.

[175] G. Mikhalkin, Amoebas of algebraic varieties and tropical geometry. In *Different Faces of Geometry*, International Mathematical Series (N. Y.), vol. 3. Kluwer/Plenum, New York, 2004, pp. 257–300.

[176] S. Mac Lane, *Categories for the Working Mathematician*, 2nd ed., Graduate Texts in Mathematics, vol. 5. Springer, New York, 1998.

[177] K. Morita, Duality for modules and its applications to the theory of rings with minimum condition. *Sci. Rep. Tokyo Kyoiku Daigaku Sect. A*, **6** (1958), 83–142.

[178] I. Mori, S. Okawa and K. Ueda, Moduli of noncommutative Hirzebruch surfaces. arXiv:1903.06457, 2019.

[179] S. Mozgovoy, Crepant resolutions and brane tilings I: Toric realization. arXiv:0908.3475, 2009.

[180] S. Mozgovoy and M. Reineke, On the noncommutative Donaldson-Thomas invariants arising from brane tilings. *Adv. Math.*, **223** (2010), 1521–1544.

[181] D. Maclagan and B. Sturmfels, *Introduction to Tropical Geometry*, Graduate Studies in Mathematics, vol. 161. American Mathematical Society, Providence, RI, 2015.

[182] D. McDuff and D. Salamon, *Introduction to Symplectic Topology*, 3rd ed., Oxford Graduate Texts in Mathematics. Oxford University Press, Oxford, 2017.

[183] S. Mukai, Duality between $D(X)$ and $D(\hat{X})$ with its application to Picard sheaves. *Nagoya Math. J.*, **81** (1981), 153–175.

[184] D. Mumford, *The Red Book of Varieties and Schemes*, expanded ed., Lecture Notes in Mathematics, vol. 1358. Springer, Berlin, 1999. Includes the Michigan lectures (1974) on curves and their Jacobians. With contributions by Enrico Arbarello.

[185] D. Murfet, Derived categories of quasi-coherent sheaves. Available at `therisingsea.org`, 2006.

[186] D. Murfet, Derived categories of sheaves. Available at `therisingsea.org`, 2006.

[187] D. Nadler, Microlocal branes are constructible sheaves. *Selecta Math. (N.S.)*, **15** (2009), 563–619.

[188] D. Nadler, Fukaya categories as categorical Morse homology. *SIGMA Symmetry Integrability Geom. Methods Appl.*, **10** (2014) Paper 018, 47 pages.

[189] D. Nadler, Cyclic symmetries of A_n-quiver representations. *Adv. Math.*, **269** (2015), 346–363.

[190] D. Nadler, Wrapped microlocal sheaves on pairs of pants. arXiv:1604.00114, 2016.

[191] D. Nadler, Arboreal singularities. *Geom. Topol.*, **21** (2017), 1231–1274.

[192] L. A. Nazarova and A. V. Roiter, A problem of I. M. Gel'fand. *Funct. Anal. Appl.*, **7** (1973), 299–311.

[193] M. S. Narasimhan and C. S. Seshadri, Stable and unitary vector bundles on a compact Riemann surface. *Ann. of Math. (2)*, **82** (1965), 540–567.

[194] Y. Nohara and K. Ueda, Homological mirror symmetry for the quintic 3-fold. *Geom. Topol.*, **16** (2012), 1967–2001.

[195] D. Nadler and E. Zaslow, Constructible sheaves and the Fukaya category. *J. Amer. Math. Soc.*, **22** (2009), 233–286.

[196] M. Olsson, *Algebraic Spaces and Stacks*, American Mathematical Society Colloquium Publications, vol. 62. American Mathematical Society, Providence, RI, 2016.

[197] S. Opper, P.-G. Plamondon and S. Schroll, A geometric model for the derived category of gentle algebras. arXiv:1801.09659, 2018.

[198] D. O. Orlov, Derived categories of coherent sheaves and equivalences between them. *Uspekhi Mat. Nauk*, **58** (2003), 89–172.

[199] D. O. Orlov, Triangulated categories of singularities, and equivalences between Landau-Ginzburg models. *Mat. Sb.*, **197** (2006), 117–132.

[200] D. Orlov, Derived categories of coherent sheaves and triangulated categories of singularities. In *Algebra, Arithmetic, and Geometry: In Honor of Yu. I. Manin. Vol. II*, Progress in Mathematics, vol. 270. Birkhäuser Boston, Boston, MA, 2009, pp. 503–531.

[201] D. Orlov, Matrix factorizations for nonaffine LG-models. *Math. Ann.*, **353** (2012), 95–108.

[202] A. Polishchuk, A field guide to A-infinity sign conventions. Preprint. `https://pages.uoregon.edu/apolish/ainf-signs.pdf`

[203] J. P. Pridham, Unifying derived deformation theories. *Adv. Math.*, **224** (2010), 772–826.

[204] J. Pascaleff and N. Sibilla, Topological Fukaya category and mirror symmetry for punctured surfaces. *Compos. Math.*, **155** (2019), 599–644.

[205] A. Polishchuk and E. Zaslow, Categorical mirror symmetry: The elliptic curve. *Adv. Theor. Math. Phys.*, **2** (1998), 443–470.

[206] D. Quillen, Projective modules over polynomial rings. *Invent. Math.*, **36** (1976), 167–171.

[207] M. Reineke, Moduli of representations of quivers. In *Trends in Representation Theory of Algebras and Related Topics*, EMS Ser. Congr. Rep. European Mathematical Society, Zürich, 2008, pp. 589–637.

[208] C. M. Ringel, The repetitive algebra of a gentle algebra, *Bol. Soc. Mat. Mexicana*, **3** (1997), 235–253.

[209] C. P. Rourke and B. J. Sanderson, Δ-sets. I. Homotopy theory. *Quart. J. Math. Oxford Ser. (2)*, **22** (1971), 321–338.

[210] A. Rizzardo and M. Van den Bergh, A note on non-unique enhancements. *Proc. Amer. Math. Soc.*, **147** (2019), 451–453.

[211] A. Rizzardo and M. Van den Bergh, A k-linear triangulated category without a model. *Ann. of Math. (2)*, **191** (2020), 393–437.

[212] D. Salamon, Lectures on Floer homology. In *Symplectic Geometry and Topology*, IAS/Park City Mathematics Series, vol. 7. American Mathematical Society, Providence, RI, 1999, pp. 143–229.

[213] R. Schiffler, *Quiver Representations*. CMS Books in Mathematics/Ouvrages de Mathématiques de la SMC. Springer, Cham, 2014.

[214] G. Segal, Lecture notes on topological field theory. Preprint, 1999. `http://web.math.ucsb.edu/~drm/conferences/ITP99/segal/`

[215] G. Segal, The definition of conformal field theory. In *Topology, Geometry and Quantum Field Theory*, London Mathematical Society Lecture Note Series, vol. 308. Cambridge University Press, Cambridge, 2004, pp. 421–577.

[216] E. Segal, Equivalence between GIT quotients of Landau-Ginzburg B-models. *Comm. Math. Phys.*, **304** (2011), 411–432.

[217] P. Seidel, Fukaya categories and deformations. In *Proceedings of the International Congress of Mathematicians, Vol. II (Beijing, 2002)*. Higher Ed. Press, Beijing, 2002, pp. 351–360.

[218] P. Seidel, *Fukaya Categories and Picard-Lefschetz Theory*. Zurich Lectures in Advanced Mathematics. European Mathematical Society (EMS), Zürich, 2008.

[219] P. Seidel, Suspending Lefschetz fibrations, with an application to local mirror symmetry. *Comm. Math. Phys.*, **297** (2010), 515–528.

[220] P. Seidel, Homological mirror symmetry for the genus two curve. *J. Algebraic Geom.*, **20** (2011), 727–769.

[221] P. Seidel, Homological mirror symmetry for the quartic surface. *Mem. Amer. Math. Soc.*, **236** (2015).

[222] J.-P. Serre, Faisceaux algébriques cohérents. *Ann. of Math. (2)*, **61** (1955), 197–278.

[223] N. Sheridan, Homological mirror symmetry for Calabi-Yau hypersurfaces in projective space. *Invent. Math.*, **199** (2015), 1–186.

[224] N. Sheridan, Versality of the relative Fukaya category. *Geom. Topol.*, **24** (2020), 747–884.

[225] I. Shipman, A geometric approach to Orlov's theorem. *Compos. Math.*, **148** (2012), 1365–1389.

[226] S. P. Smith, Non-commutative algebraic geometry. Preprint, Lecture Notes at University of Washington, 2000. `https://sites.math.washington.edu/~smith/Teaching/513nag/ch1.ps`

[227] I. Smith, A symplectic prolegomenon. *Bull. Amer. Math. Soc. (N.S.)*, **52** (2015), 415–464.

[228] D. Sorensen, Classification of vector bundles over P1. Preprint. www.derekhsorensen.com/docs/sorensen-classification-vector-bundles.pdf

[229] J. D. Stasheff, Homotopy associativity of H-spaces. I, II. *Trans. Amer. Math. Soc.* **108** (1963), 275–292, 293–312.

[230] The Stacks Project Authors. *Stacks Project.* https://stacks.math.columbia.edu, 2018.

[231] The structure of Coh $\mathbb{P}^1$. Preprint. http://people.math.harvard.edu/~hirolee/pdfs/280x-13-14-dbcoh.pdf

[232] N. Sibilla, D. Treumann and E. Zaslow, Ribbon graphs and mirror symmetry. *Selecta Math. (N.S.)*, **20** (2014), 979–1002.

[233] T. Sutherland, The modular curve as the space of stability conditions of a CY3 algebra. arXiv:1111.4184, 2011.

[234] A. Skowroński and J. Waschbüsch, Representation-finite biserial algebras. *J. Reine Angew. Math.*, **345** (1983), 172–181.

[235] R. G. Swan, Vector bundles and projective modules. *Trans. Amer. Math. Soc.*, **105** (1962), 264–277.

[236] Z. Sylvan, On partially wrapped Fukaya categories. *J. Topol.*, **12** (2019), 372–441.

[237] A. Strominger, S.-T. Yau and E. Zaslow, Mirror symmetry is T-duality. In *Winter School on Mirror Symmetry, Vector Bundles and Lagrangian Submanifolds (Cambridge, MA, 1999)*, AMS/IP Studies in Advanced Mathematics, vol. 23. American Mathematical Society, Providence, RI, 2001, pp. 333–347.

[238] D. Salamon and E. Zehnder, Morse theory for periodic solutions of Hamiltonian systems and the Maslov index. *Comm. Pure Appl. Math.*, **45** (1992), 1303–1360.

[239] G. Tabuada, Une structure de catégorie de modèles de Quillen sur la catégorie des dg-catégories. *C. R. Math. Acad. Sci. Paris*, **340** (2005), 15–19.

[240] G. Tabuada, Finite generation of the numerical Grothendieck group. arXiv:1704.06252, 2017.

[241] A. A. Takeda, Relative stability conditions on Fukaya categories of surfaces. arXiv:1811.10592, 2018.

[242] D. E. Tamarkin, *Operadic Proof of M. Kontsevich's Formality Theorem*. ProQuest LLC, Ann Arbor, MI, 1999. Thesis (Ph.D.)–The Pennsylvania State University.

[243] B. Toen, The homotopy theory of dg-categories and derived Morita theory. *Invent. Math.*, **167** (2007), 615–667.

[244] B. Toen, Lectures on dg-categories. In *Topics in Algebraic and Topological K-Theory*, Lecture Notes in Mathematics, vol. 2008. Springer, Berlin, 2011, pp. 243–302.

[245] J. Tu, Homological mirror symmetry and Fourier-Mukai transform. *Int. Math. Res. Not. IMRN*, **2015** (2015), 579–630.

[246] V. Turaev and A. Virelizier, *Monoidal Categories and Topological Field Theory*, Progress in Mathematics, vol. 322. Birkhäuser/Springer, Cham, 2017.

[247] K. Ueda, Homological mirror symmetry and simple elliptic singularities. arXiv math/0604361, 2006.

[248] R. Vakil, The rising sea: Foundations of algebraic geometry. Preprint, 2017. http://math.stanford.edu/~vakil/216blog/FOAGapr2915public.pdf

[249] M. Van den Bergh, Calabi-Yau algebras and superpotentials. *Selecta Math. (N.S.)*, **21** (2015), 555–603.

[250] M. van Garrel, D. P. Overholser and H. Ruddat, Enumerative aspects of the Gross-Siebert program. In *Calabi-Yau Varieties: Arithmetic, Geometry and Physics*. Fields Institute Monographs, vol. 34. Fields Inst. Res. Math. Sci., Toronto, ON, 2015, pp. 337–420.

[251] M. Vonk, A mini-course on topological strings. arXiv:hep-th/0504147, 2005.

[252] A. Weinstein, Symplectic manifolds and their Lagrangian submanifolds. *Adv. Math.*, **6** (1971), 329–346.

[253] C. A. Weibel, *An Introduction to Homological Algebra*, Cambridge Studies in Advanced Mathematics, vol. 38. Cambridge University Press, Cambridge, 1994.

[254] C. A Weibel, *The K-Book: An Introduction to Algebraic K-Theory*, vol. 145. American Mathematical Society, Providence, RI, 2013.

[255] C. Wendl, A beginner's overview of symplectic homology. Preprint. www .mathematik.hu-berlin.de/ wendl/pub/SH.pdf

[256] E. Witten, Supersymmetry and Morse theory. *J. Differential Geometry*, **17** (1982), 661–692.

[257] M. Wong, Dimer models and Hochschild cohomology. arXiv:1908.03005, 2019.

[258] A. Zimmermann, *Representation theory*. Algebra and Applications, Springer, Amiens, 2014.

Index

Printed in the United States
by Baker & Taylor Publisher Services